Lighting for Driving

Vehicle, road, sign, and signal lighting are provided to enable drivers to reach their destinations quickly and safely. However, the attention given to how these forms of lighting function is likely to change as new technology is introduced and understanding of ergonomics and human factors improves. **Lighting for Driving: Roads, Vehicles, Signs, and Signals, Second Edition** shows the crucial role lighting plays in road safety and examines how it could be used more effectively. With light-emitting diodes (LEDs) becoming the lighting source of choice for transport planners and vehicle designers, this book integrates information on road lighting, vehicle lighting, signs, and signals in one handy volume.

International in scope and updated for this new edition, this book features lighting examples from the USA, the UK, France, Germany, The Netherlands, Denmark, Sweden, Norway, Finland, Japan, Australia, and New Zealand. Lighting in common vehicle types including cars, vans, trucks, and motorcycles is covered as well as the visibility of pedestrians and cyclists to drivers. Coverage extends to road lighting, traffic markings, vehicle designs and internal lighting, and weather conditions. Now fully updated, the final chapter looks at the future of lighting in relation to driving. The book will help the reader in understanding how lighting systems on roads and vehicles work by explaining the thinking and scientific reasoning behind various forms of lighting and analysing their contribution to the driver's understanding of real and potential road hazards.

This book will be an ideal read for ergonomists and engineers engaged in transport and road engineering, transport planners, civil engineers, vehicle designers, and electrical engineers.

Peter Boyce has spent most of his career working in the field of lighting. From 1966 to 1990, he was a Research Officer at the Electricity Council Research Centre in England. There he conducted research on visual fatigue, the influence of age on visual performance, hue discrimination, safe lighting for emergency conditions, and security lighting. From 1990 to 2004, he was Head of Human Factors at the Lighting Research Center at Rensselaer Polytechnic Institute in the USA. There he conducted research on visual performance, visual comfort, circadian effects, emergency lighting and perceptions of safety, and directed lighting evaluations and product testing as well as teaching. From 2008 to 2019, he was the Editor of the journal *Lighting Research and Technology*. He is a Fellow of both the Society of Light and Lighting in the UK and of the Illuminating Engineering Society of North America and has received awards from both bodies for his work. He is the author of another book, *Human Factors in Lighting*, as well as numerous book chapters, papers, and articles. He has driven many different vehicle types, including cars, vans, and tractors, and ridden motorcycles, bicycles, and horses.

Lighting for Driving

Roads, Vehicles, Signs, and Signals

Second Edition
Peter Boyce

CRC Press
Taylor & Francis Group
Boca Raton London New York

CRC Press is an imprint of the
Taylor & Francis Group, an **informa** business

Front cover image: Mark Duffy/iStock

Second edition published 2025

by CRC Press
2385 NW Executive Center Drive, Suite 320, Boca Raton FL 33431

and by CRC Press
4 Park Square, Milton Park, Abingdon, Oxon, OX14 4RN

CRC Press is an imprint of Taylor & Francis Group, LLC

Library of Congress Cataloging-in-Publication Data

Names: Boyce, P. R., author.
Title: Lighting for driving : roads, vehicles, signs, and signals / Peter
Boyce.
Description: Second edition. | Boca Raton, FL : CRC Press, 2025. | Includes
bibliographical references and index. |
Identifiers: LCCN 2024009481 (print) | LCCN 2024009482 (ebook) | ISBN
9781032478289 (hardback) | ISBN 9781032484099 (paperback) | ISBN
9781003388906 (ebook)
Subjects: LCSH: Roads--Lighting.
Classification: LCC TE228 .B66 2025 (print) | LCC TE228 (ebook) | DDC
628.9/5--dc23/eng/20240404
LC record available at https://lccn.loc.gov/2024009481
LC ebook record available at https://lccn.loc.gov/2024009482

ISBN: 978-1-032-47828-9 (hbk)
ISBN: 978-1-032-48409-9 (pbk)
ISBN: 978-1-003-38890-6 (ebk)

DOI: 10.1201/9781003388906

Typeset in Times
by Deanta Global Publishing Services, Chennai, India

Contents

Preface

Since I wrote the first edition of this book in 2008, many aspects of lighting for driving have changed. It is these changes that have made a second edition necessary. The main changes have been in the fields of light sources, controls, and driver assistance systems. In 2008, the light-emitting diode was just beginning to appear as a light source in vehicles, but not on roads. Today it is ubiquitous, in both locations. Older light sources may still be found, but they are gradually disappearing from use. As for controls, the fast response speed, small size, and ease of dimming of light-emitting diodes, combined with the growth of computing and communication technology, have meant that both vehicle lighting and road lighting can now better adapt to prevailing conditions. Regarding driver assistance systems, these have grown in both number and sophistication because of the ability of cameras, sensors, and software to identify what is happening ahead and around a vehicle and to warn drivers about any problem. Ultimately, this development is likely to lead to fully autonomous vehicles which will, in turn, lead to major changes in vehicle and road lighting. But that is some way off. For the next decade or two, it is likely that most vehicles will continue to be controlled by humans who rely on their senses and brains to determine what to do. This means that vehicles and road lighting still have a major contribution to make to traffic safety.

While much has changed in the technology used for vehicles and road lighting, the failure to integrate road lighting, vehicle lighting, signs, and signals into a coherent system is still apparent. For many years, the practitioners of road lighting and vehicle lighting have studiously ignored each other. The main concern of both camps seems to be to ensure that the lighting meets existing standards rather than asking whether or not existing standards are meaningful. This does not mean that the question of the effectiveness of road and vehicle lighting is totally ignored. Rather, the effectiveness of road lighting, vehicle lighting, and signage is dealt with, at one level, through the efforts of human factors experts and, at another, by epidemiologists. This book attempts to integrate these diverse strands of evidence. The philosophy behind this integration is that the primary role of all forms of lighting for driving is to provide information to the driver about the road ahead and the presence and intentions of other people on or near the road.

All books have limitations. This book has at least two. The first is that of geography. The standards and technology relevant to lighting for driving that are discussed are those of the USA and the UK. This is because I have lived and driven extensively in both countries and am familiar with their road systems. But the literature relevant to the effectiveness of road lighting, vehicle lighting, signs, and signals comes from a wider range of countries, notably the USA, the UK, France, Germany, The Netherlands, Denmark, Sweden, Norway, Finland, Japan, Australia, and New Zealand. These countries have a number of things in common. They have a high number of vehicles relative to the population, extensive road networks, and strict regulation of people, vehicles, and other equipment allowed on the road. I have assumed that results obtained in any one of these countries will be applicable to all countries with similar characteristics. The second limitation relates to the coverage of vehicle lighting. Here, attention is focused on the most common vehicle types, such as cars, vans, trucks, and motorcycles, although the visibility of pedestrians and cyclists to drivers is also assessed.

Peter R. Boyce
Canterbury, United Kingdom

Acknowledgements

The cooperation of the following individuals and publishers in granting permission for reproduction of copyright material is gratefully acknowledged

Fabian Stahl for Figure 3.6

iStock by Getty images / Astucia for Figure 5.4

iStock by Getty images / Mediterranean for Figure 5.8

iStock by Getty images / jhorrocks for Figure 5.10

iStock by Getty Images / tornal for Figure 10.1

iStock by Getty Images / Alan Lagado for Figure 10.3a

iStock by Getty Images / LukaTBD for Figure 10.7

iStock by Getty Images / Francesca Leslie for Figure 11.3

iStock by Getty Images / Norrie3699 for Figure 11.8

iStock by Getty Images / Wirestock for Figure 11.9

iStock by Getty Images / janiebros for Figure 14.2

Lei Deng for Figure 2.5

Naomi Miller for Figure 9.7

Robert Sekuler and Randolph Blake for Figures 3.2, 3.4, 3.5, and 3.16

Society of Automotive Engineers for Figure 6.8.

The Chartered Institution of Building Services Engineers for Figure 6.5

The Illuminating Engineering Society of North America for Figures 2.4, 4.1, and 4.3.

1 Driving and Crashes

1.1 INTRODUCTION

Lighting as an aid to driving is all around us in the form of road lighting, vehicle lighting, road markings, road signs, and traffic signals. But all is not well with lighting as an aid to driving. The design principles of road lighting have changed little since the 1930s, yet vehicle lighting has changed out of all recognition. The contribution of vehicle lighting to visibility is rarely considered by the designers of road lighting, and the contribution of road lighting to visibility is rarely considered by the designers of vehicle lighting. Further, it can be argued that the driver's task has become more difficult as competition for visual attention has increased. Traffic densities are higher, traffic speeds are faster, sources of information relevant to the driver are more numerous, and sources of distraction are seldom absent. As a counterweight to these increased difficulties, many vehicles are now designed to provide greater protection for the driver in the event of a collision and are being fitted with systems based on a combination of cameras, radars, and lasers to assist the driver in lane-keeping, emergency braking, reversing, and parking, all of which make a driver's life safer and easier. Of course, such developments are moving us slowly but steadily towards the ultimate objective, the autonomous vehicle where there is no driver. However, such an achievement is some way from fulfilment. This is not just due to technical issues. There are also questions of responsibility and liability in the event of a system failure to be addressed. And even if these problems are solved, it is likely that autonomous vehicles and driven vehicles will co-exist for a considerable time. Thus, the need for road lighting, vehicle lighting, road markings, road signs, and traffic signals is likely to continue for the foreseeable future. This book has been written with the aim of providing a comprehensive review of lighting for driving today and where it might go in the future.

1.2 DRIVING AS A VISUAL TASK

Unless you have a fully autonomous vehicle, driving is a visual task. If you doubt this and are foolish enough to try it, you could attempt to drive with your eyes shut and measure how far you can go without running off the road or into something or somebody. It will not be far. This means that vision is vital to driving, but like almost all so-called visual tasks, the complete task involves a lot more than seeing. Most visual tasks have three components: Visual, cognitive, and motor. The visual component refers to the process of extracting information relevant to the performance of the task using the sense of sight. The cognitive component is the process by which sensory stimuli are interpreted and the appropriate action determined. The motor component is

DOI: 10.1201/9781003388906-1

the process by which the stimuli are explored to extract information and/or the actions decided upon are carried out. Of course, these three components interact to produce a complex pattern between stimulus and response that will be different for different tasks and for different drivers with different levels of experience (CIE, 1992a).

As examples of the visual tasks associated with driving, consider your likely responses to the onset of stop lamps on a vehicle ahead and detecting a movement at the edge of the road ahead. Your response to the former will depend on your distance from the vehicle ahead. If you are close behind, the decision is simple, you have to brake immediately. If the distance is large, no response may be necessary immediately other than to pay attention to the vehicle later. The response to detecting a movement at the edge of the road ahead will depend on the nature of the movement. If the movement is recognizable as a pedestrian, the next step is to determine the direction of movement. If the direction is into the road, there are a number of options ranging from nothing through swerving away to an emergency stop, depending on the distance to the pedestrian. If the movement is away from the road, no response is necessary. If the movement detected is a ball bouncing into the road, braking is an appropriate response because experience will tell you that a bouncing ball is likely to be followed by a child, even though the child is not yet visible.

These simple examples are enough to indicate that while vision is a necessary condition for being able to drive, but alone, it is not sufficient. The quality of driving is also influenced by cognitive skills such as learning, remembering, and decision-making, largely derived from experience; and personality variables, such as the threshold for boredom and levels of risk aversion. Further, when driving we receive information through sensors other than vision. When in motion, we receive auditory information from the noise of the vehicle itself and from the environment through which it moves, as well as information about the forces acting on the body obtained from the kinaesthetic and vestibular mechanisms. Sometimes, these other senses are the first indication of how the driver should use the sense of vision, e.g., hearing the siren of an emergency vehicle will usually initiate a visual search to locate the vehicle, but most of the time, the senses are mutually supportive. For example, accelerating into a corner will produce changes in the flow of visual information indicating increasing speed and a change of heading (see Figure 3.16); changes in the auditory stimuli produced by the engine indicating an increasing speed; and changes in the kinaesthetic and vestibular mechanisms indicating a centrifugal force.

Despite this mutual support between the senses, there can be little doubt that vision is the primary sensory input for driving. In most countries, there are no restrictions on deaf people with regard to driving, but there are restrictions on people with limited vision (see Section 13.5). Put succinctly, driving requires the driver to control the velocity, acceleration or deceleration, and direction of movement of the vehicle by seeing the line of the road ahead and the relative position and movement of other vehicles, people, animals, and objects on and around the road; detecting and understanding information presented on the vehicle's instruments and screen, and through signs and signals on the road, as well as keeping a lookout for the unexpected, such as a spilt load from a truck. Further, this has to be done dynamically at a speed that is much faster than that for which the visual system evolved which is the maximum speed of a human running (approximately 36 km.h^{-1} or 22 mph). Such is the nature of driving as a visual task.

1.3 THE ROLE OF LIGHTING IN DRIVING

The role of lighting in driving is to enable the transfer of information, either directly or indirectly. Direct transfer of information occurs when the light source itself conveys the information, as in a traffic signal or a flashing turn lamp on a vehicle. Indirect transfer of information occurs when the light is used to illuminate a surface that is then searched by the driver for the information it contains. Such surfaces may contain written information, e.g., road signs, or they may be empty but need to be searched for condition, e.g., an icy road surface. For both direct and indirect information to be transferred, the information has to be visible so the first question that needs to be considered is what makes things visible?

The answer to this question can be divided into three parts: The stimulus the object presents to the visual system, the quality of the retinal image, and the operation of the visual system. There are three variables that are always relevant to the stimulus the object presents to the visual system: Visual size, luminance contrast, and colour difference.

1.3.1 VISUAL SIZE

There are several different ways to express the size of a stimulus presented to the visual system, but all of them are angular measures. This means that an object of the same physical size will have different visual sizes depending on the distance from which it is viewed.

The visual size of a stimulus for detection is usually given by the solid angle the stimulus subtends at the eye. The solid angle is the quotient of the areal extent of the object and the square of the distance from which it is viewed and is measured in steradians (sr). The larger is the solid angle, the easier the stimulus is to detect.

The visual size of a stimulus for resolution is usually given as the angle the critical dimension of the stimulus subtends at the eye, usually measured in degrees or minutes of arc (there are 60 minutes of arc in one degree). What the critical dimension is depends on the stimulus. For two points, such as headlamps far away, the critical dimension is the separation between them. For two parallel lines that are used in road markings, it is the separation between the two lines. For a letter C on a road sign, it is the size of the gap that differentiates the letter C from the letter O. The larger is the visual size of detail in a stimulus, the easier it is to resolve that detail.

For complex stimuli, the measure used to express the size of detail is the spatial frequency distribution. Spatial frequency is the reciprocal of the angular subtense of critical detail, expressed as cycles per degree. Complex stimuli have many spatial frequencies and hence a spatial frequency distribution. The match between the luminance contrast at each spatial frequency of the stimulus and the contrast sensitivity function of the visual system determines if the stimulus will be seen and what detail will be resolved (see Section 3.4.3). Lighting can do little to change the visual size of two-dimensional objects such as lettering on a road sign, but shadows can be used to enhance the effective visual size of three-dimensional objects such as vehicles on the road.

1.3.2 Luminance Contrast

The luminance contrast of a stimulus expresses its luminance relative to its immediate background. The higher is the luminance contrast, the easier it is to detect the stimulus. There are three different forms of luminance contrast commonly used for uniform luminance targets seen against a uniform luminance background. There is no agreement on how to measure luminance contrast for complex objects when contrasts can occur within the target (Peli, 1990; Haun and Peli, 2013).

For uniform targets seen against a uniform background, luminance contrast is defined as

$$C = |L_t - L_b|/L_b$$

where C is the luminance contrast, L_t is the luminance of the target (cd.m^{-2}), and L_b is the luminance of the background (cd.m^{-2}). This formula gives luminance contrasts that range from zero to unity for targets that are darker than the background and from zero to infinity for targets that are brighter than the background.

Another form of luminance contrast for a uniform target seen against a uniform background is defined as

$$C = L_t/L_b$$

where C is the luminance contrast, L_t is the luminance of the target (cd.m^{-2}), and L_b is the luminance of the background (cd.m^{-2}). This formula gives luminance contrasts that can vary from zero, when the target has zero luminance, to infinity, when the background has zero luminance. It is often used for self-luminous targets.

For targets that have a periodic luminance pattern, e.g., a multiple chevron road sign indicating a sharp bend, the luminance contrast is given by

$$C = (L_{max} - L_{min})/(L_{max} + L_{min})$$

where C is the luminance contrast, L_{max} is the maximum luminance (cd.m^{-2}), and L_{min} is the minimum luminance (cd.m^{-2}). This formula gives luminance contrasts that range from zero to unity, regardless of the relative luminances of the target and background. It is sometimes called the luminance modulation or Michelson contrast.

Lighting can change the luminance contrast of a stimulus by producing veiling reflections on the stimulus or disability glare in the eye (see Section 6.6).

1.3.3 Colour Difference

Luminance quantifies the amount of light emitted from a stimulus but ignores the combination of wavelengths making up that light. It is the wavelengths emitted by or reflected from the stimulus that influence its colour. It is possible to have a stimulus with zero luminance contrast that can still be detected because it differs from its background in colour (Eklund, 1999). A number of different colour difference formulae have been developed over the years, but the one that has gained acceptance is the CIE 2000 formula (Luo et al., 2001; Habekost, 2013). Colour difference only becomes important for visibility when luminance contrast is low, typically below 0.3. Lighting using light sources with different spectral power distributions may alter the colour difference between the stimulus and its background and hence its visibility. It should also be noted that light source spectral power distribution affects how the colour of an object or surface is perceived. When a colour has meaning, as in some signs and signals, this may be important.

Visual size, luminance contrast, and colour difference can be used to characterize a stimulus, but that stimulus cannot be said to have been presented to the visual system until it reaches the retina of the eye. This raises the next factor involved in determining the visibility of objects: The quality of the retinal image.

1.3.4 RETINAL IMAGE QUALITY

As with all image processing systems, the visual system works best when it is presented with a sharp image. The sharpness of the image of the stimulus can be quantified by its spatial frequency distribution: A sharp image will have high spatial frequency components present, a blurred image will not. The sharpness of the retinal image is determined by the extent to which the atmosphere and materials through which light passes scatter light and the ability of the visual system to focus the image on the retina. The greater the amount of scatter and the more limited the ability to focus, the poorer the retinal image quality will be. For driving, scattering can be caused by water droplets in the atmosphere due to rain or spray and by dirt or mist on the windscreen or helmet visor. Scattering of light also occurs within the eye, increasingly so as we age (see Section 13.4.1). As for the ability to bring the object to focus on the retina, the range of distances over which we can do this becomes more restricted as we age. Lighting can do little to alter any of these factors, although it has been shown that light sources that are rich in the short wavelengths of the visible spectrum produce smaller pupil sizes for the same luminance than light sources that are deficient in the short wavelengths (Berman et al., 1992). In many circumstances, a smaller pupil size can produce a better quality retinal image because it implies a greater depth of field and less spherical and chromatic aberration.

This brings us to the operation of the visual system. There are two aspects to this: The different retinal photoreceptors active and where the retinal image of the target falls. The active photoreceptors are determined by the retinal illuminance.

1.3.5 RETINAL ILLUMINANCE

The illuminance on the retina determines the state of adaptation of the visual system and therefore alters the capabilities of the visual system (see Section 3.3.3). The retinal illuminance is given by the equation

$$E_r = e_t.\tau \, . \, (\cos \theta \, / k^2)$$

where E_r is the retinal illuminance (lx), τ is the ocular transmittance, θ is the angular displacement of the surface from the line of sight (degrees), k is a constant equal to 15, and e_t is the amount of light entering the eye (trolands). The amount of light entering the eye, measured in trolands, is given by the equation

$$e_t = L.\rho_t$$

where L is the surface luminance (cd.m^{-2}) and ρ is the pupil area (mm^2).

The amount of light entering the eye, e_t, measured in trolands, is often referred to as retinal illumination, but it does not take the transmittance of the optic media into account and therefore does not truly represent the luminous flux density on the retina. The amount of light entering the eye is mainly determined by the luminances in the field of view. For drivers in the daytime, these luminances will be determined by the extent to which the sky and sun can be seen and the reflectances of the surfaces in the field of view and the illuminances on them. For drivers at night, the relevant luminances are those of the reflecting surfaces illuminated by road lighting or vehicle forward lighting, such as the road, and of self-luminous sources, such as the headlamps of approaching vehicles.

Depending on the amount of light reaching the retina, different photoreceptors will be dominant (see Section 3.3.4). In daylight, the cone photoreceptors will be dominant. This is the photopic state. In complete darkness, only the rod photoreceptors will be working. This is the scotopic state. Under road lighting and vehicle lighting at night, both cone and rod photoreceptors are likely to be active. This is the mesopic state. Visual resolution deteriorates as the visual system passes from the photopic, through the mesopic, to the scotopic state (see Section 3.4).

The final factor determining whether a given object will be visible is where it falls on the retina. Unless the visual system is operating in the scotopic state, which is very unlikely where road lighting and/or vehicle lighting are present, the best resolution of detail is achieved at the fovea, i.e., on-axis. Further, eye movements allow the retinal images of objects that are first detected off-axis to be moved onto the fovea. However, for this to occur, the object has first to be detected off-axis. Visibility is generally worse off-axis than on-axis.

To summarize, it is the interaction between the object to be seen, the background against which it is seen, the lighting of both object and background, the atmosphere and materials through which light passes, the capabilities of the optics of the observer's eye, the operating state of the observer's visual system and where on the observer's retina the image of the object is received that determine the object's visibility. The visibility of the object influences the driver's response. But, as should be obvious from what was said earlier, on the road, visibility and response are not linked by rods of iron, but rather by bands of spaghetti. The driver's response is also influenced by the decision-making process which is, itself, dependent on the driver's knowledge, experience, attention, expectations, fatigue, and circadian state. Visibility is necessary for a response to occur but that response is dependent on many other factors.

A timely and correct response to every relevant stimulus is what is needed in order to drive safely. Timely and correct responses also make a valuable contribution to ensuring that traffic flow is smooth rather than turbulent and that the experience of driving is one of comfort rather than stress.

1.4 THE EFFECTIVENESS OF LIGHTING

Given that the fundamental purpose of lighting on vehicles and roads is to enhance the safety of road users by increasing the visibility of the road ahead and objects on and around it, one way to examine the effectiveness of such lighting is to consider the number of crashes that occur by night and day (CIE, 1992b). Different countries have different ways of recording crashes. Not all of them identify the lighting conditions at the time of the crash but three that do are the STATS19 Road Crash Injury Statistics database in the UK, the Fatality Analysis Reporting System (FARS), and the Crash Report Sampling System of the National Highway Traffic Safety Administration in the USA. Table 1.1 shows the number of motor vehicle crashes involving fatalities and the estimated number of motor vehicle crashes resulting in personal injuries and property damage alone, occurring in 2019, under different lighting conditions but in good weather, in the 50 states of the USA, the District of Columbia, and Puerto Rico (NHTSA, 2021a). The year 2019 has been used because the pandemic occurring in the next two years changed traffic flows. At first glance, the data in Table 1.1 suggest that current standards of road and vehicle lighting in the USA are more than adequate. A similar number of fatalities occur in daylight and after dark and more fatalities occur when vehicle lighting alone is being used (dark/not lit) than when road lighting is present as well (dark/lit). However, two other features of Table 1.1 suggest that such a conclusion is premature. First, there are more crashes involving personal injury and property damage during daytime than after dark, yet the visibility provided by daylight is certainly greater than anything provided by vehicle

---

TABLE 1.1

Number of Motor Vehicle Crashes Involving Fatalities and Estimated Number of Personal Injury and Property Damage Only Crashes, Occurring in Good Weather Conditions in the 50 States of the USA as well as the District of Columbia and Puerto Rico, under Different Lighting Conditions, in 2019

Consequences of Crash	Daylight	Dark/Lit	Dark/Not Lit	Dawn and Dusk
Fatality	15,815	6,719	9,047	1,330
Personal injury	1,315,359	389,833	176,918	74,448
Property damage only	3,361,940	714,924	494,886	195,033

Source: NHTSA (2021a).

lighting, alone or with road lighting. Second, although the number of crashes involving fatalities that occur after dark is fewer when the road is lit than when it is not, the numbers of personal injury and property damage crashes are greater for the dark/lit condition than for the dark/not lit condition, which implies that road lighting makes such crashes more likely.

The reason for these unexpected observations is that the number of crashes represents only one part of the quantity needed to evaluate the benefits of vehicle and road lighting. The other part is the level of exposure. The quantity needed to evaluate the benefits of any traffic safety measure is the crash risk, which is the ratio of the outcome of exposure, i.e., a crash, to the level of exposure itself (Yannis et al., 2005). This offers two possible approaches to quantifying the effectiveness of road or vehicle lighting. The simpler approach is to measure the crash risks, after dark, before and after a change in road or vehicle lighting. The difference in crash risks will establish if the modified lighting is an improvement. This approach has a limitation. It does not identify how large an improvement could be achieved. The second approach addresses this limitation by looking at the change in crash risks for night and day, separately, following a change in the road or vehicle lighting. The crash risk in the daytime provides a measure of what might be achieved by lighting. Lamm et al. (1985) reported a study using a slightly more elaborate version of the second approach to examine the effectiveness of the lighting of motorways in Germany. Specifically, the research was carried out over a period of nine years between 1972 and 1981 on a stretch of the motorway divided into three sections, A, B, and C, of lengths 1.9, 3.7, and 2.3 km (1.2, 2.3, and 1.5 miles), respectively. For the first year, all three sections were unlit. For the next five years, sections A and B were lit all night, but section C remained unlit. For the following three years, the lighting in sections A and B was usually turned off after 10 p.m., while section C remained unlit. Table 1.2 shows the crash risk by day and night in each section, measured as crashes per million vehicle kilometres of travel (Schreuder, 1998). The crash risks shown in Table 1.2 reveal a confusing picture. When all three sections of the motorway are unlit, the crash risk is higher at night than in the day, everywhere. When road lighting was introduced into sections A and B, there was the hoped-for reduction in crash risk at night. However, there was also a dramatic reduction in crash risk by day in sections A and B, a change that is difficult to relate to the presence of road lighting. When the road lighting on sections A and B is extinguished at 10 p.m., there is a decrease in crash risk for section A by night, but an increase in the corresponding measure for section B.

What these changes suggest is that the crude measure of exposure, vehicle kilometres of travel, is not an adequate representation of the difficulties facing a driver on these sections of road, and that, over the years, other changes have taken place that affect the likelihood of a crash, such as changes in road layout, changes in driver demographics, changes in traffic density, and so on. The sad fact is that crashes are influenced by many factors other than lighting. The longer the collection of data goes on, the more likely it is that some of these other factors will change, thereby increasing the level of noise in the data.

TABLE 1.2

Crash Risk, Measured as Crashes per Million Vehicle Kilometres, for Three Sections of a Motorway in Germany, by Day and Night, Lit and Unlit

Lighting Condition	Crash Risk for Section A, by Day	Crash Risk for Section A, by Night	Crash Risk for Section B, by Day	Crash Risk for Section B, by Night	Crash Risk for Section C, by Day	Crash Risk for Section C, by Night
All sections unlit	2.03	4.28	1.80	3.33	1.41	1.85
Sections A and B were lit all night. Section C unlit	0.88	1.97	1.08	1.66	1.13	2.08
Sections A and B are lit until 10 p.m. Section C unlit	0.62	1.76	0.90	2.18	0.93	1.89

Source: Schreuder (1998).

An alternative to the prolonged before and after study of the type undertaken by Lamm et al. (1985) is the relatively rapid collection of crash data from a large number of similar sites with different types and levels of lighting. This approach reduces the likelihood of changes occurring at a site but inevitably ignores the differences between sites, including different levels of exposure. This was the approach used for a study of the effect of road lighting on traffic safety undertaken in the UK (Scott, 1980). In this study, photometric measurements were taken of the lighting conditions at up to 89 different sites using a mobile laboratory (Green and Hargroves, 1979). The sites were all at least 1 km long with homogeneous lighting conditions and both the lighting and the road features had been unchanged for at least three years. The sites were all two-way urban roads with a 48 km.h^{-1} (30 mph) speed limit. The photometric measurements were made with the road dry and the crashes considered were only those that occurred when the roads were dry. Multiple regression analysis was used to determine the importance of various characteristics of the lighting on the night/day crash ratio. The average road surface luminance was found to be the best predictor of the effect of the lighting on the night/day crash ratio. Figure 1.1 shows the night/day accident ratios for the sites plotted against the average road surface luminance. The best-fitting exponential curve through the data is shown, the night/day accident ratios being weighted to give greater importance to those sites where crashes occurred most frequently. The equation for the curve is

$$N_{\mathrm{R}} = 0.66\, e^{-0.42}\mathrm{L}$$

where N_R is the night/day accident ratio and L is the average road surface luminance ($cd.m^{-2}$).

It is clear from Figure 1.1 that increasing the average road surface luminance does contribute something to a reduction in crashes at night, but the wide scatter in the individual night/day accident ratios indicates that there are many factors other than the road surface luminance provided by the lighting that matter. A similar pattern was found by Jacket and Frith (2013), who examined 7,944 crashes occurring on 270 km of roads in New Zealand over the years 2006–2010. Their regression equation for the relationship between night/day crash ratio and average road surface luminance was

$$N_R = 0.53\ e^{-0.43}L$$

where N_R is the night/day crash ratio and L is the average road surface luminance ($cd.m^{-2}$). This equation gives an expected 19% reduction in night crashes for each 0.5 $cd.m^{-2}$ increase in average road surface luminance up to an average road surface luminance of 2 $cd.m^{-2}$.

Another approach for overcoming the high level of noise evident in Figure 1.1 is to carry out a meta-analysis of multiple independent studies of the same question. Meta-analysis is useful where the independent studies are of good quality but of small size because combining them then increases the statistical power of the

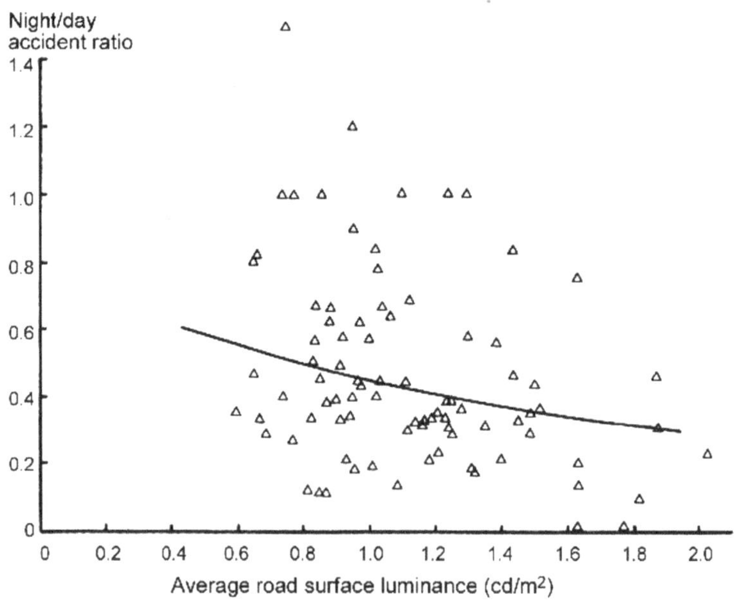

FIGURE 1.1   Night/day accident ratios plotted against average road surface luminance (cd. m^{-2}). The curve is the best-fitting exponential through the data, after weighting each ratio for the number of accidents to which it relates (after Hargroves and Scott, 1979).

analysis. It can be misleading when the underlying independent studies are of dubious quality. Elvik (1995) carried out a meta-analysis of the safety benefits of introducing road lighting to previously unlit roads. Based on results from 37 studies from different countries in which the effects had been measured in terms of the change in either the number of nighttime crashes or the number of nighttime crashes per million vehicle kilometres of travel, it was estimated that introducing road lighting should lead to a 65% reduction in nighttime fatal crashes, a 30% reduction in nighttime injury crashes, and a 15% reduction in nighttime property damage crashes. Wanvik (2009) carried out a literature review and examined data from studies in Norway and the Netherlands. He found that the presence of road lighting reduced personal injury crashes by somewhere in the range of 28%–50%.

An alternative approach to identifying the effect of darkness on crash risk is the odds ratio method (Johannson et al. 2009). This method uses national databases to get the number of crashes occurring over a year at two different hours of the day. One hour has a change from darkness to daylight over the year because of the shift in sunrise and sunset times over the year (the case hour). The other hour shows no change in daylight over the year (the control hour). The odds ratio is given by the ratio of the number of crashes in darkness and daylight in the case hour divided by the ratio of the number of crashes in the control hour when the case hour is in darkness and the number of crashes in the control hour when the case hour is in daylight. Odds ratios greater than 1.0 imply that there is an increased risk of a crash after dark. Johannson et al. (2009) applied this method to data from Norway, Sweden, and the Netherlands and found that the risk of an injury crash increased by almost 30% in urban areas, which would have had road lighting after sunset, and by nearly 50% in rural areas which would not. This method has the advantages that it does not require exposure data and crash data from the whole year is used but care is required in choosing the control hour as there may be differences in traffic density and traffic mix between the case and control hours. Current practice is to calculate odds ratios for a number of combinations of case/control hours and then to take the mean of the odds ratios to get a stable estimate of the value of road lighting (Uttley et al., 2023).

1.5   THE BENEFITS OF LIGHT FOR DIFFERENT TYPES OF CRASHES

The above discussion has served two purposes. It has demonstrated that road lighting does have some value as a crash countermeasure, and it has exposed the limitations of simple evaluations based on crash numbers or crash risks. The results discussed above are often bedeviled by noise, this being due to the multiple causes of crashes. If we wish to have a clearer picture of the role of lighting in traffic safety, a method that will reduce the amount of noise in the data is required. An elegant solution to this problem is to use the change in lighting associated with the introduction of daylight saving time (Tanner and Harris, 1956; Ferguson et al., 1995; Whittaker, 1996). In the usual daylight saving time system, the clock is moved forward by one hour in spring and back one hour in autumn. On both occasions, the effect is to suddenly change a period of driving from light to dark or vice versa. If it is assumed that activity and traffic patterns are governed by clock time, then it is likely that levels of exposure, fatigue, intoxication and driver demographics do not change substantially shortly before and shortly after the daylight saving time changeover, so any difference in crashes can plausibly be ascribed to the change in lighting conditions. Sullivan and Flannagan (2002) used data from the years 1987 to 1997 in the FARS database to determine the total number of fatal collisions involving pedestrians in 46 of the 50 states in the USA, for the hour close to the dark limit of civil twilight that showed the greatest change in light level at the daylight saving time change (Arizona, Hawaii, and Indiana were excluded because they do not have daylight saving time, and Alaska was excluded because its solar cycle is markedly different from the other states included). The dark limit of civil twilight is defined as occurring when the centre of the sun is six degrees below the horizon. The effect of the daylight saving time change on a spring morning is to move the lighting conditions from twilight to night and then back through twilight to day, as day length increases. Figure 1.2a shows the total number of fatal pedestrian crashes occurring at twilight, for the morning transition, in the nine weeks before and after the spring daylight saving change. It can be seen that in the weeks before the change, there is a steady decrease in the number of fatal crashes, but at the daylight saving change, there is a rapid return to a high level of fatal crashes, a level that then reduces with the increasing day length. Figure 1.2b shows analogous data for the spring evening twilight, for the nine weeks before and after the daylight saving change. For the evening, the effect of the daylight saving change is to change driving conditions from night to day. The dramatic decrease in the number of fatal pedestrian crashes with this transition is obvious.

This approach was adopted to examine the effect of the change from light to dark on a number of crash types using two databases (Sullivan and Flannagan, 2007). The first was the FARS database (NHTSA, 2006). The second was the North Carolina Department of Transportation Crash dataset (NCDOT). For each dataset, crashes that occurred in the one-hour time window that changed from dark to light or from light to dark in the evening when the spring or autumn daylight savings time change occurred were totalled over several years. The FARS dataset was used to examine fatal crashes of different types over 18 years (1987–2004). The NCDOT dataset was used to examine different types of fatal, personal injury, and property damage-only crashes over nine years (1991–1999). For both databases, the time window for crashes starts at the dark limit of civil twilight based on the Standard Time and extends forward by one hour. In spring, this window changes from dark to light following the daylight savings time change. In autumn, this evening window changes from light to dark following the daylight savings time change. Crashes occurring during the evenings of the five weeks either side of the daylight-saving time changes were compiled, and the ratio of crashes of each type occurring in dark and light conditions was calculated. If there is no difference between the number of crashes during dark and light periods, the dark/light ratio will be unity. Dark/light ratios greater than unity indicate that reducing the amount of light available from daylight to whatever is provided by vehicle lighting and road lighting, if present, leads to a greater likelihood of a crash. Table 1.3 shows the dark/light ratios for fatal crashes of different types that are statistically significantly different from unity ($p < 0.05$). The types of fatal crash that had dark/light ratios not statistically significantly different

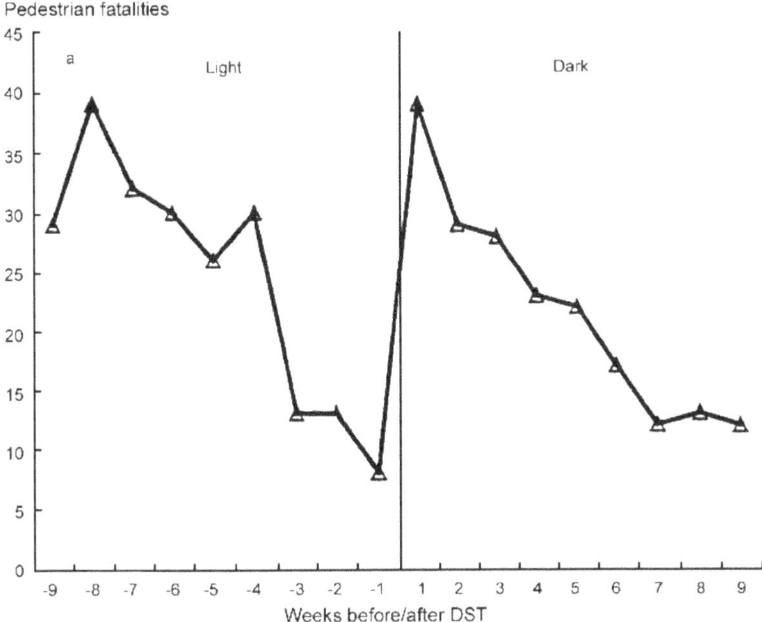

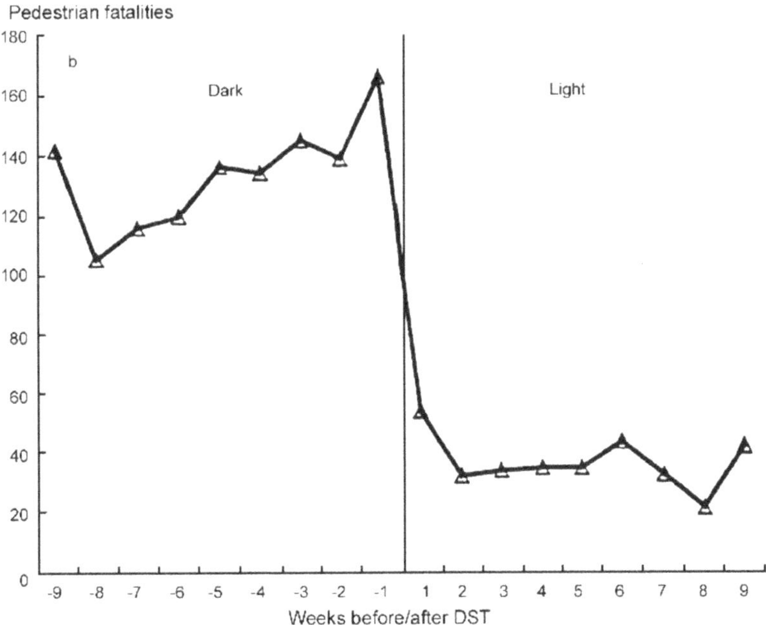

FIGURE 1.2 Cumulative number of pedestrian fatalities in 46 states of the USA, over the years 1987–1997, during twilight, for the nine weeks before and after the spring daylight saving time (DST) change (a) for morning and (b) for evening (after Sullivan and Flannagan, 2002).

from unity were a side swipe between vehicles moving in the same direction or in opposite directions, a collision with a fixed item and a collision rear to rear. There are three points to be noted from Table 1.3. The first is that some types of fatal crashes are strongly influenced by the reduction of visibility that occurs as daylight is replaced with some combination of vehicle lighting and road lighting. Others are not. Pedestrians are particularly at risk of fatal injuries after dark, a finding supported by the fact that pedestrian fatalities in Europe increase in winter (ERSO, 2007). Second, some fatal crash types are less likely to occur after dark, specifically, collisions with a fixed object off the road and a vehicle overturning. Such crashes imply a vehicle leaving the road rather than hitting something on the road. Possibly, this is due to the reduction in visibility associated with the onset of darkness making drivers more circumspect. Third, there is a large difference between the risks of pedestrians under 18 years of age being killed and for adult pedestrians. This probably has more to do with the level of exposure than anything else. Children, particularly younger children, are usually required to be indoors before dark, meaning that their level of exposure is confounded with the change in light level at the daylight saving time changeover.

Unlike the FARS database, the NCDOT database is dominated by non-fatal crashes. In the daylight savings time sample drawn from the NCDOT database, fatal crashes constituted only 0.5% of the total. 60% of the crashes were property damage

---

TABLE 1.3

Statistically Significant Dark/Light Ratios for Fatal Crashes of Different Types for the Evening Daylight Saving Time Transition, Based on the FARS Database

Crash Type	Number of Crashes in the Dark	Number of Crashes in the Light	Dark/Light Ratio
Pedestrians – 18–65 years	1,635	243	6.73
Pedestrians >65 years	845	126	6.71
Animals	61	11	5.55
Rear–end collision	440	198	2.22
Head-on collision	1,058	748	1.41
Collision with vehicle parked on road	82	58	1.41
Pedestrians <18 years	349	252	1.38
Angle collision	1,507	1,239	1.22
Miscellaneous	522	460	1.13
Collision with fixed object off road	955	1,088	0.88
Overturn	492	691	0.71

Source: Sullivan and Flannagan (2007).

only, the rest being crashes involving personal injuries. Table 1.4 shows the dark/light ratios for what are essentially non-fatal crashes that are statistically significantly different from unity ($p < 0.05$). The non-fatal crash types that had dark/light ratios not statistically significantly different from unity were colliding with elderly pedestrians, a rear-end collision when turning, a collision at an angle, a collision while turning right, a sideswipe, a collision with an object in the road, a collision with a fixed object, running off the road to the right or left, and a collision while backing.

There are a number of differences between Tables 1.3 and 1.4. Some of these differences are due to the different crash classification systems used in the two databases, but where the same crash type is considered in both databases, there is some consistency. Adult but not elderly pedestrians are at greater risk of both fatal and non-fatal crashes after dark. Both fatal and non-fatal crashes involving animals are more likely after dark. Both fatal and non-fatal rear-end and head-on collisions are more likely after dark. Both fatal and non-fatal crashes involving collision with a parked vehicle are more likely after dark. Both fatal and non-fatal crashes involving a vehicle overturning are less likely after dark.

Of course, there are also some discrepancies. The dark/light ratio for non-fatal crashes involving pedestrians under the age of 18 years is less than unity while the dark/light ratio for fatal crashes is greater than unity. This discrepancy is also probably due to the confounding of children's level of exposure with light level and the

---

TABLE 1.4

Statistically Significant Dark/Light Ratios for Non-fatal Crashes of Different Types for the Evening Daylight Saving Time Transition, Based on the NCDOT Database

Crash Type	Number of Crashes in the Dark	Number of Crashes in the Light	Dark/Light Ratio
Animals	4,656	560	8.31
Pedestrians – 18–65 years	292	115	2.54
Ran off road – straight ahead	205	96	2.14
Rear-end collision – slow	5,466	3,708	1.47
Left turn	2,265	1,819	1.25
Collision with parked vehicle	894	747	1.20
Head-on collision	205	162	1.18
Right turn cross traffic	362	310	1.17
Left turn cross traffic	1,340	1,167	1.15
Pedestrians <18 years	80	117	0.68
Overturn	52	98	0.53

Source: Sullivan and Flannagan (2007).

---

vulnerability of children when hit by a vehicle. Another anomaly involves the dark/ light ratio for crashes involving animals. The dark/light ratio for non-fatal crashes involving animals is higher than that for fatal crashes. This discrepancy is probably a matter of absolute numbers. The number of crashes associated with animals that prove fatal to humans is small but the number involving personal injury or property damage is large. Small numbers of crashes make the estimation of dark/light ratios uncertain.

Another interesting feature revealed by a comparison of Tables 1.3 and 1.4 is that for the same crash type, the dark/light ratio for fatal crashes is usually larger than for non-fatal crashes. This may be plausibly explained by the fact that fatal crashes often involve higher speeds than non-fatal crashes. Higher speeds allow less time to respond before collision, a time limit that is shortened further by low visibilities. This suggests that better road or vehicle lighting may be of greater importance for preventing fatal crashes than non-fatal crashes because they offer the possibility of increasing the time available for a response. This suggestion is supported by the findings of Plainis et al. (2006). They examined the effect of road lighting on the severity of injuries measured as the ratio of the number of crashes involving a fatality per 100 crashes. Using data from the database Road Crashes Great Britain, 1996–2005, the injury severity was calculated for day and night across a variety of road types. The mean injury severity across all road types under daylight was 1.1 but after dark it increased to 2.1 indicating that injury severity is almost doubled after dark. Injury severity was also calculated for the same road types after dark, with and without road lighting. The mean injury severity across all road types with road lighting was 1.5 but without road lighting it was 4.2, indicating the injury severity was almost three times larger on roads without lighting than on roads with lighting.

The data contained in Tables 1.3 and 1.4 are useful for three reasons. First, they indicate that some types of crashes are more sensitive to the reduction in visibility that follows the end of the day than others. If it were possible to identify where the crash types most sensitive to poor visibility were likely to happen, it would be possible to use light as a crash countermeasure more effectively. Second, the data in Tables 1.3 and 1.4 indicate that whatever the standards are for vehicle lighting and road lighting in the USA, they are capable of improvement. Ideally, vehicle and road lighting should reduce the dark/light crash ratio to unity. Third, the dark/light crash ratios can be used to assess the effectiveness of proposed lighting changes. For example, Sullivan and Flannagan (2007) used dark/light ratios for fatal and non-fatal crashes to evaluate the likely effectiveness of several innovative forms of vehicle forward lighting (see Section 6.7.2). For road lighting, the dark/light ratios combined with the frequency and cost of each crash type can be used to provide a monetary value for the benefits of road lighting to set against its undoubted cost (see Section 14.3).

Of course, the dark/light ratios derived by the daylight savings time changeover method are not without limitations. They are derived from the data of one country. Different dark/light ratios are likely to be found in other countries where different driving habits prevail. What constitutes dark will vary from site to site depending on whether road lighting is installed and at what level, so they tell us more about the absence of daylight than the value of different types of vehicle and road lighting. But the main limit is that they are based on drivers who are travelling around dusk. This may exaggerate the role of animals in crashes, because some large animals, such as deer, are crepuscular and so are most active around dusk. It may also show bias because the characteristics of drivers change through the night. Depending on the time of year and the latitude of the country dusk can range from late afternoon to late evening, clock time. People driving at dusk are much less likely to be intoxicated than those driving late at night (NHTSA, 2021a), but they are also more likely to be exposed to higher density traffic so in what direction the bias would occur is not at all clear.

Such limitations suggest that the dark/light ratios given in Tables 1.3 and 1.4 should be considered as indicative rather than definitive. Fortunately, the pattern of dark/light ratio values conforms to common sense. The crash types with the highest dark/light ratios are those involving vulnerable and unlighted objects, such as pedestrians and animals, or where objects, which may be lighted or unlighted, appear unexpectedly on the road, or where the road suddenly changes direction. Unlighted objects, such as pedestrians, will have low visibility after dark compared to lit objects, such as vehicles. Unexpected objects and unexpected road configurations require a response within a limited time. Improving visibility through better road marking, better road lighting, or better vehicle lighting allows more time to make a response. Thus, there can be little doubt that lighting has a role to play in reducing crashes, but before getting too carried away, it would be as well to remember the saying "When all you have is a hammer, everything looks like a nail." Visibility can be improved by means other than lighting. For example, a pedestrian wearing light clothing will be more visible than one wearing dark clothing and one wearing a fluorescent jacket will be even more so. Although this book is concerned with lighting, care will be taken not to ignore alternative means to achieve the desired end, an enhancement of traffic safety.

1.6 SUMMARY

Driving is a visual task but driving involves a lot more than seeing. Most visual tasks have three components: Visual, cognitive, and motor. The visual component is the process of extracting information relevant to the performance of the task using the sense of sight. The cognitive component is the process by which sensory stimuli are interpreted and the appropriate action is determined. The motor component is the process by which the stimuli are explored to extract information and/or the actions decided upon are carried out. All three components are involved in the process of driving, but lighting affects only the visual component.

The role of lighting in driving is to enable the transfer of information from the environment to the human visual system. Lighting achieves this role by making objects visible. Whether or not an object is visible will depend on the stimulus the object presents to the visual system, the quality of the retinal image, the operating state of the visual system, and where the object appears in the visual field. There are three measures that determine the stimulus: Visual size, luminance contrast, and colour difference. The quality of the retinal image is determined by the extent to which light is scattered during its passage to the retina and the ability of the visual system to focus the image on the retina. As for the operating state of the visual system, that is determined by the retinal illuminance. The significance of where the object appears in the visual field is due to the fact that visual performance deteriorates the further the object is from the fovea of the retina. It is these factors that mainly determine the visibility of the object. The visibility of the object influences the driver's response.

Given that the ultimate purpose of lighting on vehicles and roads is to enhance the safety of road users by increasing the visibility of the road ahead and objects on and around it, one way to examine the effectiveness of such lighting is to consider the number of crashes that occur by night and day. This has proved to be more difficult than would be expected. Simply counting the number of crashes occurring by night and day is not enough because it ignores the other side of the equation, the level of exposure. The quantity needed to evaluate the benefits of any traffic safety measure is the crash risk, which is the ratio of the outcome of exposure, that is, a crash, to the level of exposure. Even when this is done, there remains a lot of noise in the data. The basic problem is that road crashes have many interacting causes and visibility influences only some of them. Nonetheless, one meta-analysis of multiple studies of the effect of road lighting on crashes has led to the conclusion that introducing road lighting to previously unlit roads should lead to a 65% reduction in nighttime fatal crashes, a 30% reduction in nighttime injury crashes, and a 15% reduction in nighttime property damage crashes. Another approach, the odds ratio method, has produced an estimated increase in crash risk after dark of nearly 30% in urban areas which would have some form of road lighting and almost 50% in rural areas which would not.

Of course, these are overall figures and offer little guidance as to where road lighting might be most effectively employed. An alternative approach based on the sudden change in light level at the same clock time that occurs at the daylight savings time change has been used to examine the consequences of reduced visibility. The results indicate that some types of crashes are more sensitive to the reduction in visibility that follows the end of the day than others. For example, adult pedestrians are almost seven times more likely to be killed after dark than during the daytime, but fatalities associated with overturning the vehicle are less likely after dark. Further, the pattern of sensitivity to reduced visibility conforms to common sense. The crash types with the highest sensitivity to reduced visibility are those involving vulnerable and unlighted objects, such as pedestrians and animals, or where objects appear unexpectedly on the road, or where the road suddenly changes direction. Unlighted objects will have low visibility after dark compared to lighted objects. Unexpected objects and unexpected road configurations require a response within a limited time. Improving visibility through better road lighting and vehicle lighting allows more time to make a response. There can be little doubt that lighting has a role to play in improving traffic safety through greater visibility.

2 Light

2.1 INTRODUCTION

This book is concerned with how lighting can be used to enhance the safety of drivers and others on and near the roads. To understand the benefits and limitations of all forms of lighting, it is first necessary to understand what light is, how its characteristics can be quantified, and how it is produced and controlled. These topics are the subject of this chapter.

2.2 LIGHT AND RADIATION

To the physicist, light is simply part of the electromagnetic spectrum that stretches from cosmic rays with wavelengths of the order of femtometres to radio waves with wavelengths of the order of kilometres (Figure 2.1). What distinguishes the wavelength region between 380 and 780 nanometres from the rest of the electromagnetic spectrum is the response of the human visual system. Photoreceptors in the human eye absorb energy in this wavelength range and thereby initiate the process of seeing. Other creatures are sensitive to different parts of the electromagnetic spectrum, but light is defined by the visual response of humans.

The response of the human visual system is not the same at all wavelengths in the visible range. This makes it impossible to adopt the radiometric quantities conventionally used to measure the characteristics of the electromagnetic spectrum for quantifying light. Rather, a special set of quantities has to be derived from the radiometric quantities by weighting them by the spectral sensitivity of the human visual system.

Unfortunately, a unique spectral sensitivity curve applicable to all people in all conditions does not and cannot exist. This is because the human retina has two classes of visual photoreceptors: one class operating when light is plentiful, the cone photoreceptors, and the other operating when light is very limited, the rod photoreceptors (see Section 3.2.2). When only the cone photoreceptors are active, the visual system is said to be in the photopic state and when only the rod photoreceptors are active, it is in the scotopic state. These two photoreceptor types have very different spectral sensitivities. What these spectral sensitivities are has been the subject of international agreement. The body that organizes these agreements is the Commission Internationale de l'Eclairage (CIE). In 1924, the CIE adopted the CIE Standard Photopic Observer, based on the work of Gibson and Tyndell (1923), who took data from several experiments and proposed a smooth and symmetric spectral sensitivity curve (Viikari et al., 2005). The experiments from which the data were taken used small test fields, usually less than two degrees in diameter, viewed

DOI: 10.1201/9781003388906-2   **19**

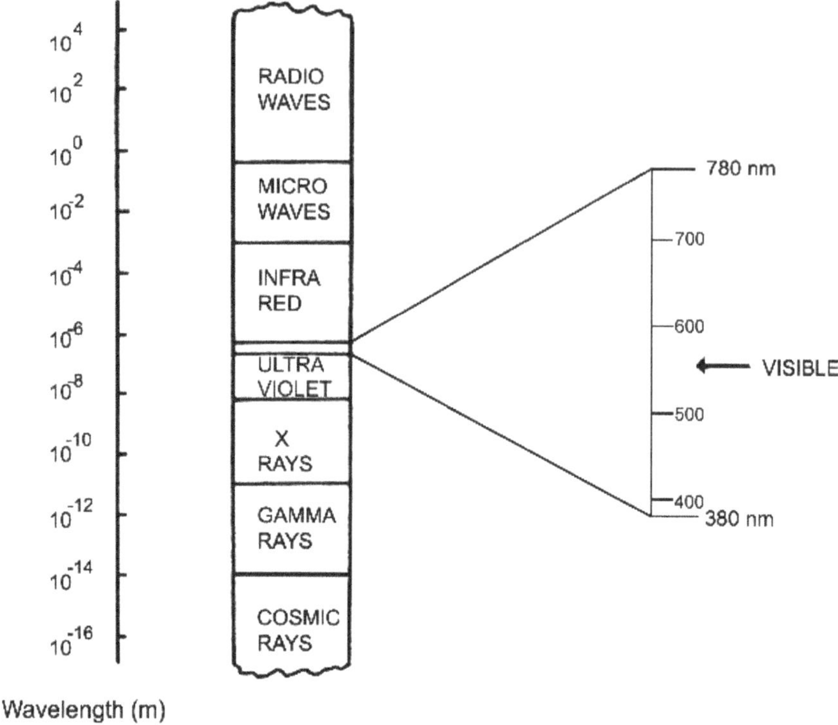

FIGURE 2.1 A schematic diagram of the electromagnetic spectrum showing the location of the visible spectrum. The divisions between the different types of electromagnetic radiation are indicative only.

directly, and the amount of light was sufficient to put the visual system into the photopic state. Later work by Judd (1951) showed that the CIE Standard Photopic Observer was too insensitive at short wavelengths, a result that eventually led the CIE to formally recognize a modified photopic spectral sensitivity curve (CIE, 1990) with greater sensitivity than the CIE Standard Photopic Observer at wavelengths below 460 nm. This CIE Modified Photopic Observer was presented as a supplement to the CIE Standard Photopic Observer, not a replacement for it. As a result, the CIE Standard Photopic Observer has continued to be widely used by the lighting industry. This is acceptable because the modified sensitivity at wavelengths below 460 nm has been shown to make little difference to the photometric properties of nominally white light sources that emit radiation over a wide range of wavelengths. It is only for light sources that emit large amounts of radiation below 460 nm that changing from the CIE Standard Photopic Observer to the CIE Modified Photopic Observer can be expected to make a significant difference in measured photometric properties (CIE, 1978). Some coloured signals, coloured signs, and narrow-band light sources, such as blue light-emitting diodes, fall into this category.

In 1951, the CIE adopted the CIE Standard Scotopic Observer, based on measurements by Wald (1945) and Crawford (1949), using an area covering the central 20 degrees of the visual field and at a light level low enough to ensure the visual system was in the scotopic state. While this is scientifically interesting because it represents the spectral response of the rod photoreceptors, until recently the CIE Standard Scotopic Observer has rarely been used by the lighting industry because the provision of almost any lighting installation worthy of the name will take the human visual system out of the scotopic state. However, interest in mesopic vision, the intermediate state where both rod and cone photoreceptors are active, has increased the use of the CIE Standard Scotopic Observer (see Section 2.4.3.3).

The CIE Standard and Modified Photopic Observers and the CIE Standard Scotopic Observer are shown in Figure 2.2 – the Standard and Modified Photopic Observers having maximum sensitivities at 555 nm and the Standard Scotopic Observer having a maximum sensitivity at 507 nm (CIE, 1983, 1990). These relative spectral sensitivity curves are formally known as the 1924 CIE Spectral Luminous Efficiency Function for Photopic Vision, the CIE 1988 Modified Two Degree Spectral Luminous Efficiency Function for Photopic Vision, and the 1951 CIE Spectral Luminous Efficiency Function for Scotopic Vision, respectively. More commonly,

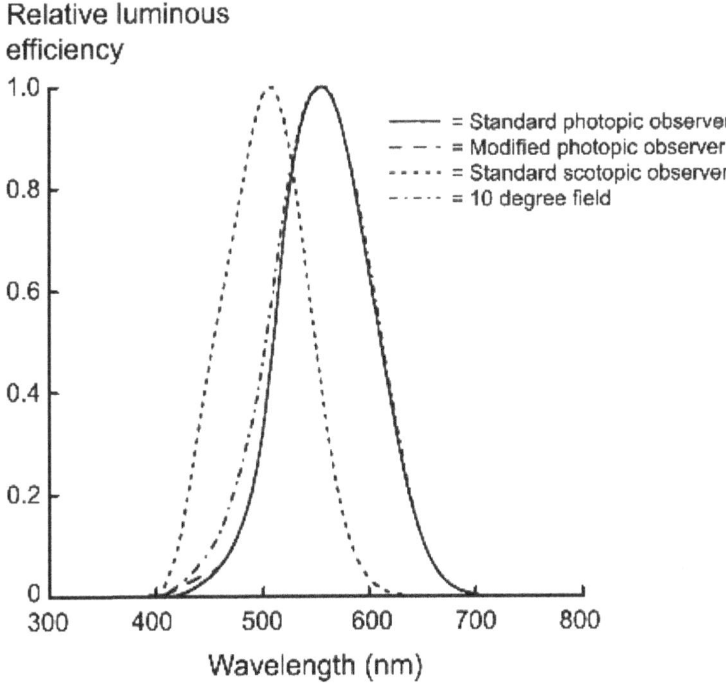

FIGURE 2.2 The relative luminous efficiency functions for the CIE Standard Photopic Observer, the CIE Modified Photopic Observer, the CIE Standard Scotopic Observer, and the relative luminous efficiency function for a 10-degree field of view in photopic conditions.

they are known as the CIE $V(\lambda)$, CIE $V_M(\lambda)$, and the CIE $V'(\lambda)$ curves. These curves are the basis of the conversion from radiometric quantities to photometric quantities, and the quantities used to characterize light.

2.3 THE MEASUREMENT OF LIGHT – PHOTOMETRY

2.3.1 DEFINITIONS

The most fundamental measure of the electromagnetic radiation emitted by a source is its radiant flux. This is a measure of the rate of flow of energy emitted and is measured in watts. The most fundamental quantity used to measure light is luminous flux. Luminous flux is radiant flux multiplied, wavelength by wavelength, by the relative spectral sensitivity of the human visual system over the wavelength range 380 nm to 780 nm. This process can be represented by the equation:

$$\Phi = K_m \sum \Psi_\lambda\, V_\lambda\, \Delta\lambda$$

where Φ is the luminous flux (lumens), Ψ_λ is the radiant flux in a small wavelength interval $\Delta\lambda$ (watts), V_λ is the relative luminous efficiency function for the conditions, and K_m is a constant (lumens/watt).

The values of K_m are 683 lm/W for the CIE Standard and Modified Photopic Observers and 1700 lm/W for the CIE Standard Scotopic Observer. These numbers arise from the decision of the CIE that one watt of radiant flux at 555 nm should produce 683 lumens, for both photopic and scotopic conditions. As 555 nm is the maximum sensitivity of the CIE Standard and Modified Photopic Observers, the constant is unchanged for the photopic condition. But for the CIE Standard Scotopic Observer, the relative spectral sensitivity is only 0.402 at 555 nm. Therefore, the constant for scotopic conditions is 683/0.402 = 1700 lm/W. It is always important to identify which of the Standard Observers is being used in any particular measurement or calculation. This requirement has led the CIE to recommend that whenever the Standard Scotopic Observer is being used, the word scotopic should precede the measured quantity, i.e., scotopic luminous flux.

Luminous flux is used to quantify the total light output of a light source in all directions. While this is important, for lighting practice, it is also important to be able to quantify the luminous flux emitted in a given direction. The measure that quantifies this concept is luminous intensity. Luminous intensity is the luminous flux emitted/unit solid angle, in a specified direction. The unit of measurement is the candela, which is equivalent to one lumen/steradian. Luminous intensity is used to quantify the distribution of light from equipment such as road lighting luminaires and vehicle headlamps.

Both luminous flux and luminous intensity have area measures associated with them. The luminous flux falling on the unit area of a surface is called the illuminance. The unit of measurement of illuminance is the lumen metre^{-2} or lux. The luminous intensity emitted per unit projected area of a source in a given direction is the luminance. The unit of measurement of luminance is the candela.metre^{-2}. Table 2.1 summarizes these photometric quantities.

TABLE 2.1

The Photometric Quantities

Measure	Definition	Units
Luminous flux	That quantity of radiant flux which expresses its capacity to produce a visual sensation	Lumens (lm)
Luminous intensity	The luminous flux emitted in a very narrow cone containing the given direction divided by the solid angle of the cone, i.e., luminous flux/unit solid angle	Candela (cd)
Illuminance	The luminous flux per unit area at a point on a surface	Lumen.metre^{-2}
Luminance	The luminous flux emitted in a given direction divided by the product of the projected area of the source element perpendicular to the direction and the solid angle containing that direction, i.e., luminous intensity/unit area	Candela.metre^{-2}
Luminance coefficient	The ratio of the luminance of a surface to the illuminance incident on it	Candela.lumen^{-1}
Reflectance	The ratio of the luminous flux reflected from a surface to the luminous flux incident on it	
For a diffuse surface	Luminance = (Illuminance x Reflectance)/π	
Luminance factor	The ratio of the luminance of a reflecting surface viewed from a given direction to that of a perfect white uniform diffusing surface identically illuminated	
For a non-diffuse surface, for a specific direction, and for lighting geometry	Luminance = (Illuminance x Luminance factor)/π	

As might be expected, there is a relationship between the amount of light incident on a surface and the amount of light reflected from the same surface. The simplest form of the relationship is quantified by the luminance coefficient. The luminance coefficient is the ratio of the luminance of the surface to the illuminance incident on the surface and has units of candela.lumen^{-1}. The luminance coefficient of a given surface is dependent on the nature of the surface and the geometry between the lighting, surface, and observer.

There are two other quantities commonly used to express the relationship between the luminance of a surface and the illuminance incident on it. For a perfectly diffusely reflecting surface, the relationship is given by the equation:

$$L = E.\rho \ /\pi$$

where L is the luminance (cd.m^{-2}), E is the illuminance (lm.m^{-2}), and ρ is the reflectance.

Reflectance is defined as the ratio of reflected luminous flux to incident luminous flux. For a non-diffusely reflecting surface, i.e., a surface with some specularity, the same equation between luminance and illuminance applies but reflectance is replaced with luminance factor. Luminance factor is defined as the ratio of the luminance of the surface viewed from a specific position and lit in a specified way to the luminance of a diffusely reflecting white surface viewed from the same direction and lit in the same way.

Both illuminance and luminance are widely used in lighting practice to quantify the end result of installing a lighting system and the stimulus to the visual system. Being able to define these quantities is useful but, in addition, it is always helpful to have an idea of what are representative magnitudes for these quantities in different situations. Table 2.2 shows some illuminances and luminances typical of commonly occurring interior and exterior lighting situations, all measured using the CIE Standard Photopic Observer.

2.3.2 Some Limitations

Although the photopic photometric quantities defined above can be calculated or measured precisely, it is important to appreciate that they only represent the visual effect of light in a particular state. Specifically, they represent the brightness response of the central two degrees of the retina in high light level conditions with a minimum contribution from the colour vision channels (see Section 3.2.5). Changing field size or light level or using coloured stimuli can change the spectral sensitivity of the visual system.

The effect of field size was recognized by the CIE in 1964 when a provisional relative spectral sensitivity curve for the central 10 degrees of the visual field in photopic conditions was approved (CIE, 1986; see Figure 2.2). This curve shows greater sensitivity to short wavelength light than the CIE Standard Photopic Observer because the visual field extends beyond the macula, an area covering the central five degrees of the retina and containing a pigment that attenuates short wavelength light, and into the area where short wavelength-sensitive cone photoreceptors are found.

TABLE 2.2

Typical Illuminance and Luminance Values for Various Lighting Situations

Situation	Typical Surface	Illuminance (lm. m^{-2})	Luminance (cd.m^{-2})
Daylight from a clear sky in summer in a temperate zone	Grass	150,000	2,900
Daylight from an overcast sky in summer in a temperate zone	Grass	16,000	300
Electric lighting for textile inspection	Light grey cloth	1,500	140
Electric lighting for office work	White paper	500	120
Electric lighting for heavy engineering work	Steel	300	20
Motorway lighting	Asphalt road	70	1.5
Residential road lighting	Asphalt road	10	0.2
Moonlight	Asphalt road	0.1	0.002

As for the effect of changing light level, in addition to the obvious change from cone-based photopic vision to rod-based scotopic vision, there is an intermediate zone where both cones and rods are active, called mesopic vision (see Section 3.3.4). For the central two degrees of the retina, the fovea, the CIE Standard Photopic Observer still applies in the mesopic state because there are only medium and long wavelength-sensitive cones present in the fovea, which is what the CIE Standard Photopic Observer is based on. In the rest of the visual field, the spectral sensitivity is in a state of continual change as the balance between rod and cone photoreceptors changes with light level, until either rods dominate, as in scotopic vision, or cones dominate as in photopic vision. Mesopic vision is important for driving because road lighting often provides visual conditions that are in the mesopic range. Nonetheless, all the photometric quantities that are used to characterize road lighting and vehicle lighting are based on the CIE Standard Photopic Observer. This practice can lead to situations where the photometric measurements bear little relation to the visual effect of the light source for off-axis vision. The absence of a CIE mesopic photometry system is not for want of trying (CIE 1989). Indeed, several different systems have been suggested, most based on brightness perception and some weighted combination of photopic and scotopic measurements (Palmer, 1968; Ikeda and Shimozono, 1981; Trezona, 1991; Sagawa and Takeichi, 1992). Others have abandoned the perception of brightness as the basis for measuring spectral sensitivity and, using reaction time, have developed a unified system of photometry that covers photopic, mesopic, and scotopic light levels (Rea et al., 2004). Yet others have developed a system of mesopic photometry based on the performance of a number of tasks associated with driving (Eloholma and Halonen, 2006; Goodman et al., 2007). The significance of these different approaches will be taken up in Section 4.6.

The presence of a coloured stimulus, such as a traffic signal, is important because such stimuli ensure that the colour vision channels as well as the achromatic channel of the human visual system contribute to the measured spectral sensitivity. At the low luminances typically found on the roads at night, the activity in the colour channels leads to spectral sensitivities that are multi-peaked with a shift in the wavelength for maximum sensitivity, relative to the conventional CIE Standard Photopic Observer (Varady et al., 2007).

Two other systematic effects that lead to different relative spectral sensitivities from the values represented by the CIE Standard Photopic Observer occur with age or with defective colour vision. As discussed in Section 13.4, as the eye ages, the transmittance of the lens decreases, particularly at the short wavelength end of the visible spectrum. This will lead to reduced sensitivity in this wavelength region for older people (Sagawa and Takahashi, 2001). For people with defective colour vision, either there are missing photopigments or the photopigments are different from the normal. In either case, if the medium or long wavelength-sensitive cone photoreceptors are involved, these individuals' relative spectral sensitivity is likely to depart from that of the CIE Standard Photopic Observer.

In addition to these systematic effects, there are inevitable individual differences in spectral sensitivity between people. This implies that although the photometric quantities can be calculated and measured precisely, there is no guarantee that they will be closely related to the visual effect produced for a specific individual. However, the CIE Standard Observers do have definite value. They lead to a system of photometry that provides a globally agreed means for regulators to specify what lighting conditions are required for specific applications, for the lighting industry to quantify the performance of its products, and for designers to state what their lighting systems will deliver. Despite the utility of such measures, whenever considering the photometric quantities for a given lighting situation, it is always important to ask whether the CIE Standard Observer being used in the calculation of the photometric quantities is appropriate to the situation. If it is not, then the apparent precision of the measurement may be misleading.

2.4 THE MEASUREMENT OF LIGHT – COLORIMETRY

The photometric quantities described above do not take into account the wavelength combination of the light received at the eye. Thus, it is possible for two luminous fields to have the same luminance but to be made up of totally different combinations of wavelengths. In this situation, and provided either photopic or mesopic conditions prevail, the two fields will look different in colour. Exactly what colour will be seen depends not only on the spectral power distribution of the radiation incident on the retina but also on such factors as the luminance and colour of the surroundings and the state of adaptation of the observer (Purves and Beau Lotto, 2010). Colour is a perception developed in the brain from past experiences and the information contained in the retinal image. Light itself is not coloured. Nonetheless, to have a means of characterizing the colour perception associated with different light sources and other stimuli to the visual system, some way had to be found to provide quantitative measures of colour. The CIE colorimetry system provides such measures.

2.4.1 THE CIE COLORIMETRY SYSTEM

The basis of the CIE colorimetry system is colour matching, an activity in which an observer is asked to determine whether two fields are the same or different in colour. From extensive colour matching measurements, the CIE Colour Matching Functions have been determined. These functions are essentially the relative spectral sensitivity curves of human observers with normal colour vision and can be considered as another form of standard observer. There are three colour matching functions, as might be expected from the fact that humans with normal colour vision can match any colour of light with a combination of not more than three wavelengths of light from the long, medium, and short wavelength regions of the visible electromagnetic spectrum.

The colour of a light source can be represented mathematically by multiplying the spectral power distribution of the light source, wavelength by wavelength, by each of the three colour matching functions $x(\lambda)$, $y(\lambda)$, and $z(\lambda)$, the outcome being the amounts of three imaginary primary colours X, Y, and Z required to match the light source colour. In the form of equations, X, Y, and Z are given by

$$X = h\Sigma\ S(\lambda).x(\lambda).\Delta\lambda$$

$$Y = h\Sigma\ S(\lambda).y(\lambda).\Delta\lambda$$

$$Z = h\Sigma\ S(\lambda).z(\lambda).\Delta\lambda$$

where $S(\lambda)$ is the spectral radiant flux of the light source (W.nm^{-1}); $x(\lambda)$, $y(\lambda)$, and $z(\lambda)$ are the spectral tristimulus values from the appropriate colour matching function; $\Delta\lambda$ is the wavelength interval (nm); and h is an arbitrary constant.

If only relative values of the X, Y, and Z are required, an appropriate value of h is one that makes Y equal to 100. If absolute values of the X, Y, and Z are required, it is convenient to take h equal to 683 since then the value of Y is the luminous flux in lumens. If the colour being calculated is for light reflected from a surface or transmitted through a material, the spectral reflectance or spectral transmittance is included as a multiplier in the above equations. For a reflecting surface, an appropriate value of h is one that makes Y equal to 100 for a reference white because then the actual value of Y is the percentage reflectance of the surface.

Having obtained the X, Y, and Z values, the next step is to express their individual values as proportions of their sum, i.e.,

$$x = X/(X + Y + Z) \quad y = Y/(X + Y + Z) \quad z = Z/(X + Y + Z)$$

The values x, y, and z are known as the CIE chromaticity coordinates. As $x + y + z = 1$, only two of the coordinates are required to define the chromaticity of a colour. By convention, the x and y coordinates are used. Given that a colour can be represented by two coordinates, then all colours can be represented on a two-dimensional surface. Figure 2.3 shows the CIE 1931 chromaticity diagram, the two axes being the x and y chromaticity coordinates. It is possible to identify a number of interesting features on the CIE 1931 chromaticity diagram. The outer curved boundary is called the spectrum locus. All pure colours, i.e., those that consist of a single wavelength, lie on this curve. The straight line joining the ends of the spectrum locus is the purple boundary and is the locus of the most saturated purples obtainable. At the centre of the diagram is a point called the equal energy point ($x = 0.33$, $y = 0.33$). This is the point where a colourless surface will be located. Close to the equal energy point is a curve called the Planckian locus. This curve passes through the chromaticity coordinates of objects that operate as a black body, i.e., the spectral power distribution of the light source is determined solely by its temperature.

The CIE 1931 chromaticity diagram can be considered as a map of the relative location of colours. The saturation of a colour increases as the chromaticity

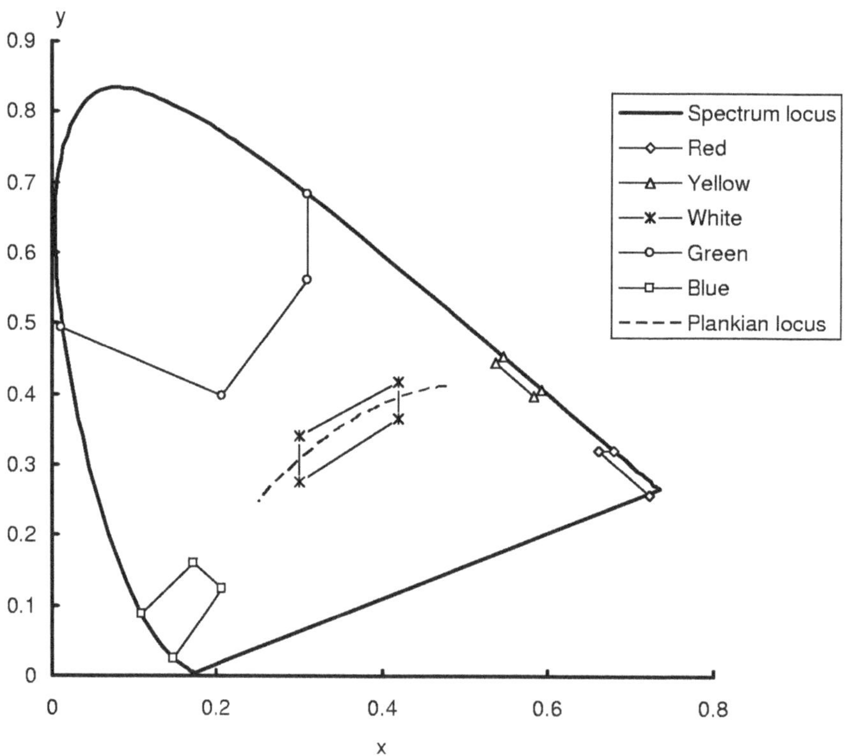

FIGURE 2.3 The CIE 1931 chromaticity diagram showing the spectrum locus and purple boundary, the Planckian locus from 2,500 K to 30,000 K and the areas defining the chromaticity coordinates suitable for red, yellow, white, green, and blue signal colours (after CIE, 1994).

coordinates get closer to the spectrum locus and further from the equal energy point. The hue of the colour is determined by the direction in which the chromaticity coordinates move from the equal energy point. Strictly, any discussion as to how a specific combination of wavelengths will appear, based on the chromaticity diagram, is nonsense. The only thing that a set of chromaticity coordinates tells us about a colour is that the colours with the same chromaticity coordinates will match. They tell us nothing about the appearance of the matched colours. But this is an argument for colour vision zealots. The fact is that a red surface lit by a nominally white light source will always plot in the bottom right corner, a green in the uppermost part, and a blue in the bottom left corner of the diagram. Thus, although the CIE 1931 chromaticity diagram is not theoretically pure, it is useful for indicating approximately how a colour will appear, a value recognized by the CIE when it specified chromaticity coordinate limits for traffic signals and traffic sign surfaces so that they will be recognized as red, yellow, white, green or blue (CIE, 1994, 1998).

Given that different colours plot at different positions on the CIE 1931 chromaticity diagram, it would seem reasonable to expect that the distance between two sets of chromaticity coordinates would be correlated with how different the two colours represented by the chromaticity coordinates appear. While this is approximately true, the correlation is very low. This is because the CIE 1931 chromaticity diagram is perceptually non-uniform. Green colours cover a large area while red colours are compressed in the bottom right corner and blue colours are compressed in the bottom left corner (Figure 2.3). This perceptual non-uniformity makes any attempt to quantify large colour differences using the CIE 1931 chromaticity diagram futile. In an attempt to improve this situation, in 1976 the CIE recommended the use of the CIE 1976 uniform chromaticity scale (UCS) diagram. This diagram is simply a linear transformation of the CIE 1931 chromaticity diagram. The axes for the CIE 1976 UCS diagram are

$$u' = 4x/(-2x + 12y + 3) \qquad v' = 9y/(-2x + 12y + 3)$$

where x and y are the CIE 1931 chromaticity coordinates.

While the 1976 uniform chromaticity scale diagram is more perceptually uniform than the CIE 1931 chromaticity diagram, it is still of limited value for determining colour differences. This is because it is two dimensional, considering only the hue and saturation of the colour. To completely describe a colour, a third dimension is needed, that of brightness for a self-luminous object and lightness for a reflecting object (Wyszecki, 1981). In 1964, the CIE introduced the U^*, V^*, W^* three-dimensional colour space for use with surface colours. This U^*, V^*, W^* system is little used now, about the only purpose for which it is routinely used is the calculation of the CIE colour rendering indices, although even here it is being replaced (see Section 2.4.3.2). For all other uses, the U^*, V^*, W^* colour space has been superseded by two other colour spaces introduced by the CIE in 1976 (Robertson, 1977, CIE, 1986). These two colour spaces are known by the initialisms CIELuv and CIELab. These two colour spaces and their associated colour difference formulas are now used to set colour tolerances for manufacture in many industries.

2.4.2 Colour Order Systems

While the CIE colorimetry system is valuable for quantifying colours, it does lack a physical presence. This need is met by a variety of colour ordering systems. A colour ordering system is a physical, three-dimensional representation of colour space. In a sense, it is an atlas of colours and like an atlas, the separation between adjacent colours is intended to be uniform in all directions. There are several different colour ordering systems used in different parts of the world (Billmeyer, 1987). One of the most widely used is the Munsell system. Figure 2.4 shows the organization of the Munsell system. The azimuthal Hue dimension consists of 100 steps arranged around a circle, with five principal hues (red, yellow, green, blue, and purple) and five intermediate hues (yellow-red, green-yellow, blue-green, purple-blue, and red-purple). The vertical Value scale contains ten steps from black to white. The horizontal Chroma scale contains up to 20 steps from grey to highly saturated. Each of the three scales is designed to provide equal steps of perception for an observer with normal colour vision looking at the samples lit by daylight, with a grey or white surround. The position of any colour in the Munsell system is identified by an alphanumeric reference made up of three terms, Hue, Value, and Chroma, e.g., a strong red is given the alphanumeric 7.5R/4/12. Achromatic surfaces, i.e., colours that lie along the vertical Value axis and hence have no hue or chroma, are coded as Neutral 1, Neutral 2, etc., depending on their reflectance. To a first approximation,

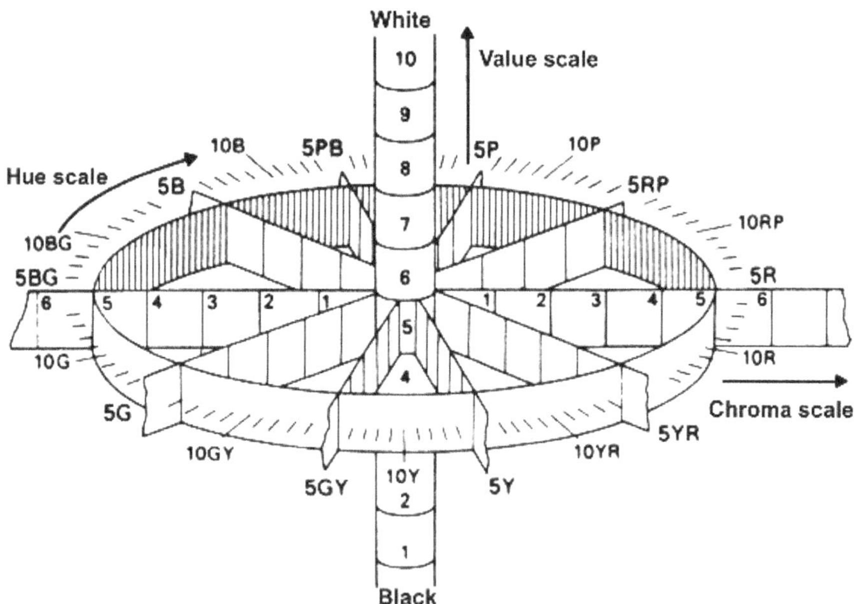

FIGURE 2.4   The organization of the Munsell colour order system. The hue letters are R = red, YR = yellow/red, Y = yellow, GY = green/yellow, G = green, BG = blue/green, B = blue, PB = purple/blue, P = purple, and RP = red/purple (from IESNA, 2000).

the percentage reflectance of a surface is given by the product of V and $(V-1)$ of the surface where V is the Munsell Value of the surface. The utility of a colour ordering system is that it makes colours manifest and hence makes it easy to communicate about colour in a more exact way than words permit, although not as accurately as numbers allow.

The existence of several different colour atlas systems used in different parts of the world, as well as the quantitative CIE colorimetry system, would seem to be a recipe for confusion. Fortunately, this is usually avoided by the fact that conversions are available between many of the colour ordering systems and the CIE colorimetry system. For example, the German DIN system provides both Munsell and CIE equivalents of its components (Richter and Witt, 1986). The name categories of the Inter-Society Color Council – National Bureau of Standards Method (Kelly and Judd, 1965) – are given in terms of the Munsell system (National Bureau of Standards, 1976). Conversions between the Munsell system and the CIE colorimetry system are given in the American Society for Testing and Materials Test Method D1535 (ASTM, 1996), based on Nickerson (1957).

2.4.3 Application Metrics

While the CIE colorimetry system is the most complete and most widely accepted means of quantifying colour, it is undeniably complex. Therefore, the lighting industry has used the CIE colorimetry system to derive two single-number metrics to characterize the colour properties of light sources: Correlated colour temperature and the CIE general colour rendering index (CRI). Correlated colour temperature is a metric quantifying the colour appearance of the light emitted by a light source. The CIE general colour rendering index is a metric quantifying the effect the spectrum emitted by a light source has on the appearance of surface colours.

2.4.3.1 Correlated Colour Temperature

Different light sources have different spectral power distributions, and hence the light emitted appears to be different in colour. In principle, the colour of the light emitted can be characterized by its chromaticity coordinates. This approach is used for signal lights (Figure 2.3) but for applications where colour is less critical and nominally white, the correlated colour temperature is used. The basis of this measure is the fact that the spectral emission of a black body is defined by Planck's Radiation Law and hence is a function of temperature only. Figure 2.3 shows the Planckian locus for a limited temperature range. The locus is the curved, dashed line in the middle of the CIE 1931 chromaticity diagram joining the chromaticity coordinates of black bodies at different temperatures. When the chromaticity coordinates of a light source lie directly on the Planckian locus, the colour appearance of that light source is expressed by the colour temperature, i.e., the temperature of the black body that has the same chromaticity coordinates. For light sources that have chromaticity coordinates close to the Planckian locus but not on it, their colour appearance is quantified as the correlated colour temperature, i.e., the temperature of the iso-temperature line that is closest to the actual chromaticity coordinates of the light source. The temperatures are usually given in degrees Kelvin (K). It is worth noting that two light sources with the same correlated colour temperature can have different colour appearances. This is because the two light sources can lie at different points along the specific iso-temperature line and hence have different chromaticity coordinates.

Nonetheless, correlated colour temperature is a very convenient and easily understandable metric of light source colour appearance, applicable to nominally white light sources. As a rough guide, such light sources have correlated colour temperatures ranging from 2,700 K to 6,500 K. A 2,700 K light source will have a yellowish colour appearance and be described as "warm" while a 6,500 K light source will have a bluish appearance and be described as "cool". It is important to appreciate that light sources that have chromaticity coordinates distant from the Planckian locus, such as the low-pressure sodium light source used for road lighting in some countries, do not have a correlated colour temperature. There is a metric that sets limits as to how far away from the Planckian locus the chromaticity coordinates of a solid-state light source can be if it is to be considered a source of white light. This metric, called Duv, is the distance between the chromaticity coordinates of the light source and the nearest point on the Planckian locus, measured on the CIE 1976 UCS diagram. Maximum allowed Duv values have been recommended, positive values for chromaticity coordinates above, and negative for below, the Planckian locus (ANSI, 2017).

2.4.3.2   CIE Colour Rendering Index

The effect the spectrum emitted by a given light source will have on the appearance of a surface colour can be estimated by calculating the chromaticity coordinates of the reflected radiation. This is reasonable if a specific set of surface colours is of interest, as is the case for road sign colours, but for most lighting applications, where many different but unspecified colours are used, more general advice is desirable. This is where the CIE colour rendering index comes in. The CIE colour rendering index measures how well a given light source renders a set of standard test colours relative to their rendering under a reference light source of the same correlated colour temperature as the light source of interest (CIE, 1995). The reference light source used is an incandescent light source for other light sources with a correlated colour temperature below 5,000 K and some form of daylight for light sources with a correlated colour temperature above 5,000 K. The actual calculation involves obtaining the positions of a surface colour in the CIE 1964, U^*, V^*, W^*, colour space under the reference light source and under the light source of interest and expressing the difference between the two positions on a scale that gives perfect agreement between the two positions a value of 100. The CIE has 14 standard test colours. The first eight form a set of pastel colours arranged around the hue circle. Test colours 9–14 represent colours of special significance, such as skin tones and vegetation. The result of the calculation for any single colour is called the CIE special colour rendering index for that colour. The average of the special colour rendering indices for the first eight test colours is called the CIE general colour rendering index. It is this latter index that is usually presented in light source manufacturers' catalogues.

Recently, a much more detailed and refined method of describing the way in which light will reveal surface colours has been developed (ANSI/IES, 2018). This method produces a metric quantifying the fidelity of surface colour reproduction and a series of measures of the saturation and hue shift of colours in different sectors of the hue circle (Boyce and Stampfli, 2019). Such information is very useful for interior lighting, particularly in retail and museum applications, but is unnecessarily detailed for road and vehicle lighting. Consequently, the CIE general colour rendering index is still dominant in these fields.

The CIE general colour rendering index has its limitations. First, it should be appreciated that just because two light sources have the same general colour rendering index (CRI), it does not mean that they render colours the same way. The general CRI is an average and there are many combinations of special CRI values that give the same average. Second, different light sources are being compared with different reference light sources. This makes the meaning of comparisons between different light sources uncertain, yet comparing light sources is what the general CRI is most often used for. Third, there is considerable argument about the method used to correct for chromatic adaptation. These limitations should be borne in mind when evaluating CIE general colour rendering indices for different light sources and when considering the improved colour metrics (ANSE/IES, 2018) that might be used where the increased complexity is considered worthwhile.

2.4.3.3 Scotopic/Photopic Ratio

Another measure of light source colour characteristic that is sometimes of value is the scotopic/photopic ratio (Berman, 1992). This is calculated by taking the spectral power distribution of the light source and weighting it by the CIE Standard Scotopic and Photopic Observers. The quotient of the resulting scotopic lumens and photopic lumens forms the scotopic/photopic ratio. The value of the scotopic/photopic ratio is that it expresses the relative effectiveness of the spectral power distribution of a light source in stimulating the rod and cone photoreceptors in the human visual system. A light source with a higher scotopic/photopic ratio will stimulate the rods more than a light source with a lower scotopic/photopic ratio when both produce the same photopic luminous flux. This can be useful when the visual system is operating in the mesopic state as it allows recommendations expressed in terms of photopic illuminance or luminance to be adjusted to more closely match the actual sensitivity of the visual system (BSI, 2013). This is of value when considering which of two light sources with very different spectral power distributions, such as high-pressure sodium and white LED, to use. Table 2.3 shows the scotopic/photopic ratios for a number of widely used light sources. As might be expected, there is a strong positive correlation ($r^2 = 0.99$) between the correlated colour temperature and the scotopic/photopic ratio of the light source (Fotios and Yao, 2019).

2.5 LIGHT SOURCES

In the early days of motoring, light sources powered by gas were used for road and vehicle lighting. Today, all the light sources used for both road and vehicle lighting

TABLE 2.3

Scotopic/Photopic Ratios for a Number of Widely Used Electric Light Sources (from Lighting Industry Association 2013)

Light Source	Scotopic/Photopic Ratio
Single-phosphor white LED (3,500 K)	1.39
Single-phosphor white LED (6,000 K)	2.18
Fluorescent (3,500 K)	1.36
Fluorescent (5,000 K)	1.97
Metal halide (5,000 K)	1.97
High-pressure sodium	0.63
Low-pressure sodium	0.25
Incandescent	1.36

are powered by electricity. The lighting industry makes several thousand different types of electric light sources. They can conveniently be divided into three classes – solid state, incandescent, and discharge. Over the last decade, solid-state light sources have taken over both road and vehicle lighting, but there are still plenty of examples of older light sources to be found on minor roads and in older vehicles.

2.5.1 Light-Emitting Diode

The light-emitting diode (LED) is a solid-state light source that is now the light source of choice for road and vehicle lighting as well as the lighting of signs and signals. This is mainly because of its high luminous efficacy and long life (see Table 2.4). The LED is a semiconductor that emits light when a current is passed through it. The spectral emission of the LED depends on the materials used to form the semiconductor. For light, the most common LED material combinations are now aluminium indium gallium phosphide (AlInGaP) and indium gallium nitride (InGaN). These combinations have replaced older combinations of gallium arsenide phosphide (GaAsP), aluminium gallium phosphide (AlGaP), and aluminium gallium arsenide (AlGaAs). LEDs typically produce narrow-band electromagnetic radiation, the spectral power distribution being characterized by the wavelength at which the maximum emission occurs (peak wavelength) and the halfbandwidth, this being half the difference in wavelengths at which the radiant flux is half the maximum radiant flux. The half-bandwidth for AlInGaP LEDs is about 17 nm while that for InGaN LEDs is 35 nm. The narrow spectrum of the light emitted by LEDs makes them an attractive proposition for traffic signals and vehicle signal lamps where a specific colour is required.

The light output of LEDs is determined by the current through the semiconductor and its temperature. LEDs made from different materials can vary widely in life, from a few thousand hours to a projected 100,000 hours, depending on the current used and the operating conditions (Narendran et al., 2001). The fact that LEDs are

TABLE 2.4

Summary of the Properties of Ssome Electric Light Sources Used for Road, Vehicle, Sign, and Ssignal Lighting

Light Source	Luminous Efficacy (lm.W^{-1})	CCT (K)	CRI	Lamp Life (hr)	Run-up Time (min)
Single-phosphor white LED	100–150	3,000–6,500	70–93	50,000	Instant
Incandescent	8–14	2,500–2,700	100	1,000	Instant
Tungsten halogen	15–25	2,700–3,200	100	1,500–4,000	Instant
Fluorescent	80–105	2,700–7,500	50–95	8,000–14,000	0.5
Compact fluorescent	45–80	2,700–4,100	80–85	8,000	0.25–1.5
Mercury vapour	33–57	3,200–3,900	40–50	8,000–10,000	4
Metal halide	65–120	3,000–6,000	60–90	2,000–10,000	0.5–8
High-pressure sodium	65–150	1,900–2,100	19–25	10,000–20,000	3–7
Low-pressure sodium	70–180	n.a.	n.a.	15,000–20,000	10–20
Induction	47–80	2,500–4,000	80	60,000	1

Notes: CCT, correlated colour temperature; CRI, CIE general colour rendering index; and n.a., not applicable.

solid-state devices and hence are less likely to fail due to mechanical damage caused by vibration has made them an attractive option for vehicle lighting. Further, the light output of LEDs can be reduced either by changing the current flowing through the semiconductor or by pulse width modulation of the power supply waveform.

It might be thought that the fact that the LED is a narrow-band source of light would preclude its use where white light is required, but this is not so. White light can be produced from LEDs by combining the light outputs from red, green, and blue LEDs in the right proportions or by adding a phosphor to the epoxy that encapsulates an InGaN LED with a peak wavelength of 470 nm (Nair and Dhoble, 2020). Figure 2.5 shows the spectral emission of both a single-phosphor white LED and three-chip RGB LED. The single-phosphor LED, also known as the phosphor-converted LED, is the one used for road and vehicle lighting. One limitation of all LEDs is the small luminous flux emitted. This means for most applications for roads and vehicles, multiple LEDs have to be assembled into a package to deliver the required light output. This necessity has advantages for vehicle design and for the control of light distribution.

The LEDs discussed above are all point sources of light and use inorganic compounds as their light-emitting elements. However, there are also organic LEDs (OLEDs) where the light-emitting element is an organic compound, which can form an area light source. OLEDs offer some interesting opportunities to the designer of vehicle interiors.

2.5.2 THE ELECTROLUMINESCENT LIGHT SOURCE

Electroluminescent light sources are a solid-state light source consisting of a sandwich in the form of a flat area conductor, a layer of dielectric–phosphor mixture, and another area conductor that is transparent. When a high, alternating voltage is applied across the two area conductors, the phosphor is excited and light is emitted. The colour of the light emitted depends on the dielectric–phosphor combination used and the frequency of the applied voltage. Spectral emissions of a wide range of colours are possible, but the most common are perceived as blue, green, or red. Electroluminescent light sources have a lower luminous efficacy than incandescent light sources. However, the fact that they have a long life and low-power requirements and can be formed as either rigid ceramic or flexible plastic sheets or tapes has made them an attractive option for self-luminous road signs, vehicle instrument panels, and for backlighting liquid crystal displays. As with LEDs, there have been developments that suggest an interesting future for the electroluminescent light source. This is the light-emitting polymer (Kwong et al., 2001). Light-emitting polymers are organic semiconducting materials that possess similar light-emitting properties to the inorganic LEDs but also have the mechanical properties of plastics (Bhuvama et al., 2018). This makes them another interesting possibility for vehicle interior lighting and displays.

2.5.3 THE LASER

Another solid-state light source is the laser. Laser stands for Light Amplification by Stimulated Emission of Radiation. The result of this stimulation is a very narrow

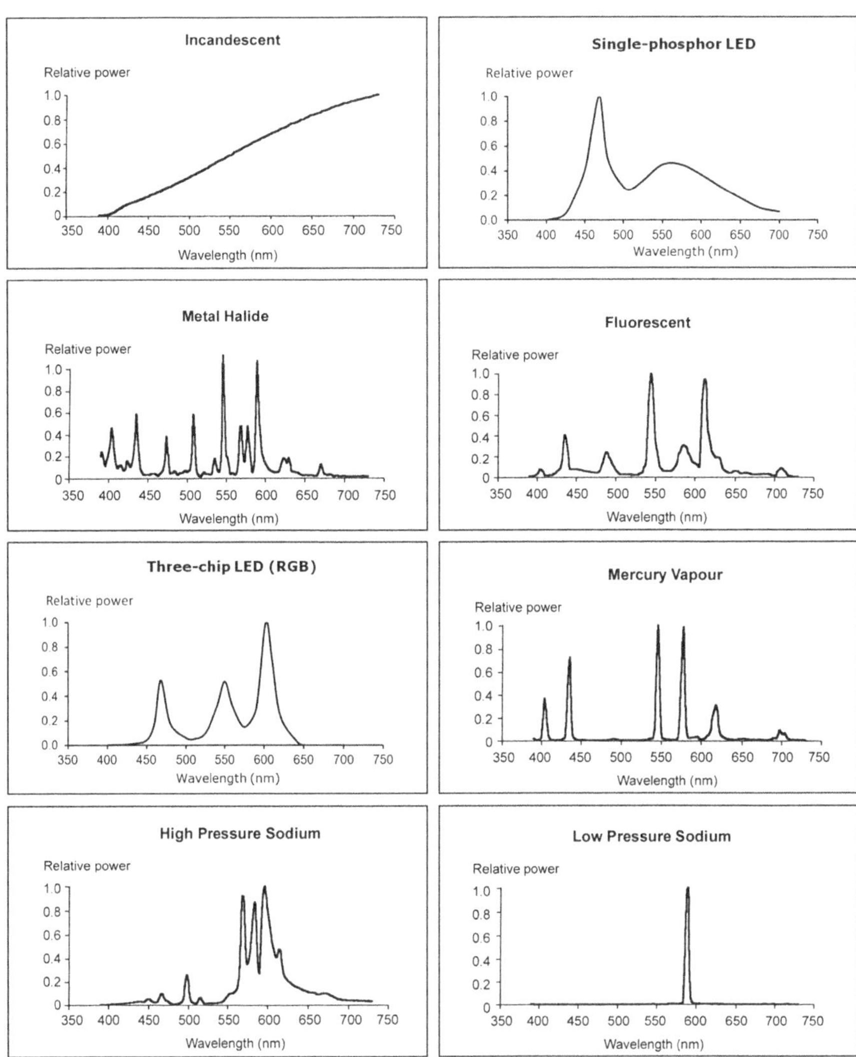

FIGURE 2.5 Relative spectral power distributions for different light sources. The light sources shown are incandescent, single-phosphor LED (also known as the phosphor-converted LED), metal halide, fluorescent, three-chip LED, mercury vapour, high-pressure sodium, and low-pressure sodium. All the spectral power distributions are normalized to unity for the wavelength with the maximum output. At various times, incandescent, fluorescent, low-pressure sodium, high-pressure sodium, and metal halide have all been used for road lighting. Today, the single-phosphor white LED is the light source of choice. These spectral power distributions are illustrative only. Different lamps of the same type can vary in their spectral power distribution, particularly fluorescent, metal halide, and three-chip LEDs (by Lei Deng).

beam of coloured light. There are several different types of laser using different materials, the actual colour depending on the material contained in the laser. It might be thought that such an intense but narrow beam would be of little value, but the laser's high efficiency and robustness mean that lasers are starting to appear in vehicle headlamp systems, vehicle signal systems, and, when connected to a fibre optic, even for vehicle interior lighting. To produce the white light necessary for vehicle headlamps, the output of the laser has to illuminate a phosphor. To control the light distribution, the output of the phosphor has to be directed via a lens, a reflector, or a set of micro-mirrors.

2.5.4 THE INCANDESCENT LIGHT SOURCE

The incandescent light source was, for many years, the light source used in traffic signals, vehicle signal lamps, and vehicle interior lighting. The incandescent light source produces light by heating a thin tungsten filament to incandescence in an inert gas atmosphere. The spectral emission of the incandescent light source is a continuum over the visible spectrum (Figure 2.5) although the light colour is determined by the temperature of the filament. This is easily seen when an incandescent light source is dimmed. Reducing the voltage reduces the current through the filament and hence the temperature of the filament. The result is that the colour appearance of the light emitted by the light source becomes more yellow and then red until, at very low voltages, no light can be seen at all, although the light source may still be emitting infrared radiation. The design of an incandescent light source is a matter of balancing luminous efficacy against life. A high light output and hence a higher luminous efficacy can be achieved by heating the filament to just below its melting point, but then the life is short. When a long light source life is desirable, as in traffic signals, the filament is heated to a lower than usual temperature by reducing the applied voltage. Incandescent light sources are sensitive to voltage fluctuations and vibration, unless specifically constructed to handle mechanical shock.

2.5.5 THE TUNGSTEN HALOGEN LIGHT SOURCE

Tungsten halogen light sources were once widely used in vehicle headlamps and some vehicle signal lamps. The tungsten halogen light source is essentially an incandescent light source with a halogen in the gas filling. The inclusion of the halogen allows the filament to be run at a higher temperature without excessive envelope blackening because, although the tungsten is evaporated off the filament faster at the higher temperature, the halogen chemically reacts with the evaporated tungsten to form a tungsten halogen compound which will not precipitate onto the light source envelope. If the tungsten halogen compound diffuses back to the filament, the higher temperature causes it to separate into tungsten and halogen, depositing some of the tungsten back onto the filament. This cycle ensures the light output is maintained at a higher level for longer than would be the case without the halogen. The spectral emission of the tungsten halogen light source is a continuum across the visible spectrum similar in form to that of the incandescent light source, as would be expected given its fundamental incandescent nature. Similarly, tungsten halogen light sources are sensitive to voltage fluctuations and vibration, unless specifically constructed to handle mechanical shock.

2.5.6 THE FLUORESCENT LIGHT SOURCE

Fluorescent light sources are still used in compact form for illuminating road signs and traffic bollards. The fluorescent light source is a discharge light source in that the physical means for producing light is the excitation of a gaseous discharge. The fluorescent light source, in either its linear or compact form, consists of a glass tube containing a mercury atmosphere. Heating the electrodes produces a stream of negatively charged electrons. These electrons are accelerated through the mercury gas by the potential difference between the electrodes. The accelerated electrons collide with the gas atoms producing two effects. The first is the ionization of the atoms into negatively charged electrons and positively charged ions. This increases the electron concentration and hence maintains the discharge. The second possible outcome is that the atom absorbs most of the energy of the colliding electron and thereby raises the energy state of its own captive electrons to a higher level. These energy levels are discrete and when the captive electrons shortly afterwards decay back to their resting level, energy is radiated at a wavelength determined by the energy-level structure of the atom. For mercury at a low pressure, which is what fills a fluorescent light source, most of the radiation emitted by the discharge is in the ultraviolet region of the electromagnetic spectrum. To produce radiation in the visible spectrum, the inner surface of the glass tube is coated with a phosphor. This absorbs the ultraviolet radiation from the discharge and emits radiation in the visible spectrum. This two-step process is evident in the spectral emission of the fluorescent light source (see Figure 2.5), which usually consists of a series of strong emission lines from the discharge superimposed on a continuous emission spectrum from the phosphor. By changing the phosphor mix, different spectral power distributions can be created so fluorescent light sources are available with a wide range of colour properties. The fluorescent light source is a discharge light source and therefore needs to have a control system to alter the electrical conditions from those necessary to start the discharge to those required to maintain it. This control gear, which is sometimes called a ballast, can be electromagnetic or electronic, the latter providing greater circuit luminous efficacy than the former. The light output of the fluorescent light source is a maximum at one ambient temperature. Higher or lower temperatures reduce the light output.

2.5.7 THE MERCURY VAPOUR LIGHT SOURCE

The mercury vapour light source is now obsolete, but it can still be found in some road lighting installations. The mercury vapour light source is similar to the fluorescent light source in that it is a discharge light source based on a mercury atmosphere in an arc tube. The difference is that the mercury vapour light source operates at high pressure. The result is that the spectral power distribution of the gas discharge is moved into the visible region, although it still consists of a series of intense spectral lines (see Figure 2.5). Today, the mercury vapour light source is usually sold, where permitted, with a phosphor coating on the inside of the envelope, the phosphor coating being used to improve its colour properties.

2.5.8 THE METAL HALIDE LIGHT SOURCE

The metal halide light source is a high-pressure gas discharge light source based on a mercury discharge, but it is different from the mercury vapour light source in that it has metal halides, such as scandium and sodium iodides, in the arc tube. When the arc tube reaches the operating temperature, the metal halides are vapourized. At the core of the discharge, the metal halides separate into the metals and halogen, with the metals emitting radiation in the visible region. At the cooler edge of the arc tube, the metals and halogen recombine and then repeat the process. The result is a spectrum consisting of many discrete spectral lines (see Figure 2.5). Over the years, the performance of the metal halide light sources has improved. The early versions used a quartz arc tube, but these suffered from poor colour stability over life. This problem was resolved in later versions by the use of a sintered aluminium arc tube and is commonly known as ceramic metal halides. Until the arrival of the light-emitting diode, ceramic metal halides were widely used for road lighting where white light was desired.

Metal halide light sources are also used for vehicle forward lighting but with one difference from those used for road lighting, the addition of xenon to the arc tube. This is added to improve the run-up time. Run-up time is the time taken to reach full light output. Ceramic metal halide light sources used for road lighting have run-up times of minutes. Such times are acceptable for road lighting but completely unacceptable for vehicle lighting. The effect of the xenon is to make a significant amount of the light output immediately available on switch-on and to make the run-up to full light output a matter of seconds rather than minutes. Such headlamps are commonly known as xenon or high-intensity discharge (HID) headlamps.

2.5.9 THE LOW-PRESSURE SODIUM LIGHT SOURCE

The low-pressure sodium light source was once widely used for road lighting in the UK, Belgium, and The Netherlands but only sparingly in other countries. The low-pressure sodium light source is a discharge light source based around sodium. Electrically, the low-pressure sodium light source operates in the same manner as the fluorescent light source, but in this case, a phosphor is unnecessary because the spectral emission from the sodium discharge is concentrated in two spectral lines, which are both close to 598 nm (see Figure 2.5). Because this wavelength is near to the maximum sensitivity of the human visual system at 555 nm, as shown by the CIE Standard Photopic Observer, the low-pressure sodium light source has the highest luminous efficacy of all the artificial light sources (up to 180 lm.W^{-1}). Unfortunately, its colour properties are what might be expected from an almost monochromatic source, non-existent.

2.5.10 The High-Pressure Sodium Light Source

Until the arrival of LEDs, the high-pressure sodium light source was the most common light source used for road lighting. Conceptually, the high-pressure sodium light source is the same as the low-pressure sodium light source, but the much higher pressure has an effect on the spectral emission. The increased pressure in the discharge leads to self-absorption of radiation within the discharge and interactions between the closely packed atoms. The combined effect of these phenomena is to reduce the power at 598 nm and to spread the spectral emission over a much wider range of wavelengths (see Figure 2.5). The result is a combination of high luminous efficacy and modest colour properties.

2.5.11 Induction Light Sources

Induction light sources are occasionally used for road lighting where access is difficult, e.g., bridges. Induction light sources differ from the discharge light sources discussed above in that they use an electromagnetic field, at either radio or microwave frequencies, to create a discharge. Induction light sources are essentially a fluorescent light source without electrodes. The electromagnetic field excites the mercury in an envelope that then emits radiation mainly in the ultraviolet. This is then absorbed by a phosphor and re-radiated in the visible region. The luminous efficacy and colour properties of induction light sources are similar to those of fluorescent light sources, their main advantage being the much longer life produced by not having any electrodes to fail.

2.5.12 Light Source Characteristics

Electric light sources can be characterized on several different dimensions. They are as follows:

- Luminous efficacy: This is the ratio of luminous flux produced to power supplied (lumens.watt^{-1}). This will vary with the wattage of the light source. If the light source needs control gear to operate, the watts supplied should include the power demand of the control gear. Note that once a light source is placed in a luminaire, the luminous efficacy will be decreased.
- Correlated colour temperature (CCT, see Section 2.4.3.1).
- CIE general colour rendering index (CRI, see Section 2.4.3.2).
- Lamp life: This is the number of burning hours until either light source failure or a stated percentage reduction in light output occurs (hours).
- Run-up time: This is the time from switch-on to full light output (minutes).

Table 2.4 summarizes these characteristics for many of the light sources used for road, vehicle, sign, and signal lighting. The values in Table 2.4 should be treated as indicative only. Details of the characteristics of any specific light source should always be obtained from the manufacturer. A more detailed discussion of the construction, operation, and properties of many of the light sources listed in Table 2.4 is available in Kitsinelis (2015).

2.6 LUMINAIRES

Being able to produce light is only part of what is necessary to produce illumination. Another part is to place the light source in a luminaire. The luminaire provides electrical, thermal, and mechanical support and protection for the light source as well as determining the light distribution. The light distribution is controlled by using reflection, refraction, or diffusion, individually or in combination (Simons and Bean, 2000; Wordenweber et al., 2007). The light distribution provided by a specific luminaire is quantified by its luminous intensity distribution. The light distributions allowed from vehicle lighting are strictly regulated, but those from road lighting luminaires are less so. All reputable road lighting luminaire manufacturers provide luminous intensity distributions for their luminaires.

Luminaires used for road lighting have to be capable of withstanding the elements. This is done by the choice of materials and careful sealing between components. The extent of protection provided by different luminaires is given by the Ingress Protection (IP) system, the degree of protection being indicated by the letters IP followed by two numbers. The first number gives the degree of protection against the ingress of foreign bodies and dust. The second gives protection against the ingress of water. Table 2.5 shows the degree of protection indicated by each

TABLE 2.5

The Ingress Protection (IP) Classification of Luminaires According to the Degree of Protection against the Ingress of Foreign Bodies, Dust, and Water

First Number	Degree of Protection	Second Number	Degree of Protection
0	Not protected	0	Not protected
1	Protected against solid objects greater than 50 mm	1	Protected against dripping water
2	Protected against solid objects greater than 12 mm	2	Protected against dripping water when tilted up to 15 degrees
3	Protected against solid objects greater than 2.5 mm	3	Protected against spraying water
4	Protected against solid objects greater than 1.0 mm	4	Protected against splashing
5	Dust - protected	5	Protected against water jets
6	Dust - tight	6	Protected against heavy seas
		7	Protected against the effects of immersion
		8	Protected against submersion to a specified depth

number. Most road lighting luminaires are IP 65 or better. The IP system is not applied to vehicle lighting, the requirements for resisting the elements being included in the relevant regulations.

2.7 THE CONTROL OF LIGHTING

The control of lighting involves the control of both light output and light distribution. The control of light output is provided by switching or dimming. Switching systems can vary from the conventional manual switch in a vehicle through simple time switches for road lighting to slightly more sophisticated photoelectric control systems that sense the level of natural light available and turn on or turn off the vehicle lighting or road lighting, as appropriate. Not all the light sources used for road and vehicle lighting can be dimmed. Solid-state light sources can be dimmed, incandescent light sources can be dimmed at the cost of a change in light colour, but most discharge light sources are difficult to dim.

More advanced control of road lighting is possible by signals transmitted down the power supply cables. This provides information to a local computer which sends the information through a wireless network to a central monitoring station. This enables the authority responsible for the road lighting to monitor its state and, through two-way communication, to adjust the light output as desired. This possibility has given rise to the idea that road lighting illuminances might be adjusted according to traffic densities and levels of pedestrian activity (see Section 14.3). Another control variable of interest for both road and vehicle lighting is the light distribution of the luminaire. Changing the light distribution is difficult when a single light source is used in the luminaire. However, where multiple light sources are used, as is the case with LEDs, there exists the possibility of changing the light output of different LEDs and hence the light distribution. The point to take from this brief discussion is that the reduction in costs and the increase in capabilities of sensors, computer networks, and communication networks have made much more sophisticated control of light output and light distribution possible.

2.8 FLUORESCENCE AND RETRO-REFLECTION

Although they are neither light sources nor luminaires, the phenomena of fluorescence and retro-reflection have important roles to play in making objects visible on and around the road. Both are part of the high-visibility clothing widely used by road workers and police officers and by some pedestrians, horse riders, and cyclists to increase their conspicuity.

Fluorescence occurs when electromagnetic radiation at one wavelength is absorbed by a chemical that then re-radiates some of it at a longer wavelength. The chemical incorporated into the material forming high-visibility clothing absorbs radiation in the ultra-violet region of the spectrum and then re-radiates some of it in the visible region. The material used in high-visibility clothing has a higher reflectance than most of the other substances against which it will be seen. This, alone, will give the clothing a high visibility for most backgrounds. However, the luminance caused by fluorescence is added to the luminance produced by reflection to produce an even higher luminance, so much so that the lightness constancy can be broken (see Section 3.5), a perception that will increase the conspicuity of the wearer. Fluorescence is effective as a means of increasing luminance during the day as daylight is a light source that is rich in ultraviolet radiation. After dark, the benefits of fluorescence are much reduced because few light sources have a lot of ultraviolet radiation. After dark, retro-reflection is a more effective means of increasing visibility (CIE, 1987).

A retro-reflector is a device that reflects light back from whence it came, over a wide range of angles of incidence. When the headlamps of a car illuminate a retro-reflective surface, the reflected light is directed back towards the car and its driver. Retro-reflectors come in a number of forms, the most common being the corner cube and the refractive/reflective combination. As its name suggests, the cube is an arrangement of three mutually perpendicular plane reflectors forming a corner, usually manufactured as a solid transparent cube. The corner cube relies on total internal reflection to redirect the light. The refractive/reflective combination has the reflective element coinciding with the image plane of the refractive element. A special case of this relevant to drivers is a transparent sphere that has a refractive index twice that of air. This will act as a retro-reflector because the back surface of the sphere forms the image plane of the refracting front surface.

Retro-reflectors are good for visibility when placed amongst diffusely reflecting surfaces, because they concentrate the reflected light in the direction from which it came, thereby giving the retro-reflector a much higher luminance than the surroundings and hence a much higher luminance contrast when viewed from that direction. Of course, this is only true for an observer sitting close, in angular terms, to a small source of high luminous intensity, such as a driver in a vehicle with the headlamps on, at night. Retro-reflectors do nothing for the visibility of objects seen by people remote from the light source, such as pedestrians.

Retro-reflectors can be small enough for large numbers to be incorporated into sheet material for use in road signs, into paint or plastic for road marking and into tape for application to clothing. Retro-reflectors can also be large enough for individual units to be embedded in the road with a slight projection above the road surface. These are the road studs familiar to drivers in many countries (see Section 5.2). Road studs are visible to drivers from a long way away and are particularly valuable where fog is frequent. Road studs are not generally used in areas that regularly experience snow during winter as snowploughs tend to damage them.

2.9 SUMMARY

This chapter is concerned with what light is, how it can be measured, and how it is produced and controlled. Light is part of the electromagnetic spectrum between 380 and 780 nm. What differentiates this wavelength range from the rest of the electromagnetic spectrum is that the human visual system responds to it. The actual human spectral response has been standardized in an internationally agreed form represented by the CIE Standard Photopic and Scotopic Observers. Using the appropriate spectral sensitivity curve, the four basic photometric quantities can be derived – luminous flux, luminous intensity, illuminance, and luminance.

These measures are all concerned with the overall amount of light and not with its colour. The CIE colorimetry system has been developed to quantify colour. By using the CIE colorimetry system, the colour appearance and colour rendering properties of light sources can be characterized, using such measures as correlated colour temperature and CIE general colour rendering index although more sophisticated metrics for colour properties are available for use where colour appearance is particularly important.

It is necessary to appreciate that while there are numerous metrics used to characterize a lighting situation or a light source, these metrics are simultaneously precise and inaccurate. The precision arises because the metrics can often be measured or calculated exactly. If they are regarded as simple physical measures, they can be considered accurate. But they are not simple physical measures. The whole reason for having photometric and colorimetric quantities is to quantify the visual effect of light. Because of the complexity and flexibility of the human visual system and the differences between individuals, any one standardized metric of visual effect is inevitably an approximation. This is particularly so for driving because all the photometric quantities used in the design and specification of vehicle lighting and road lighting use the CIE Standard Photopic Observer, but when driving at night, large parts of the visual field will be operating in the mesopic state.

Having considered how light can be measured, the physical principles and properties of the electric light sources used for roads, vehicles, signs, and signals are considered. Electric light sources used for roads, vehicles, signs, and signals fall into three types – solid state, incandescent, and discharge. Light sources differ in their physical size, light output, luminous efficacy, colour properties, life, run-up time, and sensitivity to the environment in which they have to operate. Road lighting gives priority to light output, luminous efficacy, and life. Vehicle lighting gives priority to physical size, light output, luminous efficacy, colour properties, and run-up time. Traffic signals and the lighting of traffic signs give priority to physical size, colour properties, and life.

Having chosen the light source, the next thing to consider is the characteristics of the luminaire in which it is contained. The luminaire provides electrical, thermal, and mechanical support for the light source and controls the light distribution. The light distribution is controlled by using reflection, refraction, or diffusion, individually or in combination. Lighting control systems for road and vehicle lighting are becoming much more sophisticated through a combination of sensors, computers, and communication networks.

The final optical phenomena of relevance to driving are fluorescence and retroreflection. Both are used in the high-visibility clothing commonly worn by people working on the roads. Fluorescence occurs when electromagnetic radiation at one wavelength is absorbed by a chemical that then re-radiates some of it at a longer wavelength. The chemical incorporated into the material forming high-visibility clothing absorbs radiation in the ultraviolet region of the spectrum and then re-radiates some of it in the visible region. Fluorescence is effective as a means of increasing luminance during the day as daylight is a light source that is rich in ultraviolet radiation. After dark, the benefits of fluorescence are much reduced because few light sources have a lot of ultraviolet radiation.

After dark, retro-reflection is a more effective means of increasing visibility. A retro-reflector is a device that reflects light back from whence it came, over a wide range of angles of incidence. When the headlamps of a car illuminate a retro-reflective surface, the reflected light is directed back towards the car and its driver, thereby increasing the luminance of the retro-reflector relative to the more diffusely reflecting surfaces around it. Retro-reflectors are used on road surfaces for edge and lane delineation, on road signs to make them readable at greater distances, and on vehicles and clothing to enhance conspicuity.

3 Sight Plus

3.1 INTRODUCTION

Light is essential for sight. With light, we can see, without light we cannot. But how different lighting conditions influence sight depends on the operation of the human visual system. The construction, capabilities, and constraints of the human visual system are the main subjects of this chapter. But there is another less obvious effect. Light received at the eye also influences the physiology of the human at a very fundamental level through its effect on the circadian timing system. If this system is disrupted, there are consequences for sleep and hence alertness and cognitive function, both of which may affect the ability to drive safely by day and night. This indirect effect of light exposure through the eye will also be examined.

3.2 THE STRUCTURE OF THE VISUAL SYSTEM

The human visual system is an image processing system consisting of the eye and brain working together to construct a model of the external world (Wolfe et al., 2006; Wordenweber et al., 2007). The model developed resolves the ambiguities inherent in the retinal image on the basis of probabilities derived from experience (Purves and Beau Lotto, 2010). Despite this complexity, the obvious starting point for any discussion of the visual system is the eye.

3.2.1 THE EYE

Figure 3.1 shows a section through the eye, the upper and lower halves being adjusted for focus at near and far distances, respectively. The eye is basically spherical with a diameter of about 24 mm. The sphere is formed from three concentric layers. The outermost layer, called the sclera, protects the contents of the eye and maintains its shape under pressure. Over most of the eye's surface, the sclera looks white but at the front of the eye the sclera bulges up and becomes transparent. It is through this area, called the cornea, that light enters the eye. The next layer is the vascular tunic or choroid. This layer contains a dense network of small blood vessels that provide oxygen and nutrients to the innermost layer, the retina. As the choroid approaches the front of the eye it separates from the sclera and forms the ciliary body. This element produces the watery fluid that lies between the cornea and the lens, called the aqueous humour. The aqueous humour provides oxygen and nutrients to the cornea and the lens, and takes away their waste products. Elsewhere in the eye, this is done by blood but on the optical pathway through the eye, a transparent medium is necessary.

DOI: 10.1201/9781003388906-3

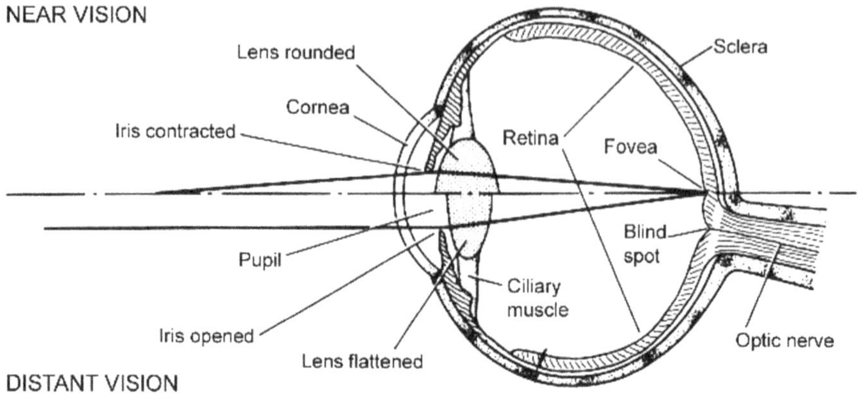

FIGURE 3.1 A section through the eye adjusted for near and distant vision.

As the ciliary body extends further away from the sclera, it becomes the iris. The iris forms a circular opening, called the pupil, that admits light into the rear of the eye. Pupil size varies with the amount of light reaching the retina but it is also influenced by the distance of the object from the eye, the age of the observer, and by emotional factors such as fear, excitement, and anger (Duke-Elder, 1944).

After passing through the pupil, light reaches the crystalline lens. The lens is fixed in position but varies its focal length by changing its shape. The change in shape is achieved by contracting or relaxing the ciliary muscles. For objects close to the eye, the lens is fattened. For objects far away, the lens is flattened.

The space between the lens and the retina is filled with another transparent material, the jelly-like vitreous humour. After passing through the vitreous humour, light reaches the retina where light is absorbed and converted to electrical signals. The retina is a complex structure (Figure 3.2). It can be considered as having three layers: A layer of photoreceptors, a layer of collector cells that provide links between multiple photoreceptors, and a layer of ganglion cells. The axons of the ganglion cells form the optic nerve which produces the blind spot where it passes through the retina out of the eye. Light reaching the retina passes through the ganglion and collector cell layers before reaching the photoreceptors, where it is absorbed. Any light that gets past the photoreceptor layer is absorbed by the pigment epithelium.

3.2.2 THE RETINA

The retina is an extension of the brain. The human visual system has four photoreceptor types in the retina, each containing a different photopigment. These four types are conventionally grouped into two classes, rods and cones, the names coming from the appearance of the photoreceptors under a microscope. All rod photoreceptors contain the same photopigment and therefore have the same spectral sensitivity. The other three photoreceptor types are all cones. Each contains a different photopigment and hence has a different spectral sensitivity. The three

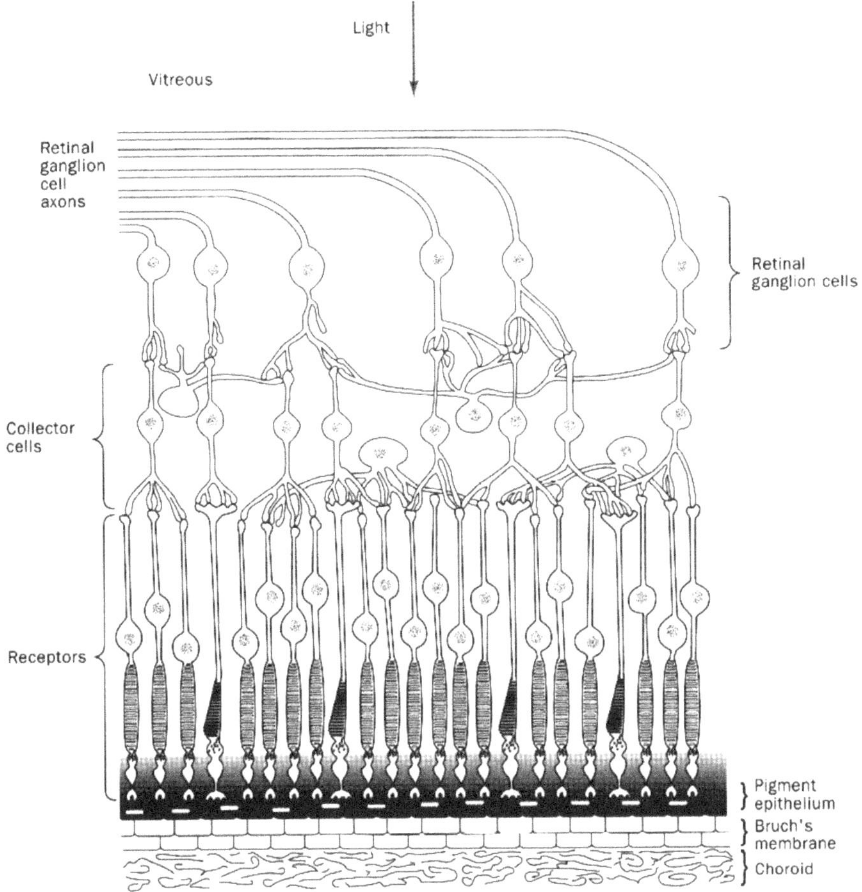

FIGURE 3.2   A section through the retina (from Sekular and Blake, 1994).

cone types are called short (S), medium (M), and long (L) wavelength-sensitive cones after the wavelength region where they have the greatest sensitivity (450, 525, and 575 nm, respectively).

Rods and cones are distributed differently across the retina (Figure 3.3). The cones are concentrated in one small area that lies on the visual axis of the eye, called the fovea, although there is a low density of cones across the rest of the retina. Rods are largely absent from the fovea and reach a maximum concentration about 20 degrees off the visual axis. The three cone types are also not distributed equally across the retina. The L and M cones are concentrated in the fovea, their density declining rapidly with increasing eccentricity. The S-cones are largely absent from the fovea, reach a maximum concentration just outside the fovea and then decline gradually in density with increasing eccentricity. The ratio of L, M, and S cones in and around the fovea is approximately 32:16:1 (Walraven, 1974).

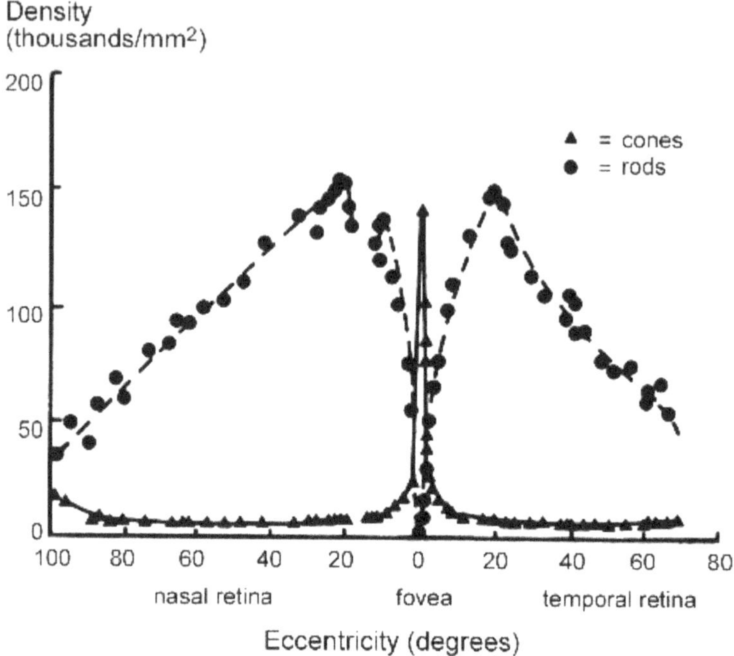

FIGURE 3.3   The distribution of rod and cone photoreceptors across the retina. Zero degrees indicates the position of the fovea.

Over the whole retina, there are many more rods than cones, approximately 110 million rods and 5 million cones. The fact that there are many more rods than cone photoreceptors should not be taken to indicate that human vision is dominated by the rods. It is the fovea that allows resolution of detail and the fovea is almost entirely inhabited by cones. There are three other anatomical features that emphasize the importance of the fovea. The first is the absence of blood vessels. For most of the retina, light passes through a network of blood vessels before reaching the photoreceptors but blood vessels avoid crossing over the central area of the retina, called the foveola, at the centre of which is the fovea. The second is the fact that even the collector and ganglion layers of the retina are pulled away over the fovea. This helps reduce the absorption and scattering of light in the region of the fovea and hence enhances the resolution of detail. The third is the fact that the outer limb of the cone photoreceptor can act as a waveguide, making cones more sensitive to light rays passing through the centre of the lens (Crawford, 1972). This characteristic, known as the Stiles-Crawford effect, compensates to some extent for the poor quality of the eye's optics.

Photons of light arriving at the retina are absorbed by the photoreceptors. This absorption results in changes to the pattern of spontaneous electrical discharges produced in the absence of light. Studies of the pattern of electrical discharges from single ganglion cells have revealed two important aspects of the operation of the retina.

The first is the existence of receptive fields. A receptive field is the area of the retina that determines the output from a single ganglion cell. The size of a receptive field is measured by exploring the retina with a very small spot of light while measuring the electrical discharges from the ganglion cell. The boundary of the receptive field is determined by the point beyond which applying the spot of light fails to alter the spontaneous electrical discharge from the ganglion cell.

A given ganglion receptive field always represents the activity of a number of photoreceptors, and often reflects input from different cone types as well as from rods. There are about 1.2 million retinal ganglion cells, about one-hundredth of the number of photoreceptors. The sizes of receptive fields vary systematically with retinal location. Receptive fields around the fovea are very small. As eccentricity from the fovea increases, so does receptive field size. The sensitivity of a receptive field to light is primarily determined by its size. Because all ganglion cells require some minimum electrical input to be stimulated, a ganglion cell with a large receptive field that receives input from a large number of photoreceptors can be stimulated by a lower retinal irradiance than can a ganglion cell with a receptive field that receives input from only a few photoreceptors. Hence, the sensitivity to light of small receptive fields is significantly less than that of larger fields. The rod photoreceptor system is characterized by large receptive fields and long integration times, a combination that makes the rod photoreceptor system significantly more sensitive to light than the cone photoreceptor system.

The second important aspect of the operation of the retina revealed by studies of electric discharges is that within each ganglion cell receptive field there is a structure. Specifically, retinal ganglion receptive fields consist of a central circular area and a surrounding annular area. These two areas have opposing effects on the ganglion cell's electrical discharge. Light falling on the central area increases the rate of electrical discharge while light falling on the annular surround decreases it, or, in other receptive fields, the reverse occurs. These types of receptive fields are known as on-centre/off-surround and off-centre/on-surround fields, respectively. If either of these two types of retinal ganglion receptive fields is illuminated uniformly, the two types of effect on electrical discharge cancel each other, a process called lateral inhibition. However, if the illumination is not uniform across the two parts of the receptive field, a net effect on the ganglion cell discharge is evident. This means that the image processing of the retina is devoted to detecting differences in luminance not absolute luminance, a process that emphasizes the contrasts in the scene and discounts absolute light level.

There are an approximately equal number of on-centre/off-surround and off-centre/on-surround receptive fields in the retina. The electrical signals from the two types of receptive fields do not cancel each other. Rather, the signals from the two types of receptive fields are kept separate, indicating that they serve different aspects of vision.

While every ganglion cell feeding the visual system has a receptive field, not all ganglion cells are the same. There are three types of ganglion cells – the magnocellular (M), the parvocellular (P), and the koniocellular (K) cells. There are a number of important differences between these cell types. First, the axons of the M-cells are thicker than the axons of the P- or K-cells, indicating that signals are transmitted more rapidly from the M-cells than from the P- or K-cells. Second, there are many more P-cells than M-cells and many more M-cells than K-cells, and they are distributed differently across the retina. The P-cells dominate in the fovea and the M-cells dominate in the periphery. The K-cells are located outside the fovea. Third, the M-cells and P-cells are sensitive to different aspects of the retinal image. The M-cells are more sensitive to rapidly varying stimuli and to small differences in illumination but are insensitive to differences in colour. The P- and K-cells are sensitive to colour. The P-cells receive input from the L- and M-cones while the K-cells receive input from the S-cones.

Overall, this brief description should have demonstrated that the retina is really the first stage of an image processing system. The retina extracts information on boundaries in the retinal image and then seeks out specific aspects of the stimulus within the boundaries, such as colour. These aspects are then transmitted up the optic nerve, formed from the axons of the retinal ganglion cells, along different channels. For a more detailed description of the operation of the retina, see Wolfe et al. (2006).

3.2.3 THE CENTRAL VISUAL PATHWAYS

Figure 3.4 shows the pathways over which signals from the retina are transmitted to the visual cortex. The optic nerves leaving the two eyes are brought together at the optic chiasm, rearranged, and then extended to the lateral geniculate nuclei. Somewhere between leaving the eyes and arriving at the lateral geniculate nuclei, some optic nerve fibres are diverted to the superior colliculus, located at the top of the brain stem, and responsible for controlling eye movements, and to the brain stem nuclei that control pupil size as well as the suprachiasmatic nuclei, the central clock of the human timing circadian system (see Section 3.8). After the lateral geniculate nuclei, the two optic nerves spread out to supply information to various parts of the visual cortex., the part of the brain where the model of the external world is constructed.

At the optic chiasm, the optic nerve from each eye is split, and then parts of the optic nerves from the same side of the two eyes are combined. This arrangement ensures that the signals from the same side of the two eyes are received together on the same side of the visual cortex.

The signals from the same side of the two eyes pass from the optic chiasm to the lateral geniculate nuclei (LGN). Anatomically, an LGN shows six broad layers. Two of these layers receive signals from the M-cells of the retina while the other four receive signals from the P-cells. Between each layer is another thin separating layer that receives signals from the K-cells. Each layer is arranged so that the location of the M-cells and P-cells on the retina is preserved. In other words, each layer preserves a map of the retina. As in the retina, electrophysiological recording of discharges from individual LGN cells hase shown the existence of receptive fields, with either an on-centre/off-surround or an off-centre/on-surround structure. The division of function found in the retina is also present in the LGN. The M-cell layers respond

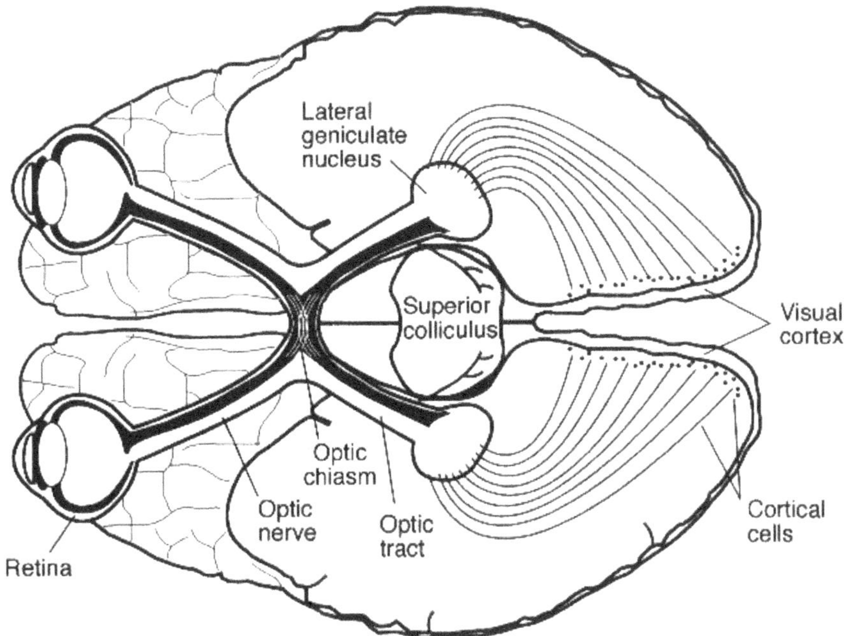

FIGURE 3.4 A schematic diagram of the pathways from the eyes to the visual cortex (from Sekular and Blake, 1994).

to movement but not to colour while the P-cell layers process detailed information about stationary targets, including their colour. Indeed, some P-cell receptive fields show strong responses when the centre is illuminated by one colour and the surround by another. The specific colour combinations are red and green or yellow and blue, this being the basis of human colour vision (see Section 3.2.5). The receptive fields of the P-layers are smaller than those of the M-layers, indicating that the P-layers will be better at resolving detail but the M-layers will respond faster to a change in the amount of light. The functions of the K layers are still the subject of study.

From the above description, it might seem that the LGN is just a relay station between the retinas of the two eyes and the visual cortex.. However, it is more than this. There are more connections from other parts of the brain to the LGN than there are connections from the LGN to the visual cortex.. Thus, the LGN is where information from other parts of the brain modifies the transfer of information to the visual cortex.. For example, the LGN receives signals from the reticular activating system, a part of the brain stem that determines the general level of arousal, and low levels of arousal have been shown to lead to an attenuated visual signal being transmitted from the LGN to the visual cortex (Livingston and Hubel, 1981). The LGN also receives signals descending from the visual cortex as well as from many other parts of the brain. Clearly, the LGN is involved in the interaction with other senses and the assumptions about incoming visual information.

3.2.4 THE VISUAL CORTEX

The visual cortex is located at the back of the cerebral hemispheres. The primary visual cortex consists of another layered array, containing many millions of neurons. At one level, the primary visual cortex is similar to the organization of the retina and the LGN. For example, the M-, P-, and K-cell channels remain separated, and signals from the different layers of the LGN are received at different layers of the visual cortex. Further, each cortical cell reacts only to signals from a limited area of the retina, and the arrangement of the cortical cells replicates the arrangement of ganglion cells in the retina. But two features separate the primary visual cortex from the LGN and the retina. The first is cortical magnification. What this means is that the number of cells allocated to each part of the retina enhances the importance of the fovea. About 80% of the cortical cells are devoted to the central ten degrees of the visual field (Drasdo, 1977), in the middle of which is the fovea. The second is the shape of the receptive fields. On-centre/off-surround and off-centre/on-surround receptive fields are found, but now they are elongated in shape rather than circular. These receptive fields show sensitivity to the orientation of a line, edge, or grating and, for the last of these three, different cells are sensitive to the different spatial frequencies of the grating. Other cells respond to moving edges and lines, but only move in one direction. Yet other cells respond more to signals from the left eye and others to the right eye, but the majority respond to signals from both eyes. It is in the primary visual cortex that the information from the two eyes is brought together. All this cellular diversity occurs at the entry level of the visual cortex. There is a much more complex structure beyond this in the higher areas of the visual cortex (Wolfe et al., 2006). Investigation of these areas has shown that different parts of the visual cortex are dedicated to specific discriminations. For example, areas have been identified in the visual cortex of monkeys that are concerned with analysing colour, motion, and even faces viewed from particular angles (Desimone, 1991). Visually, monkeys are very much like humans.

3.2.5 COLOUR VISION

The differences between rod and cone photoreceptors have been considered above, but the value of having cones with three different photopigments has not. The value of having three different cone photoreceptor types is that it allows colour vision to occur. The number of photoreceptor types used to form a colour system is a matter of compromise. A single photoreceptor type containing a single photopigment is unable to discriminate differences in wavelength from differences in irradiance and so does not support colour vision, e.g., rod photoreceptors. A system with many different photoreceptors each containing a different photopigment would be able to make many discriminations between wavelengths, but at the cost of taking up more of the neural capacity of the visual system.

The ability to discriminate the wavelength content of incident light makes a dramatic difference to the information that can be extracted from a scene. Creatures with only one type of photopigment, i.e., creatures without colour vision, can only discriminate shades of grey, from black to white. Approximately 100 such discriminations can be made. Having two photopigment types increases the number of different combinations of irradiance and spectral content that can be discriminated to about 10,000. Having three types of photopigment increases the number of discriminations to approximately 1,000,000 (Neitz et al., 2001). Thus, colour vision is a valuable part of the visual system, and not a luxury that adds little to utility.

Figure 3.5 shows how the outputs from the three cone photoreceptor types are believed to be arranged into one non-opponent achromatic system and two opponent chromatic systems. The achromatic channel receives inputs from the M- and L-cones only. The opponent blue-yellow channel produces the difference between the outputs of the S-cones and the sum of the M- and L-cones. The other opponent channel, the red-green channel, produces the difference between the outputs of the M-cones and the sum of the L- and S-cones. This opponent structure for colour vision influences the perception of colours. This was shown in an experiment by Boynton and Gordon (1965). They presented monochromatic lights of different wavelengths and asked subjects to describe the appearance of each stimulus using only the colour names

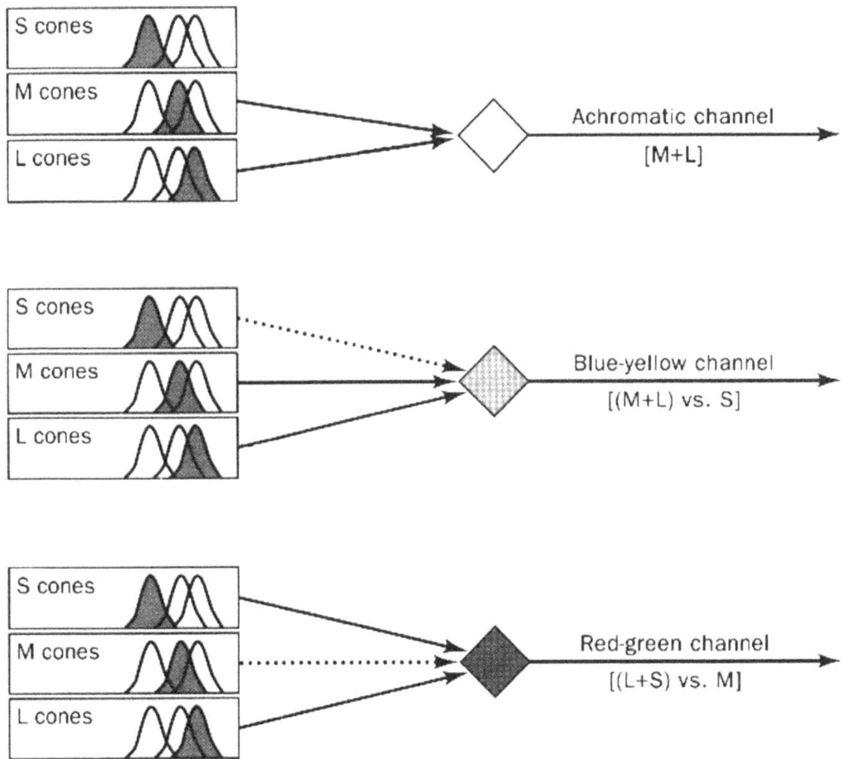

FIGURE 3.5   The organization of the human colour system showing how the three cone photoreceptor types are believed to feed into one achromatic, non-opponent channel and two chromatic, opponent channels (from Sekular and Blake, 1994).

– red, green, yellow, and blue. Either one or two names could be used for each colour presented. The interesting result was that people very rarely described a colour as red-green or yellow-blue, while yellow-red and green-blue were frequently used. Even four-month-old infants divide incident light into four categories, corresponding to what adults call red, yellow, or blue (Bornstein et al., 1976).

Physiologically, the electrical outputs from the three different cone types are organized into opponent and non-opponent classes in the retina. Outputs in the non-opponent class always give an increase in activity with increasing retinal irradiance, although the magnitude of that increase will vary with the wavelength of the incident light. Outputs in the opponent class can show either an increase or a decrease in activity depending on the wavelengths of the incident light. Cells of the non-opponent type are believed to constitute the achromatic channel shown in Figure 3.5 while cells of the opponent type form the opponent channels. The achromatic information is transmitted to the visual cortex by the M-cell channel, while the chromatic information proceeds via the P-cell channel. Much more detail on the structure and capabilities of human colour vision can be obtained from Kaiser and Boynton (1996),

By now, it should be clear that the visual system is not a camera. The retinal image is not simply transferred, lock, stock, and barrel, up to the visual cortex. Rather, the retinal image is broken down into significant elements, such as contrast, colour, and movement, while less useful information such as ambient light level is discounted. The significant elements are transmitted to the visual cortex at two different speeds, the faster providing an outline of the scene while the slower fills in the details. From these elements, and the accumulated experience of sight, a model of the outside world is constructed (Purves and Beau Lotto, 2010).

3.3 CONTINUOUS ADJUSTMENTS OF THE VISUAL SYSTEM

In everyday use, the visual system has to be capable of examining objects in different locations, at different distances, over a wide range of lighting conditions. These variable requirements indicate the need for the visual system to continually adjust its aim, focus, and sensitivity.

3.3.1 Eye Movements

The emphasis given to the fovea by the visual system makes it essential that the visual world is explored by moving the eye so that different points of interest fall on the fovea. There are several different types of eye movement. Searching the visual field is done by a series of fixations and saccades. Fixations are attempts to hold an object of interest steady on the fovea. Movement between fixations is made by saccades. Saccades are very fast with velocities ranging up to 1,000 degrees/second depending upon the distance moved. Saccadic eye movements have a latency of about 200 ms, which limits how frequently the line of sight can be moved to about five movements per second. Visual functions are substantially limited during saccadic movements.

Figure 3.6 shows the pattern of fixations made while driving on an unlit road at night using headlamps alone.

About the only situation in which saccades rarely occur is in smooth pursuit eye movements. Such movements are relatively slow, up to 40 degrees/second, and are used to keep the image of a smoothly moving object, such as a passing vehicle, on the fovea. The advantage of such movements is that, if successful, the target is always on the fovea, which means that more detailed information about it can be extracted. The smooth pursuit system cannot follow either smoothly moving targets at high velocities or slow but erratically moving targets.

These eye movements all occur in a single eye, but movements in the two eyes are not independent. Rather, they are coordinated so that the lines of sight of the two eyes are both pointed at the same target at the same time. If the lines of sight of the two eyes are not aimed at the same target at the same time, the target may be seen as double. Movements of the two eyes that keep the primary lines of sight converged on a target as it changes in distance, e.g., from the view through the windscreen to the vehicle's instrument panel or display screen, are called vergence movements. These movements are very slow, up to 10 degrees/second, but can occur as a jump movement or can smoothly follow a target moving in a fore-and-aft direction. Both types of movement involve a change in the angle between the two eyes.

3.3.2 ACCOMMODATION

The eye has a fixed image distance so to bring the images of objects at different distances to focus on the retina it is necessary to vary the optical power of the eye. This process is called accommodation. There are three optical components involved in accommodation. The first is the thin film of tears on the cornea. This film cleans the surface of the eye, starts the optical refraction process necessary for focusing objects, and smoothens out small imperfections in the surface of the second optical component, the cornea. The cornea covers the transparent anterior one-fifth of the eyeball (Figure 3.1). With the tear layer, it forms the major refracting component of the eye and gives the eye about 70% of its optical power. The crystalline lens provides the remaining 30%. The ciliary muscles have the ability to change the curvature of the lens and thereby adjust the power of the eye's optical system in response to changing target distances.

Accommodation is a continuous process and is always a response to an image of a target located on or near the fovea rather than in the periphery of the retina. Any condition, either physical or physiological, that handicaps the fovea, such as a low light level, will adversely affect the ability to accommodate. As adaptation luminance decreases below 0.03 cd.m^{-2}, the range of accommodation narrows so that it becomes increasingly difficult to focus objects near and far from the observer (Leibowitz and Owens, 1975). Reduced contrast sensitivity and poor visual acuity are consequences of inappropriate accommodation. When there is no stimulus for accommodation, as in complete darkness or in a uniform luminance visual field such as occurs in a dense fog or a whiteout snowstorm, the visual system accommodates to focus at approximately 70 cm.

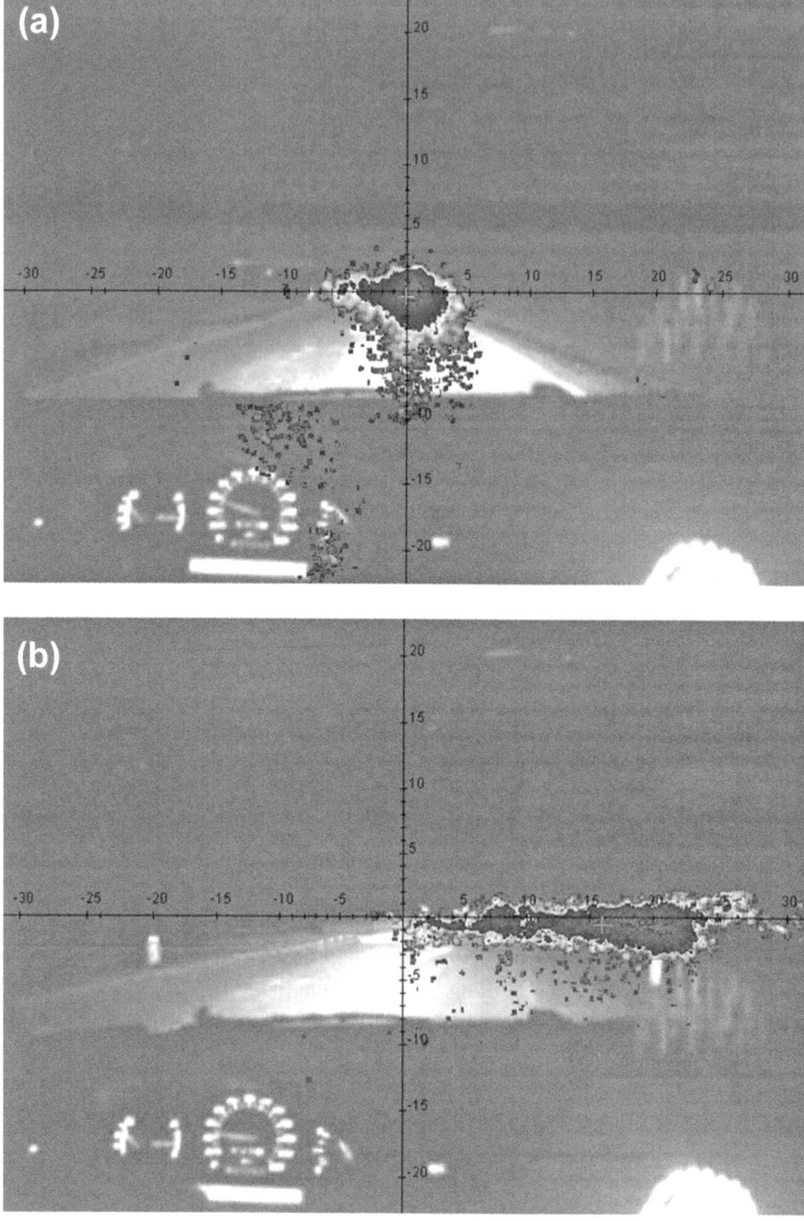

FIGURE 3.6   The pattern of fixations made while driving along an unlit road at night using headlamps (a) along a straight road and (b) around a curve. The pattern is defined into two groups based on the number of fixation points measured at each pixel. The pixel with the maximum number of fixations is given the value of 100% and is marked with a cross. All the other pixels are then allocated to one of two groups, those with 50% or more of the maximum number of fixations (dark grey) and those with 1–50% of the maximum (light grey) (from Stahl, 2004).

3.3.3 ADAPTATION

The range of luminances to which the human visual system can be exposed is very large, from 0.000001 cd.m^{-2} on a very dark night to 100,000 cd.m^{-2} on a sunlit beach. To cope with this range, the visual system continuously and unconsciously changes its sensitivity through a process called adaptation. Adaptation involves three different processes.

Change in Pupil Size – The iris constricts and dilates in response to increased and decreased levels of retinal illumination. For young people, the diameter of the pupil can range from about 2 to 8 mm. The amount of light transmitted through the pupil is proportional to its area so this range of diameters implies a maximum effect of pupil changes on retinal irradiance of 16 to 1. As the visual system can operate over a range of about 1,000,000,000,000 to 1, this indicates that the pupil plays only a minor role in the adaptation of the visual system. Iris constriction is faster (about 0.3 s) than dilation (about 1.5 s).

Neural Adaptation – This is a fast (less than 200 ms) change in sensitivity produced by synaptic interactions in the retina. It is effective over a luminance range of two to three log units.

Photochemical Adaptation – The four types of retinal photoreceptors used by the visual system contain four different photopigments. When light is absorbed by the photoreceptor, the pigment is bleached. In the dark, the pigment is regenerated and is again available to absorb light. The sensitivity of the eye to light is largely a function of the percentage of unbleached pigment. Under conditions of steady retinal illumination, the concentration of photopigment produced by the competing processes of bleaching and regeneration is in equilibrium. When the retinal illumination is changed, as happens when a vehicle with headlamps on approaches and passes, pigment is bleached and then regenerated so as to re-establish equilibrium. Exactly how long it takes to re-establish equilibrium after a change in retinal illumination depends on the magnitude of the change, the extent to which it involves different photoreceptors and the direction of the change. For changes in retinal illumination of about two to three log units, neural adaptation is sufficient so equilibrium should be reached in less than a second. For larger changes, photochemical adaptation is necessary. If the change in retinal illumination is large but still lies completely within the range of operation of the cone photoreceptors, a few minutes will be sufficient to reach equilibrium. If the change in retinal illumination ranges from cone photoreceptor operation to rod photoreceptor operation, tens of minutes may be necessary for adaptation to be completed. As for the direction of change, once the photochemical processes are involved, changes to a higher retinal illumination can be achieved much more rapidly than changes to a lower retinal illumination. It is important to note that these changes in adaptation are not linear with time. The times given above are for equilibrium to be reached, but this is approached asymptotically. The vast majority of the adaptation takes place in much shorter times.

When the visual system is not completely adapted to the prevailing retinal illumination, its capabilities are diminished. This state of changing adaptation is called transient adaptation. Transient adaptation can be significant where sudden changes from high to low retinal illumination occur, such as on entering a long road tunnel on a sunny day.

3.3.4 Photopic, Scotopic, and Mesopic Vision

The process of adaptation can change the spectral sensitivity of the visual system because at different retinal illuminances, different combinations of retinal photoreceptors are operating. The three states of sensitivity are conventionally identified as photopic, scotopic, and mesopic.

Photopic Vision –- This state of the visual system occurs at luminances higher than approximately 3 cd.m^{-2}. For these luminances, the retinal response is dominated by the cone photoreceptors, the rod photoreceptors being bleached. This means that both colour vision and fine resolution of detail are available.

Scotopic Vision – This operating state of the visual system occurs at luminances less than approximately 0.001 cd.m^{-2}. For these luminances only the rod photoreceptors respond to stimulation, the cone photoreceptors being insufficiently sensitive to respond to the low level of retinal irradiance. This means that colour is not perceived, only shades of grey, and the fovea of the retina is blind.

Mesopic Vision – This operating state of the visual system is intermediate between the photopic and scotopic states, i.e., between about 0.001 and 3 cd.m^{-2}. In the mesopic state, both cones and rod photoreceptors are active. As luminance declines from the photopic through the mesopic to the scotopic, the fovea, which contains only cone photoreceptors, slowly declines in absolute sensitivity without significant change in spectral sensitivity, until vision fails altogether as the scotopic state is reached. In the periphery, the rod photoreceptors gradually come to dominate the cone photoreceptors, resulting in a slow deterioration in colour vision and resolution and a shift in spectral sensitivity to shorter wavelengths.

The relevance of the different operating states for road and vehicle lighting varies. Scotopic vision is largely irrelevant. Any vehicle lighting worthy of the name provides enough light to at least move the visual system into the mesopic state when the driver is looking in the lit area, although parts of the visual scene outside the lit area may be at scotopic luminances. When there is road lighting present as well, there is usually enough to ensure that the visual system operates near the boundary of the photopic and mesopic states (Plainis et al., 2005)

3.4 CAPABILITIES OF THE VISUAL SYSTEM

The human visual system, like every other physiological system, has a limited range of capabilities. These limits are called the thresholds of vision. By convention, a threshold is measured as the value of the stimulus that is detected on 50% of the occasions it is presented. Threshold measurements come in many different forms and depend on many different variables, most of which interact (Boff and Lincoln, 1988). Threshold measurements are mainly of interest in determining what will not be seen rather than how well something will be seen, but, for driving at night, knowing what will not be seen is useful. The intention here is to summarize the thresholds of relevance to driving at night and how they are affected by the characteristics of the human visual system. For these threshold measurements, it can be assumed that the observers were fully adapted to the prevailing lighting conditions, that the target was presented on a field of uniform luminance, and that the observers' accommodation was correct. You should also be aware that the observer would be alert, would know there was something to be detected or identified, what it was, and, usually, where it would appear. This is not always the case in the real world which means these thresholds are truly the limits of visual capabilities.

3.4.1 Threshold Measures

The threshold capabilities of the human visual system can conveniently be divided into three types – spatial, temporal, and colour classes.

Spatial Threshold Measures – These measures relate to the ability to extract a target from its background or to resolve detail within a target. For spatial threshold measures, it is usually assumed that the target does not vary with time. Common spatial threshold measures are threshold luminance contrast and visual acuity. The luminance contrast of a target quantifies its visibility relative to its immediate background. The higher is the luminance contrast, the easier it is to detect the target. There are three different forms of luminance contrast commonly used for uniform luminance targets seen against a uniform luminance background (see Section 1.3.2). Visual acuity is a measure of the ability to resolve detail for a target with a fixed luminance contrast. Visual acuity is most meaningfully quantified as the angle subtended at the eye by the detail that can be resolved on 50% of the occasions the target is presented, expressed in minutes of arc. An alternative is decimal acuity which is the reciprocal of the angle subtended measured in minutes of arc. Using the former measure, the visual acuity corresponding to "normal" vision is taken to be 1 min arc. Unfortunately for simplicity, a relative measure of visual acuity is used by the medical profession. This is the distance at which a patient can read a given size of letter or symbol relative to the distance an average member of the population with normal vision could read the same letter or symbol. For example, if the person is said to have 6/10 vision, it means that the person can only read a given letter at 6 m that an average member of the population with normal vision can read from 10 m.

Temporal Threshold Measures – These measures relate to the speed of the response of the human visual system and its ability to detect fluctuations in luminance. For temporal threshold measures, it is usually assumed that the target is fixed in position and size.

The ability of the human visual system to detect fluctuations in luminance can be measured as the frequency of the fluctuation, in Hertz (cycles/second), and the modulation of the fluctuation, for the stimulus that can be detected on 50% of the occasions it is presented. The modulation is expressed as

$$M = (L_{max} - L_{min}) / (L_{max} + L_{min})$$

where M is the modulation, L_{max} is the maximum luminance (cd.m^{-2}), and L_{min} is the minimum luminance (cd.m^{-2}).

This formula gives modulations that range from 0 to 1. Sometimes, modulation is expressed as a percentage modulation, calculated by multiplying the modulation by 100.

3.4.1.1 Colour Threshold Measures

Colour threshold measures are based on the separation in colour space of two colours that can just be discriminated. In principle, the separation can be measured in many different ways, but the most widely used in areas other than colour reproduction has been separation on the CIE 1931 chromaticity diagram or the related uniform chromaticity scale diagram (see Section 2.4.1).

3.4.2 FACTORS DETERMINING VISUAL THRESHOLDS

There are three distinct groups of factors that influence the thresholds of vision – visual system state, target characteristics, and the background against which the target appears.

Important visual system factors are the luminance to which the visual system is adapted, the position in the visual field where the target appears, and the extent to which the eye is correctly accommodated. The luminance to which the visual system is adapted determines which photoreceptor types are operating. The position in the visual field where the target appears determines the size of the receptive field available to the visual system and the type of photoreceptors available. The state of accommodation determines the retinal image quality. As a general rule, the lower the luminance to which the visual system is adapted, the further the target is from the fovea, and the more mismatched the accommodation of the eye is to the viewing distance, the larger will be the threshold values.

Important target characteristics are the visual size and luminance contrast of the target and any colour difference between the target and the immediate background. The effect of these variables on thresholds will be modified by any movement of the target or the observer. Any one of these task characteristics can be the threshold measure of interest, but the others will interact with it. For example, the visual acuity of a target will be different for targets of different luminance contrast and colour difference. As a general rule, the closer the other target characteristics are to their own threshold, the greater will be the threshold of the measured variable. When the target is moving, the faster and the more erratic is its movement, the greater will be the increase in the measured threshold because of the difficulty of keeping the retinal image of the target on the fovea.

The important factors for the effect of the background against which the target appears are the area, luminance, and colour of the background. These factors are important because they determine the state of adaptation of the visual system and the potential for interacting with the image processing of the target. As a general rule, the larger the area around the target that is of a similar luminance to the target and neutral in colour, the smaller will be the threshold measure.

3.4.3 SPATIAL THRESHOLDS

About the simplest possible visual task is the detection of a spot of light presented continuously against a uniform luminance background, e.g., the headlamp of a distant bicycle on an unlit rural road. For such a target, the visual system demonstrates spatial summation, i.e., the product of target luminance and target area is a constant. This implies that the total amount of energy required to stimulate the visual system so that the target can be detected is the same, regardless of whether it is concentrated in a small spot or distributed over a larger area. Spatial summation breaks down when the target is above a given size, called the critical size. The critical size varies with the angular deviation from the fovea. The critical size is about 0.5 degree at five degrees from the fovea and about two degrees at 35 degrees from the fovea (Hallett, 1963). There is very little spatial summation in the fovea, the critical size being about 6 min arc.

Given that the size of the target is above the critical size, the detection of the presence of a spot of light is determined simply by the luminance contrast. When the luminance of the surround is in the photopic range, Weber's law states that the luminance difference needed for a target to be detected is a constant proportion of the background luminance. Of course, this is for a target of fixed size. A more general picture of the effect of adaptation luminance on threshold contrast for targets of different sizes is shown in Figure 3.7. The increase in threshold contrast as adaptation luminance decreases is obvious, as is the increase in threshold contrast with decreasing target size (Blackwell, 1959). These data were obtained using discs of different sizes presented, foveally, i.e., on-axis, for one second. Decreasing the presentation time below one second increases the threshold contrast, for all sizes, particularly at lower adaptation luminances (Graham and Kemp, 1938).

Figure 3.8 shows the threshold contrast necessary to detect a bar pattern subtending five degrees, stationary or moving at 24 degrees/second against a background luminance of 9.8 cd.m^{-2}, for different eccentricities (Rogers, 1972). It is evident from Figure 3.8 that a similar threshold contrast is needed to detect a moving target and a stationary target in the fovea, but in the peripheral visual field there is marked difference. The threshold contrast for the stationary target increases dramatically with increasing eccentricity but the same is not true for the moving target. For the moving target, threshold contrast changes little with increasing eccentricity. This is the basis for the common observation "I caught a movement out of the corner of my eye".

Turning now to visual acuity, Figure 3.9 shows the variation in visual acuity with adaptation luminance for the foveal viewing of a Landolt ring target. As adaptation luminance decreases, visual acuity, measured as the reciprocal of the minimum gap size detected, decreases (Shlaer, 1937).

Figure 3.10 shows visual acuity, measured as the reciprocal of the minimum gap size, measured in minutes of arc, for a Landolt ring target, plotted against eccentricity, for a range of adaptation luminances. For the adaptation luminance of 3.2 cd.m^{-2}, i.e., for the conventional photopic/mesopic border, acuity is at about 1 min arc in the fovea and declines rapidly to about 10 min arc as eccentricity increases. For adaptation luminances below 0.006 cd.m^{-2}, i.e., approaching the scotopic state

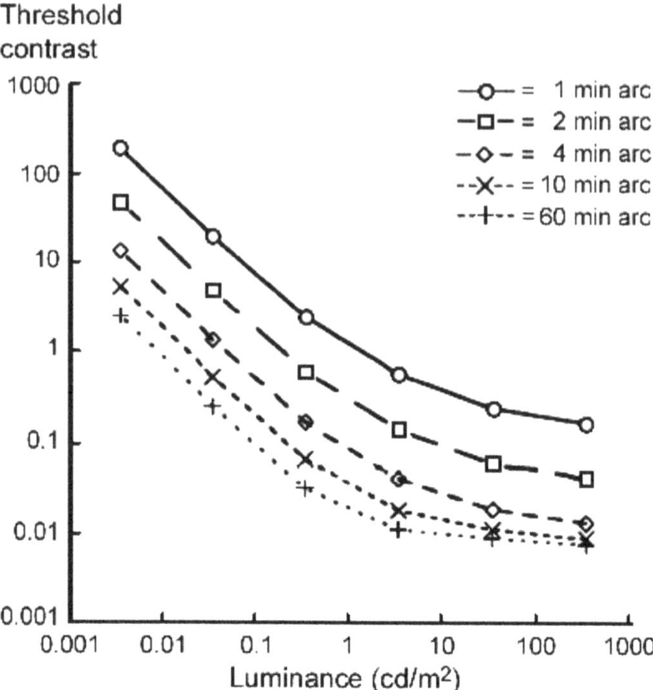

FIGURE 3.7 Threshold contrast plotted against background luminance (cd/m²) for disc targets of various diameters (min arc), viewed foveally. The discs were presented for one second and were brighter than the background (after Blackwell, 1959).

where the fovea is blind and only the rod photoreceptors are active, visual acuity is best at about 10 min arc, four to eight degrees off-axis (Mandlebaum and Sloan, 1947). For the mesopic state (0.13 cd.m^{-2}), the visual acuity is in transition between these two extremes.

Figure 3.11 shows the visual acuity, measured as the minimum gap size detected in minutes of arc, for smooth relative movement of target and observer at different velocities under different illuminances (Miller, 1958). As would be expected, visual acuity worsens as illuminance falls and angular velocity increases.

In the above discussion, threshold contrast and visual acuity have been considered separately because threshold contrast is usually measured with large-size targets, without detail, and visual acuity is measured with high luminance contrast targets, with detail. But many things of practical interest vary in both luminance contrast and size of detail, and these two target characteristics can be expected to interact. The threshold capabilities of the visual system to such targets can be expressed as the contrast sensitivity function. This is a rather grand name for what is essentially a simple piece of information, the frequency response of the visual system to spatial variations in luminance. The contrast sensitivity function of the visual system is measured using sine-wave grating targets of different spatial frequencies

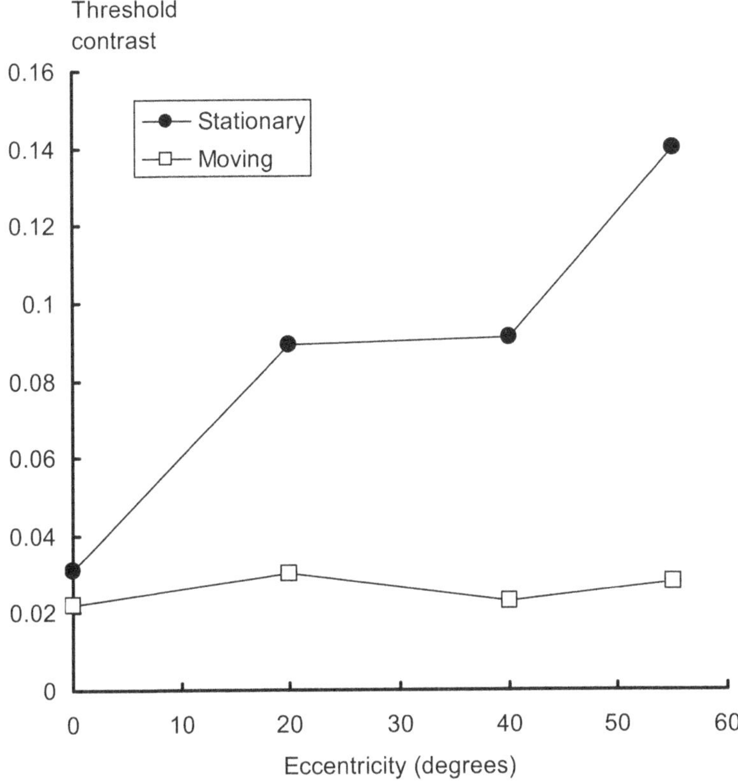

FIGURE 3.8   Threshold contrast plotted against eccentricity (degrees) for stationary and moving (24 degrees /second) bar pattern targets (after Rogers, 1972).

and adjustable luminance modulation (see Sections 1.3.1 and 1.3.2). The spatial frequency of the grating consists of the number of cycles of the grating that lie within a one-degree field of view for the observer, and hence is expressed in cycles per degree. The threshold luminance contrast condition is usually measured as luminance modulation, but it is often displayed as contrast sensitivity which is the reciprocal of luminance modulation. Figure 3.12 shows contrast sensitivity functions for different adaptation luminances (Van Nes and Bouman, 1967).

The value of this apparently esoteric piece of information is that any variation in luminance across a surface can be represented as a waveform, and any waveform can be represented as a series of sine waves of different modulation and frequency. The response of the visual system to sine waves of different modulation and frequency is given by the contrast sensitivity function. Thus, the contrast sensitivity function can be used to determine if and how a complex variation in luminance will be seen. If the luminance pattern has contrast sensitivities at all spatial frequencies that are greater than the threshold contrast sensitivities, i.e., all are above the contrast sensitivity function, the luminance pattern will be invisible. It is only when at least one spatial

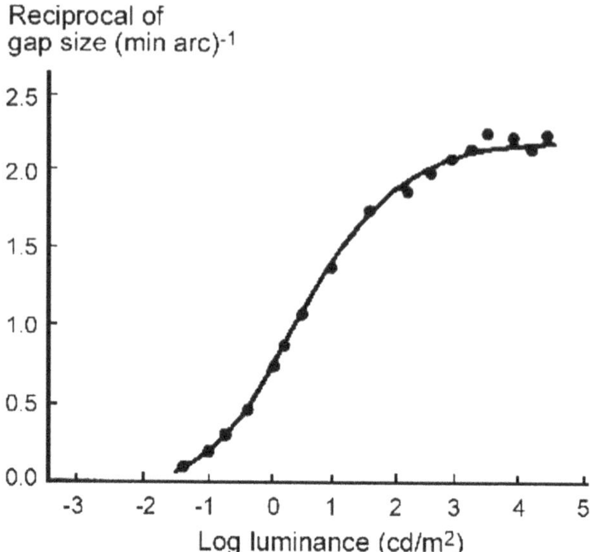

FIGURE 3.9 Visual acuity, expressed as the reciprocal of the minimum gap size ((min arc)$^{-1}$), for a Landolt ring, plotted against log background luminance (cd/m^2) (after Shlaer, 1937).

frequency has a contrast sensitivity below the contrast sensitivity function that the luminance pattern will be visible in some form. The extent to which the luminance pattern will be seen in its entirety depends on the number of spatial frequencies for which the contrast sensitivity lies below the threshold contrast sensitivity, the more spatial frequencies for which this occurs, the more complete is the perception of the luminance pattern. Contrast sensitivity functions can be used for many practical purposes. For example, they can be used to determine what size and contrast a road sign needs to be to be read from a given distance. The distance from which the observer views the luminance pattern is important because changing the viewing distance changes the spatial frequency of the pattern. As viewing distance increases, spatial frequency increases.

Returning now to Figure 3.12, it is apparent that decreasing adaptation luminance decreases both the contrast sensitivity and the maximum spatial frequency detectable, i.e., it produces a higher threshold contrast and a worse, i.e., larger, visual acuity. Also clear is the fact that the contrast sensitivity function changes rapidly below an adaptation luminance of about 30 cd.m^{-2}. The deterioration takes the form of reduced contrast sensitivities at all spatial frequencies and a shift in the spatial frequency at which maximum contrast sensitivity occurs to a lower value.

The effect of eccentricity on the contrast sensitivity function is shown in Figure 3.13. As might be expected from the increase in receptive field size with increasing eccentricity, the contrast sensitivity function shows a dramatic reduction in the highest spatial frequency visible as deviation from the fovea increases, as well as a

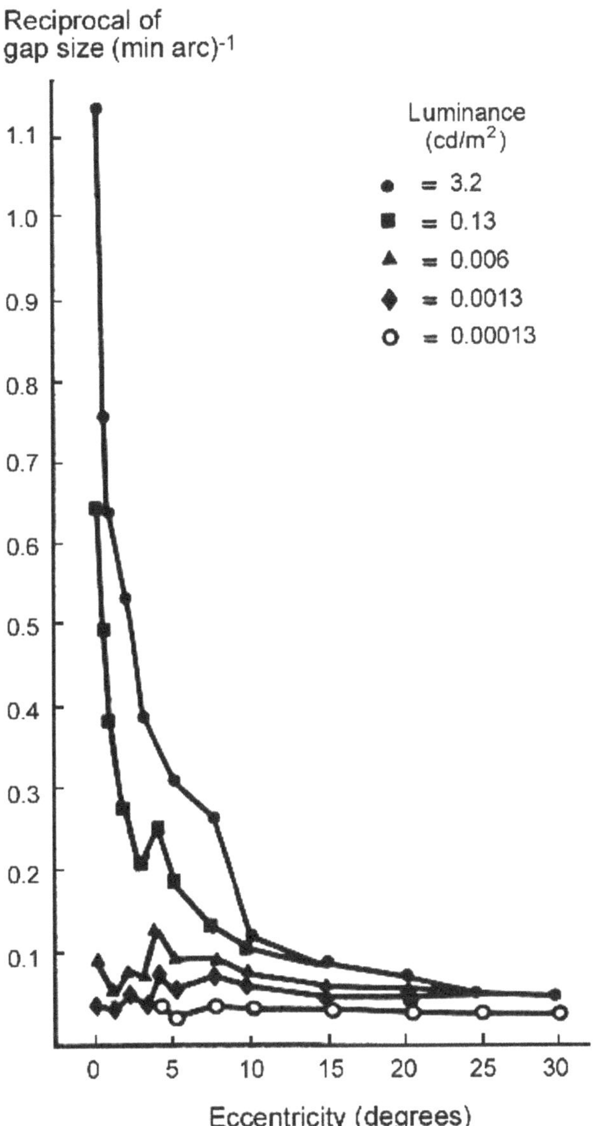

FIGURE 3.10 Visual acuity, expressed as the reciprocal of minimum gap size ((min arc)⁻¹), for a Landolt ring target presented at different eccentricities (degrees), over a range of background luminances (cd/m²) covering the photopic, mesopic, and scotopic states of the visual system (after Mandelbaum and Sloan, 1947).

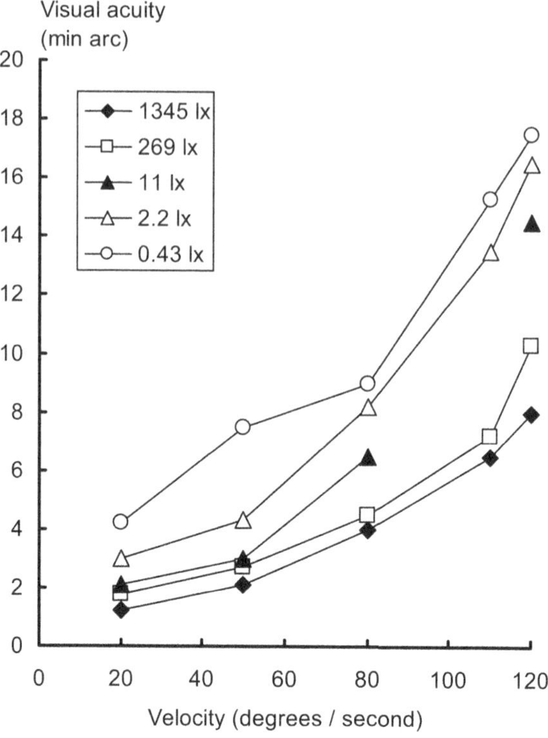

FIGURE 3.11   Visual acuity, expressed as the minimum gap size (min arc), for a Landolt ring target, illuminated to different illuminances (lx), with the observer moving at different angular velocities (degrees/second). Viewing is monocular and for a constant presentation time of 0.5 s (after Miller, 1958).

reduction in peak contrast sensitivity. What this means is that it is not possible to see fine detail more than a few degrees away from the fovea, but large objects are visible far off-axis.

3.4.4   TEMPORAL THRESHOLDS

The simplest possible form of temporal visual task is the detection of a spot of light briefly presented against a uniform luminance background. For such a target, the visual system demonstrates temporal summation, i.e., the product of target luminance and the duration of the flash is a constant. This implies that the total amount of energy required to stimulate the visual system so that the target can be detected is the same, regardless of the time for which the target is presented. Temporal summation breaks down above a fixed duration, called the critical duration. The critical duration varies with adaptation luminance, ranging from 0.1 s for scotopic luminances to 0.03 s for photopic luminances. For presentation times longer than the critical duration, presentation time has no effect, the ability to detect the flash being determined by the difference in luminance between the flash and the background.

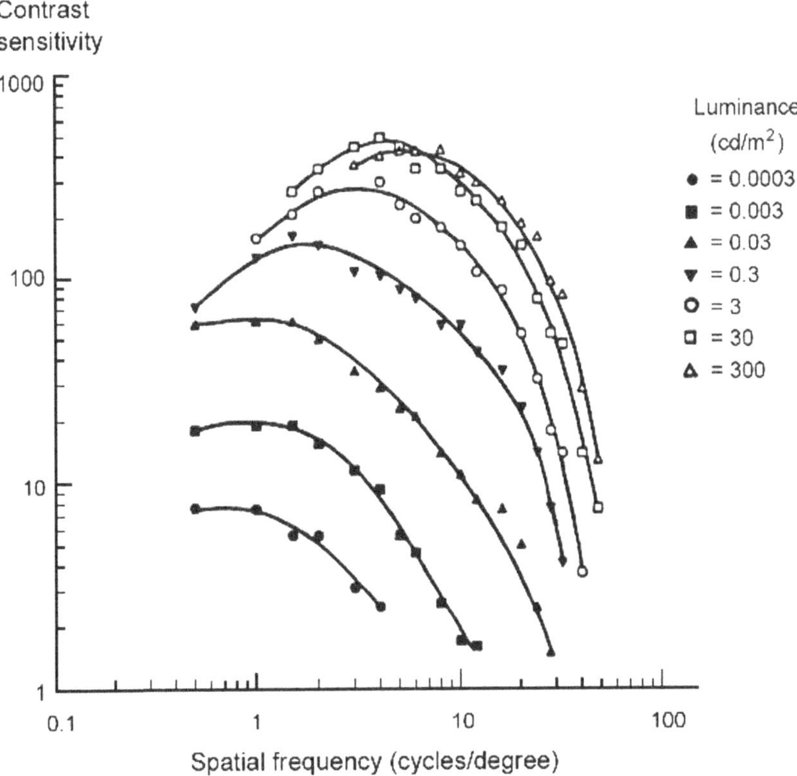

FIGURE 3.12 Contrast sensitivity functions for sine-wave gratings at different levels of background luminance (cd/m²), covering the photopic, mesopic, and scotopic states of the visual system (after van Nes and Bouman, 1967).

While the ability to detect a single flash is of interest for signalling purposes, an aspect of temporal thresholds of wider relevance to driving is the ability to detect repetitive light fluctuations, such as occur when a turn lamp is activated. Figure 3.14 shows the temporal equivalent of the contrast sensitivity function; the modulation sensitivity function, i.e., the reciprocal of the modulation threshold plotted against the frequency of the oscillation, for different adaptation luminances. These data were collected from a 68-degree diameter field, uniformly illuminated, the flicker waveform being sinusoidal. It is clear that decreasing the mean luminance decreases the sensitivity to modulation and shifts the frequency for peak sensitivity from about 20 Hz to 8 Hz. The other important point is that apart from the lowest average luminance (0.03 cd.m⁻²), the results for all the other average luminances come to a common curve at low frequencies but have different curves at high frequencies. This means that at low frequencies the ability to detect flicker is stable at a fixed modulation over a wide range of average luminances. Signalling for drivers usually involves low frequencies.

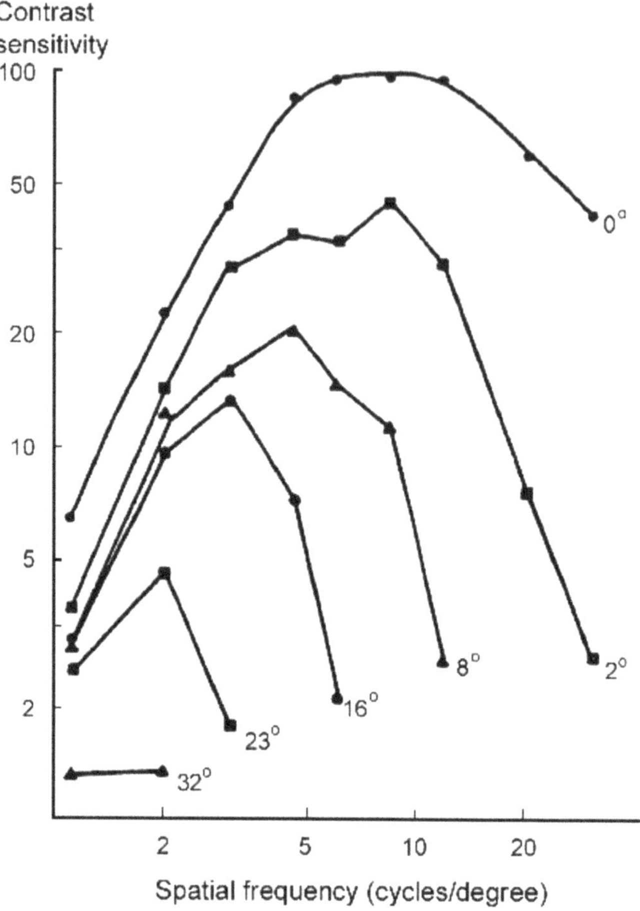

FIGURE 3.13 Contrast sensitivity functions for a 2.5-degree stimulus presented at different eccentricities (degrees) (after Hilz and Cavonius, 1974).

Figure 3.14 can be used to determine if a light fluctuation will be visible for a large area fluctuation. For a sine-wave oscillation, if the modulation at the given frequency is above the curve for the appropriate average luminance, the flicker will not be visible. If it is below the curve, it will be visible. But most vehicle signals and traffic signals are much smaller than the 68-degree target used to obtain the data shown in Figure 3.14 and have to be seen by day and by night. Figure 3.15 shows the modulation sensitivity function for 68-degree and 4-degree targets, both with an average luminance of 100 cd.m^{-2}, the 68-degree target having an edge blurred to dark over another 18 degrees, while the 4-degree target has a sharp edge to a dark background (Kelly, 1959). Also shown is the modulation sensitivity function for a 2-degree target seen against a 60-degree background with the same average luminance (de Lange, 1958). What is evident from Figure 3.15 is that peak sensitivity to fluctuations is in

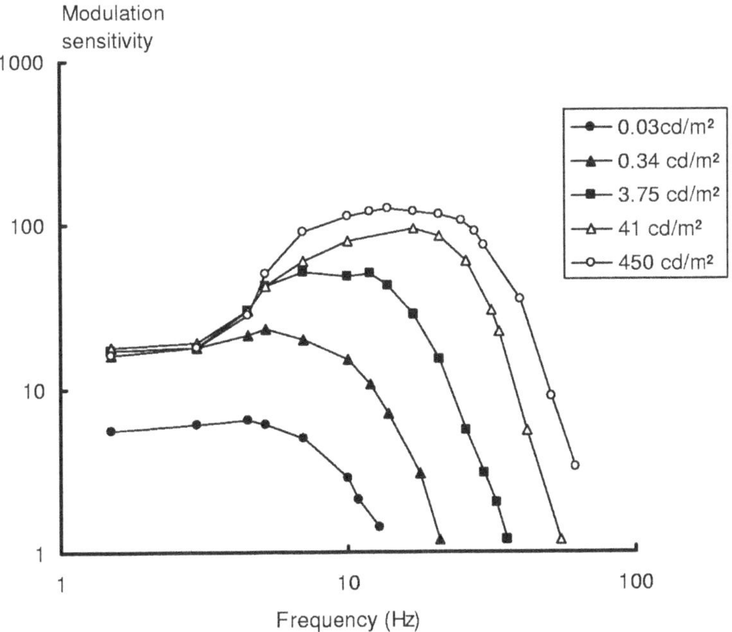

FIGURE 3.14 Modulation sensitivity functions for a large visual field at different average luminances (cd/m²). Modulation sensitivity is the reciprocal of the modulation threshold (after Kelly, 1961).

the range of 8–20 Hz for the small targets with both bright and dark backgrounds but for low frequencies, background luminance makes a large difference to sensitivity.

3.4.5 COLOUR THRESHOLDS

Both the spatial and temporal thresholds discussed above have been measured using achromatic targets lit by nominally white light. In the photopic state and to some extent in the mesopic state, the human visual system has a well-developed ability to discriminate colours. The thresholds for discriminating different colours are given by the MacAdam ellipses plotted in the CIE 1931 chromaticity diagram (MacAdam, 1942). Each ellipse represents the standard deviation in the chromaticity coordinates for colour matches made between two small visual fields with the reference field having the chromaticity of the centre point of the ellipse. The MacAdam ellipses were obtained in conditions ideal for comparison (photopic conditions with simultaneous viewing of adjacent small fields by a highly practised observer) and so represent what is likely to be the finest colour discrimination possible. Brown (1951) showed that the effect of reducing luminance to low photopic levels was to increase the size of the ellipses. Colour discrimination between targets presented successively or between targets in which there are a wide range of colours and patterns present is similarly degraded (Narendran et al., 2000).

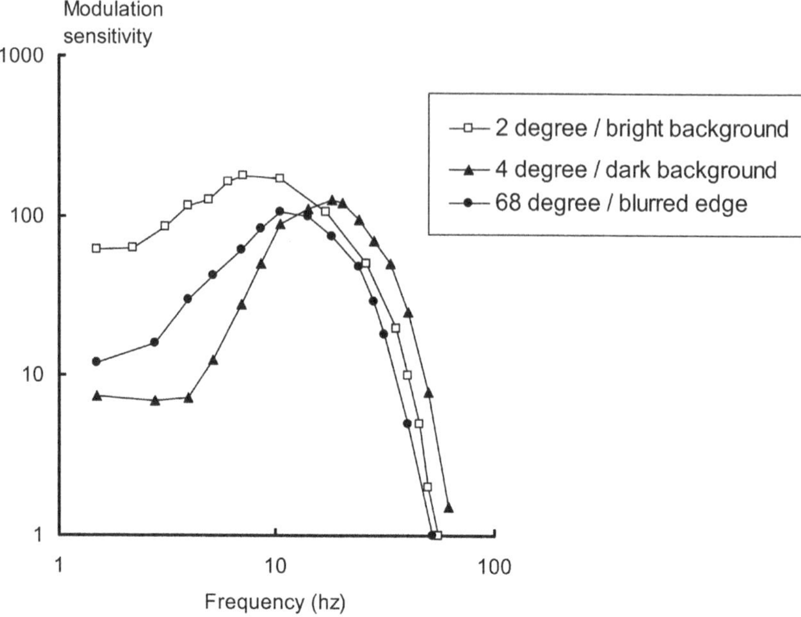

FIGURE 3.15 Modulation sensitivity functions for targets of different visual sizes, with different background brightnesses. Modulation sensitivity is the reciprocal of threshold modulation (after Boff and Lincoln, 1988).

3.5 PERCEPTION OF A SCENE

Thresholds define the limits of the capabilities of the human visual system, but how does the visual system function when presented with targets above threshold? Such stimuli are the material used in the study of perception (Purves and Beau Lotto, 2010). When considering how we perceive the world when the objects it contains are all above threshold, the overwhelming impression is one of stability in the face of continuous variation. As the eyes move in the head and the head itself moves about, the images of objects move across the retina and change their shape and size according to the laws of physical optics. Despite these variations, our perception of objects rarely changes. This invariance of perception is called perceptual constancy. There are four fundamental attributes of an object that are maintained constant over a wide range of lighting conditions. They are as follows:

Lightness – Lightness is the perceptual attribute related to the physical quantity, reflectance. In most lighting situations, it is possible to distinguish between the illuminance on a surface and its reflectance, i.e., to perceive the difference between a low-reflectance surface receiving a high illuminance and a high-reflectance surface receiving a low illuminance, even when both surfaces have the same luminance. It is this ability to perceptually separate the luminance of the retinal image into its components of illuminance and reflectance which ensures that a black cat placed in the brightest part of a headlamp beam is seen as black while a piece of white cardboard outside the beam is seen as white, even when the luminance of the cat is higher than the luminance of the cardboard.

Colour – Physically, the stimulus a surface presents to the visual system depends on the spectral content of the light illuminating the surface and the spectral reflectance of the surface. However, quite large changes in the spectral content of the illuminant can be made without causing any changes in the perceived colour of the surface, i.e., colour constancy occurs. Colour constancy is similar in many ways to lightness constancy. There are two factors that need to be separated; the spectral distribution of the incident light and the spectral reflectance of the surface. As long as the spectral content of the incident light can be identified the spectral reflectance of the surface, and hence its colour, will be stable.

Size – As an object gets further away, the size of its retinal image gets smaller, but the object itself is not seen as getting smaller. This is because by using clues such as texture, motion parallax, and masking, it is possible to estimate the distance to the object and then to compensate, unconsciously, for the decrease in retinal image size.

Shape – As an object changes its orientation in space, its retinal image changes. Nonetheless, in most lighting conditions, the distribution of light and shade across the object makes it possible to determine its orientation in space. This means that in most lighting conditions, a circular wheel that is tilted will continue to be seen as a tilted circular wheel even though its retinal image is an ellipse.

These constancies represent the application of everyday experience and the integration of all the information about the lighting available in the whole retinal image to the interpretation of a part of the retinal image that is inherently ambiguous. Given this process, it should not be too surprising that the constancies can be broken by restricting the information available coincident with the object being viewed. For example, viewing a uniform luminance surface through an aperture that restricts the view to a limited part of the surface will often eliminate lightness constancy, i.e., makes it impossible to accurately judge the reflectance of the surface. Likewise, eliminating cues to distance, such as gradients in texture, motion parallax, and overlapping of objects, will destroy size constancy; changing cues to the plane in which an object is lying will reduce shape constancy and eliminating information on the spectral content of the illuminant will reduce colour constancy.

The constancies are most likely to be maintained when there is enough light for the observer to see the object and the surfaces around it clearly, the light being provided by an obvious light source that has a spectral power distribution covering all the visible spectrum. In addition, the constancies are most likely to be maintained when there are a variety of surface colours in the scene, including some small white surfaces, and there are no large glossy areas, both factors that help with the identification of the spectral content of the light source (Lynes, 1971). Conversely, constancy is likely to break down whenever there is insufficient or misleading information available from the surrounding parts of the visual field. A breakdown of constancy is unlikely during the day, but the limited coverage of headlamps and the limited spectral power distribution of some of the light sources once used for road lighting, such as low-pressure and high-pressure sodium discharge, make a breakdown more likely when driving at night.

As examples of the problems such breakdowns can cause, consider the driver's perception of absolute and relative speed. When driving along a road, the whole retina receives a moving pattern called the optic flow in which different parts of the visual field appear to flow around the observer at different speeds in varying directions (Figure 3.16). Analysis of optic flow exposes both the structure of the world around you and your movement through that world (Gibson, 1950). But how do you know that you are moving through the world rather than the world moving past you? The answer is that you do not, at least not from optic flow alone. Either other information is necessary to conclude that you are moving through the world or you have to make an assumption. Among the other sources of information are changes in your vestibular system in your inner ear when you are accelerating or braking and vibration felt through the vehicle's contact with the road. If such information is not enough, then the usual assumption is that the larger surrounding object (the world) is stationary and the smaller enclosed object (you) is moving. This is the explanation for the common movement illusion experienced when seated on a train and the adjacent train filling the window starts to move. The initial perception is that the train you are on is moving until the absence of vestibular and vibration information tells you otherwise.

Given that you have determined that you are moving through the world, two other pieces of information are useful for the driver, his direction of movement and his speed. Gibson (1950) suggested that the direction of movement could be obtained

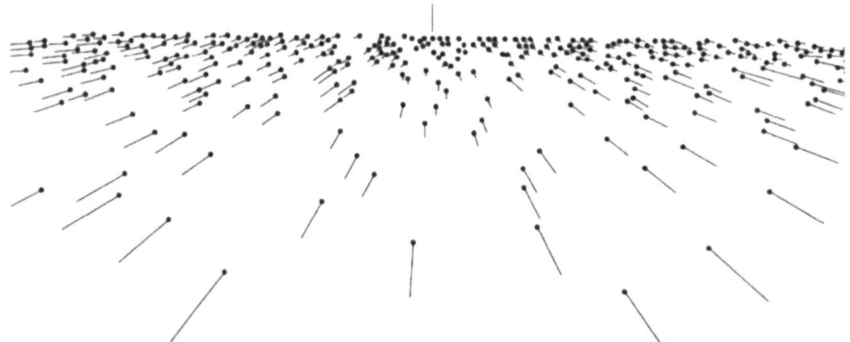

FIGURE 3.16 An optic flow pattern for an observer moving straight across and parallel to a plane in a direction given by the short line above the horizon. Each black dot represents an element on the plane, and the line attached to each dot shows the associated velocity vector. The length of the line represents the instantaneous velocity, and the direction of the line represents the direction of movement of the element (from Sekular and Blake, 1994).

by changes in the pattern of optic flow. When you are fixating the point that you are moving towards, the pattern of optic flow will be seen to expand if you are moving forward and to contract if you are moving back. Things get more complicated if you are not looking at the point towards which you are attempting to move because then the pattern of optical flow changes every time you make an eye movement (Regan and Beverley, 1982). There is no doubt that the visual system has a mechanism for extracting the effect of eye movements, but exactly what that is remains uncertain.

As for speed, absolute speed cannot be obtained from optic flow alone because to judge speed you need to be able to estimate distance, and distance cannot be obtained from optic flow. There are many other visual cues to distance in the retinal image, such as perspective relative sizes of familiar objects and texture gradients, as well as the shading and masking of one object by another. Of course, much of this information is only available when the world through which you are moving is illuminated. Driving at night on headlamps alone limits the amount of distance information available and hence makes it difficult to judge absolute speed visually.

While absolute speed is of interest when driving on an empty road, when there are other vehicles on the road relative speed is of interest. For example, if you are following another vehicle, then as long as the retinal image size is constant, you are both travelling at the same speed and you are neither increasing nor decreasing your separation. If the retinal image starts to expand, you are closing in on the vehicle ahead and the rate at which you are closing is related to the rate of expansion of the retinal image of the vehicle ahead. Of course, how much this change in relative speed matters will depend on the distance between you. This is an easy judgement to make when you are close behind a vehicle, even at night because then your headlamps will illuminate the back of the vehicle ahead, thereby providing a convenient estimate of distance. The judgement of relative speed is much more difficult when an opposing vehicle is approaching from a distance on an unlit road. Then, headlamps can be seen as two points of light: the separation between them increasing as the vehicle nears. The problem of perception is that unless you also have an estimate of distance, you cannot estimate the implication for speed of a given rate of expansion of headlamp separation. In the absence of any other lighting, your estimate of distance may have to rest on an assumed separation of the headlamps on common vehicles or on what the headlamps of the approaching vehicle illuminate. The situation gets even more difficult for a motorcycle when there is only one headlamp. Then, if you want to estimate the approach speed, you have to detect the increase in size of the single headlamp as well as judge the distance.

Clearly, lighting has a role to play in stabilizing perception when driving at night, particularly road lighting. Road lighting provides a much more extensive view of the scene than headlamps alone but its value in stabilizing perception is rarely considered. Even when road lighting is provided, enthusiasts for energy savings tend to want the light distribution restricted to the road surface. This discussion suggests that such a restriction would be unwise as providing some light on the surroundings of the road will help to set the road in context and provide a richer pattern of optic flow for the visual system to work on. Land and Horwood (1995) have shown that the edges of the road or lane markings detected in peripheral vision provide guidance for keeping the vehicle in the lane. When headlamps alone are in use, particularly low beam headlamps, the amount of information available from the visual scene is limited as is the pattern of optic flow. This suggests that a low level of light spread over a larger area with the aim of enlarging the optic flow pattern should be one objective for the designers of headlamp systems.

3.6 VISIBILITY AND CRASHES

While there is little doubt that good visibility is a necessary condition for safe driving, alone, it is not sufficient. This is evident from studies of the factors contributing to crashes. Since 2005 the UK has been collecting the opinions of police officers who attend a crash involving a fatality or personal injury that occurred on public roads about what they considered contributed to the crash. The officer is asked to provide up to six contributory factors for a crash using a form with a matrix of nine main categories, each of which can have a number of subcategories that goes under the name STATS19. The main categories, with the number of subcategories in parentheses, are road environment (10), vehicle defect (6), injudicious action (10), driver/rider error or reaction (10), impairment or distraction (10), behaviour or inexperience (7), vision affected by (10), pedestrian only (10), and special codes (5), thus giving a total of 78 possible contributing factors. In 2014, the five most frequently recorded sub-category contributory factors for such crashes were failed to look properly (44%), failed to judge person's path or speed (22%), careless, reckless, or in a hurry (17%), poor turn manoeuvre (15%), and loss of control (12%). Failed to look properly has been the most frequently reported contributory factor since 2005 (DfT, 2015a). Unfortunately, the category failed to look properly does not differentiate between failing to look in an appropriate direction at all and looking but failing to see what was there. Figure 3.17 shows some earlier data collected using a more systematic investigatory method. Specifically, these data give the percentage of crashes in which, in the opinion of the investigating team, each of the listed factors was involved (Sabey and Staughton, 1975; Hills, 1980). These percentages are based on on-the-spot investigations of 2,036 road crashes in a rural area of the UK. The percentages do not sum to 100% because crashes rarely have a single cause, usually a combination of causes is involved. Even so, Figure 3.17 declares that in 95% of the

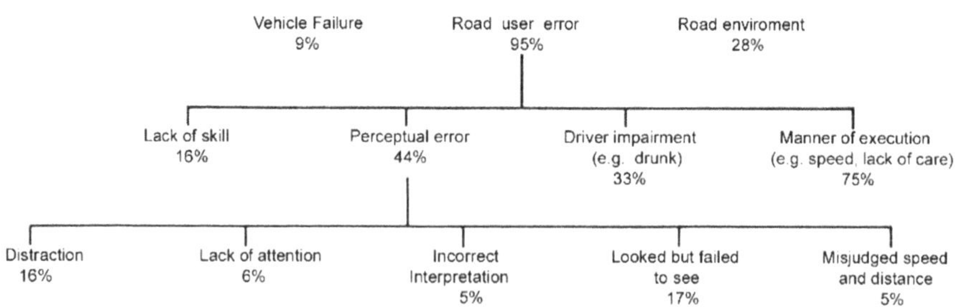

FIGURE 3.17 Apparent contributory factors in road crashes (after Hills, 1980).

crashes studied, the road user contributed to the crash a finding that is consistent with the STATS19 data. Further, 44% of the crashes involved what is termed the perceptual error. Included under perceptual error are distraction and lack of attention, incorrect interpretation, misjudgement of speed and distance, and a category intriguingly called "looked but failed to see".

A more detailed examination of the class of perceptual error can be used to explore the significance of visibility for driving. The "looked but failed to see" category makes the largest contribution, a status supported by Cairney and Catchpole (1996) who found it to be frequently involved in crashes in urban areas. A classic example of looking but failing to see is a driver who pulls out from a side road straight into the path of a motorcyclist on the main road (Wulf et al., 1989). The driver looked along the main road but failed to see the motorcyclist although measurements of luminance contrast would indicate that the motorcyclist was clearly visible. How can this be? How can a driver look in the direction of a clearly visible vehicle and not see it? The answer lies in the fact that the total amount of information available through the visual system and through the other senses at any instant is too much for the brain to deal with simultaneously (Wolfe et al., 2006). To overcome this problem, the brain uses a number of mechanisms to select which stimuli to examine in detail and which to ignore. This process of selection is unconscious and is called attention. It represents an allocation of neural resources to a specific activity at a specific time and a consequent neglect of other information arriving at that time. Thus, if the driver is paying attention to another part of the visual field or input from another sense organ, or is thinking about something other than driving, it is only too likely that a highly visible object will not be seen, even when viewed directly. This explains two of the other perceptual errors shown in Figure 3.17. Distraction occurs when attention is directed towards some other external but irrelevant source of information. Lack of attention occurs when attention is directed internally.

Given that some selection among the incoming information is necessary, how is that selection made (Trick and Enns, 2009)? Four factors have been found to influence the focus of attention: Conspicuity, expectation, mental workload, and mental capacity (Recarte and Nunes, 2009).

Conspicuity refers to the extent to which an object stands out from whatever is around it. Visibility is a necessary condition for detection, but it is not a sufficient condition to make an object conspicuous. To make an object conspicuous, it has to have some feature that easily differentiates it from other objects around it. What that feature may be can vary widely, depending on the background against which the object is usually seen. Vehicles that are likely to be stopped by the roadside in unusual places are often painted in dramatic patterns that vary in colour and reflectance to increase their conspicuity by day and equipped with flashing lights to enhance their conspicuity at night (Figure 3.18). Other examples of attempts to enhance conspicuity are the advice for motorcyclists to keep their headlamps on during the day and the requirement for slow-moving vehicles such as tractors to have flashing beacons on when on public roads. The choice of the feature or features to use to enhance conspicuity should be made on the basis of the visibility of the feature and its rarity. To be conspicuous, it is necessary to be both visible and different. The

FIGURE 3.18 The rear of a utility vehicle painted with chevrons and fitted with flashing beacons so as to be conspicuous by day and night. The painted chevrons are alternately red and yellow.

higher is the level of conspicuity, the less likely it is that a driver will look but fail to see you.

Expectation is the extent to which the presence of the object can be predicted. A clearly marked pedestrian crossing sets up an expectation that there may be pedestrians on the road and encourages the driver to look directly at it. The greater is the expectation that a specific object will be present, the less likely it is that the object will be looked at but not seen. Experience is the basis of expectation, both in general and in localities. Experienced drivers are less likely than novices to look but not see. Drivers who are familiar with a locality know what to expect and hence what sort of hazards are most likely to occur.

Mental workload refers to the amount of information that has to be handled at any moment in time. It is important to appreciate that mental workload extends beyond the information arriving at the senses to the handling of past information using memory and decision-making capabilities. Given that attention is a process for allocating neural resources and that, for an individual, such resources are fixed, it is obvious that the greater is the mental workload, the more likely it is that something of significance will be ignored, i.e., looking but failing to see is more likely as mental workload increases. Increased mental workload has been shown to be related to worse driving performance (Chapparro et al., 2005).

Mental capacity refers to the ability to handle information flow. Different people have different capacities. Further, an individual's capacity to handle information flow can be decreased by age, fatigue, alcohol, and drugs. Again, the more restricted is the individual's capacity, the more likely it is that the individual will look but fail to see.

To summarize, unexpected, inconspicuous objects that occur in front of drivers with limited mental capacity at times of high mental load are most likely to be looked at but not seen, even when they have high visibility.

The remaining two categories of crash causes classified as perceptual errors are incorrect interpretation and misjudgment of speed and distance. Incorrect interpretation is probably related to the limited time drivers have available to gather information. Frequently, drivers have only one or two seconds in which to decide whether to make a particular manoeuvre. In the few seconds available, the driver will only be able to fixate and accommodate on a few places. Limited information will sometimes lead to misinterpretation of the situation facing the driver.

Judgements of speed and distance are both involved in two manoeuvres that produce many serious crashes, turning across traffic and overtaking, These manoeuvres require the driver to move from the correct side of the road and go across or into the opposing traffic (Caird and Hancock, 2002). To do this safely, it is necessary to make accurate judgements of how far off an approaching vehicle is and at what speed it is approaching. Experimental evidence (Jones and Heimstra, 1964) suggests that people are not very good at making either of these judgements, particularly when the judgement has to be made at a long distance and the approaching vehicle is moving directly towards the driver, so the change in size and the perceived relative motion are small.

By now it will be appreciated that the drivers' task is a complex one, involving both visual and cognitive factors (Wierda, 1996). Within a very limited time, the driver has to interpret what is likely to happen on the road ahead. To do this, the driver has developed a series of expectations of other drivers' behaviour and of what are the appropriate locations to examine. The driver will be faced with objects of different degrees of visibility and conspicuity and will have to make judgements for which the visual system is not well suited. It is within this context that vehicle and roadway lighting has to operate.

3.7 SIGHT AND DRIVING

Because sight is considered an essential sense for driving, most countries require some sort of vision test before a licence to drive is issued. These vary from country to country. For most drivers, these tests involve little more than a measurement of visual acuity made under photopic conditions with unlimited time, although more stringent tests involving the size of the visual field and colour vision may be required for drivers of commercial and public service vehicles (see Section 13.5.1). Despite these requirements, attempts to find a link between simple visual functions and the accident record of drivers have proved largely fruitless. Burg (1967) measured the static and dynamic visual acuity, the field of vision, the extent of misalignment of the two eyes, and the rate of recovery from glare for 17,500 drivers who between them had had over 5,200 crashes in the previous three years. He was able to find only very weak correlations between the measured visual capabilities and the drivers accident records. Charman (1997) has confirmed that simple visual capabilities are at best weakly correlated with crash occurrence. This failure can be explained in two ways. First, it is possible that drivers with worse visual capabilities are aware of their limitations and drive within them (see Section 13.4.5). Second, there is the point that the visual capabilities measured were very simple and had no time limit. This is a problem because the driver often has to detect the presence and/or movements of a number of different obstacles, recognize them and their characteristic patterns of movement, and then relate them to each other, all the while maintaining an understanding of the more slowly changing features of the road, and all in a short time before making a manoeuvre.

Somewhat more successful have been attempts to relate limited visual function to driving performance using younger drivers with their visual capabilities artificially constricted. Troutbeck and Wood (1994) restricted the driver's field of view to 40 degrees. As a result, the drivers drove more slowly and had difficulty in detecting and identifying road signs, avoiding obstacles, and driving through narrow gaps. Brooks et al. (2005) used a driving simulator on which young drivers drove around a winding course at 55 mph (86 km.h^{-1}) and tried to detect up to six pedestrians standing at the edge of the road. While doing this, the driver's retinal image quality was either reduced by defocusing lenses, or their visual field was restricted, or the luminances of the central road markings were varied over a wide range. Figure 3.19 shows how the drivers' visual acuity, the percentage of time the whole vehicle was within the correct lane, and the percentage of pedestrians detected varied with blur, field size, and road marking luminance.

It is clear from Figure 3.19 that while visual acuity is very sensitive to a blurred retinal image, the ability to keep the car in a lane is not. What is also somewhat sensitive to blur is the ability to detect pedestrians. The same is true for reducing the luminance of road markings but not for reducing the visual field size. Both steering accuracy and the ability to detect pedestrians are sensitive to reduced visual field size. These results are consistent with the ambient–focal dichotomy. This refers to the anatomical and functional division of visual processing into two paths: One devoted to giving guidance through the environment, the ambient, and the other devoted to object recognition and identification, the focal (Scheiber et al., 2009). Another way to describe this dichotomy is to consider that it provides answers to the questions what is out there and where is it (Rea, 2018)? Such a structure is consistent with the selective degradation hypothesis (Leibowitz and Owens (1977). This hypothesis maintains that the reason for the increased probability of crashes at night is that all visual capabilities do not decline equally as luminance is reduced but that drivers are not aware of this. Specifically, while the ability to see detail in the fovea deteriorates with reduced luminance, drivers are not aware of how great the deterioration is because they experience little difficulty in steering using peripheral information and because many objects that need to be seen, such as retro-reflective road signs, are designed to be highly visible. It is only when the driver comes across an unexpected and inconspicuous object, such as a deer, that the deterioration in foveal capabilities becomes apparent.

It should now be clear why it has been difficult to establish much of a correlation between simple visual functions measured under advantageous conditions and road crashes. The basic problem is that driving involves everything from simple detection

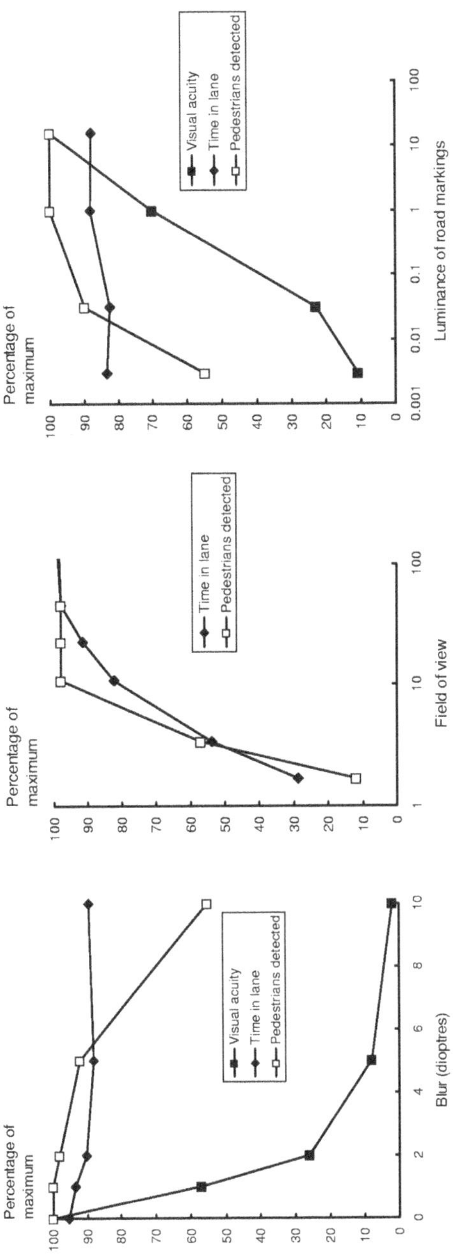

FIGURE 3.19 The percentage of the maximum possible value for visual acuity, time wholly in lane, and pedestrians detected with different degrees of blur (dioptres), different fields of view (degrees), and different road marking luminances (cd/m^2). Visual acuity measurements were not made for the restricted visual field variable (after Brooks et al. 2005).

and resolution of detail to high-level perception using different parts of the visual field, all interacting with cognitive abilities and performed under time pressure (Castro, 2009). The failure to relate simple visual functions to crashes does not mean that visual capabilities and the lighting conditions that support them are unimportant. Rather, the evidence of the effect of visual capabilities on driving performance means that visual capabilities and the lighting conditions that support them are part of the problem and part of the solution but not the whole problem, nor the whole solution.

3.8 INDIRECT EFFECTS OF LIGHT EXPOSURE

So far, this chapter has been devoted to the structure and capabilities of the human visual system because these characteristics determine visibility and hence have a direct effect on driving. However, light reaching the retina at other times and other places than the road affects many aspects of human physiology beyond vision. These effects are driven through the circadian timing system. The response of the circadian timing system to light exposure can have indirect effects on the ability to drive safely.

The circadian timing system starts from a new photoreceptor in the retina discovered more than 100 years after the rod and cone photoreceptors had been identified and called the intrinsically photosensitive retinal ganglion cell or ipRGC (Berson et al., 2002; Berson, 2007). The photopigment in the ipRGC is melanopsin, which has a maximum absorption at 480 nm. There are a relatively few ipRGCs, about 1% of all the retinal ganglion cells (Mure, 2021). Every ipRGC has an extensive dendritic arbour that spreads into the collector level of the retina and these dendrites themselves respond to incident light. Further, the ipRGC network has a high threshold and a slow but sustained output when stimulated. These characteristics make the ipRGC network well suited to signal light level to the brain. There are a range of ipRGC subtypes connecting to different parts of the brain, each with a different function (Mure, 2021; Schmidt et al., 2011).

The ipRGC subtypes of interest here are those that send signals to the suprachiasmatic nuclei (SCNs) of the hypothalamus of the brain. The SCNs are recognized as the master clock in mammals, including humans. The SCNs are responsible for synchronizing the timing of many different physiological events in the body, including DNA repair and hormone production. The whole structure is known as the circadian timing system.

The role of the circadian timing system is to establish an internal replication of external night and day so as to prepare the body for upcoming activities (Foster, 2022). The human circadian timing system involves three components: An internal (endogenous) oscillator, located in the SCN; a number of external (exogenous) oscillators that can reset (entrain) the SCN, and a messenger hormone melatonin that carries the internal "darkness" information to other melatonin sensors in all parts of the body through the bloodstream. In the absence of light, and other cues, the internal oscillator continues to operate but with a period slightly longer than 24 h (Czeisler et al., 1999). External stimuli are necessary to entrain the SCN to a 24-h period and to adjust for the seasons.

The light–dark cycle of exposure is one of the most potent of the external stimuli used for entrainment. By varying the amount of light exposure, when it is presented, and for how long, it is possible to shift the phase of the circadian timing system, either forward or backward, as required. In addition, it is possible to have an immediate alerting effect at night by suppressing melatonin. The amount of light required to cause these effects is certainly available outside in daylight and within the range of some current indoor lighting but not for most outdoor lighting at night. The spectral sensitivity of the circadian timing system peaks around 460 nm so it is very different from the CIE Standard Photopic Observer. This implies that light spectra that have a lot of power at the short-wavelength end of the visible spectrum will require lower illuminances to produce the same effect on the circadian timing system.

Although the light/dark cycle is known to be very important for entraining the circadian timing system, there are other external stimuli that can affect the system. This is because major organs in the body have their own circadian physiology; for example, human metabolism is influenced by the intake of food. If this occurs at an unusual time, the circadian timing delivered by the SCN and that identified by metabolism are in conflict with the result that the sleep/wake cycle can be disrupted (Foster, 2022).

Although many aspects of the various connections from the ipRGC to the brain still need to be investigated, one feature that is clear is that they are not totally isolated from the visual system. It is known that the ipRGCs receive inputs from the rod and cone photoreceptors and that outputs from the ipRGC project to the brain centres that control non-imaging parts of the visual system, such as pupil size (Lucas et al., 2012). This means activities in the visual and circadian timing systems fed from the retina are mingled to some degree. But there remain significant differences between the visual system and the circadian timing system. The information produced by the visual system is fast, detailed, and located. The information produced by the circadian timing system is relatively slow, limited, and not located.

But what has the circadian timing system got to do with lighting for driving? The answer is that it sets the foundations on which the driver is operating. A stable circadian timing system is necessary for sustained sleep, and a driver who has not slept well is likely to be less alert than one who has, even by day. The timing and duration of sleep is governed by the balance between two processes. One is the sleep-promoting process, the other is the wakefulness process. The balance between these two processes varies over the day. In the morning, the wakefulness process is dominant, but the longer one has been awake, the greater is the influence of the sleep-promoting process. When the balance between these two processes is disturbed, sleep disruption occurs. The causes of sleep disruption are many, ranging from lifestyle factors such as excessive caffeine consumption, work patterns such as shift work, and social responsibilities such as caring for infants. Poor quality sleep and sleep deprivation are linked with drowsiness, worse coordination, and deterioration in cognitive function, all effects that can lead to an increased number of crashes (Horne and Reyner, 1999).

Crashes where the driver falls asleep are not uncommon. Estimates of their frequency range between 10% and 20% of all crashes (European Commission, 2021).

Such crashes typically involve the vehicle leaving the road or running into the back of a vehicle ahead. Such crashes are more likely where driving is sustained yet undemanding, such as on motorways in the middle of the night. Sleep-related crashes rarely occur in urban areas when there is plenty to keep the driver busy. There are also clear time-of-day effects for the frequency of such crashes. In the UK, sleep-related crashes are most likely to occur between the hours of 02:00 and 06:00 and 14:00–16:00, this latter being consistent with what is known as the post-lunch dip in alertness (Horne and Reyner, 1995). Further, drivers of heavy goods vehicles, because of their long hours alone at the wheel, and sometimes taking their legally required rest periods in a bunk in the back of the cab while parked besides a busy road, are more likely to be involved in a sleep-related crash. Shift workers, particularly those on rapidly rotating shifts, are another group with a higher risk of a sleep-related crash, particularly when driving home at the end of their shift (European Commission, 2021).

Given that a good sleep pattern appears to be a prerequisite for minimizing sleep-related crashes is there anything that can be done with lighting to achieve this aim? Various recommendations have been made for the light exposure required in the morning and evening to ensure adequate stimulation to maintain a stable circadian rhythm (Brown et al., 2022). On the basis of these recommendations, it would appear that the amount of light received at the driver's eye from road lighting and light from headlamps reflected from the road is insufficient to interfere with a stable circadian rhythm, a conclusion confirmed by Gibbons et al. (2022). Whether adding light from the headlamps of oncoming traffic or even some short-wavelength light to the vehicle's interior lighting would produce sufficient stimulation to influence the circadian system remains to be determined. If it did, a complex balancing act is exposed. In the short term, this could be beneficial as it would suppress melatonin concentration and thereby alleviate drowsiness. In the long term, it could destabilize the circadian rhythm. Immediately, it could degrade the visual capabilities of the driver. Until this quandary is resolved, the advice that most directly deals with the immediate problem of the onset of drowsiness is to stop driving, take a break with a cup of coffee, and, ideally, a nap. Drivers are often aware that they are having to fight to stay awake and try to overcome this drowsiness by opening the vehicle window, turning up the radio, shifting about in the seat or setting the temperature controls lower. These may be effective for a short time, but if you are feeling drowsy the best thing to do is to take a break. This is the reason why, in many countries, truck and coach drivers' hours of continuous driving are limited. Another approach that is rapidly being introduced in vehicles is a system of advisory driver assistance in the form of monitoring a driver's performance in terms of lane keeping as well as frequency and magnitude of accelerating and braking (see Section 15.4). Such monitoring can lead to automated passive/aggressive suggestions about the need to take a break. Another aspect of value is interventional driver assistance in the form of automated emergency braking now available in some vehicles. This should reduce the number of rear-end collisions caused by driver drowsiness and could be extended to prevent the vehicle leaving the road. Such approaches seem more likely to address the immediate problem of sleep-related crashes than manipulating light exposure at the time, but to avoid the problem occurring in the first place, it would be better to establish a stable circadian rhythm and a good sleep pattern and then to minimize driving during the sleep phase. To establish a stable circadian rhythm, there is much to be said for following the recommended patterns of light exposure (Brown et al., 2022) and bedtime behaviour (Foster, 2022).

3.9 SUMMARY

The visual system consists of two parts: An optical system that produces an image on the retina of the eye and an image processing system that extracts different aspects of that image at various stages of its progress up the optic nerve to the visual cortex, while preserving the location where the information came from. The visual system devotes most of its resources to analysing the central area of the retina, particularly the fovea. This implies that peripheral vision is mainly devoted to identifying something that should be examined in detail by turning the head and eyes so the image of whatever it is falls on the fovea.

This emphasis on the fovea, together with the fact that the eye has a fixed image distance, means that the eye is continuously adjusting its fixation and focus. The visual system also has to operate over a wide range of luminances, from sunlight to starlight. To do this, it continually adjusts its sensitivity to light, increasing its sensitivity as the amount of light available falls. Decreasing the amount of light from daylight to darkness takes the visual system through three distinct operating states – the photopic, the mesopic, and the scotopic. In the photopic condition, fine discriminations of size and colour can be made. In the mesopic state, the ability to make these discriminations deteriorates so that by the time the scotopic state is reached, colour can no longer be seen, detail is impossible to discriminate, and the fovea is blind. Vehicle headlamps alone provide enough light to ensure that, when fixating on the road ahead, the visual system is operating in the photopic state within the beam, in the mesopic state at the edge of the beam, and, in the absence of moonlight, in the scotopic state well outside the beam. Road lighting ensures that the visual system is operating at least in the mesopic state over a much larger area.

Like every other physiological system, the visual system has a limited range of capabilities. These limits are expressed by the thresholds of vision. There are many different thresholds, one of the most common being visual acuity, i.e., the smallest size of detail that can be resolved. Others quantify the smallest luminance contrast that can be detected, the smallest colour difference that can be discriminated, and the lowest frequency of light fluctuation that can be seen as flickering. Different thresholds occur under different conditions of lighting and stimulus presentation, but, in general, vision becomes more limited as the amount of light decreases, the stimulus occurs further away from the fovea, and the speed of movement increases. Threshold measurements provide well-defined and sensitive metrics to explore the operation of the visual system and have been extensively used in the field of vision science, but for the practice of lighting, threshold measurements are mainly of interest in determining what will not be seen rather than how well something will be seen.

Given that the details of a scene are clearly visible, i.e., they are well above their threshold values, the dominant characteristic of the visual system is the stability of perception in the face of continuous variation in the retinal image. Given lighting conditions that provide enough light with a wide spectral distribution in such a way that how the space is lit can be easily understood, the lightness, colour, size, and shape of objects in the space remain constant no matter how they are viewed. It is only when the information about the space and the way it is lit is restricted or misleading that these perceptual constancies will break down.

While there is little doubt that good visibility is a necessary condition for safe driving, alone, it is not sufficient. This is evident from studies of the factors contributing to crashes. Almost half of crashes involve what is termed perceptual error, a category that includes distraction, lack of attention, incorrect interpretation, misjudgment of speed and distance, and something called "looked but failed to see". In crashes in which looked but failed to see is a factor, the driver reportedly looked at something that measurements define as being visible but failed to see it. The reason for this unexpected behaviour is that the total amount of information available through the visual system and through the other senses at any instant is too much for the brain to deal with simultaneously. To overcome this problem, the brain has to select which stimuli to examine in detail and which to ignore. This process of selection is unconscious and is called attention. Attention is influenced by conspicuity, expectation, mental load, and mental capacity. Unexpected, inconspicuous objects that occur in front of drivers with limited mental capacity at times of high mental load are most likely to be looked at but not seen, even when they have high visibility.

The belief that vision is important to driving safely is the reason why measurements of visual capability are an integral part of the test for issuing a driving licence in most countries. Despite these requirements, attempts to find a link between simple visual functions such as visual acuity and the accident record of drivers have proved largely fruitless. This may be because drivers with worse visual capabilities are aware of their abilities and drive within them. Alternatively, it may be that the visual capabilities measured are too simple. The fact is the drivers' task is a complex one, involving both visual and cognitive factors. Within a very limited time, the driver has to interpret what is likely to happen on the road ahead. To do this, the driver has developed a series of expectations of other drivers' behaviour and of what are the appropriate locations to examine. The driver will be faced with objects of different degrees of visibility and conspicuity and will have to make judgements for which the visual system is not always well suited. It is in this context that vehicle and roadway lighting has to operate.

All the above has been concerned with the visual system, but lighting conditions away from the road can have an effect on road safety indirectly. This is because light reaching the retina does more than stimulate the visual system. It also influences human physiology at a very basic level by adjusting the circadian timing system. If the circadian timing system is disrupted, the sleep/wake cycle is disrupted, and that can lead to drowsiness while driving. Drowsiness is involved in a significant minority of crashes, particularly those in which a vehicle ahead is rear-ended or a single vehicle leaves the road. Such crashes are most likely at night when driving in undemanding conditions. Until the arrival of interventionist driver assistance systems, the only realistic way to deal with the immediate problem of drowsiness while driving is to take a break. In a longer term, it would be better to eliminate the problem of drowsiness while driving as much as possible by using appropriate light/dark exposure patterns to ensure a stable circadian timing system and then to minimize driving during the sleep phase.

4 Road Lighting

4.1 SOME HISTORY

Paved roads have existed in Europe since Roman times and in South America since the time of the Inca empire, but road lighting is much more recent. Lighting designed specifically for enhancing the safety of the driver began to appear in the 1930s. Three factors converged to make road lighting possible and desirable at this time. The first was the availability of the necessary technologies in the form of an extensive electricity distribution network together with suitable lamps and luminaires. The second was the establishment of official systems for regulating the design and use of vehicles and the control of traffic. The third was the growth in the number of vehicles on the roads and the speeds those vehicles could sustain. Despite this convergence, the growth in road lighting was slow. Indeed, the autobahn system in Germany, introduced in the 1930s, and the motorway systems in Belgium and Britain, introduced in the 1950s, all designed for high-speed traffic, were opened without lighting. However, as traffic densities and traffic speeds have continued to increase, road lighting has become more extensive until it is now routinely considered as an important component of any road scheme.

The road lighting discussed above was for roads outside towns and cities, i.e., in locations where the main concern was for the safety of the driver. The lighting of roads in urban areas began much earlier for reasons of public safety (Painter, 1999, 2000). In Paris, in the 15th century, it was decreed "during the months of November, December and January, a lantern is to be hung out under the level of the first floor window sills before 6 o'clock every night. It is to be placed in such a prominent position that the street receives sufficient light" (Schivelbusch, 1988). As cities grew, there was an increased demand for some form of public lighting at night. This demand was first widely met by the introduction of gas lighting. In London, by 1823, the gas lighting system had grown to such an extent that 39,000 gas lamps provided lighting for 215 miles of urban road (Chandler, 1949). Gas was the major source for urban exterior lighting at night for about 100 years, although the first exterior electric lighting installations, using arc lamps, were installed in the 1850s. It was not until the early 20th century that gas began to give way to electricity as the primary means of providing light at night. Today, urban road lighting is designed to meet a number of objectives such as reducing the fear of crime, improving the safety of drivers, cyclists, and pedestrians, and enhancing the attractiveness of the environment. Of these, making life safer for drivers, cyclists, and pedestrians is paramount because it is always relevant while the others may or may not be, depending on the site.

DOI: 10.1201/9781003388906-4

4.2   THE TECHNOLOGY OF ROAD LIGHTING

The technology of road lighting consists of light sources and luminaires arranged in different ways and controlled by different means.

Light Sources: The light emitter originally used for public lighting was gas. However, over the last 150 years, electric light sources first became competitors to gas and ultimately, completely replaced it. The first electric light sources used for public lighting were arc lights followed by incandescent sources and then mercury vapour, fluorescent, low-pressure sodium, high-pressure sodium, and metal halide discharge sources (DiLaura, 2006) and, now, light emitting diodes (LEDs). The characteristics of these electric light sources are summarized in Section 2.5. In the UK, the light source most widely used for road lighting until recently was high-pressure sodium discharge, but it is rapidly being replaced by the LED. A similar situation prevails in the USA. The characteristics that are deemed most important for road lighting are first cost, for obvious reasons; luminous efficacy, because that affects energy costs; lamp life, because that affects maintenance costs; and colour rendering. The high luminous efficacy, long life, and reasonable colour properties of phosphor-converted LEDs have made them the light source of choice for road lighting today (Djuretic and Kostic, 2018).

Luminaires: The luminaires used for road lighting can be classified in different ways. The most complete is the full luminous intensity distribution. There are a large number of luminous intensity distributions available, some symmetrical and some asymmetrical. Further, many luminaires allow for on-site adjustments in light source position within the luminaire to modify the luminous intensity distribution. Different light distributions are necessary because roads of different widths and layouts require different light distributions if the light is to be directed onto the road surface. The exception to this concern takes the form of high mast lighting where the luminaires are mounted 30 m or more above the ground. High mast lighting is used for lighting complex road junctions where the waste inherent in illuminating large areas of grass between roads is more than offset by the cost savings produced by minimizing the number of columns and simplifying the electricity distribution network.

A simplified system of luminaire classification based on the luminous intensity distribution just above and below the horizontal plane through the luminaire is sometimes used. Limiting the amount of light emitted just above the horizontal plane through the luminaire is intended to reduce light pollution (see Section 14.7), while limiting the amount of light emitted just below the horizontal is done to control glare to the driver. The Illuminating Engineering Society of North America once used such a system, the system having four classes – full cutoff, cutoff, semi-cutoff, and non-cutoff.

A full cutoff luminous intensity distribution is defined as having zero luminous intensity at or above 90 degrees from the downward vertical and no luminous intensity in the range 80–90 degrees from the downward vertical greater than 10% of the light source luminous flux.

A cutoff luminous intensity distribution is defined as having no luminous intensity at or above 90 degrees from the downward vertical greater than 2.5% of the light source luminous flux and no luminous intensity in the range 80–90 degrees from the downward vertical greater than 10% of the light source luminous flux.

A semi-cutoff luminous intensity distribution is defined as having no luminous intensity at or above ninety degrees from the downward vertical greater than 5% of the light source luminous flux and no luminous intensity in the range 80–90 degrees from the downward vertical greater than 20% of the light source luminous flux.

A non-cutoff luminous intensity distribution is defined as a luminous intensity distribution where there is no limitation of luminous intensity above the angle having the maximum luminous intensity.

Today, in the USA, a different but more general outdoor luminaire classification system is used based on the percentage of light source luminous flux emitted in a number of zones about the luminaire (IESNA, 2020). It is colloquially known as the BUG system. Figure 4.1 shows a sphere centred on the luminaire, divided into three zones. The luminous flux emitted into the hemisphere above the horizontal plane is the uplight. The luminous flux emitted into the quarter of the sphere in front of the luminaire and below the horizontal plane is the forward light and that emitted into the quarter of the sphere behind the luminaire and below the horizontal plane is the backlight. The remaining zone not evident in Figure 4.1 is the trapped light, which is the luminous flux emitted by the light source that does not get out of the luminaire. Uplight, forward light, back light, and trapped light are all expressed as percentages of the light source luminous flux. The uplight, forward light, and back light solid angles are all subdivided into different zones according to the angle from the downward vertical from the luminaire, uplight into two classes and forward light and backlight each into four classes. These subdivisions are introduced so as to identify the percentage of luminous flux emitted by the luminaire relevant to different effects such as glare, light trespass, and sky glow (see Section 14.7).

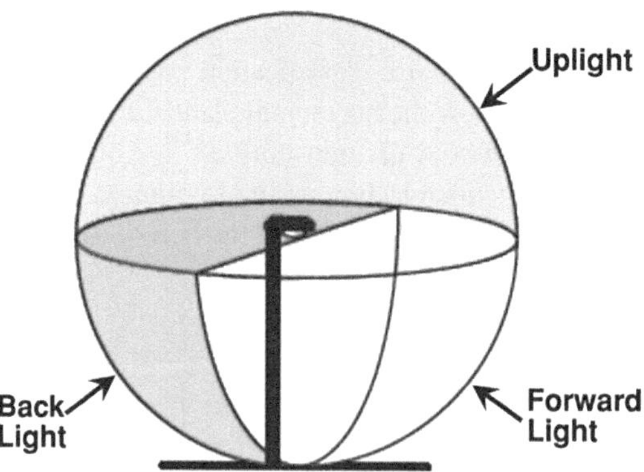

FIGURE 4.1 Three zones around a luminaire according to the IESNA luminaire classification system for outdoor luminaires (from IESNA, 2020).

The IESNA outdoor luminaire classification system is a way of systematically dividing up the luminous flux emitted by a light source, so it is useful in exposing the typical distributions for common types of road lighting luminaires. Table 4.1 gives the percentage light source luminous flux occurring in each of the subdivisions of the zones for four luminaires, two luminaires used for lighting traffic routes, and two used for lighting urban areas which include many residential streets. It is clear from Table 4.1 that the luminaires used to light traffic routes, i.e., road lighting luminaires, are very similar in their distribution of light source luminous flux although the drop lens luminaires have less trapped light and slightly more uplight than the flat lens luminaire. Where large differences occur is in the urban area lighting luminaires, particularly in the amount of upward luminous flux. It is also apparent that the road lighting luminaires provide more forward light than the urban area lighting luminaires, the latter having approximately equal forward light and backlight. It should be noted that to be effective, both classification systems assume that the luminaire is

---

TABLE 4.1

Percentage of Light Source Luminous Flux in the Different Zones of the IESNA Outdoor Luminaire Cclassification System for Two Road Lighting Luminaires and Two Urban Area Lighting Luminaires (HPS = High-Pressure Sodium, MH = Metal Halide) (from IESNA, 2007)

Zone	Zone Subdivision (Degrees from Downward Vertical)	150W HPS Drop Lens Road Lighting Luminaire (%)	150W HPS Flat Lens Road Lfighting Luminaire (%)	175W MH Glass, Tear-Drop Urban Area Lighting Luminaire (%)	175W MH Acrylic, Acorn Urban Area Lighting Luminaire (%)
Forward	0–30	11.5	5.0	3.8	0.5
light	30–60	27.6	27.3	14.9	4.0
	60–80	17.1	20.6	12.9	13.2
	80–90	1.3	0.8	0.7	8.4
	Total	57.4	53.7	32.3	26.1
Back light	0–30	5.2	4.1	3.8	0.5
	30–60	11.8	12.3	14.9	4.0
	60–80	6.0	4.4	12.9	13.2
	80–90	0.9	0.8	0.7	8.4
	Total	24.0	21.5	32.3	26.1
Uplight	90–100	1.1	0.0	0.1	8.9
	100–180	1.3	0.0	0.1	23.0
	Total	2.4	0.0	0.2	31.8
Trapped light	Total	16.2	24.8	35.2	16.0

mounted horizontally, but this is not necessarily the case. Road lighting luminaires may be tilted up as much as 15 degrees from the horizontal.

The only luminaire classification system used in the UK (BSI, 2015) has six classes defined by the luminous intensity of the luminaire, in candelas/1000 lumens of the luminaire luminous flux, at 70, 80, and 90 degrees and above, from the downward vertical, in any direction, and the luminous intensity above 95 degrees, in any direction (see Section 4.4.2). This system is aimed primarily at controlling disability glare to drivers and pedestrians.

The luminous flux distribution is not the only important characteristic of road lighting luminaires. Others are the light output ratio, the protection provided against the ingress of dirt and moisture, and the physical size. The luminaire light output ratio, or the sum of uplight, forward light, and backlight or 100 minus the percentage trapped light in the IESNA outdoor luminaire classification system quantifies the proportion of light emitted by the light source that gets out of the luminaire. Light output ratios for road lighting luminaires tend to be about 0.8. The level of ingress protection is given by the IP number (see Table 2.5). Typically, road lighting luminaires are IP65, meaning they are strongly protected against the ingress of dust and driving rain. This implies that the interior of the luminaire should remain clean although the outside will still require regular cleaning if a marked deterioration in light output is to be avoided. As for physical size, this is relevant to the amount of leverage applied to a lighting column in high winds. The smaller the physical size of the luminaire and the more aerodynamic its shape, the less will be the leverage.

Columns: Most road lighting installations use columns to carry the luminaire or luminaires. Columns vary in height and materials used. Column heights above ground level typically range from 3.5 m to 14 m although high mast installations use columns of 30 m and taller. Materials are usually aluminium, steel, or composites, such as glass-reinforced polyester, although concrete was used in the past. Most columns allow the luminaire mounted at the top of the column to be changed. In the last decade, this process has been extensively used for replacing high-pressure sodium discharge road lighting with LED road lighting. In rural areas where the electricity network uses overhead distribution, the wooden poles carrying the electricity lines are often used for mounting the luminaires, regardless of the distortions this imposes on the lighting conditions achieved.

Lighting columns themselves are a hazard for the driver, which is one reason why concrete columns have fallen into disuse. Measurements in The Netherlands suggest that 20% of crashes causing death or injury involve single vehicle crashes where the vehicle collides with objects on or near the road. In 11% of such crashes, the object collided with is a lighting column (Schreuder, 1998). There are three approaches to reducing the hazard of the lighting column: To set the column back from the road edge so as to provide an obstacle-free zone alongside the road (AASHTO, 2011; BSI, 2020), to shield the column with a crash barrier of some sort, or to design the column to be passively safe by either absorbing the energy of the crash or bending or breaking away on impact (BSI, 2019). Each of these approaches has limitations. The setback required on high-speed roads is large and may not always be available. Crash barriers are expensive although they may be necessary for other reasons. Columns that bend or break away are restricted to shorter columns, usually less than 7 m and that are made of aluminium or a composite material. Columns that break away on impact may cause more crashes because it is not possible to predict where they will fall, although this has been estimated to be a minor risk (Williams et al., 2008).

The choice of light source, luminaire, and column will be guided by many factors, among them the desired layout of luminaires (van Bommel, 2015). The layout of the luminaires for single-carriageway roads is usually single-sided, staggered, or opposite. In a single-sided installation, all the luminaires are located on one side of the carriageway. The single-sided layout is used when the width of the carriageway is equal to or less than the mounting height of the luminaires. The luminance of the lane on the far side of the carriageway is usually less than that on the near side. In a staggered layout, luminaires are arranged alternately on opposite sides of the carriageway. Staggered layouts are typically used where the width of the carriageway is between 1 and 1.5 times the mounting height of the luminaires. With this layout, care is necessary to ensure that the luminance uniformity criteria are met. In the opposite layout, pairs of luminaires are located symmetrically opposite each other on the sides of the carriageway. This layout is typically used when the width of the carriageway is more than 1.5 times the mounting height of the luminaires.

The layout of luminaires for dual carriageways and motorways is usually central twin, or central twin and opposite. In a central twin layout, pairs of luminaires are located on a single column in the central reservation. This layout can be considered as a single-sided layout for the two carriageways. Where the overall width of the road is wider, either because the central reservation is wide or there are several lanes, the central twin and opposite layout can be used. In this, the central twin luminaires alternate with the opposite luminaires to form a staggered layout. Alternatively, where there are several lanes in each direction and there is a central reservation available, a catenary mounting system can be used. This involves a series of closely spaced luminaires mounted on a wire between widely spaced columns. The luminaires distribute light across the road rather than along the road. A catenary system can provide good uniformity across multiple lanes with less disability and discomfort glare. Unfortunately, a catenary system is usually more expensive than other layouts which limits its use.

The above layouts apply to continuous linear stretches of roads, not to sharp bends or conflict areas such as intersections and roundabouts. Layouts for these areas are bespoke, although some typical layouts are given in road lighting recommendations (BSI, 2020; IESNA, 2021b). Another situation where bespoke design is required is in urban areas, particularly in narrow streets where there is no room for columns. Then, luminaires suspended from wires stretched between buildings can be used.

Controls: The control of road lighting is usually based either on time or on the amount of daylight available and is applied to individual luminaires or to groups of luminaires. In the past, most road lighting was controlled by time switches to be on from half an hour after sunset to half an hour before sunrise although enthusiasm for financial or energy savings have meant that some installations were partially or totally switched off at midnight. However, the use of time switches makes it difficult to deal with unexpected meteorological conditions. Today, the most common control system consists of a photoelectric cell mounted on the luminaire, this being used to detect the amount of daylight available and thus to ensure that the road lighting is only used when necessary. However, there are now more sophisticated systems available. For example, individual luminaires can be fitted with circuitry that identifies their GPS locations and monitors their performance. Wireless communication between luminaires is possible and communication from a group of luminaires to a remotely located server can be arranged. Further, where LEDs are the light source, dimming of light output can be achieved. This makes it possible to control many individual luminaires from a remote site and hence to manage their operation and maintenance. These technical advances are behind the current interest in adaptive road lighting according to which the light output of the luminaires is adjusted to match the traffic flow and weather conditions (Guo et al., 2007; Ozadowicz and Grela, 2016). In addition, the ability to dim road lighting can be used to compensate for the decline in light output that occurs over light source life. At present, designers tend to anticipate this decline so that, when new, most road lighting installations produce more than the required road surface luminance. This excess can be eliminated by dimming: The amount of dimming being reduced as the light output of the light source decreases (Valetti et al., 2023).

4.3 METRICS OF ROAD LIGHTING

Given the need to make a choice between a wide variety of light sources, luminaire types, column heights, road layouts, and control systems, it is necessary for the designer of road lighting to have a target to aim at. What will be discussed here are the attempts that have been made to provide single-number metrics suitable for quantifying the effectiveness of the choices made.

The primary purpose of road lighting is to make people, vehicles, and objects on the road visible without causing discomfort to the driver. Road lighting makes people, vehicles, and objects on the road visible by producing a difference between the luminance of the person, vehicle, or object and the luminance of its immediate background, which is usually the road surface or its edges (BSI, 2020). In principle, this may present a problem because depending on the luminance of the people, vehicle, or object and its immediate background, the luminance difference can be positive or negative, above, or below the threshold. Both positive and negative luminance differences that are below the threshold are unlikely to be seen. In practice, this is seldom a problem because people, vehicles, and objects rarely have a single, uniform luminance and neither do their backgrounds. This is a good thing because it means that it is unlikely that the whole person, vehicle, or object will disappear.

While the existence of multiple luminance differences is useful for detection, they may make identification difficult, particularly when the pattern of luminance differences changes dramatically as a vehicle moves. However, road lighting and vehicle lighting provide some stability to the pattern of luminance differences. Specifically, the road lighting produces a pattern of shadow on the road around the vehicle while the headlamps of the vehicle increase the luminance of the road immediately in front of the vehicle. The shadow pattern around the vehicle fluctuates as the vehicle moves beneath successive road lighting luminaires, but is always present. Similarly, the luminance differences between the vehicle profile and the road immediately ahead will vary as the vehicle moves under the road lighting, but they are always present.

Despite this variability in luminance differences for real objects, the metrics that have been developed to quantify the ability of road lighting to make objects visible have been based on the luminance of planar surfaces. The first such metric was called revealing power (Waldram, 1938). To calculate the revealing power of a road lighting installation at a specific point requires four pieces of information: The road surface luminance, the vertical illuminance on the critical object, the luminance difference threshold of the critical object, and the cumulative frequency distribution of the reflectances associated with critical objects. The last two requirements depend on the definition of the critical object. The critical object is usually taken to be a diffusely reflecting, square plate of 20 cm side, placed vertically on the road surface (Narisada, 1995). The justification for the choice of an object of this size is that an object 20 cm high will just pass beneath most vehicles without hitting the underside. While the physical dimensions of the critical object are 20 cm for determining the luminance difference threshold, it is necessary to define the visual size of the critical object, i.e., the angle the critical object subtends at the eye. This, in turn, requires the specification of a distance at which the critical object is to be seen. The distance usually specified is 100 m on the basis that if the critical object can be seen at 100 m, it can be avoided (de Boer et al., 1959). Given the visual size of the critical object and the luminance of the road surface forming the background to the object, the luminance difference threshold can be determined from the psychophysical literature (Boff and Lincoln, 1988). Given the luminance difference threshold and the road surface luminance, it is possible to identify the upper and lower luminances of the critical object necessary for it to be above threshold and hence visible. Once these luminances have been calculated, then with knowledge of the vertical illuminance on the critical object, the upper and lower reflectances that are necessary for the critical object to be visible can be calculated. These are known as the critical reflectances. The final step is to apply these critical reflectances to the cumulative frequency distribution of reflectances that are likely to be seen on the road. Figure 4.2 shows such a cumulative frequency curve for pedestrian clothing reported by Smith (1938). As an example, assume that the upper and lower critical reflectances are 0.15 and 0.20, respectively. Critical objects with reflectances of 0.15 and below will be seen in negative contrast against the road while critical objects with reflectances of 0.20 and above will be seen in positive contrast. Examination of Figure 4.2 reveals that 80% of pedestrian clothing will fall below the lower critical reflectance and 10% will be above the upper critical reflectance, making the total revealing power at the point considered to be 90%. Of course, to cover the whole road, it is necessary to make these calculations for a regular array of points. This will give a distribution of revealing power values. The general level can then be expressed as the area ratio, this being the percentage of the road area where the revealing power is above a fixed level, such as 70% (Narisada et al., 2003).

This somewhat convoluted procedure has the advantage of quantifying the effectiveness of road lighting in terms of visibility and emphasizing the importance of

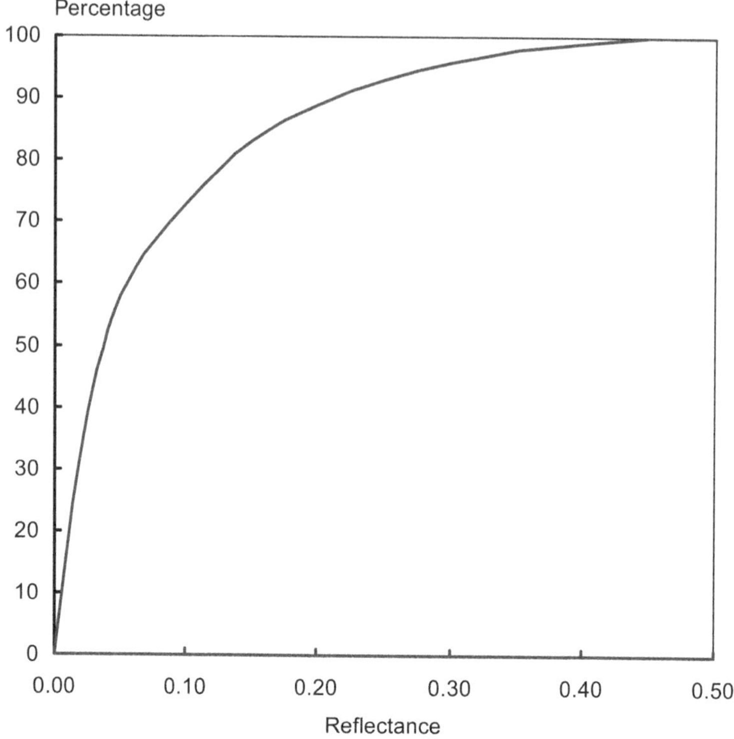

FIGURE 4.2   The cumulative frequency of the reflectance of clothing worn by pedestrians (after Smith, 1938).

vertical illuminance. It has been used to generate equi-revealing power contours for different road lighting luminaire layouts (Narisada and Karasawa, 2001) and for identifying significant design parameters (Narisada et al., 2003). However, it does suffer from some limitations. One is that there is limited and maybe outdated data on the cumulative frequency distribution of reflectances (Smith, 1938; Hansen and Larsen, 1979) likely to be found on roads. More seriously, the luminance increment threshold data in the literature has usually been obtained using uniform luminance fields to which the observers were fully adapted. The luminance distribution in front of the driver is rarely uniform and may even be diverse enough for the driver not to be fully adapted following some changes on fixation, although Narisada (1995) claims to have found a way around this problem. These limitations, together with the assumptions made about the critical object, have been enough to ensure that revealing power has been a subject of academic interest rather than a topic of practical relevance.

Revealing power has not been the only attempt to quantify the effectiveness of road lighting using visibility. In the USA, an approach called small target visibility has been developed. Small target visibility is defined as the weighted average of the visibility levels of a number of flat, 18 cm side, square targets, all with a diffuse reflectance of 0.5, mounted vertically on the road surface (IESNA, 2021a). The targets are arranged in a regular matrix of points across the road surface. The target is always 83 m ahead of the observer, orientated normal to the observer-to-target sightline, which is parallel to the centre line of the road. These distance and viewing direction conventions maintain both target size and target/observer geometry constant. The observer is assumed to be 60 years of age, with normal vision and using a fixation time of 200 ms. The visibility level of the target is defined as the ratio of the actual luminance difference of the target to the luminance difference threshold of the target; the higher is the ratio, the more visible is the target. Adrian (1989) has provided a quantitative model that allows the calculation of luminance difference thresholds for different sizes of targets as a function of background luminance, for positive and negative contrast. The actual luminance difference that will occur under a proposed lighting installation can be obtained by calculating the road surface luminance and the target luminance, including, for the latter, both the light received on the target directly from the luminaires and after reflection from the road surface (Adrian et al., 1993). A complete calculation procedure is given in IESNA (2021a). Janoff (1990, 1992a) has been able to show that there is a clear relationship between the visibility of the target and the distance at which it can be detected and recognized, and a link between the visibility level of small targets and subjective ratings of their visibility.

It is clear the revealing power and small target visibility metrics are similar in some respects but different in others. Both metrics use critical objects, both involve the use of luminance difference thresholds and both use a grid of measurement points placed across the road surface. Beyond this, they diverge. For revealing power, the luminance difference threshold is used together with the vertical illuminance on the object, to identify the critical reflectances that are then applied to the cumulative frequency distribution of reflectances to obtain an estimate of the percentage of objects that will be visible. For small target visibility, the critical object has a fixed reflectance and the merit of the road lighting installation is judged simply on the weighted average of the calculated visibility levels. This eliminates the concerns expressed above about the relevance of the distribution of reflectances but leaves unanswered concerns about basing the assessment of road lighting on a single reflectance at the extreme of the cumulative frequency of pedestrian clothing (Figure 4.2). Small target visibility has been mentioned as a metric for the quality of road lighting in the USA (IESNA, 2021b) but only half-heartedly in the sense that it is one of three possibilities, the other two being average road surface luminance and average illuminance on the road. Average road surface luminance is intended for use where a clear direction of view can be defined, for example, from the driver to the road. Average illuminance on the road is used where there are many different directions of view, as at a roundabout or intersection. The main use of small target visibility is now likely to be to compare road lighting installations with the same average road surface luminance or illuminance by examining the distribution of small target visibilities for the array of targets.

Both revealing power and small target visibility are attempts to develop metrics for road lighting that are closer to its ultimate purpose than conventional photometric measures, namely to make objects visible. Unfortunately, both are probably doomed to failure because the visual task of driving is multi-faceted and both revealing power and small target visibility address only one facet. The driver certainly has to be able to detect small targets on-axis but also needs to be able to detect objects on the road but off-axis, to detect movement at the edges of the roadway, to judge the relative speeds and direction of movement of other vehicles and to identify many different signs and signals. Further, the revealing power and small target visibility metrics are only really applicable in low traffic densities where the driver can see a considerable distance down the road. In heavy traffic, detection of small obstacles at a distance is not an issue, much more importance being attached to the relative movements of nearby vehicles. Even in the absence of other traffic where the driver can see far down the road, there can be no guarantee that all objects will be small and when they are, they are of marginal significance for crashes. Data from The Netherlands suggest that only 0.6% of crashes causing death or injury involve collision with small, loose objects on or near the road (Schreuder, 1998).

This is a practical objection to the use of revealing power or small target visibility as metrics for road lighting effectiveness. A more fundamental objection arises from the fact that most objects on and near roads that have to be seen have multiple luminance differences, both within the object and between the object and the background. This poses a problem for the use of a single visibility level for an object because when it has multiple luminance differences, the object has multiple visibility levels, a difficulty experienced by Blackwell et al. (1964) in their early work on visibility measurement. Attempts have been made to overcome this problem by using multi-faceted hemispherical objects rather than flat plates (Lecocq, 1993). This approach has been shown to produce good predictions of perceived visibility on the road, but so has the small target visibility approach (Bacelar et al., 1999). As a result, the use of hemispherical targets for visibility calculations has not been pursued. This is a pity because they are more representative of the complexity of real targets. It is not until this complexity is recognized and the very diverse nature of the driver's tasks is accepted that the visibility approach to quantifying the quality of road lighting will be appreciated.

So far, this discussion of the metrics of road lighting has been focused on the effects on visibility of the light that illuminates people, vehicles, and objects on and near the road. However, light emitted by road lighting luminaires can have other effects on visibility, as well as visual discomfort when they deliver significant amounts of light directly to the eyes of the driver. When these effects are noticeable, they are called glare. There are two forms of glare that are relevant to road lighting – disability glare and discomfort glare.. Disability glare, as its name implies, produces a measurable change in visibility. This change is caused by light scattered in the eye, the scattered light forming a luminous veil over the retinal image of the road (Vos, 1984). The amount of disability glare produced in a given situation can be measured by comparing the visibility of an object seen in the presence of the glare source with the visibility of the same object seen through a uniform luminous veil. When the visibilities are the same, the luminance of the veil is a measure of the amount of disability glare produced by the glare source, and it is called the equivalent veiling luminance. Numerous studies have led to several different empirical methods for predicting the equivalent veiling luminance (Holladay, 1926; Stiles, 1930; Stiles and Crawford, 1937). Based on this work, an equation has been developed to predict the equivalent veiling luminance from directly measurable variables (Adrian and Bhanji, 1991). It is

$$L_v = k.E_n.\Theta_n^{-2}$$

where L_v is the equivalent veiling luminance (cd.m^{-2}), k is a multiplier that varies with age, E_n is the illuminance at the eye from the nth glare source (lx), and Θ_n is the angle between the line of sight and the nth glare source (degrees). The exponent of Θ_n is 2 when the glare source is more than 2 degrees from the line of sight which will be the case for road lighting but not for headlamps.

This formula is adequate for glare sources between about 1 and 30 degrees from the line of sight and for young people. However, a series of modifications of the formula have been suggested to extend the range over which accurate predictions can be made from 0.1 to 100 degrees, for age ranges up to 80 years and for eye iris colour (Vos, 1999; CIE, 2002).

The BS 13201 Part 2 recommendations for road lighting limit disability glare by setting a maximum percentage increase in the luminance difference threshold occurring under the roadway lighting when the equivalent veiling luminance is allowed for. This criterion is known as the threshold increment. The percentage threshold increment (TI) can be obtained from the formula:

$$\text{TI} = 65 \, (L_v \, / \, L^{0.8})$$

where TI is the threshold increment, L_v is the equivalent veiling luminance (cd.m^{-2}), and L is the average road surface luminance (cd.m^{-2}). This metric is used in the UK and the rest of Europe.

The IESNA (2021a and b) recommendations for road lighting attempt to control disability glare by setting maximum values for the ratio of the equivalent veiling luminance to the average road surface luminance. Where the small target visibility recommendations are followed, equivalent veiling luminance is included in the estimation of adaptation luminance used to determine the luminance difference threshold. Regardless of whether the percentage threshold increment or the veiling luminance ratio metrics are used, the importance of disability glare diminishes rapidly as the angle between the line of sight and the luminaire increases. Disability glare from road lighting luminaires is a minor matter compared to the disability glare produced by vehicle headlamps.

The other aspect of glare from road lighting luminaires that needs to be considered is discomfort glare. The study of discomfort glare from road lighting has a long and tortuous history, starting with Hopkinson (1940), and continuing until the present day, but without reaching any very definite conclusion. Three alternative metrics have been developed at various times; the luminaire luminance corresponding to the boundary between comfort and discomfort (de Boer et al., 1959), the Glare Mark (Adrian and Schreuder, 1970), and the Cumulative Brightness Evaluation (IESNA, 1980). Adrian (1991) examined each of these formulations and showed that despite their apparent differences they actually had a very similar form. Specifically, discomfort glare from road lighting increases with increasing luminaire luminance and the solid angle subtended by the luminaire at the driver's eye and decreases with increasing road surface luminance and increasing angle between the road lighting luminaire and the line of sight. There are differences in the exponents used in the three metrics, but Adrian concluded that they all had the same characteristics and would yield comparable results when used to evaluate different road lighting installations.

A different approach has been taken by Bullough (2022). He uses three illuminances received at the eye, from the luminaire, from the area immediately surrounding the luminaire, and from the ambient environment, together with the maximum luminance of the luminaire. Then,

$$DG = \log (E_{\mathrm{L}} + E_{\mathrm{S}}) + 0.6 \log (E_{\mathrm{L}}/E_{\mathrm{S}}) - 0.5 \log (E_{\mathrm{A}})$$

where

DG is an interim value of discomfort glare.
E_{L} is the illuminance at the eye direct from the luminaire (lx).
E_{S} is the illuminance at the eye from the area immediately surrounding the luminaire (lx).
E_{A} is the illuminance at the eye from the ambient illumination (lx).

The interim discomfort glare value (DG) can be converted to a value on the de Boer discomfort glare scale by the equation

$$DB = 6.6 - 6.4 \log (DG) + 1.4 \log (50,000/L)$$

where

DB is the value on the de Boer scale.
DG is the interim discomfort glare value.
L is the maximum luminance of the luminaire (cd m^{-2}).

If the small size of the luminaire makes it difficult to measure its maximum luminance, the equation for conversion of the interim discomfort glare value (DG) to a value on the de Boer scale (DB) can be reduced to

$$DB = 6.6 - 6.4 \log (DG)$$

This is the equation developed in earlier work (Bullough et al., 2008) that has been shown to make accurate predictions of discomfort (Abboushi et al, 2024). It has also been incorporated into the OSP "shoebox" method for limiting light pollution (see Section 14.7).

The de Boer scale. expresses the level of glare the observer is experiencing on a 9-point scale from 1 unbearable, 3 disturbing, 5 just permissible, and 7 satisfactory to 9 just noticeable or no glare (de Boer, 1967). It has been widely used in studies of glare from outdoor lighting.

The great advantage of this approach is that all the variables can be measured in the field using illuminance and luminance meters with appropriate baffles and switching of the luminaire or calculated using luminaire photometric data and an assumption about the ambient illuminance (Bullough, 2022).

Despite this interesting development, the present British road lighting recommendations (BSI, 2015) use the product of the luminous intensity at 85 degrees from the downward vertical and the area of the luminous part of the luminaire in the direction of the luminous intensity to limit discomfort glare, stating that this is mainly of concern to pedestrians and cyclists. Going even further, the IESNA (2021a and b) road lighting recommendations have not adopted any specific system for limiting discomfort glare. This lack of enthusiasm for limiting discomfort glare from road lighting suggests that, in practice, such glare is considered insignificant compared to that experienced when facing oncoming headlamps.

4.4   ROAD LIGHTING STANDARDS

Road lighting standards vary in detail from country to country, but they do have some common features. One such feature is the division of the road network into different classes, according to the type of users and the road geometry (Schreuder, 1998). Different lighting metrics are used to form standards for the different classes, and these metrics are then varied to match the sub-divisions of each class. Another common feature is the aspects of road lighting dealt with. Most lighting standards contain metrics for the amount of light on the road, the distribution of light on and adjacent to the road, and the extent of disability glare produced by road lighting. Two rather different sets of road lighting standards will be examined here, namely, those used in the USA and the UK. These recommendations owe much to the work of the Road Lighting Committee of the Illuminating Engineering Society of North America (IESNA) and Division 4: Transportation and Exterior Applications of the Commission Internationale de l'Eclairage (CIE). The lighting recommendations considered here are the latest issued but are likely to change as more research on their effectiveness is completed (Fotios and Gibbons, 2018).

4.4.1   ROAD LIGHTING STANDARDS USED IN THE USA

Road lighting standards in the USA vary from state to state, but many states have adopted the IESNA Recommended Practice RP-8-21 as a basis for their standards. The road classes used in IESNA (2021b) and their defining characteristics are summarized in Table 4.2. The classes are divided into two types – highways which are primarily used by vehicles, and streets where pedestrians and cyclists as well as

TABLE 4.2

The Road Classification Used in Many Parts of the USA (IESNA, 2021b)

Class	Characteristics
Freeway A	A divided highway with full control of access; a road with high visual complexity and high traffic volumes, typically found in major urban areas. It is used only by vehicles and operates at design capacity during the morning and evening rush hours
Freeway B	A divided highway with full control of access, used only by vehicles
Expressway	A divided highway for through traffic with partial control of access but has some intersections. It is used mainly by vehicles
Major (arterial)	Roads connecting areas of traffic generation and roads entering and leaving a city; the part of the road network that serves most of the through traffic flow.
Collector	Roads servicing traffic between major and local class roads. These roads are used for traffic within residential, commercial, and industrial areas. Vehicles, bicycles, and pedestrians will all be users
Local	Roads for access to residential, commercial, and industrial property. Vehicles, bicycles, and pedestrians will all be users

vehicles may be found. IESNA (2021b) uses road surface luminance as the preferred measure for both highway and street lighting.

Another classification used in IESNA (2021b) is the pedestrian activity area. This has three classes – high, medium, and low. Assignment to a class is based on the highest average number of pedestrians present in a 100 m section of sidewalk, over an hour, after dark. Specifically, a low activity area will have ten or fewer pedestrians, a medium activity area will have 11–99 pedestrians and anywhere with 100 or more pedestrians will be a high pedestrian activity area.

There are a number of different metrics recommended for the design of continuously lit, straight roads. They are the minimum average luminance of the road as seen by the driver, two minimum luminance uniformity ratios, and a maximum veiling luminance ratio. These values should be maintained over the life of the installation. Table 4.3 gives the recommendations for the different combinations of road and pedestrian activity classes. There are no pedestrian activity classes for freeways and the expressway because pedestrians should be absent from such highways. The recommendations are based on the luminances calculated or measured at a regular array of test points spread over each lane of the road. The array should have at least two points per lane across the road and at least 10 points with a maximum separation between points along the road of 5 m for one luminaire cycle, which is the distance between two adjacent luminaires on the same side of the road. The assumed position of the driver used for the calculation of luminance is taken to be 1.45 m above the road surface with a line of sight one degree down, resulting in a constant viewing distance for each test point of 83 m. The two uniformity criteria are the average

TABLE 4.3

Luminance Criteria for Different Road Classes Used in Parts of the USA (IESNA, 2021b)

Road Class	Pedestrian Activity Class	Average Road Surface Luminance L_{avg} (cd.m^{-2})	Average Uuniformity Ratio (L_{avg}/L_{min})	Maximum Uniformity Ratio (L_{max}/L_{min})	Maximum Veiling Luminance ratio ($L_{v.max}/L_{avg}$)
Freeway A	-	0.6	3.5	6.0	0.3
Freeway B	-	0.4	3.5	6.0	0.3
Expressway	-	1.0	3.0	5.0	0.3
Major	High	1.2	3.0	5.0	0.3
Major	Medium	0.9	3.0	5.0	0.3
Major	Low	0.6	3.5	6.0	0.3
Collector	High	0.8	3.0	5.0	0.4
Collector	Medium	0.6	3.5	6.0	0.4
Collector	Low	0.4	4.0	8.0	0.4
Local	High	0.6	6.0	10.0	0.4
Local	Medium	0.5	6.0	10.0	0.4
Local	Low	0.3	6.0	10.0	0.4

uniformity ratio, which is the average road surface luminance to the minimum luminance at any test point, and the maximum uniformity ratio, which is the maximum road surface luminance to the minimum luminance at any test point. Disability glare is limited by the maximum veiling luminance ratio which is the maximum veiling luminance to the average road surface luminance.

There are three interesting points about Table 4.3. The first is that the highest luminance is recommended, not for freeways or expressways, where traffic speeds are likely to be highest, but for major roads with high pedestrian activity areas. The second is that the recommended luminances for street lighting decrease as the pedestrian activity class changes from high to low. The third is that the luminance uniformity is relaxed as the road becomes more residential.

Table 4.3 has one caveat that should be noted. The luminance recommendations should not be used when lighting sharp bends (less than 600 m radius) or steep hills (more than 6% gradient) because of the limitations of the road reflectance tables (see Section 4.5). For these locations, the illuminance on the road should be adopted, using the equivalances: 1 cd.m^{-2} is equivalent to 10 lx on an R1 road surface; 1 cd.m^{-2} is equivalent to 15 lx on an R2 or R3 road surface; 1 cd.m^{-2} is equivalent to 13.3 lx on an R4 road surface;

Another area requiring special attention is intersections. These fall into two types – partial, where only the area of the intersection is lit, and the roads leading to it are

TABLE 4.4

Illuminance Recommendations for Partial Intersections on Otherwise Unlighted Roads Designed Solely for Vehicular Traffic and with Ddifferent Road Surface Classes, Used in Parts of the USA (IESNA, 2021b)

Road Class	R1 Road Surface (lx)	R2 Road Surface (lx)	R3 Road Surface (lx)	Maximum Uuniformity Ratio (E_{avg}/E_{min})
Freeway class A	6.0	9.0	8.0	3.0
Freeway class B	4.0	6.0	5.0	3.0
Expressway	6.0	9.0	8.0	3.0

TABLE 4.5

Illuminance Recommendations for Full Intersections on Continuously Lighted Roads and Different Pedestrian Activity Levels, Used in Parts of the USA (IESNA, 2021b)

Intersection	Minimum Maintained Average Road Surface Illuminance (lx)			Maximum Illuminance Uniformity Ratio (E_{ave}/E_{min})
	High Pedestrian Activity Area	Medium Pedestrian Aactivity Area	Low Pedestrian Activity Area	
Major/major	34	26	18	3.0
Major/collector	29	22	15	3.0
Major/local	26	20	13	3.0
Collector/collector	24	18	12	3.0
Collector/local	21	16	10	3.0
Local/local	18	14	8	3.0

unlit, and full, where two roads intersect and both are continuously lit. The lighting recommendations for intersections use illuminance as the appropriate measure because there are multiple viewing directions. Table 4.4 gives the recommended illuminances for partial intersections with different road surface classes. Table 4.5 sets out the recommended average maintained illuminances for full intersections in low, medium, and high pedestrian activity areas, involving all possible combinations of local, collector, and major roads together with the maximum illuminance uniformity ratio. The principle behind these recommendations is that the average maintained illuminance should be equal to the sum of the recommended values for the two intersecting roads.

It should also be noted that regardless of whether luminance or illuminance measures are used, they both relate to the lighting of the road surface. But there is also some benefit to lighting the area adjacent to the road, known as the surround area (National Academies of Sciences, Engineering and Medicine, 2020). Lighting the surround area leads to more distant detection of objects. The recommendation is that the average illuminance on the 3.6-m-wide strip of land beside the road should be 0.8 times the average illuminance of the nearest lane.

4.4.2 ROAD LIGHTING STANDARDS USED IN THE UK

The road lighting recommendations used in the UK (BSI, 2015) identify three distinct situations – traffic routes where vehicles are dominant, conflict areas where streams of vehicles intersect with each other or with pedestrians, and cyclists or residential roads where the lighting is primarily intended for pedestrians and cyclists. These last are sometimes called subsidiary roads.

The criteria used to define lighting for traffic routes are the average road surface luminance, the overall luminance uniformity, the longitudinal luminance uniformity, the threshold increment, and the edge illuminance ratio. These metrics are based on calculations or measurements at a grid of test points arranged across and along the road. The average road surface luminance is the average of the luminances of the road surface at all the test points. The overall luminance uniformity is the ratio of the lowest luminance at any test point to the average road surface luminance. The longitudinal luminance uniformity is the ratio of the lowest to the highest luminance found at test points on a line along the centre of a single lane. For the whole carriageway, the value taken is the lowest longitudinal luminance uniformity found for any of the lanes of the carriageway. The purpose of the longitudinal luminance uniformity metric is to reduce the perception of a regular fluctuation in road surface luminance as the driver moves past successive luminaires. The edge illuminance ratio is the average illuminance on a strip of land at the edge of the road with a width matching the width of the nearest driving lane, relative to the average illuminance on that driving lane.

Traffic routes are divided into six different classes (M classes). The different classes are based on such aspects as the road layout, traffic flow, and ambient environment. Advice on the choice of class applicable to different locations is given in BSI (2020), the choice being based on traffic speed, traffic density, traffic composition, intersection density, parked vehicles, ambient luminance, and visual guidance. Details of the recommended lighting criteria for dry roads of each class are given in Table 4.6. Where roads are likely to be damp or wet for a significant part of the night, a looser requirement for an overall luminance uniformity ratio of 0.15 can be used for all classes. The minimum edge illuminance ratio criterion is only applied where there are no areas with their own criteria adjacent to the carriageway, e.g., a separate cycle path. Disability glare from road lighting is limited by the use of a maximum threshold increment. Where threshold increment cannot be calculated an alternative method to limit disability glare is to select a luminaire according to the luminaire classes given in Table 4.7. The different classes are defined by the luminous intensity

TABLE 4.6

Lighting Recommendations for Dry Traffic Routes in the UK and Europe (BSI, 2015)

Lighting Class	Minimum Maintained Average Road Surface Luminance (cd. m^{-2})	Minimum Overall Luminance Uniformity Ratio (Minimum/ Average)	Minimum Longitudinal luminance Uniformity Ratio for the Carriageway (Minimum/ Maximum)	Maximum Threshold Increment (%)	Minimum Edge Illuminance Ratio
M1	2.00	0.40	0.70	10	0.35
M2	1.50	0.40	0.70	10	0.35
M3	1.00	0.40	0.60	15	0.30
M4	0.75	0.40	0.60	15	0.30
M5	0.50	0.35	0.40	15	0.30
M6	0.30	0.35	0.40	20	0.30

TABLE 4.7

Luminaire classes for the control of disability glare on traffic routes used in the UK and Europe (BSI, 2015)

Lighting class	Maximum luminous intensity / 1000 lumens at 70 -79° (cd/1000 lm)	Maximum luminous intensity / 1000 lumens at 80-89° (cd/1000 lm)	Maximum luminous intensity / 1000 lumens at 90-95° (cd/1000 lm)	Luminous intensity above 95° (cd)
G*1	-	200	50	-
G*2	-	150	30	-
G*3	-	100	20	-
G*4	500	100	10	0
G*5	350	100	10	0
G*6	350	100	0	0

of the luminaire/1000 lumens of light source luminous flux, at 70, 80, and 90 degrees and above from the downward vertical, in any direction, and the luminous intensity above 95 degrees, in any direction. This method can also be used where there is concern about light pollution.

All traffic routes eventually reach a conflict area. A conflict area is a place where traffic flows merge or cross, e.g., at intersections or roundabouts, or where vehicles, cyclists, and pedestrians are in close proximity, e.g., at a roundabout or on a shopping street. Lighting for conflict areas is intended for drivers, cyclists, and pedestrians.

The criteria used for the lighting of conflict areas are the average horizontal illuminance and the minimum overall horizontal illuminance uniformity, which is the ratio of the lowest illuminance at any point on the lit area to the average illuminance of that area. The lit area can be limited to the road or to the road and its surroundings. The recommendations for the five different lighting classes for conflict areas (C classes) are given in Table 4.8. Again, advice on selecting an appropriate class is given in BSI (2020).

So far, the lighting recommendations used in the UK have been concerned primarily with the needs of drivers, but there are a large number of locations where the needs of pedestrians and cyclists take priority. Such locations include many residential streets as well as cycle lanes and sidewalks. The lighting recommendations for these areas are covered by six classes (P classes) and are shown in Table 4.9. Advice on the choice of class is given in BSI (2020). The measures used are the average and minimum horizontal illuminance over the lit area. The lit area includes the road, the sidewalks, and any cycle lanes. To ensure sufficient illuminance uniformity, it is recommended that the actual average horizontal illuminance should not exceed 1.5 times the minimum maintained average horizontal illuminance specified in Table 4.9. Interestingly, the lighting recommendations also include additional criteria where the ability to see the faces of people approaching is considered to be important for the safety of pedestrians. The measures used for this purpose are the minimum vertical illuminance and the minimum semi-cylindrical illuminance. The vertical illuminance is taken on a vertical plane orientated so that the normal to the plane points in the direction of movement. The semi-cylindrical illuminance is the luminous flux falling on the curved surface of a small semi-cylinder divided by the area of the curved surface, located 1.5 m above the road and orientated so that the inside normal of the flat back of the semi-cylinder points in the direction of movement.

It is obviously important that the lighting of traffic routes, conflict areas, and residential streets should be coordinated. Table 4.10 shows the compatible lighting

TABLE 4.8

Lighting Recommendations for Conflict Areas in the UK and Europe (BSI, 2015)

Lighting Class	Minimum Maintained Average Horizontal Illuminance (lx)	Minimum Overall Illuminance Uniformity Ratio (Minimum/Average)
C0	50	0.4
C1	30	0.4
C2	20	0.4
C3	15	0.4
C4	10	0.4
C5	7.5	0.4

TABLE 4.9

Lighting Recommendations for Residential Roads, Sidewalks, and Cycle Paths in the UK and Europe (BSI, 2015)

Lighting Class	Minimum Maintained Average Horizontal Illuminance (lx)	Maintained Minimum Horizontal Illuminance (lx)	Maintained Minimum Vertical Illuminance (lx)	Maintained Minimum Semi-cylindrical Illuminance (lx)
P1	15.0	3.0	5.0	5.0
P2	10.0	2.0	3.0	2.0
P3	7.50	1.5	2.5	1.5
P4	5.0	1.0	1.5	1.0
P5	3.0	0.6	1.0	0.6
P6	2.0	0.2	0.6	0.2

TABLE 4.10

Compatible Lighting Classes for Traffic Routes, Conflict Areas and Subsidiary Streets in the UK and Europe (BSI, 2020)

Traffic Routes, M Class	Conflict Area, C Class	Subsidiary Streets, P Class
–	C0	–
M1	C1	–
M2	C2	–
M3	C3	P1
M4	C4	P2
M5	C5	P3
M6	–	P4
–	–	P5
–	–	P6

classes for traffic routes, conflict areas, and residential streets. It should be noted that the lighting classes of comparable levels should not be used in place of the appropriate M, C, or P classes related to the road type. This is particularly important for M-class lighting as there are no comparable luminance uniformity requirements for the C and P classes.

4.4.3 SIMILARITIES AND DIFFERENCES

A comparison of the road lighting standards used in parts of the USA and in the UK will reveal that there are some similarities but also some differences. The similarities are most evident in the classifications used for different road types. The US classifications of freeways, expressways, major, collector, and local roads can be roughly matched to the UK classifications of traffic routes, conflict areas, and subsidiary roads, although the multiple classes available in the UK recommendations suggest a greater flexibility. Further similarities lie in the measures used for the lighting recommendations, both using road surface luminance for traffic routes and illuminance for conflict areas and subsidiary roads. The reason for this is that in conflict areas there will be multiple directions of view as drivers, cyclists, and pedestrians attempt to negotiate the intersection or roundabout. A similar argument can be used to explain the use of illuminance as a metric for residential roads. Luminance is only useful as a metric for road lighting where there are a limited number of directions of view. This is the situation for traffic routes where there are few pedestrians and all the drivers are, hopefully, on the road and looking along it. Another similarity occurs in recommendations to light the edge of the road, although different names are used for what is essentially the same measurement.

The differences between the USA and UK recommendations become more apparent when the actual values are compared. A comparison of Tables 4.3 and 4.6 shows that the UK recommendations for traffic routes cover a wider range (0.3–2.0 cd.m^{-2}) than the USA recommendations (0.3–1.2 cd.m^{-2}). Interestingly, the highest average road surface luminance in the UK recommendations occurs for motorways where traffic moves at the highest speed and pedestrians are absent, while in the US recommendations, the maximum occurs for major roads with high pedestrian activity areas. In fact, the average road surface luminance for freeways in the US (0.6 cd.m^{-2}) is less than a third of the average road surface luminance for motorways recommended in the UK (2 cd.m^{-2}). A similar difference occurs in the illuminances recommended for the lighting of conflict areas (c.f. Tables 4.5 and 4.8). The range of illuminances for intersections in the US recommendations is 8–34 lx, but for conflict areas in the UK, which include intersections, the illuminance range is 7.5–50 lx.

Another comparison can be made on the basis of overall luminance uniformity for traffic routes, when defined as the minimum to average luminance ratio for both sets of recommendations (c.f. Tables 4.3 and 4.6). The UK recommendations call for more uniform lighting than the US recommendations, the range of overall luminance uniformity for traffic routes being 0.35 to 0.40 for the former and 0.17 to 0.33 for the latter. A similar difference is evident for the lighting of conflict areas in the two sets of recommendations (c.f. Tables 4.5 and 4.8).

Other differences between the UK and US recommendations lie in the various metrics used for what are the same phenomena. There are additional but different luminance uniformity criteria in both sets of recommendations, as well as different approaches to limiting disability glare. There are also differences in the extent to which recommendations for traffic routes, adjacent areas, and conflict areas are coordinated. In addition, the general US recommendations make no mention of changing values when roads are frequently wet, although individual states may, whereas the UK recommendations specifically consider wet roads. Also, the UK recommendations for subsidiary streets consider the need for pedestrians to be able to see the faces of people approaching. The US does not.

In summary, road lighting recommendations in the UK and USA are similar in concept but differ in detail. The UK recommendations should lead to brighter, more uniform lighting on roads with faster traffic than the US recommendations. However, the allocation of the highest average road surface luminance to the major roads in high pedestrian conflict areas rather than freeways in the USA is consistent with putting most light where it will have the most beneficial effect in reducing fatalities and personal injury crashes (see Section 1.5).

4.5   ROAD LIGHTING IN PRACTICE

The extent to which road lighting standards are achieved in practice depends on the assumptions that have to be made (Hargroves, 1981). The lighting criteria requiring the fewest assumptions are those based on illuminance. Given knowledge of the position of a luminaire relative to the road and the luminous intensity distribution of the luminaire, the horizontal illuminance on the road surface can be calculated. Figure 4.3 shows the relevant geometry for an observer looking down the long axis of the road.

The formula for calculating the illuminance at a specific point, P, on the road surface received from a specific luminaire is

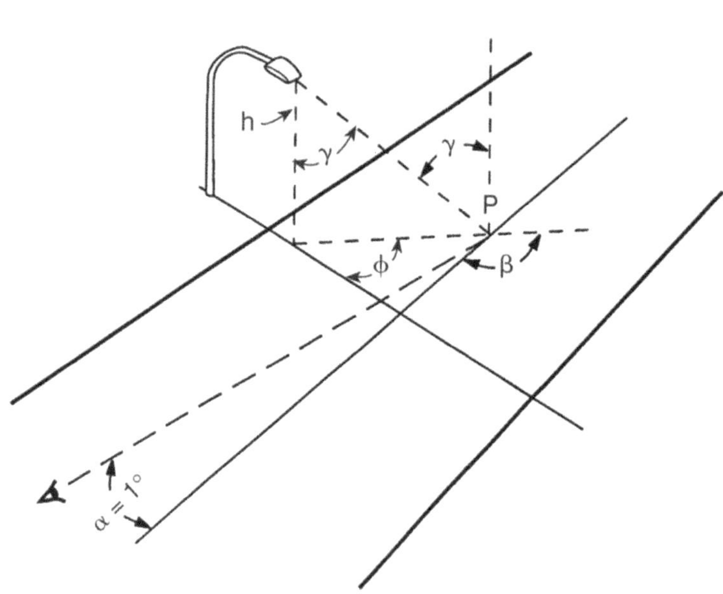

FIGURE 4.3   The luminance coefficient at point P for an observer looking straight down the road axis is dependent on four angles. These are α = angle of observation from the horizontal, β = angle between the vertical planes of incidence and observation, γ = angle of incidence from the upward vertical, and ϕ = angle between the transverse axis across the road and the line joining the downward vertical at the luminaire and point P (from IESNA, 2000).

$$E_n = I\,(\phi,\gamma)\cos\gamma^{\,3}/h^2$$

where E_n is the horizontal illuminance at point P from the n'th luminaire (lx)., $I\,(\phi,\gamma)$ is the luminous intensity in the specified direction for the n'th luminaire (cd), ϕ is the angle between the transverse axis across the road and the line joining the intersection of the downward vertical through the luminaire with the road and point P, γ is the angle of emission from the downward vertical through the luminaire, and h is the height of n'th luminaire above the road surface (m).

The illuminance contributions at point P from all the surrounding luminaires are summed. Similar calculations are then made for a grid of points arranged across and along the road so that the average road surface illuminance and the illuminance uniformity criteria can be calculated.

Today, these calculations are always made using software but assumptions are still present. For a start, it is always assumed that the road is level. Then there is the fact that the average road surface illuminance recommended is a maintained value. This means the average road surface illuminance should not fall below the recommended value at any time during the life of the installation. This calls for assumptions about the change in the luminous flux emitted by the light source and the collection of dirt on the luminaire over time. Guidance on suitable values for these variables is given by manufacturers and in design guidance (CIE, 2003), but measurements with different luminaires located on roads with different levels of pollution for one and three years have shown that the recommended maintenance factors are often too conservative (Sanders and Scott, 2008). In practice, determining the maintenance factor for a specific location seems to be a matter of judgement that itself requires further assumptions despite authoritative advice (ILP, 2020a).

Road lighting criteria based on luminance include all these assumptions plus others concerned with the road surface reflection properties. The luminance of any point on a road surface lit by a single luminaire and as seen by a driver is given by the formula:

$$L_n = q.E_n$$

where L_n is the road surface luminance at point P produced by the n'th luminaire (cd. m^{-2}), E_n is the illuminance at point P produced by the n'th luminaire (lx)., and q is the luminance coefficient of the road surface at point P (cd.lm^{-1})

The road surface luminance depends on the position and luminous intensity distribution of the luminaire, the road surface reflection properties, and the geometry of the observer and the luminaire relative to the point under consideration. In principle, five angles are needed to define the relevant geometry. Figure 4.3 shows four of them. They are α, the angle of observation from the horizontal; β, the angle between the vertical planes of incidence and observation; γ, the angle of incidence from the upward vertical at point P, and ϕ, the angle between the transverse axis across the road and the line joining the intersection of the downward vertical through the luminaire with the road and point P. A fifth angle not shown in Figure 4.3 but which has an influence on the luminance at P is the angle between the vertical plane of observation and the road axis. This angle is ignored because the reflection properties of most road surfaces are almost completely isotropic and the range of possible angles is limited.

To simplify matters further, one of the angles in Figure 4.3 is fixed. The angle to be fixed is α, the angle of observation from the horizontal. This angle has a limited range in practice. For a driver's eye height ranging from 1 m to 3 m, which covers both sports cars and heavy trucks, α ranges from 0.35 to 2.86 degrees. Road surface luminance coefficients within this range show little variation (Moon and Hunt, 1938; De Boer et al., 1952). The angle α is conventionally fixed at one degree.

With these assumptions, it is possible to describe the reflection properties of a point on a road surface by a two-dimensional array of luminance coefficients, the dimensions of the array being β, the angle between the vertical plane of incidence and the vertical plane of observation, and γ, the angle of incidence from the upward vertical at point P. However, such a table is not convenient for use in calculation because the fundamental photometric data available for road lighting luminaires consists of a luminous intensity distribution which will be expressed in terms of angles ϕ and γ. This can be allowed for by replacing the illuminance in the formula for road surface luminance by the luminous intensity and using the inverse square law. The result is an expression for the luminance of a point on the road surface due to a single luminaire of the form

$$L_n = (q(\beta,\gamma) . I(\phi,\gamma)/h^2) \cos^3\gamma$$

where L_n is the road surface luminance at point P produced by the n'th luminaire (cd.m^{-2}), $I(\phi,\gamma)$ is the luminous intensity of the n'th luminaire towards point P (cd), $q(\beta,\gamma)$ is the luminance coefficient of the road surface at point P (cd.lm^{-1}), h is the mounting height of the n'th luminaire (m), and γ is the angle of incidence at point P of light from the n'th luminaire (degrees).

The term $q(\beta,\gamma).\cos^3\gamma$ is called the reduced luminance coefficient (r) and is the metric conventionally used in what are called the r-tables that characterize the reflection properties of road materials. The two dimensions of the r-table are the angle β, the angle between the vertical plane of incidence and the vertical plane of observation, and the tangent of the angle γ, the angle of incidence from the upward vertical at point P (see Figure 4.3). Each cell in an r-table contains a value for the reduced luminance coefficient multiplied by 10,000.

With an r-table matched to the road material and the luminous intensity distribution for the luminaire, the luminance produced by a single luminaire at any point on the road surface as seen from a specified position can be calculated. This process can then be repeated for adjacent luminaires and the contributions from all luminaires summed to get the luminance at that point for the whole lighting installation. This process can then be repeated over an array of points on the road so as to get the luminance metrics used in road lighting standards (BSI, 2015). In practice, there is a limit to how large an area of road surface should be included in the array. Calculation is conventionally limited to the road surface between two adjacent luminaires on the same side of the road. By convention, in the UK, the observer is placed 60 m in front of the first transverse row of calculation points and 1.5 m above the road. In the USA, the corresponding assumptions are 83 m and 1.45 m. As regards the other dimension needed to define the observation position, for the average luminance and overall luminance uniformity metrics, the observation position is taken to be one quarter of the road width from the near side of the road.

While the above process is possible in principle, it is rarely used in practice. This is because, strictly, every piece of road has a unique r-table and that itself will change over space and time as different parts of the road wear differently. This implies that before designing a road lighting installation, measurements should be made of the reflection properties of samples taken from the road to be lit. Such measurements are difficult and time consuming although attempts have been made to automate the process (Muzet et al., 2021). Given the current difficulty in measuring actual road surfaces, a road reflection classification system has been developed by which many different road surfaces are approximated by a single r-table.

The first step in building a classification system is to identify some descriptive parameters for the measured r-table. Several different attempts have been made to do this (de Boer and Westermann, 1964a, 1964b; Roch and Smiatek, 1972; Range, 1972; Massart, 1973; Erbay, 1974). After consideration, the CIE decided that most r-tables could be described by three parameters, one concerned with lightness and two concerned with specularity (CIE, 1976).

The parameter adopted for lightness is the average luminance coefficient, Q_0, which is the solid angle-weighted average of the reduced luminance coefficients in the r-table. The solid angle weighting ensures that the reduced luminance coefficient values, corresponding to large γ angles, do not have an overwhelming influence on the value of Q_0. The average luminance coefficient, Q_0, can be calculated from the r-table using a weighting factor procedure developed by Sorensen (1974). The average luminance coefficient, Q_0, has been shown to be highly correlated to the average luminance produced on the road surface (Bodmann and Schmidt, 1989)

As for the parameters relating to specularity, there are two, defined as ratios.

$$S1 = r(0, 2)/r(0, 0)$$

where $r(0, 2)$ is the reduced reflection coefficient for $\beta = 0$ degrees and $\tan \gamma = 2$, $r(0, 0)$ is the reduced reflection coefficient for $\beta = 0$ degrees and $\tan \gamma = 0$.

$$S2 = Q_0/r(0, 0)$$

where Q_0 is the average luminance coefficient and $r(0, 0)$ is the reduced reflection coefficient for $\beta = 0$ degrees and $\tan \gamma = 0$.

Frederiksen and Sorensen (1976) have shown that changes in the two parameters of specularity, $S1$ and $S2$, have somewhat different effects on the luminance patterns produced. Specifically, increases in $S1$ lead to rapid decreases in the overall luminance uniformity ratio but have little effect on the longitudinal luminance uniformity ratio. Increases in $S2$ lead to increases in both overall and longitudinal luminance uniformity ratios. As might be expected, there is evidence that the use of these two parameters of specularity together gives more accurate predictions than either one alone (Frederiksen and Sorensen, 1976). However, data provided by Sorensen (1975) and Frederiksen and Sorensen (1976) from numerous measurements taken on road samples in Europe show that $S1$ and $S2$ are highly correlated ($r = 0.93$). As a result, $S2$ has been dropped from the CIE classification system, leaving just Q_0 as a parameter for the diffuse reflectance and $S1$ as the parameter for the specular reflectance (CIE, 2001).

Having identified two descriptive parameters that can be used to characterize any r-table, the next step in developing a classification system is to decide on how many classes to use and where the boundaries should be. In 1976, the CIE recommended the use of two different four-class classification systems, the R system and the N system, the latter being recommended for countries where artificial brighteners are used in pavement materials to give very diffusely reflecting surfaces (CIE, 1976). The boundaries of classes in both the R and N systems are determined by the value of $S1$. Table 4.11 shows the boundary values for the four classes in each system. Different countries opted to use either the R or the N system, both the USA and the UK choosing the R classification system.

However, there were doubts about the value of having four classes. Figure 4.4 shows the relationship between the descriptive parameters, Q_0 and $S1$, for the 286 measured road surfaces reported in Sorensen (1975).

It is apparent from Figure 4.4 that there is little change in Q_0 for classes R2, R3, and R4 (see Table 4.11). A similar relationship between Q_0 and $S1$ is shown in van Bommel and de Boer (1980). Further, Burghout (1979) demonstrated that combining classes R2, R3, and R4 would make little difference to the predicted road surface luminance patterns. In this work, the average luminance, overall luminance uniformity ratio, and longitudinal luminance uniformity ratio were calculated for 413 road surfaces lit by two different lighting systems using five different luminaires at three

---

TABLE 4.11

The Boundaries and Standard Values for the Classes in the R, N, and C Road Reflection Classification Systems (from CIE, 1999)

System	Class	$S1$ Boundaries	Standard $S1$	Standard Q_0
R	R1	$S1 < 0.42$	0.25	0.10
	R2	$0.42 < S1 < 0.85$	0.58	0.07
	R3	$0.85 < S1 < 1.35$	1.11	0.07
	R4	$S1 > 1.35$	1.55	0.08
N	N1	$S1 < 0.28$	0.18	0.10
	N2	$0.28 < S1 < 0.60$	0.41	0.07
	N3	$0.60 < S1 < 1.30$	0.88	0.07
	N4	$S1 > 1.30$	1.55	0.08
C	C1	$S1 < 0.40$	0.24	0.10
	C2	$S1 > 0.40$	0.97	0.07

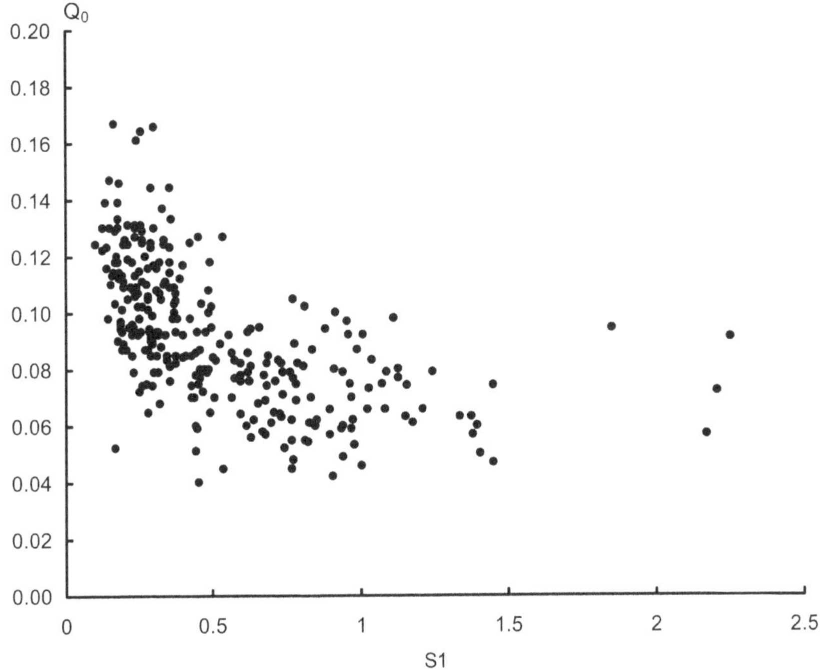

FIGURE 4.4   Relationship between descriptive parameters Q_0 and $S1$ for 286 measured road surfaces (after Sorensen, 1975).

different spacings, from the measured r-tables for each road surface. These calculations showed that the average road surface luminances for class R1 are markedly different from those of classes R2, R3, and R4, which show little difference. There are clear differences between the predicted mean overall luminance uniformity ratios for the R2, R3, and R4 classes and smaller differences for the mean longitudinal luminance uniformity ratios. Burghout (1979) recommended the use of a two-class classification system for road surfaces. As a result of this work, CIE (1984) recommended the adoption of a two-class classification system, called the C system. The boundary values of $S1$ and the standard values of $S1$ and Q_0 for each of the two classes in the C classification system are given in Table 4.11.

Given that a representative road surface r-table is used in the calculation of luminance rather than an r-table of the actual road surface, it is interesting to consider the magnitude of the consequent differences. CIE (1984) reports the errors in terms of the mean and standard deviation of the percentage deviation between the luminance metric calculated using the standard r-table. and the actual r-table, after re-scaling the standard r-table. for the actual Q_0 value. The expression for percentage difference for the luminance metric is

$$D = ((L_1 - L_2)/L_1) \times 100$$

where D is the percentage deviation, L_1 is the luminance metric calculated for the actual r-table, and L_2 is the luminance metric calculated for the standard r-table.

The mean and standard deviation of percentage deviation in the three luminance metrics for 44 widely different luminaire light distributions and 113 different dry road surfaces for a single-sided lighting installation and for spacing/mounting height ratios of 4.0 and 5.5 are given in Table 4.12. Examination of Table 4.12 shows that the mean percentage deviations and the associated standard deviations are the least for the average road surface luminance metric and generally greatest for the longitudinal luminance uniformity metric. The directions of the mean percentage deviations are such that the use of the representative r-table is likely to underestimate all the luminance metrics. Similar results have been obtained by Bodmann and Schmidt (1989) and Dumont and Paumier (2007). It is clear from Table 4.12 that the convenience of using the standard r-tables rather than the actual r-table comes at a cost in accuracy, even when care is taken to rescale the Q_0 value.

Another possible source of error is the change in reflection characteristics of a road surface over time. Bodmann and Schmidt (1989) measured the reflection characteristics of a number of different road surface samples, from new on roads in Germany, over a period of approximately three years. Most of the change in reflection properties occurred over the first six months of exposure. Dumont and Paumier (2007) measured the change in reflection properties in France over three years. They found that some road materials became less specular and more diffuse in reflection over time, but others showed the reverse trend. Another cause of variability is the difference in reflection properties across the road. An observation of any well-travelled lane will reveal the existence of five distinct areas: Two polished wheel tracks, a darker oil track in the centre of the lane, and two unworn tracks at the edges

---

TABLE 4.12

Mean Percentage Deviation and the Associated Standard Deviation (in Brackets) for Calculations of Average Road Surface Luminance, Overall Luminance Uniformity Ratio, and Longitudinal Luminance Uniformity Ratio Using Actual and Standard r-Tables after Rescaling the Q_0 Value of the Standard r-Table

Class	Spacing/Mounting Height Ratio	Average Road Surface Luminance	Overall Luminance Uniformity Ratio	Longitudinal Luminance Uniformity Ratio
C1	4.0	5.3% (3.6%)	6.6% (4.4%)	10.0% (8.1%)
C1	5.5	5.3% (3.6%)	9.0% (6.2%)	13.4% (9.5%)
C2	4.0	7.0% (5.0%)	10.8% (8.4%)	9.9% (8.0%)
C2	5.5	7.0% (5.1%)	11.0% (8.7%)	12.9% (9.8%)

Note: Calculations are done for dry roads of two classes with luminaires at two different spacing/mounting height ratios (from CIE 1984).

of the lane. Each of these areas will have a different balance of reflection properties although they are all represented by a single r-table.

In summary, despite the precision of the standards used for road lighting, the assumptions made in the design process mean that the reality is likely to be much more variable from site to site. This variability could be identified and then reduced if on-site measurements were routinely taken (ILP, 2007). Unfortunately, such measurements are not routinely taken. Rather, the usual approach is to over-design the installation so as to exceed the relevant metric. Using a higher wattage light source and hence a higher light source luminous flux, without changing the luminaire luminous intensity distribution, will result in a higher average road surface illuminance and luminance without changing the uniformity. Reducing the spacing between luminaires without changing the light output of the luminaire or the luminous intensity distribution will increase the average road surface illuminance and luminance and improve uniformity. Increasing luminaire mounting height without changing luminaire light output or luminous intensity distribution will decrease average road surface illuminance and luminance but improve uniformity. Increasing the luminaire overhang without changing luminaire light output or luminous intensity distribution will increase average road surface illuminance and luminance but the effect on uniformity is unpredictable. Another possibility is to change the layout of the luminaires along the road. By changing from a staggered layout to an opposite layout and doubling the spacing between luminaires without changing the light source or luminaire, the uniformity can be improved without changing the average road surface illuminance and luminance. By changing from a staggered layout to a centre-mounted layout and doubling the spacing between luminaires without changing the light source or luminaire, the uniformity will be degraded with a small change in the average road surface illuminance and luminance. A more technological solution can be available when LED luminaires are used. This is to continually adjust the light output of the luminaires to match the change in road surface reflection properties so as to maintain the specified lighting conditions (Muzet et al., 2019).

Yet another cause of uncertainty is the fact that the measurements of road surfaces from which the r-tables were developed were made during the late 1960s and 1970s. It is widely believed that these measurements are no longer representative of modern road surfaces (Fotios et al., 2005a). The obvious conclusion is that, like a stopped clock, road lighting recommendations will, at best, only be met at one point during an installation's life, if at all. Whether or not the variation in visibility that results is a matter of practical significance for traffic safety is doubtful.

4.6 SPECTRAL EFFECTS

One choice in the design process that has not been considered so far is the spectral power distribution of the light source. The perceived colour of road lighting varies greatly from the monochromatic yellow of low-pressure sodium, through the orange of high-pressure sodium, to the white of metal halide and phosphor-converted LED light sources. Other colours produced by such relics of the past as incandescent or mercury vapour light sources can sometimes be seen. Several studies have been made of the effectiveness of these light sources for making largely achromatic objects on the carriageway visible, without any clear conclusions, suggesting that any effects are small (Eastman and McNelis, 1963; de Boer, 1974, Buck et al., 1975). One common feature of these evaluations is that all the measurements were taken fixating on the object, i.e., the retinal image fell on the fovea. More recent measurements of the effect of light spectrum on the detection of off-axis targets suggest that there may be a significant effect of light colour relevant to road lighting. Specifically, He et al. (1997) carried out a laboratory experiment in which high-pressure sodium and metal halide light sources were compared for their effects on the reaction time to the onset of a two-degree diameter disc with the centre either on-axis or 15 degrees off-axis, for a range of photopic luminances from 0.003 to 10 cd.m^{-2}. The luminance contrast of the disc against the background was constant at 0.7. The same light source was used to produce both the background luminance and the stimulus so there was no colour difference between the stimulus and its background. Figure 4.5 shows the median reaction time to the onset of the stimulus, on-axis and off-axis, for a range of photopic luminances, for two practised observers.

From Figure 4.5, it is evident that reaction time increases as photopic luminance decreases from the photopic to the mesopic state, for both on-axis and off-axis detection. There is no difference between the two light sources in the change of reaction time with luminance for on-axis detection, but for off-axis detection, the reaction times for the two light sources begin to diverge as vision enters the mesopic region. Specifically, the reaction time is shorter for the metal halide light source at the same photopic luminance, and the magnitude of the divergence between the two sources increases as the photopic luminance decreases further. These findings can be explained by the structure of the human visual system. The fovea, which is what is used for on-axis vision, contains only cone photoreceptors, so its spectral sensitivity does not change as adaptation luminance decreases until the scotopic state is reached, at which point the fovea is effectively blind. The rest of the retina contains both cone and rod photoreceptors. In the photopic state, the cones are dominant but as the mesopic state is reached, the rods begin to have an impact on spectral sensitivity until in the scotopic state, the rods are completely dominant. Given the different balances between rod and cone photoreceptors in different parts of the retina and under different amounts of light, it should not be surprising that the metal halide light source produces shorter reaction times for off-axis detection than the high-pressure sodium in the mesopic range because it is better matched to the rod spectral sensitivity as shown by its scotopic/photopic ratio (see Section 2.4.3.3). It is also evident why there is no difference between the two light sources for on-axis reaction times.

Lewis (1999) has obtained similar results. Figure 4.6 shows the mean reaction time to correctly identify the vertical or horizontal orientation of a large, achromatic, high contrast, 13-degree by 10-degree grating, where the grating was lit by one of five different light sources: Low-pressure sodium, high-pressure sodium, mercury vapour, incandescent, and metal halide, plotted against the photopic luminance. As long as the visual system is in the photopic range, i.e., at 3 and 10 cd.m^{-2}, there is no difference between the different light sources provided they produce the same

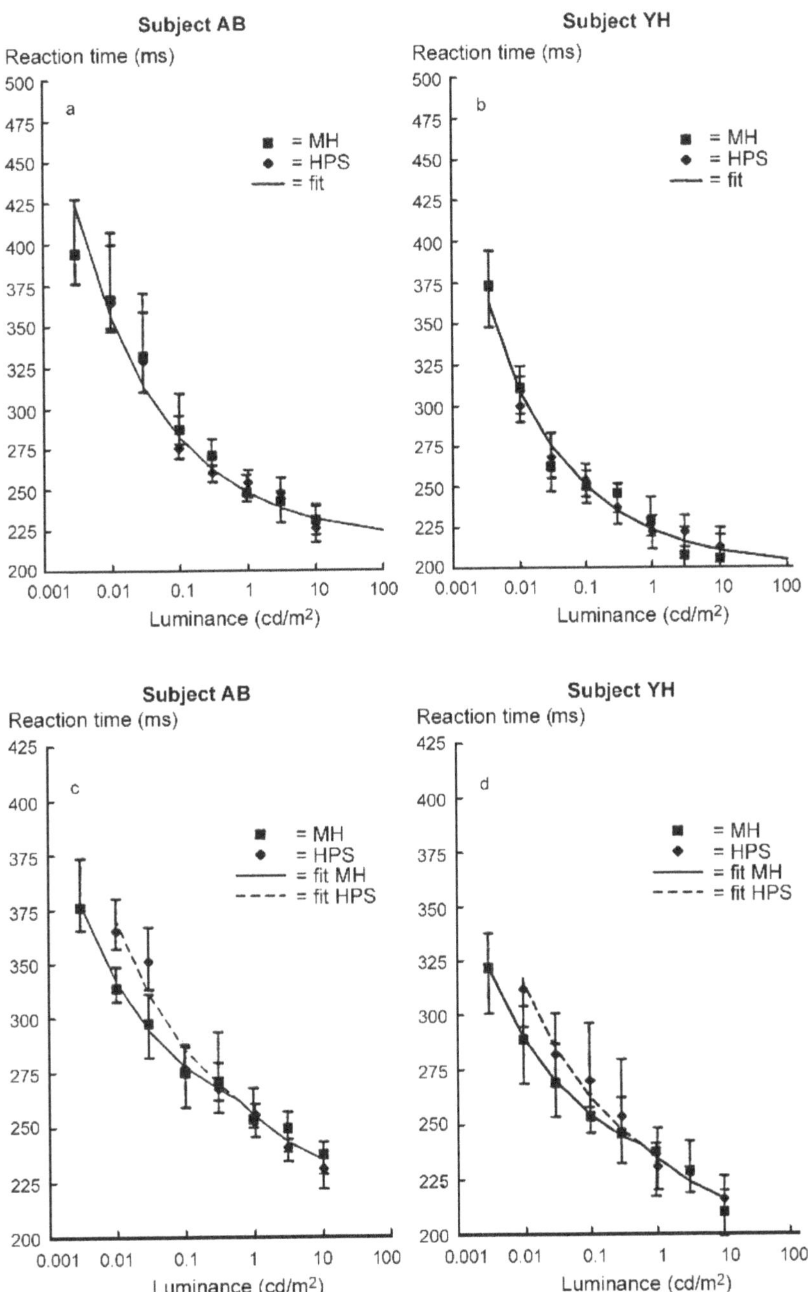

FIGURE 4.5 Median reaction times (ms), and the associated interquartile ranges, to the onset of a 2-degree, high-contrast target seen either (a and b) on-axis or (c and d) 15 degrees off-axis, and illuminated using either high-pressure sodium (HPS) or metal halide (MH) light sources, for photopic luminances in the range 0.003 to 10 cd/m² (after He et al., 1997).

photopic luminance. However, when the visual system is in the mesopic state, i.e., at 1 and 0.1 cd.m^{-2}, the different light sources produce different reaction times, with the light sources that better stimulate the rod photoreceptors (incandescent, mercury vapour, and metal halide) giving shorter reaction times than the light sources that stimulate the rod photoreceptor less (low- and high-pressure sodium).

Such measurements of the time to detect the onset of abstract targets under different light sources may seem irrelevant to the task of driving, but, in fact, driving often requires the visual system to extract information from the peripheral visual field. Lewis (1999) verified that the spectral power distribution of a light source does have an effect on the time taken to extract information of relevance to driving by repeating the experiment described above but replacing the gratings with a transparency of a female pedestrian standing at the right side of a roadway in the presence of trees and a wooden fence. In one transparency, the woman was facing towards the road, in the other she was facing away from the road. The observer's task was to identify which way the woman was facing. Figure 4.6 also shows the mean reaction times for this task under the different light sources over a range of photopic luminances. Again, there is no difference between the light sources as long as the visual system is in the photopic state, but once it reaches the mesopic state, the light sources that more effectively stimulate the rod photoreceptors show shorter reaction times.

Another approach to evaluating the effect of light spectrum in mesopic conditions measured the probability of detecting the presence of a target off-axis. Bullough and Rea (2000) used a simple driving simulator based on a projected image of a road, controlled through computer software. The subject could control the speed and direction of the vehicle along the road through a steering wheel and accelerator

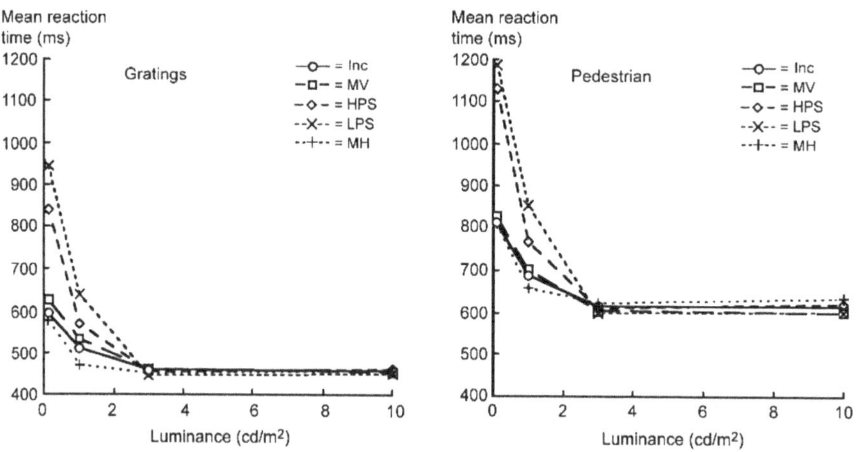

FIGURE 4.6 Mean reaction time (ms) to correctly identify the vertical or horizontal orientation of a grating and the direction a pedestrian located adjacent to a roadway is facing, plotted against the photopic luminance (cd/m²) produced by five different light sources (Inc = incandescent, MV = mercury vapour, HPS = high-pressure sodium, LPS = low-pressure sodium, and MH = metal halide) (after Lewis, 1999).

pedal. The computer monitored the time taken to complete the course and the number of crashes occurring. Filters were applied to the projected image of the course to simulate the spectrum of both high-pressure sodium and metal halide light sources and more extreme red and blue light for a range of luminances. Interestingly, there was no effect of light spectrum on the time taken to complete the course, i.e., on driving speed, but there was a marked effect on the ability to detect the presence of a target near the edge of the roadway. The light spectra that stimulate the rod photoreceptors more (blue and metal halide) led to a greater probability of detection than light spectra that did not stimulate the rod photoreceptors so effectively (red and high-pressure sodium).

Of course, the consequences of making an error while driving a simulator are not quite the same as driving a real vehicle on the road and may lead subjects to pay less attention to what they are doing. Fortunately, Akashi and Rea (2002) have extended this work into the field, having people drive a car along a short road while measuring their reaction time to the onset of targets 15 and 23 degrees off-axis. The lighting of the road and the area around it was provided by either high-pressure sodium or metal halide road lighting, adjusted to give the same amount and distribution of light on the road, and seen with and without the vehicle's halogen headlamps. There was a statistically significant difference between the high-pressure sodium and metal halide lighting conditions. Specifically, the mean reaction time to the onset of the targets was shorter for the metal halide road lighting than for the high-pressure sodium road lighting at both angular eccentricities (Figure 4.7). Using the same experimental site and equipment, Akashi and Rea (2002) also examined the effect of disability glare, provided by halogen headlamps from a stationary car in the adjacent lane, on the ability of a stationary driver to detect off-axis targets at 15 degrees and 23 degrees when the road lighting was provided by metal halide and high-pressure sodium lighting. Again, the mean reaction times to the onset of the targets were longer for the high-pressure sodium road lighting than for the metal halide road lighting. As might be expected, the mean reaction times were also longer when the headlamps in the opposing vehicle were switched on.

Given the results discussed above, there can be little doubt that light spectrum is a factor in determining off-axis visual performance, but how important is it? It could be argued that the increase in reaction times in the luminance range of interest is small and would make little difference to traffic safety. For example, an increase of 100 ms in reaction time would mean a vehicle moving at 80 km.h^{-1} (50 mph) would travel only 2.2 m further because of the longer reaction time. Fortunately for those committed to the importance of light source spectrum, there is some evidence that longer reaction times are associated with more missed events. Rea et al. (1997) measured observers' responses to a change of high contrast character on a changeable message sign located 15 degrees off-axis. The setting was a roadway lit by either high-pressure sodium or metal halide light sources to give an average road surface luminance of 0.2 cd.m^{-2}. An effective average road surface luminance of 0.02 cd.m^{-2} was achieved by asking the observers to wear glasses with a transmittance of 0.1. Figure 4.8 shows the measured reaction times to the changes that were detected and the percentage of off-axis signals missed, plotted against the photopic luminance

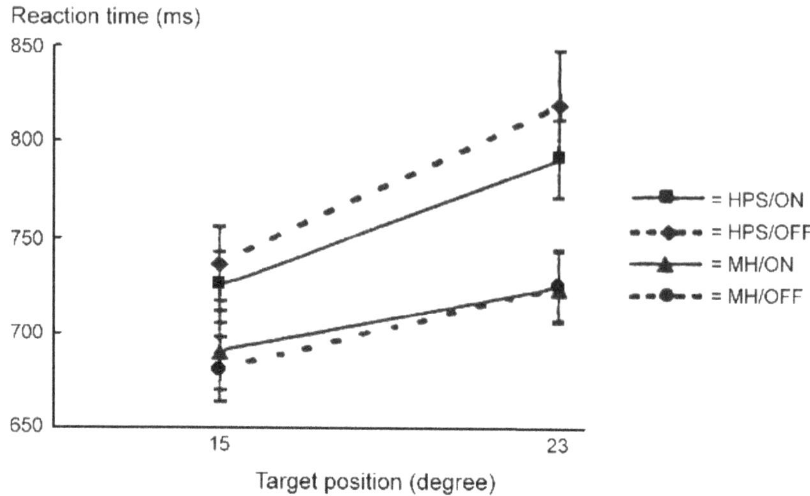

FIGURE 4.7 Mean reaction times (ms) (and the associated standard errors of the mean) to the onset of a target at 15 degrees and 23 degrees off-axis while driving, with high-pressure sodium (HPS) and metal halide (MH) road lighting, and with halogen headlamps turned on and off. The road lighting using the two light sources was adjusted to give similar illuminances and light distributions. The rectangular target subtended 3.97×10^{-4} steradians for the $15°$ off-axis position and 3.60×10^{-4} steradians for $23°$ off-axis position. Both targets had a luminance contrast against the background of 2.77 (after Akashi and Rea, 2002).

of the road surface. It is evident that high-pressure sodium lighting leads to longer reaction times and more missed signals than metal halide lighting, differences that increase as road surface luminance decreases further into the mesopic. Similar increases in missed off-axis changes under high-pressure sodium relative to metal halide illumination have been obtained in other experiments (Bullough and Rea, 2000; Lingard and Rea, 2002). While the importance of increases in reaction time of the order shown are debatable, there can be little doubt about the importance of missing off-axis changes altogether.

A number of other studies have been made of the effect of light spectrum on visual performance in mesopic conditions (IESNA, 2006). These studies have produced a consistent pattern in which tasks done on-axis, such as measurements of visual acuity (Eloholma et al., 1999) and the visibility distance of small targets (Janoff and Havard, 1997) show no effect of light spectrum at the same photopic luminance in the mesopic range, while tasks requiring off-axis activity, such as measurements of effective field size (Lin et al., 2004) and identifying the direction of movement of an off-axis target while driving (Akashi et al., 2007), do. The effect is that light sources that provide greater stimulation to the rod photoreceptors, i.e., with a higher *S/P* ratio, ensure better off-axis visual performance in mesopic conditions.

It is now necessary to consider the relevance of this to road lighting practice. Until recently, the most widely used light source for road lighting was high-pressure sodium but today the phosphor-converted LED is preferred. The relative S/P ratios

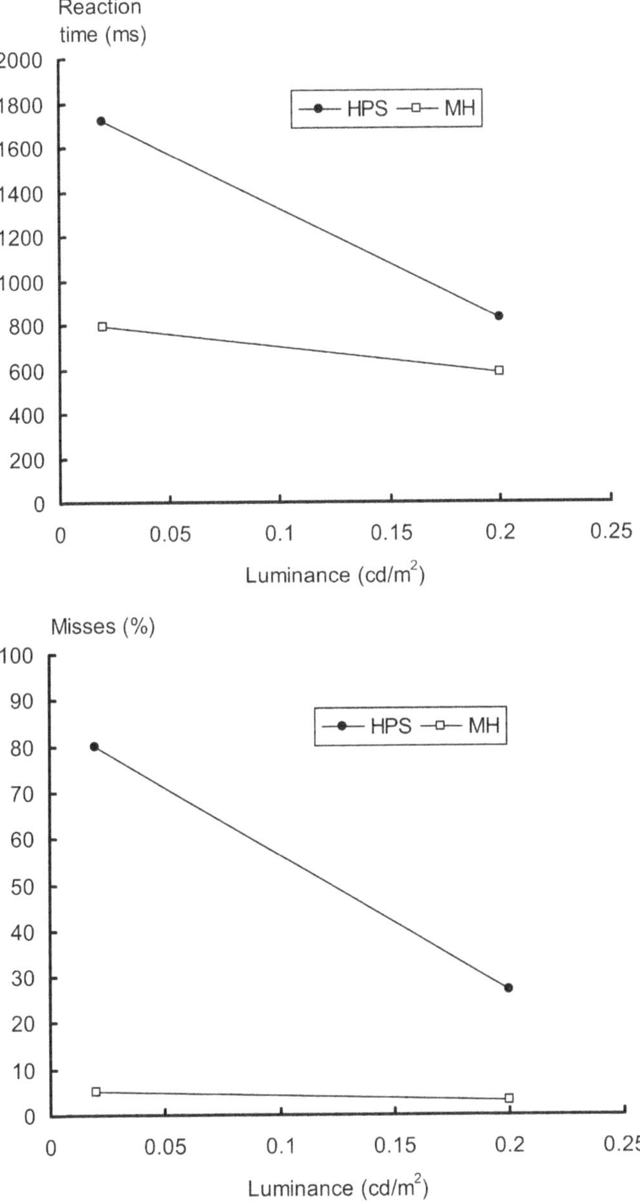

FIGURE 4.8 Mean reaction times (ms) and percentage of misses for changes in a single high-contrast character in a message sign located 15 degrees off-axis for observers looking down a road lit to the same average luminance (cd.m^{-2}) by either high-pressure sodium (HPS) or metal halide (MH) lighting (after Rea et al., 1997).

of these two light sources are very different and predict that the white LED should allow better off-axis visual performance than the orange high-pressure sodium. However, the implications of the results are rather more complex than such a simple statement suggests. In reality, the benefit of choosing a light source that stimulates the rod photoreceptors depends on the driver's adaptation luminance, the balance between on-axis and off-axis tasks and the nature of those tasks. Provided the adaptation luminance is such that the visual system is operating in the photopic state, say 3 cd.m^{-2} and above, there is no effect of light spectrum on off-axis visual performance so the choice can be made on the other factors such as luminous efficacy, life, and colour rendering. If the adaptation luminance is in the high mesopic, say about 1 cd.m^{-2}, the effect of light spectrum is slight. It is only when the adaptation luminance is well below 1 cd.m^{-2} that the choice of spectral power distribution is likely to make a significant difference to off-axis visual performance. How often this occurs is open to question. If the standards described in Section 4.4 are met then the road classes most likely to benefit from a careful choice of light spectrum are those where pedestrians are present and traffic speeds are slow, local roads in the US and the P class in the UK. However, an average luminance masks a wide range, from the area lit by the headlamp beam, to the lit road, and to the luminance of the scene beyond the reach of either the road lighting or the vehicle headlamps. Further, discussion of a single value of adaptation luminance serves to hide the truth. The fact is the concept of adaptation luminance is a convenient fiction. It was originally developed to describe the effects of luminance on basic visual functions. Its use for this purpose was not unreasonable as such measurements are usually made on a uniform luminance field but where the visual field has a wide range of luminances, the adaptation of different parts of the retina will be different, depending on where and for how long the eye is fixated. If the driver has one dominant fixation point, such as might be the case of a driver approaching a tunnel entrance, then the average luminance within about 20 degrees of the fixation point is a reasonable estimate of the adaptation luminance (Adrian 1976). If the observer has many fixation points, i.e., the observer is moving his eyes around a lot, as might occur when approaching a roundabout, then the concept of adaptation luminance is meaningless and specific luminances need to be considered (EPRI, 2005). However, Mortimer and Jorgeson (1974) found that when driving at night, eye fixations tended to be confined to the front edge of the headlamp beam, an observation also made by Stahl (2004), at least for straight roads (see Figure 3.6). Unfortunately, the luminances of the visual field will then vary widely, ranging from the darkness of the far periphery through the luminances of the road ahead of the headlamp beam to the luminances of the road surface lit by both the road lighting and the vehicle headlamps.

Of course, all these luminances are photopic luminances, calculated using the CIE Standard Photopic Observer. It might be thought that the use of the CIE Standard Photopic Observer for the measurement of light when the visual system is operating in the mesopic state is a fundamental problem. There is no doubt that light sources that more effectively stimulate the rod photoreceptors enhance the performance of off-axis detection tasks when the visual system is operating in the mesopic state but at what luminance the mesopic state begins is the subject of controversy. A unified model of photopic, mesopic, and scotopic photometry based on reaction times has mesopic vision starting at 0.6 cd.m^{-2} (Rea et al., 2004) while a model of mesopic effects, based on the performance of tasks claimed to be important to driving, shows mesopic vision having an impact up to 10 cd.m^{-2} (Elohoma and Halonen, 2006; Goodman et al., 2007). Most road lighting for traffic routes is designed to produce average road surface luminances between these two limits. Fortunately, comparisons of the predictions of the two models show only small differences (Rea and Bullough, 2007) (Figure 4.9), which suggest that either could be used to evaluate the role of spectral power distribution in road lighting. As a result, the CIE has produced a consensual model of mesopic photometry (CIE, 2010a) and a set of interim recommendations for using it (CIE, 2017a).

Given that road lighting does produce conditions in which the visual system is operating in the mesopic state while driving at night or walking on a residential street, it is also necessary to consider the nature of the driver's or pedestrian's task and the balance between on-axis and off-axis tasks. These can vary widely, both in the stimuli presented to the driver and the information that needs to be extracted from them. This is important because the magnitude of any spectral effect on off-axis

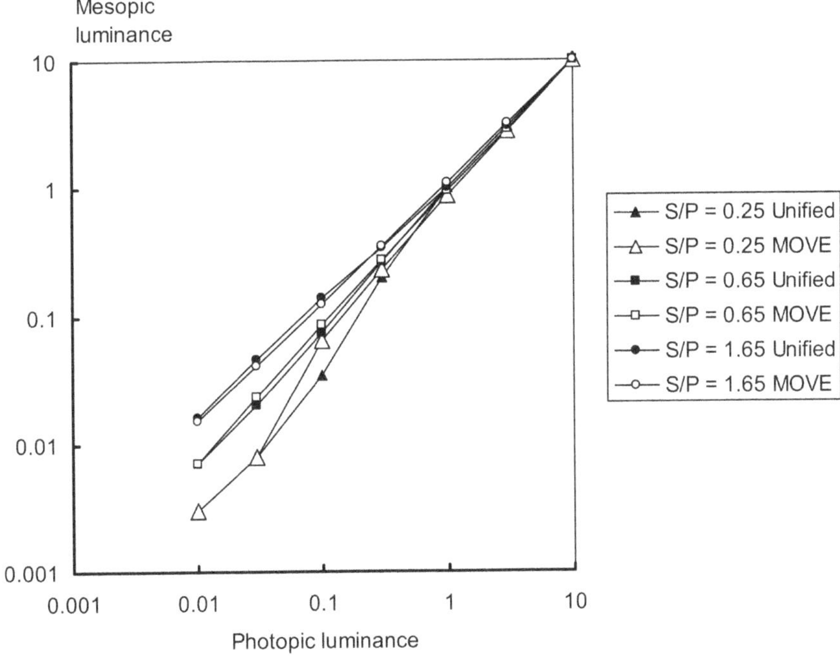

FIGURE 4.9 Calculated mesopic luminances plotted against photopic luminance for three different scotopic/photopic (*S/P*) ratios, using the unified system of photometry of Rea et al. (2004) and the MOVE model of Eloholma and Halonen (2006). The three *S/P* ratios approximate to low-pressure sodium (*S/P* = 0.25), high-pressure sodium (*S/P* = 0.65), and metal halide (*S/P* = 1.65) light sources, all of which have been used for road lighting.

visual performance will depend on the exact task (IESNA, 2006). For stimuli close to threshold, the spectral effects can be large, but for stimuli well above threshold, the spectral effects may be insignificant. What can be said is that using a light source that better stimulates the rod photoreceptors at a given photopic luminance will not make off-axis visual performance worse and may make it better, except where accurate colour discrimination is required, and in this case, a worsening of performance is possible.

As for the balance between on-axis and off-axis tasks, this could be important because it is sometimes argued that the metal halide or phosphor-converted LED light sources which produce white light can be used at a lower road surface luminance than the high-pressure sodium light source, without penalty. Certainly, the results discussed above suggest that, in mesopic conditions, the same off-axis visual performance can be achieved at a lower photopic luminance or illuminance with a metal halide or LED light source than a high-pressure sodium light source. However, in principle, for an on-axis target, a lower photopic luminance will produce worse visual performance for both light sources. This implies that it is only if off-axis detection is assumed to be the more important task in driving or walking that a reduction in road surface luminance for white light sources can be justified. There cannot be many drivers who would be willing to deny the importance of both on-axis and off-axis vision for driving. Fortunately, Ohashi et al. (2020) have found that people could recognize in foveal vision almost all the targets that peripheral vision detected and that the S/P ratio of the light sources used did not change this capability. Likewise, Akashi et al. (2023) have found reducing light levels when implementing mesopic vision adjustments in street lighting have little effect on foveal visual performance.

It is interesting to note that this concern with mesopic vision was most evident at a time when the competition between light sources for use in road lighting was between high-pressure sodium and metal halide. Indeed, in BSI (2013), where light sources with a CIE general colour rendering index of sixty or greater are used in residential roads, the recommended average illuminance for those areas can be reduced by one P class (see Table 4.9). The motivation for this recommendation is not stated but a colour rendering index of 60 does nicely separate metal halide from high-pressure sodium. The luminous efficacy of high pressure sodium is typically higher than that of metal halide. A cynic would observe that in order to equalize the energy consumption of the same installation, the metal halide would have to provide a lower road surface luminance than the high-pressure sodium and applying a mesopic correction would ensure that this would occur. The cynic would also note that since phosphor-converted LEDs, which have a similar luminous efficacy but much longer life and better colour properties than either high-pressure sodium or metal halide, have become the light source of choice, interest in mesopic vision has diminished, so much so that recent lighting recommendations (IESNA, 2021a, 2021b ; BSI, 2020) do not suggest any correction of the recommended illuminances, even for residential streets where lower light levels are used and the interests of pedestrians are given priority. Rather, there is now a lot more interest in other aspects of lighting for pedestrians, especially brightness, perceived safety and reassurance (Fotios and Cheal, 2007; Fotios et al., 2015), and the ability to recognize the faces of people approaching and even to identify their intent from facial expression (Willis et al., 2011; Fotios et al., 2015). Provided a light source with good colour rendering is used, which means a light source producing white light, these concerns have a lot more to do with the amount of light and its distribution than with the light source spectrum.

4.7 BEYOND VISIBILITY

While the emphasis in the specification and design of road lighting is undoubtedly on its impact on visibility, it is important to appreciate that there are other benefits associated with road lighting. These benefits come in two forms, the visual and the behavioural. The visual benefits are directly relevant to the driver. They are an increase in the amount of time the driver has before a response is essential, a reduction in the amount of discomfort and disability glare produced by opposing vehicles' headlamps, and guidance on the direction of the road far ahead. The behavioural effects are both direct and indirect. The direct behavioural effect is on the speed at which drivers choose to travel. The indirect effects of road lighting are evident in the level of crime occurring and the possibility of economic development.

Anyone who has made the transition from an unlit to a lit section of road while driving using low-beam headlamps will be aware of the immediate sense of relaxation that results. The reason for this relaxation is the greater distances over which objects on and near the road can be seen and hence the longer times available for selecting an appropriate response. This benefit of road lighting will be most evident on high-speed traffic routes where the amount of additional information revealed by the road lighting is likely to be modest and, without road lighting, the response times are short. Where the amount of additional information revealed by the road lighting is large, as may be the case in urban areas, the sense of relaxation may be less because of the additional information that has to be dealt with.

Road lighting itself will produce some disability and discomfort glare but, given that the standards discussed in Section 4.4 are met, the amount of glare produced by road lighting will be much less than that produced by the headlamps of approaching vehicles. The formulae for disability and discomfort glare produced by vehicle headlamps are discussed in Section 6.6. For discomfort glare from headlamps, road lighting will tend to increase the adaptation luminance with the result that discomfort glare is reduced. For disability glare from headlamps, road lighting will not change the equivalent veiling luminance but the impact of the equivalent veiling luminance on luminance contrast will be diminished as the luminance of the background, which is usually the road surface, is increased. Thus, road lighting will always tend to diminish both disability and discomfort glare from headlamps, an achievement that makes driving more comfortable.

As for guidance, the view of road lighting luminaires stretching away into the distance provides easily understandable clues to the run of the road far ahead, further than is possible with retro-reflective road markings. Such guidance is most obvious when the road lighting is in a central twin or single-sided layout. Double, staggered, or mixed luminaire layouts can be more difficult to interpret.

The behavioural effects of road lighting can be positive or negative. A negative effect is based on observations of risk compensation in human behaviour. The idea behind risk compensation is that drivers will adjust their behaviour to a constant level of risk. This implies that if road lighting is installed, visibility of hazards is improved so the risk to drivers is reduced, and drivers will increase their speed. Assum et al. (1999) found that drivers in Norway did indeed increase their speed at night after road lighting was installed, by about 5% on straight roads and about 0.7% on curves. Presumably, the reason for the difference between straight and curved roads is that road lighting is more effective in enabling the driver to see further ahead on straight roads than on curved roads. Despite this finding, the results from the meta-analysis of road lighting as an accident countermeasure conducted by Elvik (1995) suggest that such increases in speed are not enough to nullify the value of road lighting, but it is undeniable that they may go some way to diminish its impact.

A possible positive effect of road lighting in urban and suburban areas through behaviour is its use as a crime prevention measure. For this to occur, the road lighting has to be designed not just to light the road but also the surroundings. Even then, there can be no guarantees. This is because lighting, per se, does not have a direct effect on the level of crime. Rather, lighting can affect crime by two indirect mechanisms (Boyce, 2014). The first is the obvious one of facilitating surveillance by people on the street after dark, by the community in general, and by the authorities. If such increased surveillance is perceived by criminals as increasing the effort and risk and decreasing the reward for a criminal activity, then the incidence of crime is likely to be reduced. Where increased surveillance is perceived by the criminally inclined not to matter, then better lighting will not be effective. The second indirect mechanism by which an investment in better lighting might affect the level of crime is by enhancing community confidence and hence increasing the degree of informal social control. This mechanism can be effective both day and night but is subject to many influences other than lighting.

Finally, urban lighting, which includes the lighting of roads and pedestrian areas, has a role to play in economic development. The purposes of such lighting are to attract people into the area and to make their experience safe and enjoyable. Given this purpose, lighting has to be both bright and interesting (Hargroves, 2001). Brightness is essentially a means to an end, namely providing enough light so that visitors can see well enough to feel safe in the space (Boyce et al., 2000a). For a given illuminance, brightness perception is influenced by the spectral power distribution of the light source (Fotios et al., 2015; Besenecker et al., 2016). Lighting using sources with a higher S/P ratio will be seen as brighter at the same photopic illuminance. Interest can be produced by many different means ranging from lighting a historic area using "historic" luminaires through floodlighting iconic buildings to filling the surroundings with continuously changing advertising, as in Times Square, New York. Most large-scale developments involve a lighting master plan intended to ensure that the development is seen as coherent rather than muddled. Such plans cover many types of lighting, but road lighting is usually one of them.

4.8 SUMMARY

Road lighting, as opposed to urban street lighting, first appeared widely in the 1930s. Incandescent, mercury vapour, fluorescent, low-pressure sodium, high-pressure sodium, metal halide, and phosphor-converted LED light sources have all been used for road lighting at different times, the last in this list being the current light source of choice. The luminaires used for road lighting are classified according to their luminous intensity distribution. There are a large number of luminous intensity distributions available, some symmetrical and some asymmetrical. Different luminous intensity distributions are necessary because roads of different widths and layouts require different light distributions if the light is to be directed onto the road surface and not wasted. Luminous intensity distribution is also important for limiting light pollution and controlling glare to drivers.

Road lighting luminaires are typically mounted on columns beside the road. Today, columns are mostly made of steel, aluminium, or composites such as glass-reinforced polyester. The layout of the luminaires for single carriageway roads can be single-sided, staggered, or opposite. The layout of luminaires for dual carriageways and motorways can be central twin, central twin, and opposite or catenary arranged along the axis of the road. These layouts apply to continuous linear stretches of roads, not to sharp bends or conflict areas such as intersections and roundabouts. Layouts for these areas are bespoke. It is worth noting that lighting columns represent a hazard for the driver. The magnitude of the hazard can be reduced by setting the columns back from the road or shielding them with a crash barrier or designing them to bend or break away on impact.

For many years, the most common control system for road lighting was the photoelectric switch mounted on the luminaire, this being used to detect the amount of daylight available and thus to ensure that the road lighting was only used when necessary. Today, light source dimming, mains signalling, and wireless communications are being used to make road lighting installations more sensitive to prevailing conditions and easier to manage and maintain.

Given the need to make a choice between a wide variety of light sources, luminaire types, column heights, road layouts, and control systems, it is necessary for the designer of road lighting to have a target to aim at. The primary purpose of road lighting is to make objects on and near the road visible without causing discomfort to the driver. Road lighting makes objects on the road visible by producing a difference between the luminance of the object and the luminance of its immediate background, which is usually the road surface or its edges. Two single-number metrics have been developed to quantify the visibility provided by road lighting, revealing power, and small target visibility. Revealing power is the average percentage of pedestrian clothing reflectances that are above threshold at a matrix of points spread across and along the road. Small target visibility is defined as the weighted average of the visibility levels of a number of flat, 18 cm side, square targets, all with a diffuse reflectance of 0.5, mounted vertically in a matrix across and along the road. The visibility level of the target is defined as the ratio of the actual luminance difference of the target to the luminance difference threshold of the target; the higher is this ratio, the more visible is the target.

Neither revealing power nor small target visibility is much used because the visual task of driving is multi-faceted and both metrics address only one facet, on-axis detection. The driver certainly has to be able to detect small targets on-axis but also needs to be able to detect objects on the road off-axis, to detect movement at the edges of the road, to judge the relative speeds and direction of movement of other vehicles, and to identify many different signs and signals. Further, the revealing power and small target visibility metrics are only really applicable in light traffic where the driver can see a considerable distance up the road. In heavy traffic, the detection of small obstacles is not an issue, much more importance being attached to the relative movements of nearby vehicles. This is probably why most current road lighting standards are expressed in conventional photometric terms.

Road lighting standards vary in detail from country to country but they do have some common features. One is the division of the road network into different classes, according to traffic speeds and density, the type of users, and the road geometry. Different lighting metrics are used to form standards for the different classes and these metrics are then varied to match the sub-divisions of each class. Another common feature is the aspects of road lighting dealt with. Most lighting standards contain metrics for the amount of light on the road, the distribution of light on the road, and the extent of disability glare produced by the road lighting. In the UK, the amount of light and its distribution across and along the road for traffic routes is specified in terms of maintained levels of average road surface luminance, overall luminance uniformity, and longitudinal luminance uniformity while the amount of disability glare is restricted by setting a maximum threshold increment, this being the percentage increase in luminance difference threshold that occurs because of disability glare. For conflict areas, such as intersections and roundabouts, and residential areas, the amount and distribution of light is specified in terms of road surface illuminance. The change from a luminance to an illuminance metric is rationalized as being due to the fact that for traffic routes there are a limited number of directions of view, but in conflict areas and residential streets, there are multiple directions of view.

In the USA, standards use road surface luminance as the primary measure of interest. Again, this metric is used to specify different maintained average and uniformity values for different traffic routes and residential areas, although maintained illuminance is used for intersections. For the luminance recommendations, an explicit metric is used to limit disability glare, this being the maximum veiling luminance ratio which is defined as the ratio of the equivalent veiling luminance to the average road surface luminance.

A comparison of the road lighting recommendations used in the UK and USA shows they are similar in concept but differ in detail. The UK recommendations should lead to a higher level, more uniform lighting on roads with faster traffic than the US recommendations. However, the allocation of the highest average road surface luminance to major roads in high pedestrian conflict areas in the USA, rather than high-speed traffic routes as is the case in the UK is consistent with putting most light where it will have the most beneficial effect in reducing fatalities and personal injury crashes.

Setting standards is one thing, but the reality may be different. The extent to which road lighting standards are achieved in practice depends on the assumptions that have to be made. Standards based on illuminance are the easiest to achieve in practice because they require no assumptions about the reflection characteristics of the road surface. Standards based on luminance do. In principle, the reflection properties of a specific road surface can be measured, but in practice they rarely are. Rather, the detailed reflection properties of a small number of what are claimed to be representative road surfaces are used in the calculation of road lighting installations, in the form of an *r*-table. Measurements of the difference between the calculated luminance metrics using the representative reflection characteristics and the actual reflection characteristics for the road suggest that the use of the representative reflection characteristics is likely to underestimate the luminance metrics.

It should also be noted that the reflection characteristics of a road surface change over time due to wear. An observation of any well-travelled road lane will reveal the existence of five distinct areas: Two polished wheel tracks, a darker oil track in the centre of the lane, and two unworn tracks at the edges of the lane. Each of these areas will have different reflection properties although they are all represented by a single *r*-table. It should be apparent that, despite the precision of the standards used for road lighting, the assumptions made in the design process mean that the reality is likely to be much more variable from site to site.

Another choice in the design process that needs to be considered is the spectral power distribution of the light source used. The perceived colour of road lighting can vary greatly from the monochromatic yellow of low-pressure sodium lighting, through the orange of high-pressure sodium lighting, to the white of phosphor-converted LED lighting. Measurements in the laboratory and on the road have shown that as the visual system enters the mesopic state, light sources that provide more stimulation to the rod photoreceptors produce faster reaction times and fewer missed signals for off-axis detection at the same photopic luminance. Attempts to provide a mesopic photometry system have been made but have not been implemented in standards. Fortunately, the phosphor-converted LED can provide good stimulation to the rod photoreceptors along with high luminous efficacy, good colour rendering, and long life, which explains why it is now the light source of choice for road lighting.

While the emphasis in the specification and design of road lighting is undoubtedly on its impact on visibility, there are other benefits associated with road lighting. These benefits come in two forms, the visual and the behavioural. The visual benefits are directly relevant to the driver. They are an increase in the amount of time the driver has before a response is essential, a reduction in the amount of discomfort and disability glare produced by opposing vehicles' headlamps, and guidance on the direction of the road far ahead. The behavioural effects are both direct and indirect. The direct behavioural effect is on the speed at which drivers choose to travel. The indirect effects of road lighting are evident in the level of crime occurring and the possibility of economic development in an urban area.

Clearly, road lighting is not as simple as it might seem. The design of road lighting involves many assumptions that influence the photometric conditions achieved.

Even when these assumptions are reasonable, doubts must remain as to the suitability of the photometric conditions for driving. The fundamental problem is that our understanding of how lighting affects the driver's ability to extract information of different types on- and off-axis is limited. Until this deficiency is rectified, road lighting will continue to be based more on experience than science.

5 Markings, Signs, and Traffic Signals

5.1 INTRODUCTION

One of the earliest signals associated with the use of motor vehicles was introduced by the British Parliament in the form of the 1865 Locomotives on Highways Act. This called for each self-propelled, steam-powered vehicle to have a crew of three, one driver, one stoker, and a man to walk 60 yards in front of the vehicle carrying a red flag to warn of its approach, the purpose being to avoid frightening the horses. Today, horse-drawn transport and steam-powered vehicles have almost vanished from the developed world, but signals have proliferated dramatically. Today, drivers are faced with a plethora of markings, signs, and signals designed to inform and regulate their behaviour, some fixed, some changeable, some unlit, some lit, but all needing to be seen by day and night. The form and location of markings, signs, and signals are strictly controlled so as to ensure consistency across road networks, although different countries have different rules (DfT, 2022a, FHWA 2023). The factors considered in designing markings, signs, and signals are the distances from which they need to be visible; their shapes and colours, shape and colour being used as cues to meaning as well as being important for visibility; the advantages and disadvantages of pictograms rather than text in a specific language for signs; and the need for some means to attract attention to the sign or signal. The aspect of design that will be considered here is the use of light, either as an inherent part of the marking, sign, or signal or as a means to make the marking, sign, and signal conspicuous, visible, and legible by day and night.

5.2 FIXED ROAD MARKINGS

Some of the simplest means used to guide drivers are markings on the road. Such markings usually consist of different patterns of lines on the road surface, although sometimes solid shapes, texts, or pictograms are used and, occasionally, the road surface itself is the sign, the colour of the surface indicates a restricted class of use. Road markings are used to indicate lane boundaries, bends in the road, areas where overtaking is prohibited or limited to a specific lane, edges of junctions, parking restrictions, vehicles allowed to use a lane, mini-roundabouts, etc. (Figure 5.1).

The lighting variables important to road markings are colour and luminance. Colour is important because it often carries part of the message. For example, in the UK, road markings that restrict parking in some way are yellow while markings that indicate the road ahead are white. There is rarely a problem in discriminating

FIGURE 5.1 Three examples of road marking: (a) a mini-roundabout without road lighting, (b) lane markings with road lighting, and (c) a lane restricted to use by buses, taxis, and bicycles, all by day.

colours by day, but at night there can be. In theory, under low-pressure sodium road lighting, yellow and white road markings will be difficult to discriminate, except in the area illuminated by the vehicle's headlamps where small amounts of light from light sources with a wide spectral power distribution will make colour discrimination possible (Boynton and Purl, 1989). In practice, the inability to separate yellow and white lane markings at night is not much of a problem because the different locations of the markings related to parking and driving make the meaning obvious. However, where the colour of the whole lane is used to indicate a restricted use, e.g., buses and taxis only, there can be a problem depending on the colouring material used. Often, this is a dark red, with the result that if low-pressure sodium road lighting is used, the lane marking is indistinguishable from adjacent conventional asphalt lanes. This is why such lanes usually have text or pictograms indicating their use set out in white retro-reflective paint at regular intervals. Fortunately, the use of low-pressure sodium as a light source for road lighting, or any other application, is largely a distant memory, despite its high luminous efficacy.

The most common role of road markings is to provide visual guidance and lane definition. Drivers need both long-range guidance (more than 5 s preview time) (Schieber et al., 2009) and short-range guidance (less than 3 s preview time) (Rumar and Marsh II, 1998). Long-range guidance is accessed intermittently and consciously, using foveal vision. Short-range guidance is accessed continuously and unconsciously, using peripheral vision. Road markings can provide both short and long-range guidance. Road markings usually consist of a thermoplastic paint containing spherical retro-reflective beads (see Section 2.8). The paint material is a high-reflectance, diffuse reflector, which ensures the mark will be seen in positive luminance contrast against the low-reflectance road surface during daylight. At night, the luminance of the white paint has two components – the diffuse reflected component mainly from any road lighting and the retro-reflected component from the vehicle headlamps (CIE, 2001). Where there is no road lighting, the luminance depends almost entirely on the retro-reflective materials. The greatly enhanced reflection of these materials in the direction of the vehicle means that the luminance of the markings will be much greater than the luminance of the adjacent road surface, the resulting luminance contrast making the markings visible at a distance (Figure 5.2). The

FIGURE 5.2 Painted retro-reflective road markings and road studs lit by high beam headlamps alone (by Mick Stevens).

main limitation of such markings is that they tend to lose their retro-reflective properties with wear (Owusu et al., 2018). As a result, visual guidance is much reduced (Rumar and Marsh II, 1998). Recently, the Federal Highways Administration has issued a requirement for maintaining a minimum retro-reflectivity for longitudinal road markings. For roads with a speed limit of 35 mph (56 km.h^{-1}) and a traffic flow of 6000 vehicles per day or more, the minimum retro-reflectivity is set at 50 mcd. m^{-2}.lx^{-1}. For roads with a speed limit of 70 mph (113 km.h^{-1}), the minimum retro-reflectivity is set at 100 mcd.m^{-2}.lx^{-1}. Advice on methods of maintaining the retro-reflectivity of road markings is given in FHWA (2022a).

Retro-reflective road markings are also subject to reductions in visibility in certain weather conditions. For example, they tend to disappear when the road is covered with water, the water surface forming a specular reflector above the markings which reflects the grazing-incident light from the vehicle's headlamps along the road away from the driver before it reaches the retro-reflectors (see Section 12.3). They are also less effective in fog because the light from the vehicle's headlamps is scattered, reducing the amount of light reaching the markings and the retro-reflected light is also scattered, reducing the contrast of the markings as seen by the driver. Retro-reflective road markings are also of little use when the road is covered in snow, even after plowing. To provide visual guidance where snow is common, vertical posts of about 1 m height are placed at regular intervals along the edges of the road. These posts often have small panels of retro-reflective material fixed near the top. Such posts define the edges of the road well in rain, snow, and fog but do not give any guidance as to the lanes.

An alternative approach to retro-reflectors involves photoluminescent materials, such as strontium aluminate pigments, producing what is popularly known as glow-in-the-dark markings (Bacero et al., 2015). These materials are excited by electromagnetic radiation received during the day and continue to emit light for a time after the sun sets. Depending on the amount, spectrum, and duration of the incident radiation, useful amounts of light can be emitted for times ranging from minutes to hours although the light output decays over time (Villa et al., 2023). The benefit of photoluminescent materials for road marking is that they will provide guidance to the driver beyond the range of the headlamps something that retro-reflectors cannot do. So far, this approach has been little used which is a pity because it has the potential to increase visibility distances.

A mechanism used to provide both lane definition and guidance as to the run of the road ahead, in rain and fog but not in snow, is the individual retro-reflector, originally known as a "cats-eye" but now commonly called a road stud (see Figure 5.2). The original "cats-eye" consisted of a cast-iron frame holding a rubber housing itself containing four refractive/reflective retro-reflectors, two facing each way down the road. A series of such assemblies were set into the centre of the road at approximately 10 m intervals to provide visual guidance. The rubber housing deflected when a vehicle's wheel passed over it, the deflection acting to clean the front surface of the retro-reflectors when wet. There are now a number of simpler road studs available designed to withstand the impact of wheels without deflection (Figure 5.3). All road studs place the retro-reflectors high enough above the road surface to stand above water on the road although this can make them prone to damage by snowploughs. Retro-reflective road studs can be fitted with filters so that colour can be used to carry a message. For example, road studs acting as lane dividers are usually white while those acting as road edge markers are conventionally red on the nearside and amber of the offside of the road. Where the edge of a road can be crossed, as at a slip road off a major road, the colour of the road studs is changed from red to green.

FIGURE 5.3 A retro-reflective road stud fitted with a red filter to mark the edge of the road.

Retro-reflective road studs depend on light for their visibility from a vehicle's headlamp. This inevitably limits the distance over which guidance is delivered to less than 100 m. An alternative is an active road stud. This contains an LED making it self-luminous. Such road studs can be powered from a battery fed by a built-in solar panel or by induction from an electricity cable buried in the road. By installing active road studs at regular intervals along a road, visual guidance is available over much longer distances, up to 1,000 m, and around curves in the road before the vehicle reaches the curve (Figure 5.4). Active road studs are more effective than retro-reflective road studs in wet and foggy conditions but require their luminous intensity to be tuned to the environmental condition to avoid causing discomfort glare (Villa et al., 2015). They are also more expensive although it is also claimed they have a longer life. Shahar et al. (2018) undertook a study of driver's behaviour at night on a winding road without road lighting, with road lighting, and with active road studs marking the edges of the road and lanes round the curves, using a driving simulator. They found better lateral control of the vehicle in the active road studs condition relative to the unlit and road lighting conditions. The mean speed of the vehicle was very slightly higher when the road was lit, either by road lighting or by active road studs (101 km.h^{-1}) relative to the unlit road (98 km.h^{-1}). There was no difference in the mean speed for the road lighting and the active road studs conditions. Subjectively, both the road lighting and the active road stud conditions were considered safer and more comfortable,thus allowing better control of the vehicle than the unlit condition. Such findings have led to suggestions that active road studs could be an alternative to road lighting on rural roads (Llewellyn et al., 2020).

FIGURE 5.4 An arrangement of powered road studs providing guidance around a bend. The studs at the edge of the road are red and those marking the centre line are white (iStock by Getty Images/Astucia).

Earlier studies of the effects of conventional road marking on drivers' behaviour are more mixed. Adding lines marking the edges of a road where previously there had been no marking results in increased driving speeds with the lateral position of the vehicle being closer to the edge of the road (Rumar and Marsh II, 1998; Davidse et al., 2004). When edge lines are added to an existing centre line, there is no overall change in speed but when a centre line marking is replaced with edge lines, there tends to be a decrease in speed (Davidse et al., 2004). The rationale for such changes in behaviour lies in the driver's confidence about what lies ahead. Providing edge markings, or a centre line, on a previously unmarked road will increase the amount of visual guidance and confidence in where the road goes, hence the increase in speed. Adding edge markings to a road with a central line marking adds little to visual guidance, so a change in speed is unlikely. Removing central marking and replacing it with edge markings may have the effect of making the road appear narrower, hence the reduction in speed. These examples serve to make a basic point that providing better visual guidance to the driver at night may not result in safer driving. There are two opposing views on the value of road marking. One view holds that better visual guidance leads to smoother and safer driving. The other is that better visual guidance leads to over-confidence resulting in an increase in speed. The problem this conflict exposes is that while some visual guidance is certainly necessary and road markings and road studs are a convenient way to provide it, markings only address one part of the driver's task. An overemphasis on visual guidance and a neglect of the other aspects of the driver's task may diminish traffic safety rather than improve it.

5.3 FIXED SIGNS

Another common feature of roads are fixed signs mounted beside or over the road giving information on directions, lane changes, speed limits, etc. The size, shape, colour, and content of such signs have been extensively studied (Forbes, 1972; Mace et al., 1986) and regulated (DfT, 2022a, FHWA, 2023). The form of road signs used in the UK and in many other parts of the world are based on the Vienna Convention of Road Signs and Legends, a product of the United Nations intended to bring some uniformity to road signs and signals in different countries. This convention is not followed in North America where versions of the Manual on Universal Traffic Control Devices for Streets and Highways are used. Despite this variation, the first question of interest here is whether or not such signs should be illuminated and, if so, how? The decision on whether or not to illuminate a sign depends first and foremost on the distance at which the sign needs to be detected and the distance at which it needs to be legible. These distances depend on the speed and density of traffic approaching the sign and whether the driver has to carry out some manoeuvre in response to the sign. High speeds, dense traffic, and the need for a manoeuvre all increase the distances at which the sign needs to be detected and legible. Other factors to be considered are the complexity and brightness of the background against which the sign has to be seen, the location of the sign relative to the driver, the size of the sign, the amount of information contained in the sign, and the retro-reflection properties of the material from which the sign is constructed. The more complex the background, the brighter the ambient light level, the further the sign is from the driver's normal direction of fixation, the larger the sign, and the greater the amount of information the sign contains the more likely it is that lighting should be provided. The usual interpretation of this complex picture is that road signs in rural areas do not need lighting, the retro-reflective nature of these sign material being sufficient to ensure their visibility. On lit roads in suburban and urban areas, sign lighting is necessary because the alternative sources of light, road lighting and headlamps, are inadequate. Road lighting and headlamps are designed to illuminate the road surface, not signs remote from it. Attempts to use road lighting or headlamps for sign lighting risk producing increased glare to other road users. The ranges of luminance of the white parts of externally lit signs when lit according to the UK regulations and with the nature of the reflective material taken into account are 9–53 cd.m^{-2} for medium background luminances and 35-123 cd.m^{-2} for high ambient background luminances (Cooper et al., 2008).

The lighting of the sign can be external or internal. Where external lighting is provided for individual signs, the luminaire is usually positioned close to and either above or below the sign (Figure 5.5). Luminaires mounted above the sign will collect less dirt on the emitting face and will produce less light pollution but may cause glare to drivers approaching from the back of the sign, may cast shadows on the sign in daylight, and may be more difficult to reach for maintenance. Luminaires

FIGURE 5.5 Two different externally illuminated road signs on a single post. The light is provided by compact fluorescent light sources.

mounted below the sign will collect more dirt and will cause more light pollution but will not cast any shadows on the sign. Regardless of the mounting position, all luminaires need to be positioned so that they do not obstruct the view of the sign from any meaningful direction and so as to ensure that any specular reflections of the luminaire do not occur in the direction of approaching traffic. Further, the light source must be chosen to accurately reveal the colours of the sign, colour often being an important element in identifying the sign before any printing or pictogram is readable. This implies that the best light sources for lighting signs are now white phosphor-converted LEDs, with their high luminous efficacy, long life, and good colour properties.

Internal lighting of a sign can take two forms. One is to incorporate an electroluminescent element into the sign. The other consists of an inter-reflecting box containing a light source, with the front face of the box providing the information. Both the reflection and transmission properties of the front face are important because the sign has to be legible by both day and night and should look the same under both conditions. It is the reflection properties that dominate the appearance of the sign by day and its transmission properties that dominate by night. The great advantage of the internally illuminated sign is that, compared with external sign lighting, it produces much less light pollution. The risk with internally illuminated signs is that, at night, the luminance is so high that the sign itself becomes a glare source. The ranges of luminance for the white part of an internally illuminated sign in medium and high background luminance conditions when lit according to the UK regulations are 40–150 cd.m^{-2} and 150–300 cd.m^{-2}, respectively (Cooper et al., 2008). Power for internally illuminated signs can be provided by a solar panel/battery system or by connection to the mains.

A special road sign is the bollard placed in the middle of the road to indicate an obstruction such as a pedestrian island. These can be retro-reflective (Figure 5.6a) or internally illuminated (Figure 5.6b). The retro-reflective bollard has no light source or electrical power and is designed to be self-righting in the sense that when struck by a vehicle at moderate speed, the sign will spring back to the vertical position. When the bollard is internally illuminated with power supplied from the mains, the light source and associated electrics are usually placed below road level. The whole of the bollard above road level is made of translucent plastic which emits light in all directions, making the bollard visible from all directions. Vehicles colliding with the bollard may destroy the plastic upper part usually without damaging the below-road part, making replacement relatively inexpensive.

Where neither external nor internal sign lighting is provided, the luminance of the sign at night will be dependent on the illumination provided by the headlamps of approaching vehicles, the effectiveness of the retro-reflective treatment of the material from which the sign is constructed and the angular separation of the driver from the headlamps (Sivak and Olson, 1985). Olson et al. (1989a) examined how the detection distances for differently coloured retro-reflective sign materials varied with the effectiveness of the retro-reflective material expressed as the specific intensity/unit area of material. Specific intensity is the luminous intensity emitted by the retro-reflector per unit of illuminance received at the retro-reflector. The distances were

FIGURE 5.6 Two types of bollard to warn drivers of a pedestrian island in the middle of the road, (a) a self-righting retro-reflective bollard and (b) an internally illuminated bollard.

obtained from observers driving along unlit roads at night using headlamps alone. There were two linear relationships between the logarithm of the specific intensity/ unit area and detection distance, one for yellow, white, blue, and green materials and one for red and orange materials. For all colours, the higher is the specific intensity/ unit area for the material, the greater is the detection distance (Figure 5.7).

As for the role of headlamps, the two questions of interest are, can the colour of the sign be correctly recognized and is there sufficient light to read the sign? McColgan et al. (2002) examined the effect of different light sources designed to be used in headlamps on the ability to correctly identify the colour of road sign materials at high (63 cd.m$^{-2.}$) and low (12 cd.m^{-2}) sign luminances. They showed that tungsten halogen, tungsten halogen with a neodymium coating, tungsten halogen with a blue coating, and xenon (HID) light sources all enabled very accurate identification of the sign colours specified in the then US Manual on Uniform Traffic Control Devices (FHWA, 2003), at both high and low sign luminances. Given the colour properties of phosphor-converted LEDs, there is little doubt that these would also reveal road sign colours correctly at these luminances.

The angular separation of the driver from the headlamps matters because the retro-reflective materials used in signs reflect the incident light back along its own path, i.e., light received at a sign from a headlamp will be reflected back to the head-lamp. Of course, such a material is not perfect so there will always be some spread in the reflected light distribution. The position of the driver relative to the headlamps is not usually a problem for cars, but for large trucks it can be. Sivak et al. (1993) have shown that the luminance of retro-reflective signs can be much less for truck drivers than for car drivers and that such reduced sign luminances will seriously reduce the detection distances of signs. Similar conclusions were reached in static and dynamic

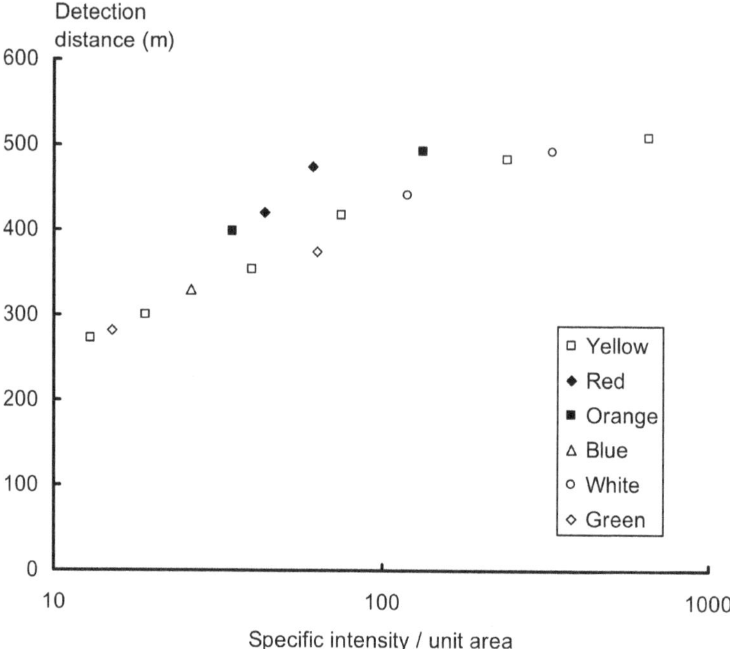

FIGURE 5.7   Mean detection distances (m) for road sign materials of various colours plotted against the efficiency of the retro-reflective material, measured as the specific intensity/unit area of the material (cd.lx^{-1} m^2) (after Olson et al., 1989).

trials of sign legibility by day and night, with and without road lighting, undertaken for the UK Highways Agency in 2010. It is also important to note that the efficiency of retro-reflectors is much reduced when water and dirt cover the sign, suggesting that regular cleaning is required, particularly for bollards which are positioned close to the road surface. However, such cleaning will ultimately diminish the retro-reflector efficiency through wear.

Another area of concern for signs, both lit and unlit, is the background against which the sign is seen. The background can be important for two different reasons. The first is the presence of a very bright light source close to the sign. Such a source can produce enough disability glare to make the sign invisible. The classic example of this is a sign with the setting sun immediately adjacent to it. This problem is usually solved by surrounding the sign with a low reflectance screen that cuts off the view of the sun within a few degrees of the sign. The second reason is where the background against which the sign is seen is visually complex so that the sign is just one sign amongst many. This often occurs in city centres where there are a multitude of advertising signs of high luminance to compete with the road sign. Schwab and Mace (1987) examined the detection and legibility distances for signs seen against backgrounds of different complexity. They found that the more complex was the background, the shorter was the detection distance, but there was little effect on legibility distance. This is not surprising because legibility is primarily dependent on the details within the sign when the sign is fixated while a sign is usually first detected off-axis. The effectiveness of off-axis detection during visual search will be influenced by the presence of competing visual information. Schwab and Mace suggest that in the presence of a complex background either the luminance of the sign or its size should be increased, or an advanced warning sign should be used. An alternative would be to separate the sign from the background by using a large, low-reflectance surround. It would also be possible to attract attention to the sign by means of a simple strobe light.

5.4 VARIABLE ROAD MARKINGS

The most common form of variable road marking is active road studs that can change their state on instruction from a remote control centre. The change can be one of luminous intensity, colour, or fluctuation. Installations of these studs along roads have been used to change lane usage patterns and to warn of hazards ahead, the former by changing colour and the latter by changing from continuous to flashing light output. Glare at night can be avoided by reducing the luminous intensity of the stud from that necessary to maintain visibility by day. Essentially, installations of variable road studs form part of what is called an intelligent road system, the aim of this being to maximize the throughput of the road network while maintaining safe travel. Variable road studs tend to be used mainly at choke points in the road network, i.e., in areas of very high traffic density.

5.5 VARIABLE MESSAGE SIGNS

Another element in the intelligent road system that is found with increasing frequency is the variable message sign (BSI, 2014) (Figure 5.8). These signs are used to provide information about temporary road conditions, such as the presence of road works, variable speed limits, and traffic congestion.

Variable message signs use a series of pixels to display a text message or pictogram. The pixels may be light reflecting or light emitting although the latter are now much more common than the former. Light-reflecting pixels are usually of low reflectance on one side and high reflectance on the other, electromechanical means being used to flip each pixel as required. During the day, the reflectances are enough to ensure the visibility of the message. At night, some illumination of the pixels is required; illumination that sometimes uses ultraviolet radiation to excite phosphors in the high-reflectance material. For the self-luminous pixels, the light sources used are LEDs. The legibility and readability of such signs depend on many factors, including the pixel shape, letter width/height ratio, font, and letter separation. Collins and Hall (1992) showed that words with a regular pixel, a letter width/height approaching unity, an upper case font, and a letter separation of two pixels are most legible. Collins and Hall worked with light-reflecting signs and so had no need to consider the luminance of the display. Padmos et al. (1988) carried out field evaluations of self-luminous message signs mounted above the road so that the immediate

FIGURE 5.8 A gantry over a UK motorway carrying conventional blue direction signs, variable speed limit signs over each lane with the speed limit displayed inside a red circle, and a variable text message sign giving the time predicted to reach the next two junctions (iStock by Getty Images/Mediterranean)

background was the sky. Figure 5.9 shows the mean message luminances set for three different formats of the number 5, for two different visibility criteria, plotted against the horizon luminance. The number 5 was viewed from 100 m. The message luminance is given by the equation:

$$L_{mes} = 10^6 . I_{px}/d^2$$

where L_{mes} is the message luminance (cd.m^{-2}), I_{px} is the pixel luminous intensity (cd), and d is the distance between pixels (mm).

Figure 5.9 shows that the visibility of the message varies with the horizon luminance, the higher the horizon luminance the higher the message luminance required for the message to be visible. By using other visibility criteria, Padmos et al. (1988) were able to show that the message luminances necessary for a rating of "optimum" on a bright day would be rated as "glaring" at night. This finding implies that some degree of luminance control is necessary to ensure comfortable and effective viewing of the message by day and night. Padmos et al. (1988) suggest that a sufficiently legible but not too bright message can be obtained by a two-step message luminance, 4,000 cd.m^{-2} by day and 100 cd.m^{-2} by night, although three steps (4,000, 400, and 40 cd/m^{-2}) would be better.

The number of words that can be displayed on a variable message sign is limited, and the aspect ratio of the letters forming the words is adjusted according to the

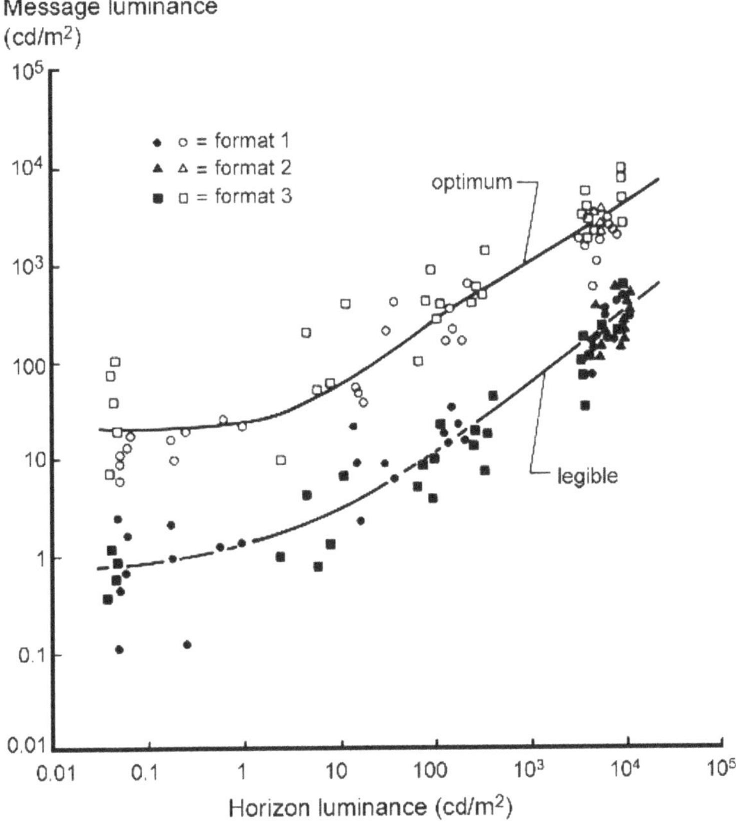

FIGURE 5.9 Message luminances (cd.m⁻²) set by individual subjects for the number 5 presented in three different formats on a self-luminous message sign, seen from 100 m, plotted against horizon luminance (cd.m⁻²). The luminances were set to match two different visibility criteria at different times of day and night and hence for different horizon luminances. The criterion "optimum" is based on the perception that the display is conspicuous but not glaring. The criterion "legible" is based on the perception that the display is just recognizable. The three formats used different numbers of pixels to form the number 5. Specifically, format 1 = 23 pixels, format 2 = 50 pixels, and format 3 = 141 pixels (after Padmos et al., 1988).

speed of vehicles approaching (DfT, 2015b). Variable message signs can also be used to display information without words, such as a lane closure or a current speed limit on a multi-lane road (Figure 5.8) or as a warning to a driver that they are exceeding the speed limit (Figure 5.10).

In another variation, flashing lights may be incorporated into a variable message sign although not in the message itself. For example, on motorways in the UK, messages can be of two types – those informing the driver of problems ahead and those giving expected time to a destination or advocating safe driving behaviour (Figure 5.8). Those indicating problems ahead have flashing lights at the corners of

FIGURE 5.10 Variable message sign warning an approaching vehicle that it is travelling faster than the speed limit. The approaching vehicle's speed is measured by radar (iStock by Getty Images/jhorrocks).

the display to make the sign more conspicuous. Similarly, message signs warning an approaching driver of excessive speed often have flashing lights in the corners of the display.

The content of variable message signs on main traffic routes (Figure 5.8) is usually determined remotely from a control room that contains multiple screens showing views of traffic flow along the road. Where the number of possible messages is limited and there is no central control, as is the case for speed messages (Figure 5.10), a radar detection system built into the sign will indicate the measured speed of an approaching vehicle and may change the colour of the indicated speed to red if it is above the limit or green if below, or produce the message "Slow Down" with separate flashing lights if the approach speed is too fast.

Clearly, the potential for variable message signs to influence driver behaviour is large but whether or not that potential will be fulfilled depends on the message displayed (Chatterjee and Mcdonald, 2004). Messages fall into three classes, those that are valuable, those that are wrong, and those that are already obvious to the driver. A valuable message is one that gives information that is not otherwise available to the driver, such as an accident causing congestion some miles ahead that can be avoided by taking a different road or a change of speed limit made at times of heavy congestion with the intention of maintaining smooth traffic flow. Help in avoiding long delays and the evident benefit to the driver of being in heavy traffic moving smoothly

TABLE 5.1

Levels of Information Provided to Drivers in a Study of the Effect of Message Content on Driver Behaviour (from Alm and Nilsson, 2000)

Message Level	Information Provided	Example of Message
0	None	–
1	Warning (flashing red light)	Warning
2	Warning, nature of incident	Warning, congestion
3	Warning, nature of incident, distance to incident	Warning, road works, 1 km ahead
4	Warning, nature of incident, distance to incident, recommended action	Warning, accident, 1 km ahead, use left lane

rather than in a stop and start manner will ensure that such messages are appreciated and obeyed. But, if the message is wrong, e.g., a message about the presence of fog being displayed when visibility is clear, or a statement of the obvious, e.g., a message telling drivers to slow down because of heavy spray when the windscreen wipers have already been set to maximum speed, the message may be ignored because drivers will already have taken what they consider to be the appropriate action. Ensuring that the message is relevant to the driver and matches the prevailing conditions is the responsibility of the control room.

Where the message is relevant as regards event, location, and timing, variable message signs can have a beneficial effect on traffic safety. Alm and Nilsson (2000) conducted an experiment looking at the effect of different message content on drivers' behaviour. In a driving simulator, drivers were faced with three incidents: A queue of cars moving at 30 km.h^{-1}, road works requiring a lane change, and an accident requiring a lane change. Five levels of information were provided at a distance of 1,000 m from the incident (Table 5.1).

Figure 5.11 shows the mean speed plotted against distance from the slow-moving traffic queue. It is evident that all forms of message result in a slower approach to the slow-moving traffic queue than when no warning is given. Indeed, one of the drivers who did not receive a warning failed to slow soon enough and collided with the back of the queue. Figure 5.12 shows the mean speed of approach for the accident. Again the speed of approach is reduced when any form of warning is given. Interestingly, when the message contains a recommended action, namely to use the left lane, lane changing occurs earlier, but the speed past the accident is faster than when less information is given. There can be little doubt that messages that are correct in describing event, location, and timing are helpful to drivers (Nygardhs and Helmers, 2007).

5.6 TRAFFIC SIGNALS

A ubiquitous feature of roads in urban and suburban areas is the traffic signal. Traffic signals are placed at intersections to identify priorities for both vehicular and pedestrian traffic (Figure 5.13).

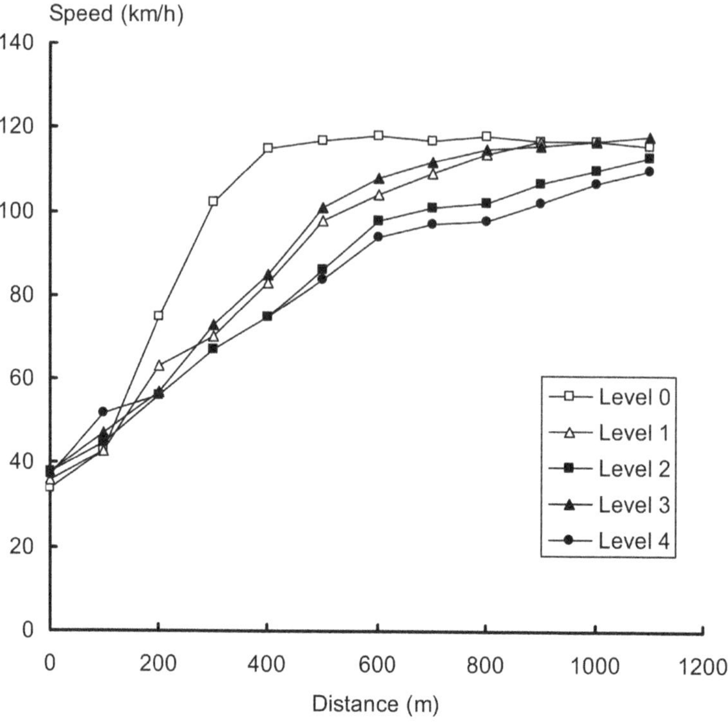

FIGURE 5.11 Mean speed (km/h) plotted against distance (m) from a slow-moving queue of traffic for the five levels of message content listed in Table 5.2 (after Alm and Nilsson, 2000).

The photometric and colorimetric characteristics of traffic signals are closely regulated in terms of their luminous intensity distributions and colour, the latter because the meaning of the signal is given by its colour (ISO, 1999. ITE, 2005, 2008 and 2011; European Committee for Standardisation, 2006). The recommendations are consensus decisions made by a committee, but those decisions are based, at least in part, on studies of the reaction time to the onset of the signals and the number of signals that are not detected under different conditions. Bullough et al. (2000) have reported an extensive series of measurements of reaction time and missed signals using a tracking task requiring continuous fixation and simulated traffic signals occurring a few degrees from the fixation point, the traffic signals being provided by both incandescent and LED light sources. Reaction times for all three signal colours tended to become shorter as signal luminance increased until a minimum was reached. However, small changes in reaction time are of little significance for traffic signals because of the delays built into the sequencing of the signals. Much more important are signals that are missed altogether. Figure 5.14 shows the percentage of missed signals for three traffic signal colours, over a range of signal luminances, seen against a 5000 cd.m^{-2} large-area background, i.e. against a simulated

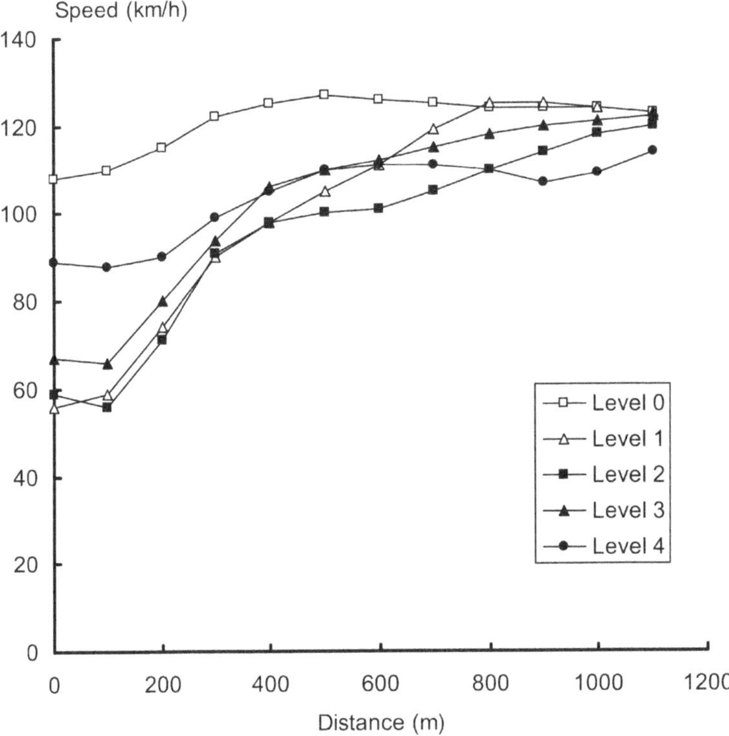

FIGURE 5.12　Mean speed (km/h) plotted against distance (m) from an accident requiring a change of lane for the five levels of message content listed in Table 5.2 (after Alm and Nilsson, 2000).

daytime sky. A missed signal was one that was lit for more than one second without a response from the observer. It is evident that increasing the signal luminance reduces the percentage of missed signals until a minimum level is reached.

The practical implication of Figure 5.14 is that different traffic signal colours need to have different luminances to achieve the same percentage of missed signals unless the luminances are so high that virtually no signals are missed. This observation explains differences in practice in different countries. The USA has different luminous intensities for different signal colours (ITE, 2005), while Europe recommends a higher luminous intensity for all traffic signal colours (European Committee for Standardization, 2006).

Until the turn of the century, the only light source used in traffic signals was the incandescent lamp. However, since then coloured LEDs have completely taken over in the traffic signal market. These traffic signals have a much lower power demand and a longer maintenance interval. Measurements have shown that there is no difference in the percentage of missed signals between LED and incandescent traffic signals of the same luminous intensity and the same nominal colour

FIGURE 5.13   A complex traffic signal layout with roundels and arrows for specific directions of travel.

(Bullough et al., 2000). However, these measurements were taken against a uniform background. Where the background to the traffic signal is complex and filled with other lights, there may be an advantage for LED traffic signals in that the colour produced tends to be more saturated than when an incandescent lamp and filter are used to produce the required colour. Saturation matters because, for the same hue, more saturated colours are perceived as brighter than less saturated colours at the same luminance (Ayama and Ikeda, 1998). This difference in perception is associated with the stronger signal produced through the colour channels of the human visual system by more saturated light. What this means in practice is that, at the same luminance, traffic signals using LEDs will be seen as brighter than those using an incandescent light source and hence will be more conspicuous (Bullough et al., 2007a). As for the ability to correctly identify the signal colour, Boyce et al. (2000b) examined this question for both incandescent and LED signals in daytime, both light sources producing signals with chromaticity coordinates inside the Institute of Transportation Engineers colour boundaries for signal lights (ITE, 1985). They found that errors in colour identification were rare but, when they did occur, were most likely to occur for off-axis viewing, at signal luminances below about 8,000 cd.m^{-2}, for the incandescent light source and the yellow signal, the yellow signal being identified as red. Taken together, these results suggest that, visually, the LED light source has a slight perceptual advantage over the incandescent light source for use in traffic signals.

One general conclusion from the results discussed above is that the higher is the luminous intensity of the signal, the fewer are the number of missed signals and the

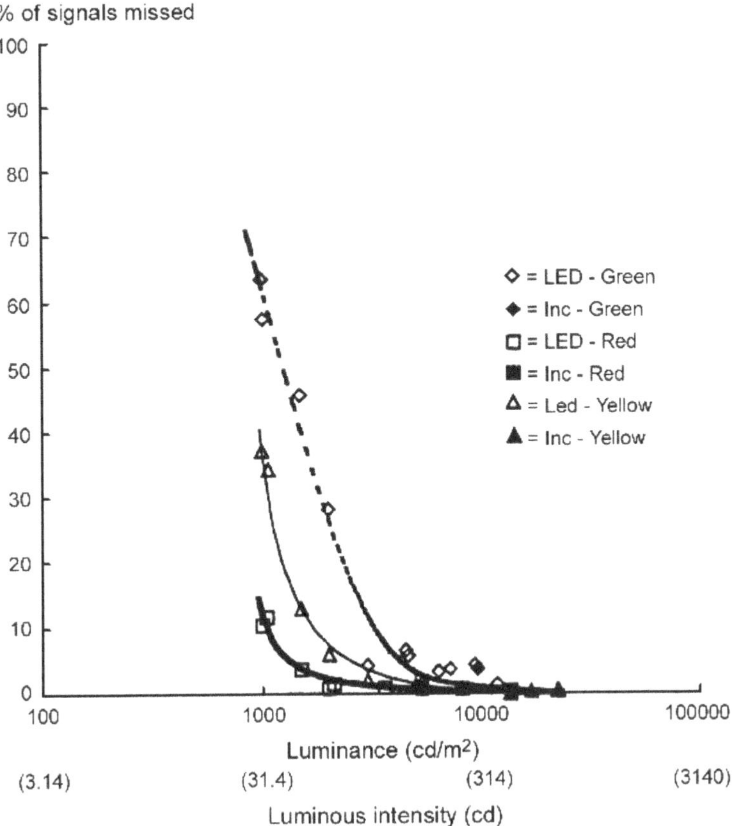

FIGURE 5.14 Percentage of signals missed for each signal colour plotted against signal luminance (cd/m²). The signals were provided by either LED or filtered incandescent light sources. To be counted as a missed signal, the signal had to have been on for one second without a response from the observer. The second horizontal axis is the luminous intensity (cd) corresponding to the signal luminance (cd/m²) for a 200-mm-diameter signal (after Bullough et al., 2000).

more likely it is that the signal colour will be correctly identified. This suggests that the higher is the luminous intensity, the better is the signal, but there is a limit to how far the luminous intensity of a signal can be taken. A traffic signal has to be seen both day and night. A higher luminous intensity is of value during the day because it will tend to increase the conspicuity of the signal, but by night a high luminous intensity can become a source of discomfort and even disability glare, particularly for drivers of heavy vehicles where the driver is much higher above the road and closer to the mounting height of traffic lights. Bullough et al. (2001a) measured the percentage of people considering traffic signals of different luminances uncomfortable when viewing them directly (Figure 5.15). Such data could be used to set desirable traffic signal maximum luminances at night, which should be lower than the maximum allowed by day. Unfortunately, there appears to be little interest in this

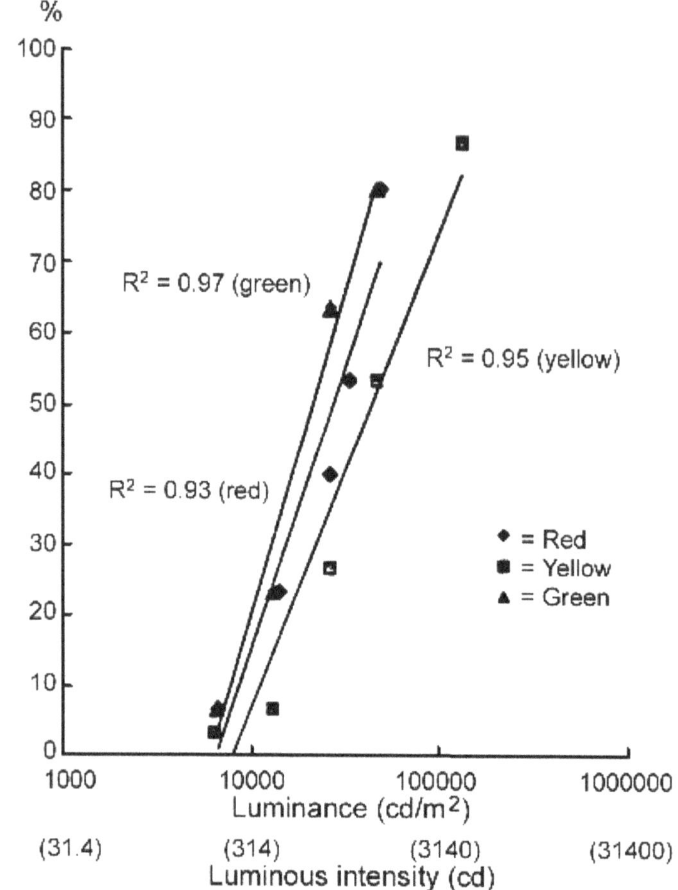

FIGURE 5.15 Percentage of observers considering a signal uncomfortable for the three signal colours seen in darkness, plotted against signal luminance (cd/m²). The signal simulates a 200-mm-diameter signal seen from a distance of 20 m. The second horizontal axis is the luminous intensity (cd) corresponding to the signal luminance (cd/m²) for a 200-mm-diameter signal (after Bullough et al., 2001a).

possibility, most research on traffic signal systems being concerned with constructing models of system timing to optimize traffic flow (Eom and Kim, 2020) or to make such systems adaptive to traffic flow (Wang et al., 2018)

5.7 SUMMARY

Road markings, road signs, and traffic signals are designed to inform and regulate drivers' behaviour. Some markings, signs, and signals are fixed, some are changeable, some are unlit, some are lit, but all need to be seen by day and night. The aspect of the design of markings, signs, and signals that is considered here is the use of light, either as an inherent part of the marking, sign, or signal or as a means to make the mark, sign or signal conspicuous, visible, and legible by day and night.

Some of the simplest means used to guide drivers are markings on the road. Road markings usually consist of white thermoplastic paint containing some retro-reflective materials. The white material is a diffuse reflector, which ensures the mark will be seen in positive contrast against the road surface during daylight. At night, the luminance of the white paint has two components, the diffuse reflected component mainly from any road lighting and the retro-reflected component from the vehicle headlamps. Where there is no road lighting, the luminance depends almost entirely on the retro-reflective materials. The retro-reflective materials produce a greatly enhanced reflection in the direction of the vehicle which means that the luminance of the markings will be much higher than the luminance of the adjacent road surface, the resulting positive luminance contrast making the markings visible at a distance. The main limitation of such markings is that they tend to lose their retro-reflective properties with wear and they tend to disappear when the road is covered with water or snow. As a result, visual guidance is much reduced at the time when it is most required.

One way to overcome this problem is to use road studs. These devices consist of an assembly containing one or two large retro-reflectors. The assembly is usually high enough to stand above any water on the road. Road studs depend for their visibility on light from a vehicle's headlamps. This inevitably limits the distance over which guidance is delivered. An alternative now available is a self-luminous road stud containing a white LED. Power for such studs can be provided by integrated solar cells or by induction from a power cable buried in the road. Studs of this type can provide visual guidance over a much longer distance than is possible using light from headlamps alone.

Another common feature of roads is fixed signs mounted beside or over the road giving information on directions, lane changes, speed limits, and so on. The decision on whether or not to illuminate a sign depends on the distance at which the sign needs to be detected. High speed, dense traffic, and the need for a manoeuvre all increase the distance at which the sign needs to be detected and hence the need for lighting of the sign. To be sure of being detected, the sign also needs to be conspicuous. The minimum sign luminance necessary to ensure conspicuity is influenced by the complexity of whatever is around the sign, the sign colour, and the driver's age. Individual sign lighting is most required where the necessary detection distances are large, the background is complex, and the amount of light reaching the sign from the headlamps of approaching vehicles is small.

Where individual lighting is provided for a sign, there are two options, external or internal lighting. For external lighting, the luminaire is positioned close to and either above or below the sign. For internally illuminated signs, both the reflection and transmission properties of the front face are important because the sign has to be legible by both day and night and should look the same under both conditions. It is the reflection properties that dominate the appearance of the sign by day and its transmission properties that dominate them by night. The great advantage of the internally illuminated sign is that, compared with external sign lighting, it produces much less light pollution.

Where individual sign lighting is not provided, the illumination of the sign is dependent on the headlamps of approaching vehicles. The need to control the light distribution from headlamps in order to control glare to other vehicles limits the amount of light that can be expected to reach a sign. To increase the amount of light from headlamps reflected towards the driver, signs contain retro-reflective materials.

Road markings and signs are usually invariant but there are variable versions of both. Variable road markings are powered road studs that can change their colour or state on instruction from a control unit. Variable message signs are used to provide information about temporary road conditions, such as the presence of road works, variable speed limits, traffic congestion, and travel times. A variable message sign is only as useful as the message it displays. Messages fall into three classes, those that are valuable, those that are wrong, and those that are already obvious to the driver. Only messages in the first class are likely to be obeyed. To contribute to traffic safety, the message has to be correct about the event, its location, and its timing.

Another ubiquitous feature of roads in urban and suburban areas is the traffic signal. Traffic signals are placed at intersections to identify priorities for both vehicular and pedestrian traffic. The photometric and colorimetric characteristics are closely regulated in terms of their luminous intensity and colour, the latter because the meaning of the signal is given by its colour. Until the turn of the century, incandescent light sources were used in traffic signals but since then LED light sources have taken over the market. The most important response measure to a traffic signal is missing the signal altogether. Different traffic signal colours need to have different luminances to achieve the same percentage of missed signals unless the luminances are so high that virtually no signals are missed. This would seem to suggest that all that is necessary to ensure an adequate signal is to use high luminance, but this may make the signal itself a glare source at night.

Markings, signs, and signals can be considered as a balancing act on a number of levels. There is a need to balance the level of information so that enough information is provided to be useful, but not so much that the driver is overwhelmed. There is a need to balance conspicuity and visibility against visual comfort and light pollution, by day and night. Lighting and retro-reflection are the means used to make the information provided by markings, signs, and traffic signals accessible to the driver, but care is necessary to achieve the proper balance.

6 Vehicle Forward Lighting

6.1 INTRODUCTION

The exterior lighting of vehicles can be conveniently divided into two types – forward lighting and signal lighting. Forward lighting is lighting designed to enable the driver to see after dark. Signal lighting is lighting designed to indicate the presence of or give information about the movement of a vehicle to others, by night and day. Forward lighting includes headlamps and fog lamps. Often, these are of such a high luminous intensity that they cause glare to approaching vehicles. Signal lighting includes front, side and rear position lamps, turn lamps, stop lamps, rear fog lamps, reversing lamps, daytime running lamps, hazard flashers, and licence plate lamps. These vary in size and are of limited luminous intensity. Forward lighting is there to see by. Signal lighting is there to be seen. One exception to this crude classification is reversing lamps. Reversing lamps provide both visibility to the rear and a signal to others around the vehicle. This chapter is concerned with the forward lighting fitted to all vehicles, i.e., headlamps. Fog lamps are dealt with in Section 12.5. Forward lighting is now an essential part of vehicle design, serving functional, aesthetic, and marketing demands. This chapter is concerned only with the functional aspects.

6.2 THE TECHNOLOGY OF VEHICLE FORWARD LIGHTING

The technology of vehicle forward lighting involves light sources, optical systems, and structure (Wordenweber et al., 2007).

6.2.1 LIGHT SOURCES

To be suitable for vehicle forward lighting, light sources have to have sufficient light output to meet the legal requirements that specify the minimum luminous intensities of headlamps. They also have to meet the customer's expectations about the amount of light immediately available on switch-on. Further, they have to meet the manufacturer's need to operate in a wide range of climates as well as to withstand frequent vibration and to last as long as the vehicle, assuming common patterns of use. For the first motor vehicles, the light source used in headlamps was a flame produced by burning acetylene gas generated by dripping water onto calcium carbide. It was only in the 1910s that electric lighting began to be widely used and not until the 1950s that vehicle headlamps became the subject of quantitative legal requirements. The first electric light source used in vehicle headlamps was the basic incandescent. This served until the 1960s when the tungsten halogen light source was introduced but that, in turn, was challenged by the arrival of the xenon discharge (HID) light source in the

1990s. Today, the light-emitting diode (LED) is rapidly replacing all the others. These light sources differ in their light spectrum, luminous efficacy, and life (see Section 2.5). All are used for the conventional lighting of buildings but when used in vehicles they are modified to be able to withstand the rigours of the vehicle environment that include extremes of heat, cold, moisture, and vibration. The main modification for the incandescent and tungsten halogen light sources is to increase the strength of the filament assembly. For the HID light source used in headlamps, the main modification is the addition of xenon to what is basically a metal halide light source so as to provide a significant amount of light available immediately on switch-on and to make the run-up time to full light output much shorter. For headlamps, the LED light source comes in two forms. One is the chip-on-board form in which several LED semiconductors are grouped close together on a circuit board to form an effective point source. This allows conventional methods of optical control to be used. The other is a number of LEDs arranged to meet optical requirements with different LEDs providing different parts of the beam. Regardless of form, LED systems designed to be used in vehicles are modified to provide better thermal control.

6.2.2 OPTICAL CONTROL

To provide the required light distribution, the light source is housed in a structure, which itself is either mounted on the vehicle or, more usually, integrated into the vehicle body. At least two headlamp light distributions are required by law: High beam for use when there is no other vehicle on the road ahead, either approaching or leading; and low beam for use when there is another vehicle approaching or leading (see Section 6.3). In some countries, high beam is referred to as the main beam or driving beam while low beam is also called the dipped beam, meeting beam, or passing beam.

There are several systems of optical control used in headlamps to produce the specified luminous intensity distribution, based on the physical phenomena of reflection and refraction, either individually or combined (Wordenweber et al., 2007). In most reflector systems, the light distribution is determined by the position of the light source relative to the reflector, the shape of the reflector, and any optical patterning of the front cover glass. The oldest and simplest headlamp consists of a parabolic reflector with the filament of the light source at the focus and with a shading cap over the light source. The presence of the shading cap means that all the light emitted from the headlamp comes from the reflector. The parabolic reflector produces a parallel beam of light so the desired light distribution is achieved through cylindrical or prism lenses built into the cover glass. Designers prefer smooth clear cover glasses, so in many vehicles, the parabolic reflector has been replaced by smooth, segmented, or faceted reflectors designed to produce the desired light distribution without any lenses on the cover glass (Figure 6.1a).

A more recent form of reflector system uses a micro-mirror. This consists of up to several million tiny mirrors. Each mirror can be individually addressed. Its orientation can be changed by an electrostatic field. Effectively, each mirror forms a pixel of a display. When illuminated by an LED, the micro-mirror can be used to deliver a wide range of luminous intensity patterns from a headlamp and therefore is a useful part of some adaptive forward lighting systems (see Section 6.7)

FIGURE 6.1 Two systems of headlamp optical control: (a) A reflector headlamp on a Suzuki Jimny and (b) a projector headlamp on a BMW X5.

Headlamps known as projectors have at least three components: a light source, a near ellipsoidal reflector, and a condensing lens. Because the reflector is nearly ellipsoidal, it has two foci. The light source is placed at one focus so the reflector produces an image of the source at the second focus. The light distribution after the second focus is strongly divergent, so a condensing lens is used to collimate the beam (Figure 6.1b). Projector headlamps typically use LED or HID light sources.

There are also LED headlamps using multiple light sources to form the beams. Unlike tungsten halogen or HID light sources, a single LED does not have enough light output to form a headlamp. Therefore, LED headlamps all have multiple LEDs. The desired light distributions can be achieved by means of reflectors or lenses together with switching or dimming of individual LEDs.

Although these systems of optical control have been discussed separately, in practice they are often integrated into one assembly (Figure 6.2).

Various methods have been used to provide the two main luminous intensity distributions required – high beam and low beam. The simplest is to provide two different headlamps, one for high beam and one for low beam, and to switch between them, thereby creating a four rather than a two headlamp vehicle. The same or different optical control methods can be used in the high-beam and low-beam headlamps. Rumar (2000) concluded that four headlamp systems are better than two headlamp systems in all respects except cost and size. If a two-headlamp system is required, and they often are, some means has to be found to get two different light distributions from the same headlamp (Ying et al., 2021). For reflector headlamps, this can be done by using a single light source and moving the reflector to change the light distribution. An alternative approach suitable for the tungsten halogen light source is to keep the reflector fixed but to use two filaments in the same light source. By placing the two filaments at different positions relative to the reflector, two different light distributions can be achieved. For projector headlamps, a moveable baffle within the headlamp can be used to cut-off part of the high beam when low beam is required. The change from high to low beam is made by raising the baffle across the second focal plane. For multiple-source headlamps, different light distributions can be achieved by switching or dimming different combinations of light sources.

FIGURE 6.2 A headlamp assembly on a Jaguar XF250. Both reflector and projector optics are used to provide low-beam and high-beam lighting as well as a daylight running light, a front position lamp, and a turn indicator.

6.2.3 HEADLAMP STRUCTURE

Headlamp structure can be considered in four areas – mechanical, electrical, environmental, and supplementary. The mechanical elements are concerned with ensuring that the light source and optical control elements are consistently related to each other and the whole is mounted securely on or in the vehicle body. It is also necessary to provide some means of adjusting the aiming of the headlamp. The electrical elements involve the connection of the light source to the vehicle's electrical system, either directly or through any necessary control gear, and the means to change the light source in the event of premature failure. The environmental elements require careful consideration of the materials used and the mechanical structure so that the sealing of the headlamp is sufficient to prevent water access, yet there is adequate ventilation to prevent fogging. The supplementary aspects refer to the systems used for automatic washing of the cover glass, automatic levelling of the headlamp, and automatic change from high beam to low beam and vice versa. Details of these matters are given in Wordenweber et al. (2007)

This brief consideration of the light sources, optical control systems, and structure of headlamps is sufficient to suggest that the design of headlamps is complex and that headlamps can take many different forms. However, all have to meet the relevant regulations.

6.3 THE REGULATION OF VEHICLE FORWARD LIGHTING

Headlamp luminous intensities, light distribution, and placement on the vehicle are closely regulated. The purpose of these regulations is to bring some order to the potential conflict between drivers caused by the fact that headlamps increase the visual capabilities of the driver sitting behind them but simultaneously decrease the visual capabilities of the driver facing them. These regulations are implemented through adherence to them by vehicle manufacturers and by annual government inspections of used vehicles, designed to ensure continued roadworthiness. These regulations can take different forms in different countries, but the vast majority follow either the recommendations of the USA Federal Motor Vehicle Safety Standard or the Economic Commission for Europe (ECE). Regardless of which set of standards is followed, there are two features they have in common. The first is that the headlamps fitted to a vehicle have to produce two different luminous intensity distributions, here called high beam and low beam. High-beam headlamps are for use when there is no other vehicle on the road ahead so there is no need to limit glare. Low-beam headlamps are for use when there is an approaching vehicle or a vehicle immediately in front and it is necessary to limit the headlamp luminous intensity so as not to compromise the vision of the other driver. The second is that the colour appearance of the light emitted by headlamps must be white, defined as emitting light with chromaticity coordinates that fall within a specified region of the CIE 1931 chromaticity diagram (see Figure 2.3 for an example). Insisting on a similar colour appearance for all headlamps is an example of using colour as a means to signal the direction of movement of the vehicle. This principle also explains why reversing lights, which are mounted on the rear of the vehicle, are also white, reversing lights only being lit when the reverse gear is engaged and the vehicle is ready to move backwards.

6.3.1 FORWARD LIGHTING IN NORTH AMERICA

In the USA, the design, performance, and installation of all motor vehicle lighting equipment are regulated by the Federal Motor Vehicle Safety Standard (FMVSS) 108, which incorporates the technical standards of the Society of Automotive Engineers (SAE). A 2022 amendment of FMVSS 108 permits the use of adaptive headlamp systems (see Section 6.7.3) as well as separate low and high beams. Daytime running lights are legal in the USA but are not required. Both Canada and Mexico, which have land borders with the USA, generally follow the same standard, although Canada requires the installation and use of daytime running lights (see Section 7.14). The required photometric properties of headlamps are given as minimum and maximum luminous intensity values for specific directions relative to the beam axis. Interestingly, there are very few points where both minimum and maximum luminous intensities apply. For most points, the limit is either a minimum or a maximum but not both. The points where a minimum luminous intensity is required are those associated with providing good vision for the driver. The points where a maximum is set are those most likely to cause glare to an approaching driver. The outcome of having different low- and high-beam luminous intensity distributions is to produce different illuminances at different locations along the road ahead. Figure 6.3 shows contours of the median vertical illuminance for pairs of headlamps used on the twenty best-selling passenger vehicles in the USA for the 2000 model year, measured at road level for both low and high beams (Schoettle et al., 2002). Other requirements relate to the positioning of the headlamps. Mounting heights of headlamps are limited because the higher is the mounting height, the more likely it is that

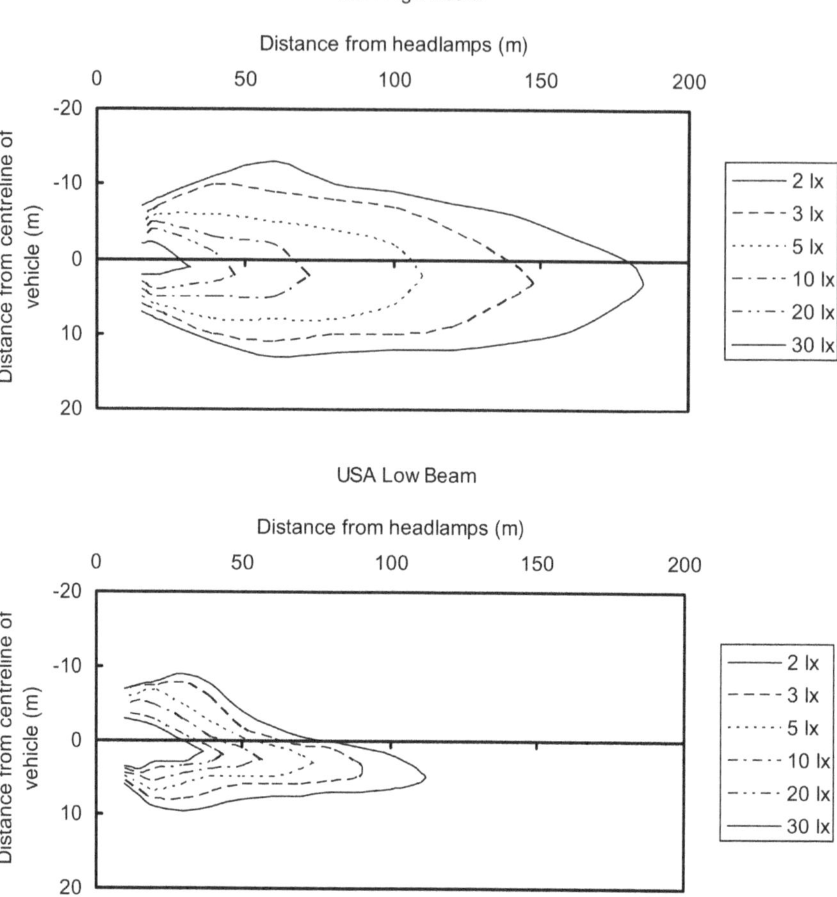

FIGURE 6.3 Contours of median vertical illuminances (lx) produced by pairs of halogen headlamps operating on high beam and low beam for the twenty best-selling passenger vehicles in the USA in the 2000 model year. The vertical illuminances were calculated at road level. Vehicles in the USA are driven on the right (after Schoettle et al., 2002).

approaching drivers or drivers being followed will experience disability and discomfort glare (Akashi et al., 2008). Left and right headlamps should be equidistant from the vehicle centre line and at the same height above the road surface.

6.3.2 Forward Lighting Elsewhere

The headlamp regulations used in virtually all other industrialized nations follow the ECE recommendations. The photometric requirements are specified as the minimum and maximum illuminances at different positions on a vertical plane positioned 25 m from the headlamp, normal to the beam axis. Figure 6.4 shows the contours of median vertical illuminance for pairs of headlamps used on the 20 best-selling passenger vehicles in Europe for the 1999 model year, measured at road level for both low and high beams (Schoettle et al., 2002). The ECE regulations also set limits on mounting heights and require that a levelling device is fitted for all headlamps on vehicles in Europe.

6.3.3 SIMILARITIES AND DIFFERENCES

A comparison of Figures 6.3 and 6.4 reveals some similarities and some differences between the North American and ECE standards, at least as regards vertical

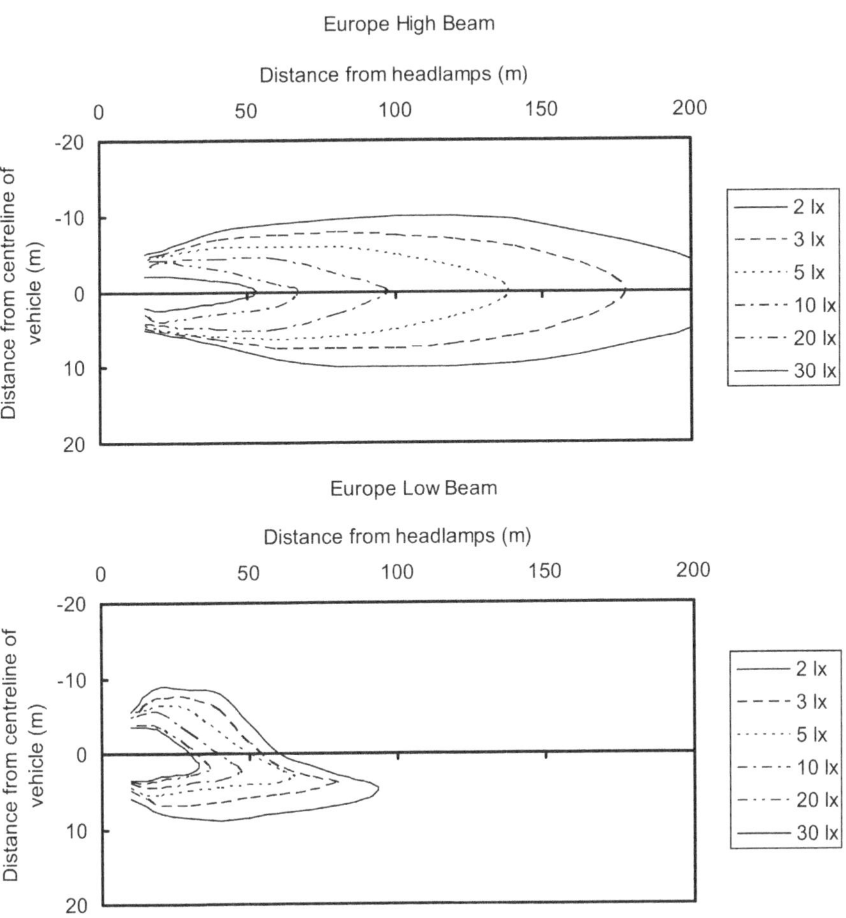

FIGURE 6.4 Contours of median vertical illuminances (lx) produced by pairs of halogen headlamps operating on high beam and low beam for the 20 best-selling passenger vehicles in Europe in the 1999 model year. The vertical illuminances were calculated at road level. Vehicles in most countries of Europe are driven on the right (after Schoettle et al., 2002).

illuminance distributions. Specifically, the European high beam is narrower and provides more light further down the road than does the American high beam. This is consistent with the fact that the luminous intensity values for high beams differ as regards the maximum luminous intensity allowed. In the American regulations, the maximum luminous intensity is 75,000 cd while in the ECE recommendations, the maximum is 140,000 cd. Rumar (2000) examined this difference on the basis of visibility and glare and concluded that the US maximum should be increased to the then ECE level. For low beams, both American and ECE regulations are similar in that they show an emphasis on the near side of the road and a limitation in the vertical illuminance produced towards vehicles in the opposing lane. However, the ECE low beam has a sharper cut-off than the North American low beam, indicating the greater emphasis given to controlling disability glare. Conversely, the American low beam provides more light down the road and more on the edge of the road, indicating a greater emphasis on visibility.

For several years there were moves to resolve this difference in low beam luminous intensity distributions by the production of a harmonized vehicle headlamp specification (SAE, 1995; GTB, 1999). Sivak et al. (2001) carried out an evaluation of the proposed harmonized low-beam luminous intensity distribution in terms of the change in luminous intensities in directions where significant elements in the road environment are to be found, e.g., pedestrians at the edge of the road, road edge markings, road signs, glare towards oncoming drivers, glare towards the mirrors of a vehicle immediately ahead, and glare reflected from wet road surface towards oncoming drivers. As might be expected from a luminous intensity distribution based on a compromise between the American approach, which emphasizes forward visibility, and the ECE approach, which emphasizes control of glare, headlamps meeting the harmonized specification provide less light for road signs and vehicle retro-reflectors, and less light towards mirrors on the vehicle ahead than headlamps meeting the American luminous intensity requirements. Conversely, headlamps meeting the harmonized specification provide more light for pedestrians, road signs, and vehicle retro-reflectors, more glare to oncoming drivers and more light towards mirrors on the vehicle ahead than headlamps meeting the ECE requirements. The harmonized specification also provides more foreground illumination and more glare from a wet road towards oncoming drivers than for headlamps that conform to either the North American or the ECE requirements. Despite these well-meaning attempts at harmonization, there has been little progress in resolving the different priorities in different countries (American Automobile Association, 2019). The fact is the recommended luminous intensity distributions used in different parts of the world are all compromises between the need for visibility and the need to control glare.

6.4 HEADLAMPS IN PRACTICE

The process of determining whether or not a headlamp design meets the relevant regulations involves careful measurements taken in a laboratory under very specific conditions. However, headlamps in a vehicle on the road may produce different luminous intensities in important directions for a number of reasons. Some are transient and inherent in the road layout or the nature of the vehicle. An example of the former is the reduction in illumination of the road ahead and the increase in glare to opposing drivers that occur when breasting a hill. An example of the latter is the reduction in the illumination of the road ahead and the increase in glare to opposing drivers produced by motorcycles when cornering to the right on right-hand drive roads due to the tilting of the machine (Konyukhov et al., 2006). Others are permanent and occur because the vehicle is not level, or the headlamp is incorrectly aimed, or the headlamp is dirty. Yerrel (1971) reported a set of roadside measurements of headlamp luminous intensities in Europe and found a very large range of luminous intensities for the same direction despite a common standard. Alferdinck and Padmos (1988) found similar results from roadside measurements in The Netherlands. They also examined the importance of aiming, dirt, and lamp age on the luminous intensity in a series of laboratory measurements. Figure 6.5 shows the cumulative frequency distributions of luminous intensity in a direction important for forward visibility and in a direction important for glare to an oncoming driver, for 50 cars taken from a parking lot. The luminous intensity measurements of the headlamps, as found, but taken in the laboratory, agreed with measurements taken at the roadside.

From Figure 6.5 it can be seen that the headlamps, as found, tend to produce less forward visibility and more glare than new headlamps. The forward visibility is most improved by correcting the aiming. Cleaning the headlamps and operating them at 12 V also increases the luminous intensity for forward visibility a little and brings it closer to that of new headlamps. For the direction important for glare, correcting the aiming makes things slightly worse, but cleaning the headlamps reduces the luminous intensity causing glare and again brings it close to that of new headlamps. Some of these effects will have been diminished by the introduction of washing and automatic levelling systems on vehicles as well as the annual testing of older vehicles required for continued use on the road. However, the range of luminous intensities evident in Figure 6.5 implies that even when headlamps are correctly aimed, new, and clean, there will be a wide variation in how effective they are. This variation is evident from the measurements that have been made of the distances at which targets can be detected when driving on low beams on an unlit road. Perel et al. (1983) reviewed 19 studies in which observers had been driven along an unlit road at a constant speed in vehicles equipped with standard North American or ECE headlamps. The observers were asked to press a button when they detected small (typically 0.5 m square) or large (man-sized), low reflectance, low-contrast targets placed at the edge of the road. The mean detection distances for the large target ranged from 51 m to 122 m, while for the small target, the mean detection distances ranged from 45 m to 100 m. It is likely that these detection distances are overestimates of reality because the observers were told to look for the targets and were not distracted from that task by having to drive. Roper and Howard (1938) have shown that an unexpected target is seen at about half the distance of an expected target. Little seems to have changed since then. Green (2000) observed that the response time to an unexpected target was about 1.50 s when, for an expected target, the time taken to detect the target and move the foot from the accelerator to the brake was about 0.75 s.

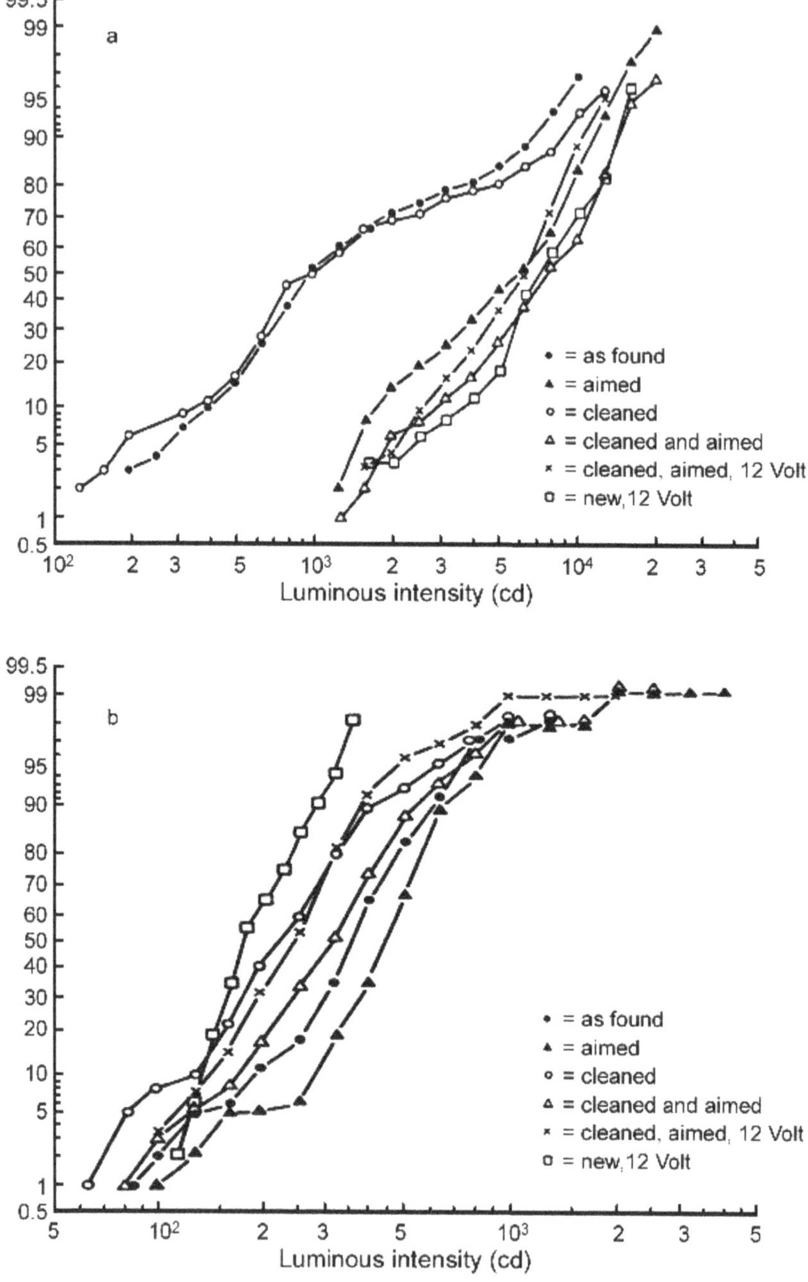

FIGURE 6.5 Cumulative frequency distributions of luminous intensities (cd) in directions (a) important for visibility of the near side of the road and (b) important for glare to oncoming drivers, for headlamps on 50 cars as found; aimed; cleaned; cleaned and aimed; cleaned, aimed, and operated at 12 V; and for new headlamps (after Alferdinck and Padmos, 1988).

How significant such detection distances are can be revealed by comparing detection distance with stopping distance for a particular speed. Olson et al. (1984) have calculated stopping distances for cars and trucks when making an emergency stop from different speeds on a wet road with worn tyres, assuming a driver reaction time of 2.5 s. Two types of emergency stops were considered: One where the driver locked the wheels and hence lost control of the vehicle and one where the driver adjusted the braking so as to avoid locking the wheels (Figure 6.6). If it is assumed that a necessary condition for traffic safety is that the detection distance should be greater than the stopping distance, it is possible to use Figure 6.6 to calculate the maximum speed for safe driving. Using the bottom of the range of detection distances found by Perel et al. (1983), such calculations suggest that the safe speed for driving on low-beam headlamps alone is about 30 mph (48 km.h^{-1}).

Another measure of the effectiveness of low-beam headlamps is given by Olson and Sivak (1983). They also measured the distance at which observers could detect a pedestrian wearing a dark or light top while being driven along an unlit road in a car using low beam headlamps, as well as calculating the stopping distance for a

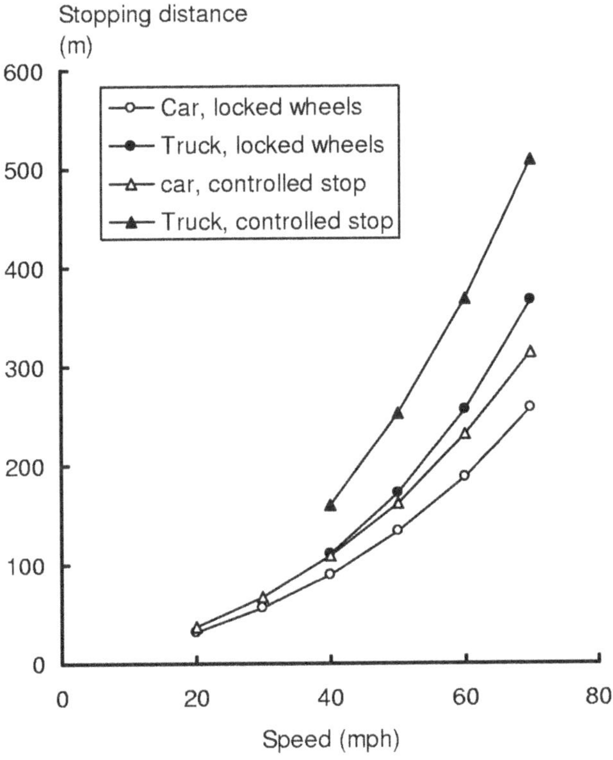

FIGURE 6.6 Stopping distances (m) for cars and trucks with worn tyres making an emergency stop on a wet road surface, with and without locked wheels, plotted against speed (mph) and assuming a driver reaction time of 2.5 s (after Olson et al., 1984).

vehicle moving at 55 mph (89 km.h^{-1}). The percentage of trials in which the detection distance was less than the stopping distance ranged from 3%, which was for young observers with the pedestrian wearing the light top to 83% for old observers with the pedestrian wearing the dark top. Similar estimates of the percentages of people able to stop within the distance at which they can detect low-contrast targets have been made by Yerrel (1976).

More recently, the Insurance Institute for Highway Safety has undertaken a different approach to evaluating the effectiveness of headlamps for passenger cars (IIHS, 2018). Cars that have travelled less than 1000 km were driven, at night, along an unlit road with a straight section and left and right bends of radii 150 and 250 m. The visibility provided by the high beam was measured by the distance over which an illuminance of 5 lx was maintained at the edges of the road and in the middle of the driven lane up to a distance of 10 m from the measurement point for the right edge and lane positions and 15 m for the left edge. For the straight section, the approach to the measurement points started from 250 m while for the curves it started from 120 m. The glare produced by the headlamps on low beam was assessed by two criteria. The first was that the illuminance at the eye of an approaching driver should not exceed 10 lx at separation distances of 5–10 m. The other was the cumulative distance over which the illuminance received at the eyes of an approaching driver exceeded a specified threshold value while approaching from 250 to 10 m for the straight section and 120 m to 10 m for the curves. The threshold illuminance increases as the separation between the two vehicles reduces starting from about 1 lx at 100 m separation rising to 3.5 lx at 10 m separation for the straight and left curves and 7.5 lx for right curves. The distances for each headlamp set were then graded according to a system of demerits, the shorter the visibility distance and the longer the glare distance, the greater the demerit. The IIHS claims that as a result of this evaluation process, since 2016 manufacturers have improved the headlamps fitted to passenger vehicles sold in America and that the grade assigned to headlamps is correlated with nighttime single-car crash statistics (Brumbelow, 2022).

Taken together, these distance comparisons go some way to explain the increase in pedestrian deaths around the change in daylight saving time (see Section 1.5). They also indicate that driving at night on low beams on unlit road is very much an act of faith unless the speed is very limited. Of course, this might not matter so much if drivers used high beams whenever possible, i.e., whenever there was no vehicle approaching and no vehicle immediately ahead. Unfortunately, field measurements have shown that this is not what happens. Sullivan et al. (2004a) observed drivers' use of low and high beams on unlit rural roads. Figure 6.7 shows the percentage of drivers using high beams when there was no approaching vehicle and none immediately ahead plotted against traffic density. As might be expected, the percentage of vehicles using high beams decreases with increased traffic density, but it is not until traffic density falls below about 50 vehicles/hour that high beams are used by more than about 50% of drivers. Similar results have been found by Reagan et al. (2017).

These results imply two basic truths. First is that low beams alone do not provide adequate visibility of the road ahead, except at low speeds. Second is that high beams, which do increase visibility, are not used as frequently as they could be. Why

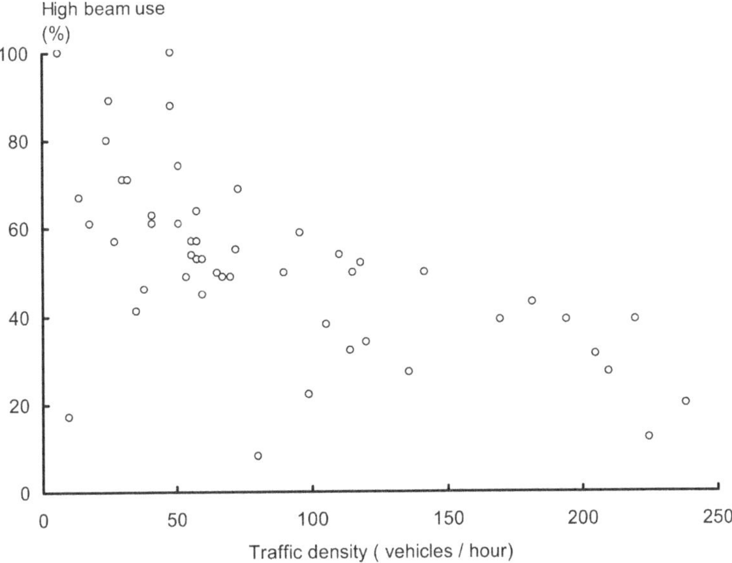

FIGURE 6.7   The percentage of drivers on unlit rural roads using headlamps on high beam when there was no approaching vehicle and none immediately ahead plotted against traffic density (vehicles/hour) (after Sullivan et al., 2004a).

high beams are not used as frequently as they could be has been the subject of some theorizing. One hypothesis is that drivers drive faster than they should because they do not recognize the extent to which their vision has been degraded at low light levels because the degradation is not uniform across all visual capabilities (Leibowitz and Owens, 1977). Specifically, the ability to steer the vehicle using peripheral vision is maintained at low light levels but the ability to recognize small, low-contrast objects is not (Owens and Tyrrell, 1999). Further, the widespread use of high-contrast, retro-reflective signs and markings may give drivers an unrealistic estimate of their ability to detect and recognize detail. The problem with this selective degradation hypothesis is that it is difficult to believe there are many drivers who are not aware that they can see much better in all respects with high beams than low beams. Certainly, the professional drivers of heavy trucks seem to be so aware. An early study of high beam use (Hare and Hemion, 1968) showed that where there were no opposing vehicles or vehicles immediately ahead, only 25% of car drivers used their headlamps on high beam. Under the same conditions, 67% of truck drivers used their headlamps on high beam.

An alternative explanation for the failure to use high beams as frequently as possible is based on a combination of probability and inconvenience. The idea is simply that given the low probability of there being a pedestrian on an unlit rural road at night, the increase in visibility that is achieved by changing from low beam to high beam is not worth the trouble, particularly if there is a high probability of having to revert to low beam shortly after. There has been no study of this probability/

inconvenience hypothesis. One way to do this would be to examine the pattern of use of headlamp beams on unlit rural roads where there was and was not a high probability of there being large unlit obstacles on the road, such as deer.

The selective degradation hypothesis for the under-use of high beams treats drivers as rational beings and excuses their behaviour. The probability/inconvenience hypothesis treats them as emotional humans and condemns the implicit gamble. While the argument between these two hypotheses is interesting, it may soon be moot because the technology to change between low and high beams automatically is already available (Wordenweber et al., 2007). With this technology, called high beam assist, the default state is high beam. Sensors detect the presence of approaching vehicles or vehicles immediately ahead and automatically change to low beam, reverting to high beam as soon as the approaching vehicle has passed or the vehicle ahead has moved on. Adaptive forward lighting systems are also becoming available (see Section 6.7). In these, a combination of camera and radars with artificial intelligence software controls the luminous intensity distribution of the headlamp so that an effective high beam is maintained apart from that part that would be seen by the approaching driver. If such systems are successful in what they seek to do and become common, the simple dichotomy between headlamp high and low beam will become irrelevant.

6.5 HEADLAMPS AND LIGHT SOURCES

The most dramatic change in headlamps since the start of the 21st century has been the steady replacement of the conventional sealed-beam halogen headlamp, first by the xenon discharge (HID) headlamp, identifiable by their blue-white colour appearance, and now by the LED headlamp, also white. Both LED and HID headlamps have a much higher system efficiency than halogen headlamps. However, LED headlamps are more expensive than HID headlamps and both are more expensive than halogen headlamps. A major advantage of LED headlamps is that they have a much longer life than HID headlamps although both have a longer life than halogen headlamps. Finally, HID light sources are mainly used in projector headlamps while LED light sources are used in both reflector and projector headlamps. Questions of system efficiency, life, cost, and design freedom are certainly relevant for the vehicle manufacturer, but the three characteristics that are important for visibility are the amount of light produced, the luminous intensity distribution, and the spectral power distribution. LED and HID headlamps typically produce two to three times more luminous flux than halogen headlamps. The recommended minimum and maximum luminous intensities used in regulations apply regardless of the light source used, a fact that raises the question of how the additional luminous flux should be distributed. The maximum luminous intensities specified in regulations are mainly restricted to parts of the beam that cause glare to opposing drivers or drivers immediately ahead. In other parts of the beam, the regulations specify minimum values but not maxima. Consequently, the additional luminous flux produced by LED and HID headlamps tends to be directed to the parts of the beam where no maximum is specified.

Van Derlofske et al. (2001) report a method for quantifying the benefits of this different light distribution by comparing HID to halogen headlamps. They measured reaction times to the onset of a change in reflectance of targets at various angles off-axis when illuminated by a HID headlamp set and two halogen headlamp sets, all conforming to ECE regulations and used on low beam. Figure 6.8 shows the geometry of the experiment as set out on an unused and unlit asphalt runway.

The targets were placed on an arc of radius 60 m from the headlamps. Each target consisted of a 178 mm square grid of 12.7-mm-diameter flip dots, each dot being a disc painted black on one side and white on the other. By applying current to the target, the dots are flipped over within 20 ms, thereby changing the target from a black square to a grey square, grey because each dot is surrounded by a black frame and at 60 m, the dots and the frame merge and together appear grey with an average reflectance of 0.4. The luminance of the target varies with position because the illuminance on the target also varies with position. Figure 6.9 shows the illuminances on each target produced by the three headlamp sets. Note the similar maximum illuminance for all three light sources but the much wider light distribution of the HID headlamps.

The subjects performed a continuous tracking task, designed to maintain fixation directly ahead, and released a press switch as soon as they detected the change in reflectance of any of the targets. The reaction time to the onset of the target, i.e., the change from black to grey, was measured. Any response longer than one second was taken as a miss although each such miss was included in the data from which mean reaction time was calculated at an assumed reaction time of 1,000 ms. Figures 6.10 and 6.11 show the mean reaction times to the onset of the target and the percentage of missed signals, respectively, plotted against angular deviation from the line of sight, for the three headlamp sets. An examination of Figures 6.10 and 6.11 show there is little difference between the three headlamp sets for less than 7.5 degrees deviation, but beyond this, the HID headlamps give statistically significantly shorter values of mean reaction time and a lower percentage of missed signals than either of the halogen headlamp sets.

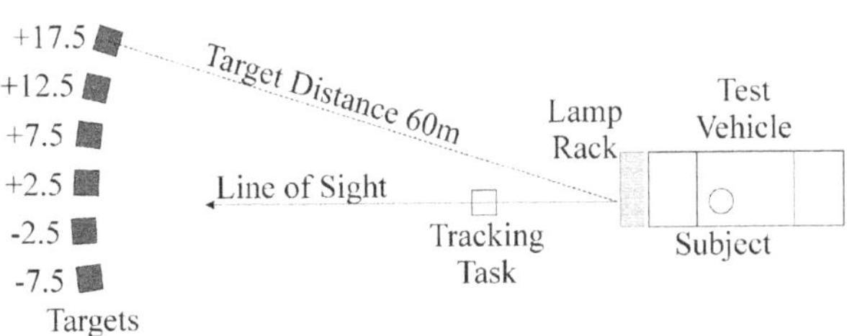

FIGURE 6.8 Geometry of the experiment conducted by Van Derlofske et al. (2001). The subject sat in the test vehicle and did the continuous tracking task. The headlamps were mounted on the lamp rack at the front of the vehicle. The flip-dot targets were positioned on an arc 60 m away from the headlamps spaced at five-degree intervals (Reprinted with permission from SAE Paper 2001-01-0298 © 2001 SAE International)

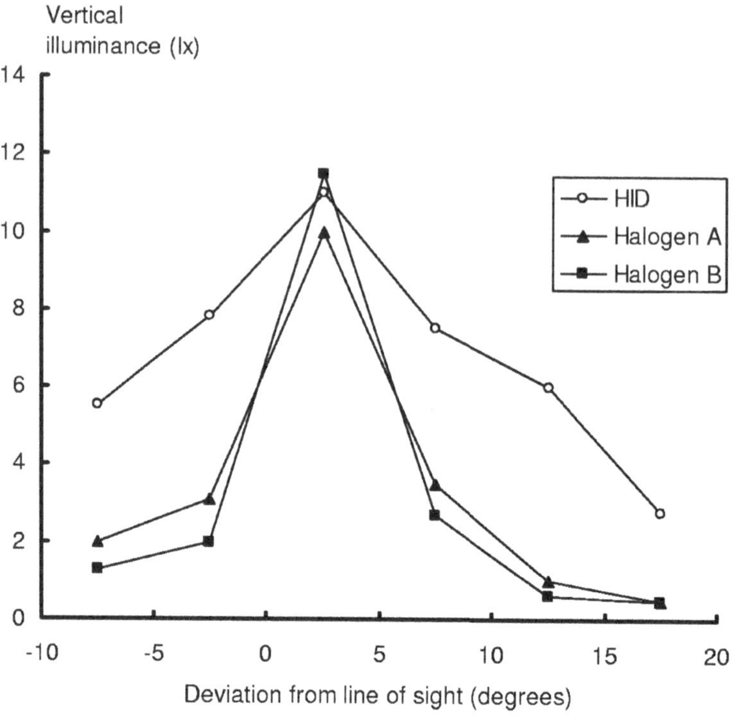

FIGURE 6.9 Illuminance (lx) on the flip-dot target produced by one HID headlamp set and two different halogen headlamp sets, all conforming to ECE regulations and used on low beam, plotted against deviation from the line of sight in degrees (after Van Derlofske et al., 2001).

This difference between headlamp sets is what might be expected from the illuminances they produce on the targets (Figure 6.9). However, the HID and halogen headlamp sets differ in spectral power distribution as well as illuminance, and there is evidence that at low light levels light sources that provide greater stimulation to the rod photoreceptors allow faster reaction times off-axis than light sources that do not (see Section 4.6). This implies that HID headlamps should allow faster reaction times for off-axis detection than halogen headlamps, even when both provide the same photopic illuminance. Van Derlofske and Bullough (2003) have examined this possibility with the same equipment and protocol as that described above but using a filtered HID headlamp set. The headlamp set conformed to North American regulations for low beam. The filtering changed the spectral power distribution of the light but not the luminous intensity distribution or the illuminances on the targets. For each position, two target average reflectances, 0.4 and 0.2, were created by having the target viewed with and without a neutral density filter in front of it. Four different spectral power distributions were examined, the relative efficiency of each at stimulating the rod and cone photoreceptors being quantified by the *S/P* ratio (see Section 2.4.3.3). Figures 6.12 and 6.13 show the mean reaction times and the percentage of

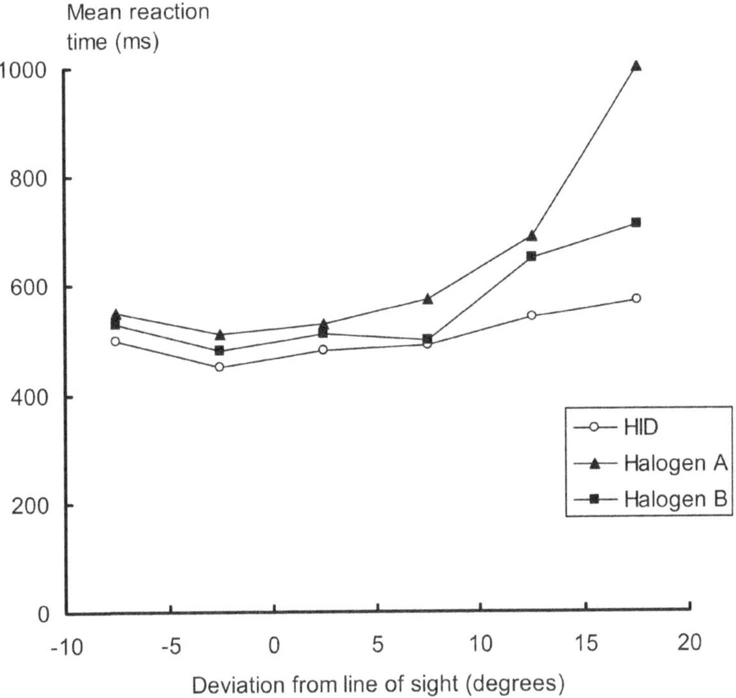

FIGURE 6.10   Mean reaction times (ms) to the onset of the targets for one HID headlamp set and two different halogen headlamp sets, all conforming to ECE regulations and used on low beam, plotted against deviation from the line of sight in degrees (after Van Derlofske et al., 2001).

missed signals, respectively, for the 0.2 average reflectance target, plotted against deviation from the line of sight, for each spectral power distribution.

A common pattern is evident in Figures 6.12 and 6.13. Mean reaction times and the percentages of missed signals increase with increased deviation from the line of sight, for all spectral power distributions. These increases are certainly due to the decrease in illuminance on the target as deviation from the line of sight increases (Figure 6.9). However, at the extreme deviations, where the illuminances on the targets are least, an effect of spectral power distribution is evident. Specifically, the mean reaction time and the percentage of missed signals both decrease as the *S/P* ratio of the light source increases. This is what would be expected from what is known about how the spectral sensitivity of off-axis vision changes in the mesopic state (see Section 3.3.4) and indicates that HID headlamps have an additional advantage for off-axis visibility over and above the greater illuminances produced. Although real, this advantage is small relative to the impact of the greater illuminances. This became apparent when the data for the 0.4 average reflectance targets were examined. For these data, the increases in mean reaction times and percentages of missed signals with increasing deviation from the line of sight were present and of

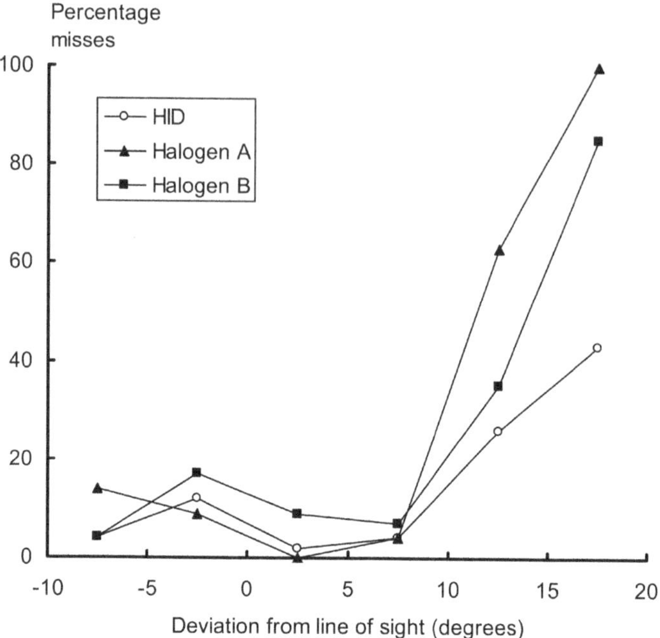

FIGURE 6.11 Percentage of missed signals for one HID headlamp set and two different halogen headlamp sets, all conforming to ECE regulations and used on low beam, plotted against deviation from the line of sight in degrees (after Van Derlofske et al., 2001)

similar magnitude to those obtained for the 0.2 average reflectance targets, but there was no statistically significant effect of S/P ratio.

All the above tests have compared HID headlamps with halogen headlamps because at the time these were the two available light source technologies, but now there are LED headlamps. Flannagan (2019) reports the luminous intensity distributions of 48 LED headlamps and 62 halogen headlamps when on low beam, both types intended for the US market. Examination of the data shows that LED headlamps will deliver similar maximum illuminances to the halogen headlamps at the centre of the beam but higher illuminances off-axis. Further, the spectral power distribution of LED headlamps has a lot of power at the short wavelength end of the visible spectrum (see Figure 2.5) so they have an S/P ratio of about 2.0 which is similar to that of HID headlamps. Consequently, LED headlamps should produce similar effects to those shown for S/P ratio of 2.04 in Figures 6.12 and 6.13.

Using the data on reaction times presented above, Bullough and Skinner (2009) developed a simple equation to predict reaction time for the detection of off-axis targets. The equation was

$$RT = 8.98\Theta - 115\log E - 332\log S + 409\log D - 49.3\log p - 23.4R + 291$$

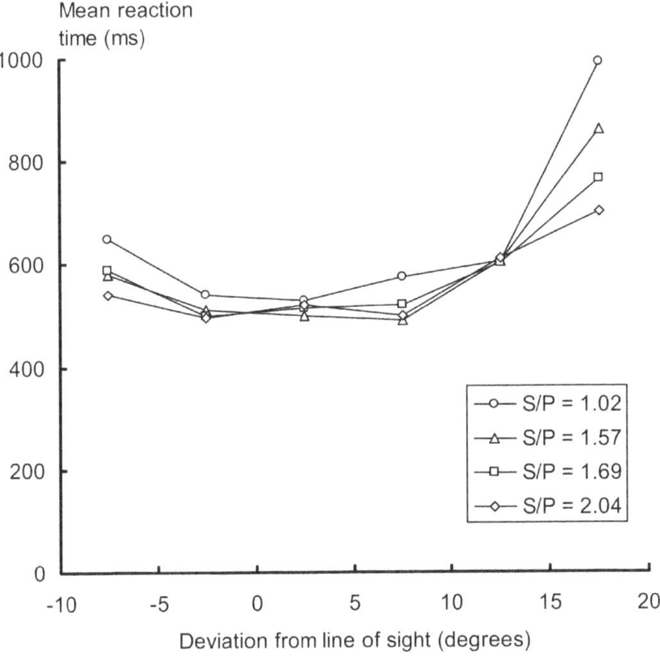

FIGURE 6.12 Mean reaction times (ms) to the onset of the low reflectance (0.2) target for four different light spectra specified by the scotopic/photopic (S/P) ratio, plotted against deviation from the line of sight in degrees (after Van Derlofske and Bullough, 2003).

where RT is the predicted reaction time in milliseconds, Θ is the target's angular deviation from the line of sight in degrees, E is the vertical illuminance on the target in lux, S is the target size in centimetres, D is the distance to the target in metres, p is the target reflectance, and R is the *S/P* ratio of the light source.

To test this model, a field experiment was conducted in which the subject drove a car at not more than 30 mph (48 km.h^{-1}) along an unlit, slightly curved road at night. The car was fitted with either halogen or HID headlamps. Two low reflectance (p = 0.05) plywood cutouts of a pedestrian were placed randomly at six possible locations, three on each side of the road. They were arranged so that sometimes they faced into the road and sometimes they faced away from the road. The driver was instructed to stop the car as soon as a pedestrian facing the road was detected. As would be expected, the HID headlamp produced higher illuminances on the pedestrians than the halogen headlamps, particularly on the left side of the road. The stopping distances from the pedestrians, called the safety margin, were about 20 m for the HID headlamps and 10 m for the halogen headlamps. To compare the results with the reaction time model, the identification distance had to be estimated. This was done by adding the difference between the locations when the brake pedal was activated and the vehicle came to a stop to the safety margin. This identification distance was the distance from the pedestrian at which the driver determined that the pedestrian

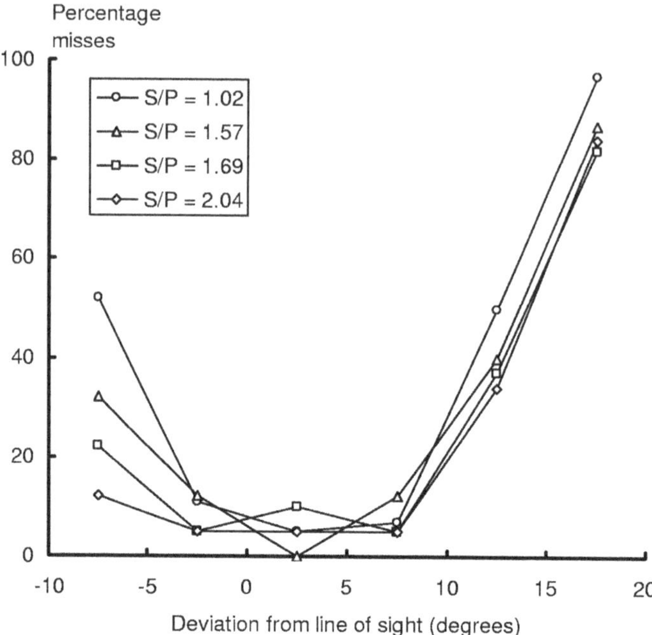

FIGURE 6.13 Percentage of missed signals for the low reflectance (0.2) target for four different light spectra specified by the scotopic/photopic (*S/P*) ratio, plotted against deviation from the line of sight in degrees (after Van Derlofske and Bullough, 2003).

was present and facing the road. Again, the identification distance was nearly 10 m longer for the HID headlamps relative to the halogen headlamps. It was concluded that the reaction time model could be applied in a constructive and practical way to real-world situations such as stopping a vehicle when a pedestrian about to enter the road is detected.

There can be little doubt that LED and HID headlamps will give better off-axis detection than halogen headlamps but this is primarily because of the higher light output and wider light distribution rather than their spectral power distributions. The question then becomes: Do they also produce more glare?

6.6 GLARE FROM HEADLAMPS

6.6.1 THE INCIDENCE OF GLARE

It is not difficult to find complaints about glare from headlamps. The media, both electronic and print, are full of them. Equally, it is not difficult to find advertisements for brighter headlamps to retrofit to your vehicle which can only exacerbate the situation. But why has this tide of complaints occurred? There is no doubt that LED and HID headlamps emit more light than the previously dominant halogen headlamps, but there are many other factors involved in the level of glare experienced by drivers facing an approaching vehicle at night. For a start, there is the irresponsible approaching driver who persists in using high-beam rather than low-beam headlamps. Then, there is the possibility that the headlamps are misaimed, unadjusted for vehicle load, or mounted higher above the road surface on the approaching vehicle. There are also road situations that differ from the level state assumed in the regulations that headlamps have to meet before they can be fitted to a vehicle. There are basically minimum luminous intensities for most positions. It is only positions where it is assumed an approaching driver will be present that have both minimum and maximum luminous intensities. This gives headlamps that produce more light output freedom to spread the light more widely than before. Unfortunately, there are a number of common situations where such wide light distributions are inappropriate, such as meeting a vehicle rounding a bend where the higher illuminance part of the low beam illuminates the opposing lane. More generally, given a wider and higher luminous intensity distribution, even on a straight, flat road, a driver facing an approaching set of such headlamps will be exposed to glare for longer. Then there are the characteristics of drivers themselves. In both the UK and USA and in many other countries, people are living longer. As they age, their visual capabilities degrade so that their contrast sensitivity and visual acuity deteriorate and more light is scattered in the eye (see Section 13.4). These changes make older people more sensitive to glare. All this suggests that the regulations covering headlamps need to be revised to keep up with developments in lighting technology and so rebalance the relationship between the visibility and glare. Until that happens, or technology comes to the rescue, it would be as well to have some understanding of what glare is and what its consequences can be.

6.6.2 Forms of Glare

Glare occurs because while the human visual system can operate over about twelve log units of luminance in total, it can only operate over about three log units simultaneously. Any luminance more than about two log units above the average luminance of the scene will be considered glaringly bright. Vos (1999) has classified glare into eight different forms. Of these eight, only four are experienced while driving: Saturation glare, adaptation glare, disability glare, and discomfort glare.

Saturation glare occurs when a large part of the visual field is bright for a long time. The behavioural response is to shield the eyes by wearing low transmittance glasses. Such behaviour is common when driving in very sunny climates during the day.

Adaptation glare occurs when the visual system is exposed to a sudden, large change in luminance of the whole visual field, e.g., on exiting a long road tunnel during a sunny day. The perception of glare is due to the visual system being misadapted. On exiting the tunnel the visual system is adapted to the low light level of the tunnel but is exposed to the brightness of sunlight. Adaptation glare is temporary in that the processes of visual adaptation will soon adjust the visual sensitivity to match the new conditions (see Section 3.3.3). It is to be hoped that the design of tunnel lighting takes the possibility of such adaptation glare into account (see Section 11.2).

Disability glare, as its name implies, disables the visual system to some extent. This disabling is caused by light scattered in the eye (Vos, 1984). The scattered light forms a luminous veil over the retinal image of adjacent parts of the scene, thereby reducing the luminance contrasts of the image on those parts on the retina. Disability glare is experienced most frequently on the roads at night when facing the headlamps of an approaching vehicle.

Discomfort glare is said to occur when people complain about visual discomfort in the presence of bright light sources. There is no known cause for discomfort glare, although suggestions have been made ranging from fluctuations in pupil size (Fry and King, 1975) through distraction (Lynes, 1977) and muscle tension around the eye (Murray et al., 2002) to hyperexcitability in the visual cortex (Bargary et al., 2015). This distinction between disability and discomfort glare does not mean that disability glare does not cause visual discomfort nor that discomfort glare does not affect task performance. Headlamps at night can certainly be both visually disabling and visually uncomfortable. In essence, these two forms of glare, disability glare and discomfort glare, are two different outcomes of the same stimulus pattern, namely a wide variation of luminance across the visual field.

6.6.3 THE QUANTIFICATION OF GLARE

Disability glare can be quantified by comparing the visibility of an object seen in the presence of the glare source with the visibility of the same object seen through a uniform luminous veil. When the visibilities are the same, the luminance of the veil is a measure of the amount of disability glare produced by the glare source, and it is called the equivalent veiling luminance. The CIE has developed a disability glare formula suitable for use with vehicle headlamps. This is more elaborate than that used for road lighting (see Section 4.3) because headlamps are usually seen much closer to the line of sight. The CIE formula applies at all angles from the line of sight in the range 0.1 to 30 degrees and to either young or old people (CIE, 2002). This equation takes the form:

$$L_v = \Sigma(10E_n/\Theta_n^3 + (1 + (A/62.5)^4)\ 5\ E_n/\Theta_n^2)$$

where L_v is the equivalent veiling luminance produced by the n'th glare source (cd. m^{-2}), E_n is the illuminance (lx) at the observer's eyes from the n'th glare source, θ_n is the angular deviation of the n'th glare source from the line of sight (degrees), and A is the age of the observer (years). The effect of the equivalent veiling luminance on the luminance contrast of an object can be estimated by adding it to the luminance of both the object and the immediate background.

The only photometric quantity relevant to equivalent veiling luminance is the illuminance from the glare source received at the eye. There is little evidence for other aspects of exposure, such as the illuminated area of the headlamp and the light spectra influencing disability glare (Bullough, 2022). Van Derlofske et al. (2004) examined the impact of different illuminances at the eye on the ability to detect off-axis targets using the same equipment and protocol as that described earlier (see

Figure 6.8) but using another HID headlamp set positioned 50 m ahead and 5 degrees to the left of the driver's line of sight. The HID headlight set was tilted slightly to produce three different illuminances at the subject's eyes, 0.2, 1.0, and 5.0 lx. Each flip dot target was presented with and without a neutral density filter placed in front, the result being that the average reflectance of the target was, when presented, either 0.4 or 0.2. Again, the subjects performed a continuous tracking task to control fixation and released a button when they detected a change in one of the targets. Any change that was not detected within one second was counted as a missed target and included in the data used to calculate mean reaction time as a reaction time of 1,000 ms. Figures 6.14 and 6.15 show the mean reaction times and percentage of missed signals plotted against the target position relative to the line of sight, for the three illuminances at the eye and the two target average reflectances. Also shown are the predicted percentages of missed targets when no glare source is present, based on the model of Bullough (2002b).

The first point to note about Figures 6.14 and 6.15 is that the mean reaction times for targets of both reflectances positioned at –2.5 degrees and 17.5 degrees are concentrated at 1000 ms. This is because for these two positions, virtually all the targets were missed. The targets at –2.5 degrees were closest to and those at 17.5 degrees were furthest from the glare source. The reason for the missed targets at –2.5 degrees

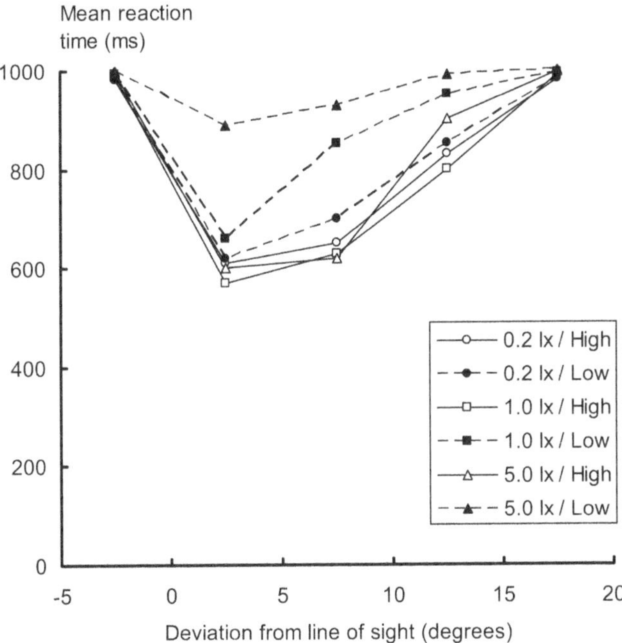

FIGURE 6.14 Mean reaction times (ms) to the onset of the high (0.4) and low (0.2) reflectance targets for three different levels of disability glare specified by the illuminance (lx) received at the driver's eyes, plotted against deviation from the line of sight in degrees (after Van Derlofske et al., 2004).

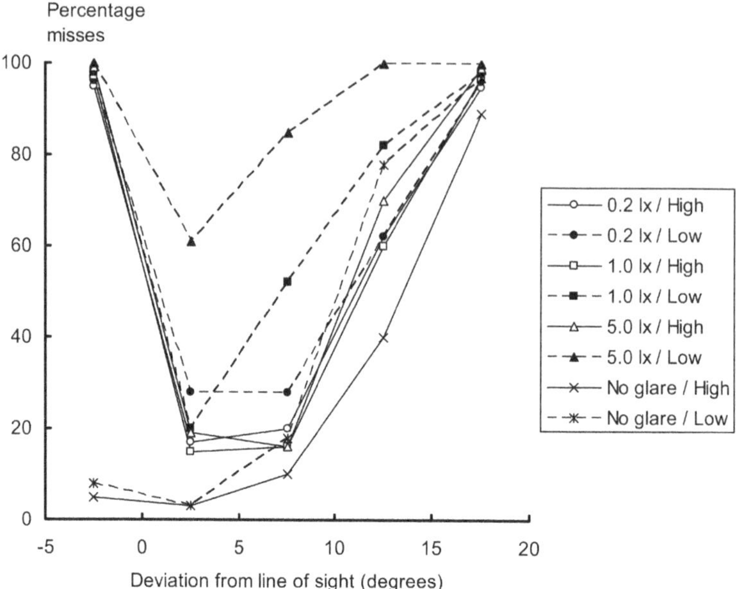

FIGURE 6.15 Percentage of missed high (0.4) and low (0.2) reflectance targets for three different levels of disability glare specified by the illuminance (lx) received at the driver's eyes, plotted against deviation from the line of sight in degrees. Also shown are the percentages of missed targets in the absence of glare predicted by the model of Bullough (2002b) (after Van Derlofske et al., 2004).

is the reduction in luminance contrast caused by the disability glare produced by the glare source, an observation supported by the low level of misses predicted for the –2.5 degrees position in the absence of opposing headlamps. Even an illuminance at the eyes as low as 0.2 lx ensures the target at –2.5 degrees will be missed. This is bad news for any pedestrian caught behind two opposing vehicles when attempting to cross the road. The reason for the missed targets at 17.5 degrees is not disability glare but rather the failure of the observer's headlamps to illuminate the target. For the other positions, 2.5, 7.5, and 12.5 degrees from the line of sight, it is clear that reaction times increase and the percentage of misses increase with increasing deviation from the line of sight and that these increases are much greater for the low average reflectance than for the high average reflectance targets. The difference between the low and high average reflectance targets is to be expected because the effect of a given equivalent veiling luminance on visibility will depend on the luminance contrast of the target. The low average reflectance target will have a lower luminance contrast with its immediate background in the absence of glare so the addition of the veiling luminance will take the low average reflectance target closer to threshold than it will the high average reflectance target. The effect of the illuminance at the eyes is only evident at 2.5, 7.5, and 12.5 degrees from the line of sight for the low reflectance ($p = 0.2$) target. For this target, a glare illuminance of 0.2 lx has hardly any effect on mean reaction time and the percentage of missed targets but glare illuminances of 1.0 lx and 5.0 lx both cause increases in mean reaction times and percentage of targets missed, the increases for 5.0 lx being much greater than for 1.0 lx.

As for discomfort glare, Schmidt-Clausen and Bindels (1974) have produced an equation relating the illuminance at the eye to the level of discomfort produced by headlamps, expressed on the de Boer scale. The equation is

$$W = 5.0 - 2\log(E/0.003(1+\sqrt{(L/0.04)}) \times \phi^{0.46})$$

where W is the discomfort glare rating on the de Boer scale, E is the illuminance at the observer's eyes (lx), L is the adaptation luminance (cd.m^{-2}), and ϕ is the angle between the line of sight and glare source (min. arc). The de Boer Scale is a nine-point glare scale with five anchor points labelled 1 = unbearable, 3 = disturbing, 5 = just admissible, 7 = acceptable, and 9 = unnoticeable. Note that on this scale, lower values are more uncomfortable. Conditions producing ratings of 4 or less are usually considered uncomfortable.

Unfortunately, people have a tendency to look directly at a glare source, even if only temporarily but for longer periods if it contains information, as is the case for some roadside video displays (Vos, 2003). For this condition, Bullough (2022) has developed a simple equation to predict the level of discomfort glare experienced. It is

$$DG = \log(E_L + E_S) + 0.6\log(E_L.E_S^{-1}) - 0.5\log(E_A)$$

where DG is the level of discomfort glare, E_L is the illuminance from the glare source (lx), E_S is the illuminance from the immediate surround of the glare source (lx) a suggested definition of the surround being the area within 7 degrees of the glare source, and E_A is the illuminance from the ambient environment (lx).

For a small glare source, such as a headlamp, the level of a discomfort glare (DG) can be converted to a rating on the de Boer scale (DB) by the equation

$$DB = 6.6 - 6.4\log(DG)$$

The form of the de Boer scale is explained above.

Figure 6.16 shows the mean ratings of discomfort glare on the de Boer scale plotted against the illuminance at the eye from the HID headlamps in the experiment of Van Derlofske et al. (2004) described above. Also shown are the ratings predicted by the Schmidt-Clausen and Bindels discomfort glare equation for the same experimental situation.

The predictions of the discomfort glare equation show a broad agreement with the findings for the low reflectance target of Van Derlofske et al. (2004). More interesting is the finding that there is a clear difference between the low and high reflectance targets. This implies that the perception of discomfort glare depends not only on the stimulus to the visual system produced by the glare source but also on what the observer is trying to do. This should not be too surprising given that it is well known that the same photometric stimuli can be considered as comfortable or

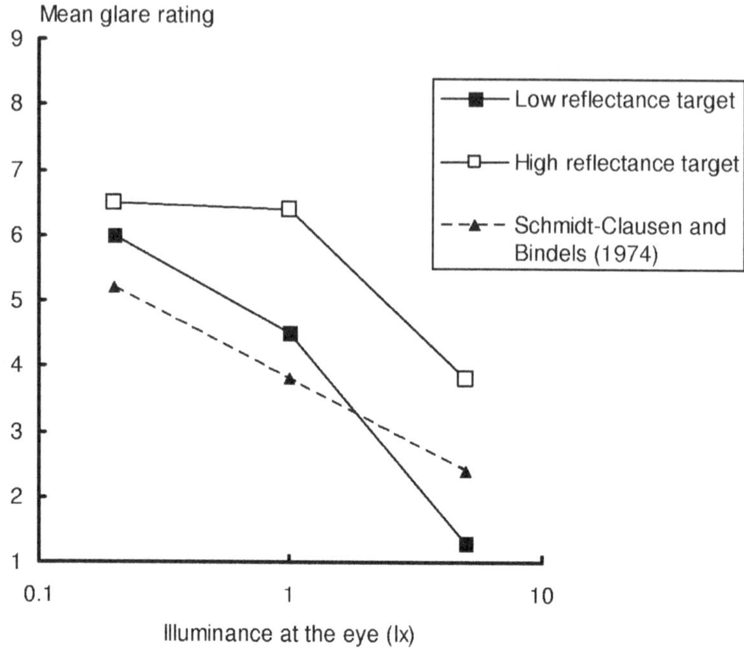

FIGURE 6.16 Mean discomfort glare (de Boer) ratings collected when the observer was attempting to detect the onset of low (0.2) and high (0.4) reflectance off-axis targets, plotted against the illuminance (lx) received at the driver's eyes. Also shown are the predicted glare ratings derived from the discomfort glare equation of Schmidt-Clausen and Bindels (1974) (after Van Derlofske et al., 2004).

uncomfortable depending on the task being performed (Sivak et al., 1991), the context in which the stimuli are presented (Boyce, 2014) and the information contained in the scene (Tuaycharoen and Tregenza, 2005).

The discomfort glare equation produced by Schmidt-Clausen and Bindels (1974) involves three components – the illuminance at the eye, the adaptation luminance, and the angle between the glare source and the line of sight. However, there is evidence that other factors have an effect. While the perception of discomfort from headlamps is dominated by the illuminance at the eye (Sivak et al., 1990; Alferdinck, 1996, Van Derlofske et al., 2004), light spectrum (Flannagan et al., 1989; Van Derlofske et al., 2004; Van Delofske and Bullough, 2006), and headlamp size (Sivak et al., 1990; Alferdinck and Varkevissar, 1991; Van Derlofske et al., 2004) both have small effects, although the evidence for the latter is mixed. What this means is that for the same illuminance at the eye, light spectra with more energy at the short wavelength end of the visible spectrum produced by smaller size headlamps will tend to cause more discomfort. There is also some evidence that the magnitude and frequency of changes in the illuminance at the eye as the vehicle moves along the road impact the muscle tension around the eye, conditions that have been associated with driver discomfort and fatigue (Murray et al., 2002).

6.6.4 Performance in the Presence of Glare

The most obvious and best-understood consequence of exposure to disability glare is a reduction in the luminance contrasts of the scene around the glare source. This leads to a reduction in the ability to detect targets that are close to threshold in the absence of glare (see Figure 6.15). However, the measurements on which this conclusion is based were taken in a static situation but glare is most usually experienced in a dynamic situation as two vehicles approach and pass each other. Mortimer and Becker (1973), using both computer simulation and field measurements, have shown that the distances at which targets of reflectances 0.54 and 0.12 become visible diminish as opposing cars close and then start to increase rapidly (Figure 6.17). The separation at which the visibility distance is a minimum depends on the relative luminous intensity distribution of the headlamps, the relative positions of the two vehicles, the obstacles to be seen, and the physical characteristics of the obstacles.

Helmers and Rumar (1975) measured visibility distances for flat, dark-grey 1.0 m by 0.4 m rectangles with a reflectance of 0.045. Observers were driven towards a parked car with its headlamps on and asked to indicate when they saw the obstacles. It was found that for the dark-grey obstacle, a headlamp system with the maximum high beam luminous intensity gives a visibility distance of about 220 m when no opposing vehicle is present. This is the same as the stopping distance for a vehicle

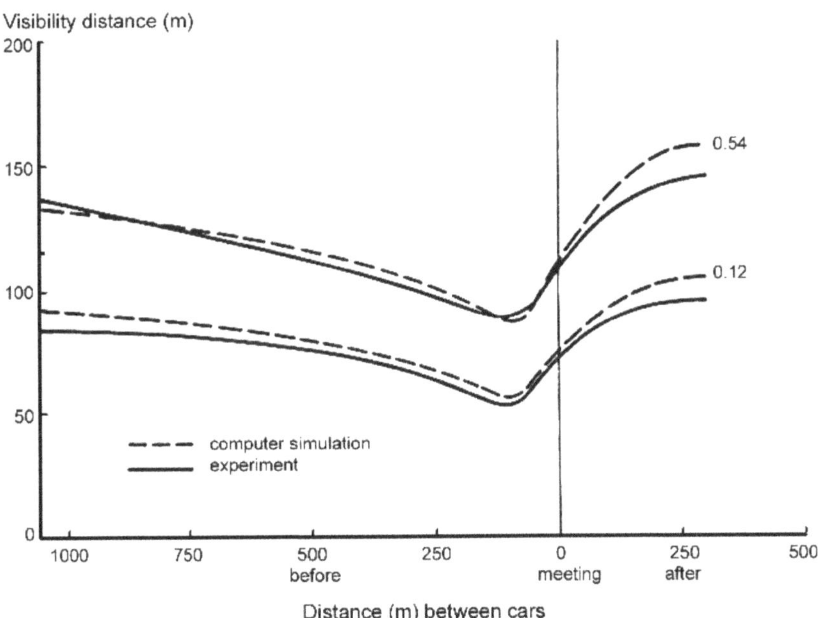

FIGURE 6.17 Visibility distance (m) for targets of reflectance 0.54 and 0.12, plotted against the distance (m) between two cars approaching each other, with headlights of equal luminous intensity (after Mortimer and Becker, 1973).

moving at 110 km.h^{-1} (68 mph) on wet roads (AASHTO, 2011). However, when two opposing vehicles have equal luminous intensity headlamps, the visibility distance is reduced to about 60–80 m, which is much less than the stopping distance and when the opposing vehicle had a luminous intensity about three times more than the observer's vehicle, the visibility distance was reduced to about 40–60 m. Again, it is clear that driving at high speeds against opposing traffic at night approaches an act of faith.

Kimlin et al. (2017) had older drivers drive around a closed circuit track with various items to be detected on and near the road. A glare source was mounted on the bonnet of the car so that it was ten degrees off the driver's normal line of sight and produced 13 lx at the driver's eye when viewed directly which would have been both disabling and uncomfortable. The glare source was illuminated intermittently to simulate the appearance of an approaching vehicle. Driver performance was statistically significantly degraded when glare was present, particularly for detecting pedestrians (48% detected without glare, 10% with glare), although this dramatic reduction was probably enhanced by the pedestrians appearing on the same side as the glare source. There was also a small decrease in the percentage detection and avoidance of low-contrast hazards (82% without glare, 78% with glare).

6.6.5 Recovery from Glare

The discomfort experienced when exposed to headlamps is replaced by a feeling of relief almost immediately after the other vehicle passes. Likewise, the light scattered in the eye disappears with the glare source, but that does not mean that vision is immediately restored to the state that existed before exposure to glare. The additional light that has reached the retina of the driver from the approaching headlamps will have had an effect on the state of adaptation of the photoreceptors so immediately after the other vehicle passes the driver's vision will be misadapted. The process of adjusting adaptation is called recovery from glare.

Van Derlofske et al. (2005) examined what factors determined the time taken to recover from glare. The subject was exposed to four different glare stimuli (Figure 6.18) differing in maximum illuminance and light dose; this latter being the product of illuminance and time of exposure.

Specifically, illuminance profiles 1 and 2 have different maximum illuminances but the same light dose. Illuminance profiles 3 and 4 also have different maximum illuminances but the same light dose although the light dose was twice that of illuminance profiles 1 and 2. Immediately after exposure, the subject was presented with a square target the contrast of which was a fixed ratio of the individual's threshold contrast, i.e., at a fixed visibility level. The subject's task was to indicate when the target could first be detected. Figure 6.19 shows the mean detection times for different target contrast ratios and for the different glare exposure profiles. From Figure 6.19, it is evident that detection times are shorter for the higher contrast target ratios and that the detection time is determined by the light dose and not the maximum illuminance. It is interesting to note that in the same experiment, it was shown that

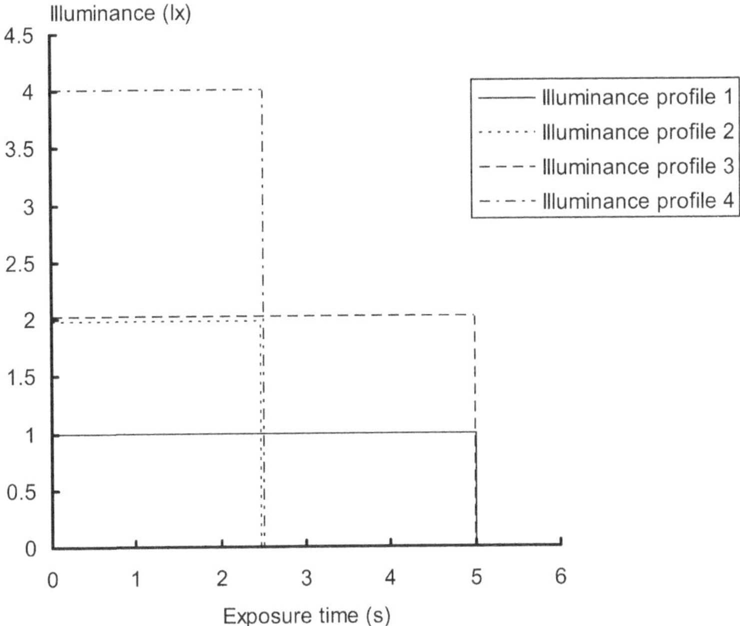

FIGURE 6.18 Profiles of the four glare stimuli used by Van Derlofske et al. (2005) showing the illuminance at the eye (lx) and the duration of exposure (s). Illuminance profiles 1 and 2 have different maximum illuminances but the same light dose. Illuminance profiles 3 and 4 also have different maximum illuminances but the same light dose although the light dose was twice that of illuminance profiles 1 and 2.

ratings of discomfort on the de Boer scale were more closely related to the maximum illuminance at the eye than the light dose.

6.6.6 Behaviour When Exposed to Glare

Different illuminances received at the eye can be associated with different driving behaviours. The range of illuminance received at the eye during normal driving is from 0 to 10 lx (Alferdinck and Varkevisser, 1991). Illuminances of 3 lx and more are likely to be considered very uncomfortable (Bullough et al., 2002). Illuminances of the order of 1 to 3 lx are sufficient to cause drivers to request dimming from the approaching vehicle (Rumar, 2000). Given that the approaching driver does not respond to a request for dimming, how does the requesting driver respond? Theeuwes and Alferdinck (1996) had people drive over urban, residential, and rural roads at night, with a glare source simulating the headlamps of an approaching vehicle mounted on the bonnet of the car. They found that people drove more slowly when the glare source was on, particularly on dark winding roads where lane-keeping was a problem. Older drivers showed the largest speed reduction. The presence of glare also caused the drivers to miss many roadside targets, particularly older drivers.

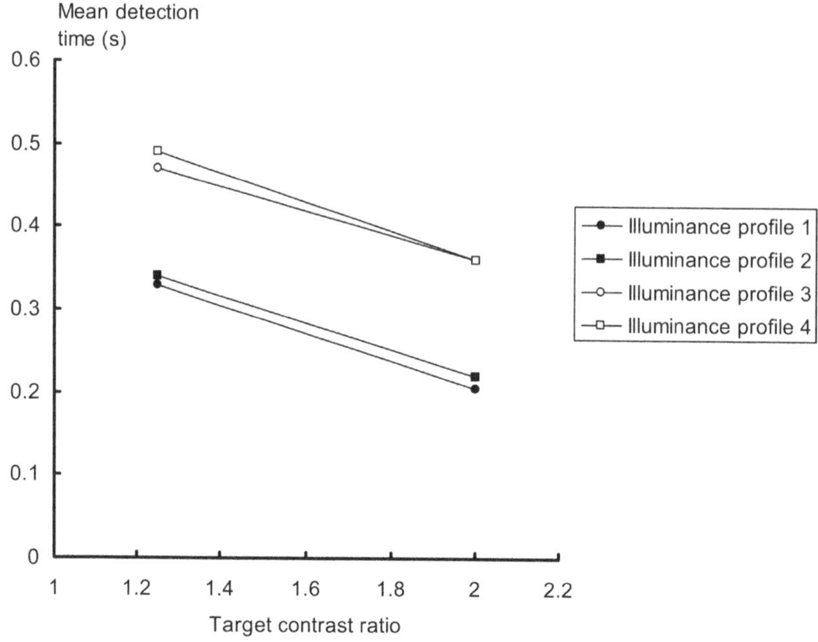

FIGURE 6.19 Mean detection time (s) for targets after exposure to the four glare stimuli shown in Figure 6.18 plotted against target contrast ratio. Target contrast ratio is the ratio of the actual contrast to the individual's threshold contrast without glare (after Van Derlofske et al., 2005).

6.6.7 LED, HID, AND HALOGEN HEADLAMPS

LED and HID headlamps differ from conventional halogen headlamps in two respects relevant to glare; the luminous intensity distribution from the headlamp and hence the illuminance at the eye and the spectral power distribution of the light emitted.

LED and HID headlamps typically have a brighter and wider luminous intensity distribution than halogen headlamps (see Figure 6.9). These differences in the amount and distribution of light from LED and HID headlamps, together with the variability introduced by aiming, dirt, and the different geometries that can occur between two approaching vehicles, as well as the existence of an after-market that allows LED headlamps to be fitted to vehicles without self-levelling systems, are probably enough to explain the widespread anecdotal complaints of glare from drivers meeting vehicles equipped with LED headlamps. The differences in the amount and distribution of light from LED, HID, and halogen headlamps also imply that a driver meeting a car equipped with LED or HID headlamps is likely to be exposed to higher illuminance for longer than if the vehicle was using halogen headlamps. Consequently, the time for recovery from glare should often be longer after meeting a vehicle fitted with LED or HID headlamps.

These differences between LED, HID, and halogen headlamps are all to do with differences in the illuminances produced at the eye but there is another difference between these light sources. The spectral power distributions of LED and HID headlamps have much more power at the short wavelength end of the visible spectrum than the halogen headlamp. This alone will tend to lead to greater discomfort from the LED or HID headlamp for the same illuminance at the eye (Bullough et al., 2003; Sivak et al., 2005b; Van Derlofske and Bullough, 2006). Figure 6.20 shows the mean ratings of discomfort on the de Boer scale for halogen and HID headlamps positioned at five and ten degrees from the line of sight and plotted against the illuminance at the eye (Bullough et al., 2002). As would be expected, both the illuminance at the eye and the deviation from the line of sight show statistically significant effects on the magnitude of discomfort glare but so does light spectrum.

To evaluate the effect of any specific headlamp light spectrum on discomfort glare, Dee (2003) has proposed a spectral sensitivity curve as follows:

$$V_{dg}(\lambda) = V_{10}(\lambda) + (0.19\ SWC(\lambda))$$

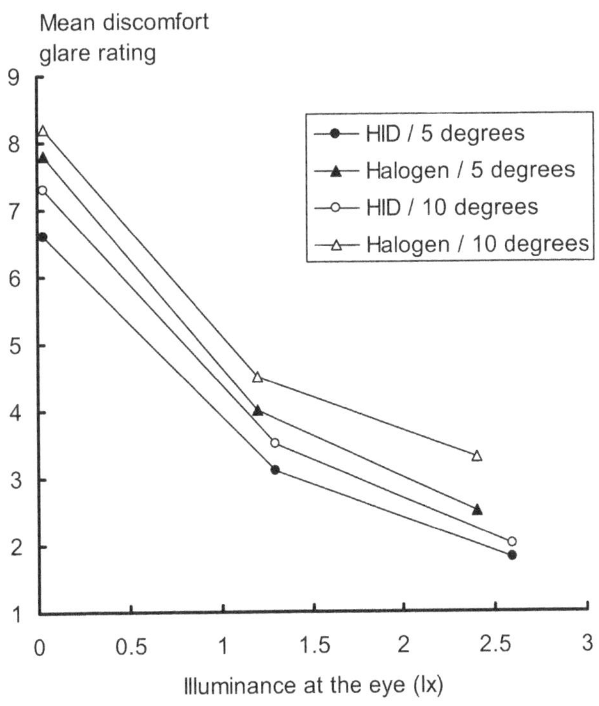

FIGURE 6.20 Mean discomfort glare (de Boer) ratings for exposure to halogen and HID headlamps at five and ten degrees from the line of sight plotted against the illuminances (lx) at the driver's eyes (after Bullough et al., 2002).

where $V_{dg}(\lambda)$ is the discomfort glare spectral sensitivity, $V_{10}(\lambda)$ is the photopic spectral sensitivity for a ten-degree field, and SWC(λ)) is the short wavelength cone spectral sensitivity. This discomfort glare spectral sensitivity has been shown to rectify discomfort glare ratings for conditions simulating exposure to headlamps from both white light and monochromatic glare sources (Watkinson, 2005).

After the above discussion, there can be little doubt that glare, particularly disability glare, is a major constraint on the visibility available to drivers at night. The detection distances achievable in the presence of opposing vehicles are insufficient unless speeds are dramatically reduced, and everyday experience shows that this rarely happens. There is a clear need to find a better way to deal with glare than the two-beam approach used at the moment.

6.7   INNOVATIONS

So far, this chapter has concentrated on the headlamps that are fitted to the vast majority of vehicles on the roads today, headlamps that use either LED, halogen or, decreasingly, HID light sources, that have two possible states, high and low beam, each producing luminous intensity distributions conforming to either the ECE or American regulations. However, the last two decades have seen the introduction of a number of innovative ideas in headlamp design, covering light sources, light spectrum, and luminous intensity distributions and the control thereof. Some of these have achieved lift-off and are gradually being introduced on new vehicles. Others have failed to achieve lift-off and have been dropped into the bin of history. Yet others are still being evaluated.

6.7.1   Light Sources

One light source from which much is expected is the solid-state laser diode. When used in a headlamp, a number of such lasers emit blue light towards a set of micro-mirrors which focus the beam onto a lens containing a phosphor. The result is white light which can then be directed into a powerful high beam using a reflector, projector, or hybrid optical system. Headlamps using this technology are already available on very upmarket vehicles but only for a high beam designed to deliver visibility at much greater distances than usual. The advantages of the laser diode are its small size and high system efficiency relative to both LED and HID headlamps. The disadvantages are its high cost and its use being restricted to high beam applications. It may well be that with further technological development of the associated control systems, the laser headlamp will be the next step after LED headlamps, particularly for electric vehicles where reducing power demand is desirable to extend range.

6.7.2   Light Spectrum

There have been a number of innovations in the spectrum emitted by headlamps. An earlier one was the mercury-free HID headlamp. The reason for eliminating mercury from the HID headlamp is nothing to do with the efficiency or effectiveness of the light source and everything to do with the environmental damage caused by mercury when such lamps are discarded. A consequence of removing mercury is a change in the spectral power distribution to give greater emphasis to short wavelengths. Sivak et al. (2006a) examined the effects of this change in spectrum for discomfort glare and the colour rendering of red retro-reflective materials. They concluded that discomfort glare for the mercury-free HID headlamps would be similar to that produced by the bluest of the then conventional HID headlamps and that the colour rendering and brightness of red traffic signs would be acceptable. Since the advent of LED headlamps and the consequent reduction in the use of HID headlamps, this innovation has disappeared from view.

Another innovation is a headlamp in which different parts of the beam have different spectral power distributions. Specifically, a projector-type headlamp has been designed which uses a dielectric coating to filter and redirect short wavelength light away from an opposing vehicle and towards the nearside edge of the road. One outcome is less light going towards an opposing vehicle and what does go in this direction is spectrally shifted to longer wavelengths resulting in a more yellow appearance. From Section 6.6, it would be expected that such a spectrally tuned headlamp would result in less glare to opposing drivers and better detection of off-axis targets at the edge of the road. Van Derlofske et al. (2007) confirmed these expectations in a study using the same methodology as described in Section 6.5. What they found was that the spectrally tuned headlamps had no effect on disability glare but did reduce discomfort glare by about one unit on the de Boer scale from that produced by standard HID headlamps. Such a difference is comparable with the difference between halogen and HID headlamps producing the same illuminance at the eye. As for off-axis performance, the spectrally-tuned headlamp set reduced reaction times to the onset of the targets by about 70 ms relative to the standard HID headlamp set at 12.5 degrees from the line of sight. This improvement is brought about by both the increase in illuminance and the shift in the spectrum. Again, the advent of the LED headlamp has caused this innovation to be abandoned.

A more dramatic attempt to resolve the conflict between visibility and glare involves the use of headlamps producing wavelengths outside the visible (Mace et al., 2001). An example of this has been the development of ultra-violet (UV) emitting headlamps. These headlamps emit radiation in the wavelength range 320–380 nm and are designed to supplement conventional low-beam headlamps. When these UV headlamps are combined with fluorescing materials in road markings, the distances at which the road markings can be seen increase dramatically, as do the distances at which pedestrians who are wearing clothing washed in detergents containing a fluorescent whitening agent can be detected (Turner et al., 1998). The UV radiation emitted by such headlamps does not pose a health threat to people exposed to it (Sliney et al., 1995). The combination of UV headlamps with fluorescing materials is considered by some to be a very cost-effective approach to improving the safety of driving at night (Lestina et al., 1999) but this innovation has not been developed.

A variation in this approach is to use an infrared night vision system. These come in two forms – active and passive. Active night vision systems use a headlamp emitting infrared radiation in the wavelength range 800–1,000 nm coupled to a camera sensitive to these wavelengths and software to identify salient details of the camera image. The output is then delivered to a display, either a head-up display or an LCD display, as part of the information system available to the driver (Holz and Weidel, 1998; Wordenweber et al., 2007). Because the wavelengths used are outside the visible range, the infrared headlamp can be kept on high beam without blinding any approaching driver. A high level of infrared radiation together with the fact that many materials that have a low reflectance in the visible wavelengths have a much higher reflectance in the near–infrared means that active infrared night vision systems can be effective in revealing people and animals at much greater distances than would be possible using low beams alone.

An alternative night vision system is passive in that it simply detects infrared radiation emitted by surfaces at different temperatures, usually in the wavelength range 8–14 um. There is no additional radiation emitted from the vehicle. Such a system is effective in detecting warm objects with a distinct temperature difference from the background, such as pedestrians and other animals, but not objects at a similar temperature as the background, such as lane markings.

Tsimhoni et al. (2005) had people perform a simulated steering task while viewing video recordings of a drive along a road where pedestrians were to be seen and the same trip as seen by active and passive night vision systems. The mean distances at which the pedestrians were detected were about three times greater for the passive night vision system than for the active night vision system. Hankey et al. (2005) report measurements of the distances at which drivers were able to detect pedestrians wearing black or white clothing, crossing or standing beside the road, with and without a passive infrared system using a head-up display. The drivers were also asked to detect a pedestrian wearing white clothes standing behind a stationary vehicle with its headlamps on and standing behind a crash barrier on a curve as well as a piece of tyre tread on the far side edge line. For the pedestrians crossing and standing by the road, the drivers were not aware of their location, but they were for the pedestrian near the glare source. The pedestrian standing behind the crash barrier on the curve was outside the field of view of the passive system and the tyre tread had a similar temperature to the road. Table 6.1 shows the mean detection distance for each detection task, with and without the passive system. In addition, the percentage of misses is given for the pedestrian. Any detection of pedestrians crossing or standing beside the road at less than 150 m was counted as a miss and given a detection distance of zero metres. Similarly, any detection of pedestrians beside the glare vehicle or behind the crash barrier, or the tyre tread that occurred at less than 50 m was counted as a miss and given a detection distance of zero metres.

An examination of Table 6.1 reveals large, statistically significant increases in mean detection distances as well as zero misses for pedestrians crossing and standing beside the road and standing beside the glare vehicle when the passive system is operating. When it is not operating, the increased mean detection distances for pedestrians wearing high reflectance clothing are clear. Another interesting point is that the mean detection distance for the pedestrian standing behind the crash barrier on a curve is statistically significantly shorter and there are more misses when the passive system is operating than when headlamps alone are used. This pedestrian

TABLE 6.1

Mean Detection Distance (m) and Percentage of Misses for Detecting Pedestrians and Tyre treads at Night Using Headlamps Alone Or Headlamps with a Passive Infrared Night Vision System (from Hankey et al., 2005)

Target	Headlamps only		Headlamps and passive infrared night vision system	
	Mean detection distance (m)	Misses (%)	Mean detection distance (m)	Misses (%)
Pedestrian in black crossing road	61	31	455	0
Pedestrian in white crossing road	119	3	444	0
Pedestrian in black standing beside road	42	26	414	0
Pedestrian in white standing beside road	137	0	409	0
Pedestrian in white standing near glare vehicle	87	0	379	0
Pedestrian in white standing behind crash barrier on curve	50	12	36	29
Tyre tread on far side edge line	49	6	44	23

is outside the field of view of the passive system, suggesting that when the passive system is used, attention is focused on the area it covers. Nonetheless, the greatly increased detection distances for pedestrians on or close to the road, who are much more at risk than a pedestrian standing behind a crash barrier, are clear. Infrared night vision systems are still confined to the top end of the market but, given the advantages shown by the studies discussed above, are likely to become more widely available in the future.

6.7.3 LUMINOUS INTENSITY DISTRIBUTIONS

The most dramatic innovation in vehicle forward lighting in recent years has been the introduction of adaptive forward lighting systems (AFS) that change the luminous intensity distribution of the headlamp. This has occurred in two stages. The first was the introduction of the bending light designed to increase visibility around a curve. There are two forms of bending light. Either the headlamps are swiveled to better illuminate the curve in the road without changing the luminous intensity distribution (dynamic) or the headlamps are fixed but the luminous intensity distributions are changed by switching on additional light sources to increase the illuminance around the curve (static). The movement of the swiveling headlamp beam is automatically determined by some combination of signals from sensors providing information on the vehicle's motion. This movement increases visibility distances around the bend, particularly for low radius bends but may also increase glare to oncoming vehicles

(Sivak et al., 2005a) which may be why NHTSA decided that steerable headlamps should not be used in the US if the whole beam was swiveled.

The next stage in the development of AFS involved systems to produce different luminous intensity distributions for a number of different commonly occurring driving situations. Adaptive forward lighting consists of three elements, a surround sensing system that searches what lies ahead of the vehicle, software that evaluates the results of the search so as to detect other vehicles and hazards and then to decide how to alter the light distribution of the headlamps and a headlamp system that is capable of a wide range of luminous intensity distributions. The surround sensing systems can be based on microwave radar, infrared laser radar or artificial vision using video cameras, the different systems varying in their range, resolution, field of view, sensitivity to weather conditions and the amount of processing power required. One such headlamp has multiple LED light sources so that by adjusting the current supplied to each LED, the luminous intensity distribution can be varied over a wide range. The great advantage of such a system is that it has no moving parts. An alternative system based on digital micro-mirrors does have moving parts. In this, light from a single source is reflected from an array of micro-mirrors. Information from the surround sensing system is used to adjust each of the micro-mirrors so as to shape the headlamp luminous intensity distribution to the required conditions. Matrix headlamps using these technologies are accepted in many countries but they have only recently been approved for use in the US. Matrix headlamps are being installed in luxury vehicles but given the tendency for expensive options first introduced in up-market vehicles to gradually spread into more modest vehicles, it seems likely that AFS in various forms will eventually become much more widely available. If so, then according to Sullivan and Flannagan (2007) there is a potential to significantly reduce crashes involving pedestrians, particularly on high-speed roads.

Modified luminous intensity distributions have been proposed for use in towns where speeds are low, for use on motorways and divided carriageways where speeds are high and there is a large separation between traffic streams, and for use on wet roads, where the increased specular reflection leads to more glare to opposing drivers (Wordenweber et al., 2007). The transition between these beams is automatic, determined by some combination of signals from sensors giving the vehicle's speed and direction, and the ambient light level. The town beam is wider than the conventional low beam and extends a shorter distance up the road, thereby emphasizing the visibility of pedestrians, road signs, and road markings. In addition, the light output of the headlamp is halved. This town beam is activated when speeds are below 50 $km.h^{-1}$ (31 mph) or if the road surface luminance is higher than 1 $cd.m^{-2}$. The motorway beam is created by moving the low beam up a quarter of a degree to extend the beam further down the road. This beam is activated when the vehicle is moving over 110 $km.h^{-1}$ (68 mph). The wet road beam involves a reduction in the illumination just in front of the vehicle and increased light to the sides of the vehicle. This beam is activated when either rain is detected on the road or when the windscreen wipers are switched on. As if this were not enough, predictive AFS systems for lighting bends have been suggested. In these, GPS information is used to predict road curvatures immediately ahead with the result that the forward lighting is moved to illuminate a curve before the vehicle enters it. Evaluations of bending lighting, motorway lighting, and town lighting by drivers indicate that bending lighting is considered the most valuable, followed by motorway lighting with town lighting having the least value (Hamm, 2002).

But this is not the latest development in AFS systems. One recent development has been to directly address the problem of glare for approaching drivers (Mazzae et al., 2015; Reagan and Brumbelow, 2015; Hamm et al., 2016). To do this, a matrix headlamp is linked to a camera and image processing software that enables hazards and information in and near the road ahead to be identified. Such hazards include approaching and preceding vehicles, pedestrians on or near the road, cyclists, obstacles, and detritus in the road as well as parked vehicles, bends in the road, road markings, and road signs. The software will then automatically modify the beam appropriately. For example, in order to avoid causing glare to the driver of an approaching vehicle or the driver of a preceding vehicle, the part of the beam that illuminates that driver can be dimmed or extinguished, the area being changed as the situation changes. Bullough et al. (2016) have shown that with LED adaptive forward lighting, the distance at which a pedestrian at the side of the road and facing into the road could be detected is equivalent to that achieved by the high beam and about twice that achieved using low beam. But for discomfort glare, the de Boer ratings for the LED adaptive forward lighting were the same as that of the low beam and much less than that caused by the high beam. As a consequence, high beam becomes the default state for headlamps.

Another recent development is a system designed to adjust the range of the luminous intensity distribution of the headlamp according to the road lighting class (Vogel et al., 2024). Different road lighting classes deliver different road surface luminances. Using LED light sources for both road and vehicle lighting, measurements of visibility level for vertical, 200 mm side, square targets of different reflectance at different distances from the vehicle showed that depending on the road surface luminance, the target reflectance and the distance of the target from the vehicle, adding light from a vehicle's headlamps sometimes increased and sometimes decreased the visibility level of the target. The concept associated with this innovation is to only provide light from headlamps when and where it will produce a higher visibility level.

6.7.4 Controls

As should be evident from the discussion above, there have been rapid advances in the sophistication of controls used to change forward lighting. In the beginning, control of when to turn on headlamps and when to shift from low beam to high beam, and vice versa, were matters for the driver using a simple switch. The first step in providing automatic control was a photosensor that monitored the available ambient light level and turned on the headlamps when the ambient light level dropped below a specified threshold. The next step was a camera system that identified a vehicle approaching or just ahead and automatically switched from high beam to low beam. It restored high beam after the approaching vehicle had passed or the vehicle just ahead moved on. This was followed by the delivery of a much more detailed display of what lay ahead of the driver and what items to pay attention to. This was achieved using a camera, radar, and software but the decision of what action to take was left to the driver. The final development has been to reduce the driver's workload by automatically adjusting the luminous intensity distribution of the headlamps to better match what lays ahead. Whether or not this trend has reached a conclusion or will proceed to the point that a truly autonomous vehicle becomes a practical proposition remains to be seen, but there is little doubt that the nature and performance of forward lighting is likely to become much more automated in the future.

By now it should be apparent that this is an exciting time to be involved in the design of vehicle forward lighting. After decades of little change, there are now many possibilities for enhancing the ability of the driver to see the road ahead without blinding those approaching. Some of these possibilities are evolutionary in that they involve introducing more beam types and more automation to the existing high-beam/low-beam options. Others are revolutionary in that they offer additional information based on parts of the electromagnetic spectrum outside the visible. Presently, these possibilities are used to supplement human vision but they may ultimately replace it. It will be interesting to see which of these possibilities thrive and which decline.

6.8 SUMMARY

Forward lighting of vehicles is designed to enable the driver to see after dark. The most common form of forward lighting, headlamps, is fitted to all motor vehicles. Headlamps in modern vehicles use either tungsten halogen, LED, or xenon discharge (HID) light sources. At least two headlamp light distributions are required by law – high beam for use when there is no other vehicle on the road ahead, either approaching or leading; and low beam for use when there is another vehicle approaching or leading. High beam is sometimes called the main beam or driving beam while low beam is sometimes called the dipped beam, meeting beam, or passing beam.

Headlamp light distributions and placement on the vehicle are closely regulated. The purpose of these regulations is to bring some order to the potential conflict between drivers caused by the fact that headlamps increase the visual capabilities of the driver sitting behind them but simultaneously decrease the visual capabilities of the driver facing them. The vast majority of countries follow either the recommendations of the Economic Commission for Europe (ECE) or the USA Federal Motor Vehicle Safety Standard (FMVSS). A comparison between these standards shows some similarities and some differences. For high beams, the maximum luminous intensity values allowed are higher in the ECE regulations than in the regulations used in America. For low beams, both American and ECE regulations are similar in that they show an emphasis on the near side of the road and a limitation in luminous intensity directed towards vehicles approaching in the opposing lane. However, they differ in the luminous intensity distribution towards the approaching driver. The ECE low beam has a much sharper cut-off than the American low beam, indicating the greater emphasis given to controlling disability glare.

While the regulations applied to headlamps are exact, headlamps in a vehicle on the road may depart from the regulations for a number of reasons. Some are transient and inherent in the road layout or the nature of the vehicle. Others are permanent and occur because the vehicle is not level, or the headlamp is incorrectly aimed or dirty. Even when headlamps are correctly aimed, new, and clean, there is still a question about how effective they are, especially on low beam. Comparisons between detection distances using low beam and stopping distances suggest that the maximum safe speed when using low beams is about 30 mph (48 km.h^{-1}). This might not matter so much if drivers used high beams whenever possible. Unfortunately, they do not.

The most dramatic change in headlamps over the last decade has been the increasing use of LED rather than halogen or HID light sources. LED and HID headlamps typically produce two to three times more luminous flux than halogen headlamps and emit light with a spectrum better suited to off-axis detection. Consequently, LED and HID headlamps conforming to the same regulations allow objects to be detected at greater distances and over a wider range of angles than halogen headlamps. However, the same photometric characteristics lead to more glare to the drivers of opposing or preceding vehicles.

When driving, glare is commonly experienced in two forms – disability glare and discomfort glare. Disability glare reduces the capabilities of the visual system due to light being scattered in the eye, with the important variable being the illuminance received at the eye. Discomfort glare is said to occur when people complain about visual discomfort in the presence of bright light sources. The illuminance at the eye is also important in determining discomfort glare, although light spectrum and headlamp size also have small influences.

Drivers most commonly experience glare in a dynamic situation as two vehicles approach and pass each other. In this situation, the visibility distance depends on the relative luminous intensity distribution of the headlamps, the relative positions of the two vehicles, and the physical characteristics of the objects to be seen. When two opposing vehicles have equal luminous intensity headlamps, the visibility distance for low-reflectance objects is about 60–80 m. When the opposing vehicle has a luminous intensity about three times more than the observer's vehicle, the visibility distance for low-reflectance objects is reduced to about 40–60 m. Clearly, driving at high speeds against opposing traffic at night approaches an act of faith. Further, it should not be thought that once the opposing vehicles have passed each other, the effects of glare are over. They are not. This is because after exposure to glare the driver's vision will be misadapted. The time taken to recover from exposure to glare depends on the light dose, i.e., the product of illuminance at the eye and the exposure time.

Over the last two decades, a number of innovations in headlamps have occurred. These range from the use of night vision systems using ultraviolet or infrared radiation to deliver an image of the road head on a display, to adaptive forward lighting systems. All aim to enable the driver to see far ahead without blinding the driver of an approaching vehicle by glare. The latest innovation is an adaptive forward lighting system based on an array of individually addressable LEDs linked to a system of camera, radar, and software that can automatically dim or turn off the LEDs as required. This enables the high beam to be modified to have different luminous intensity distributions for different commonly occurring driving situations while maintaining good visibility without glare.

This is an exciting time to be involved in the design of vehicle forward lighting. After decades of little change, there are now many possibilities for enhancing the ability of the driver to see the road ahead without blinding those approaching. Some of these possibilities are evolutionary in that they involve introducing more beam types and more automation to the use of the existing high-beam/low-beam options. Others are revolutionary in that they offer additional information based on non-visual parts of the electromagnetic spectrum. Presently, some of these possibilities are being used to supplement human vision but they may ultimately replace it.

7 Vehicle Signal Lighting

7.1 INTRODUCTION

Vehicle signal lighting has two functions. The first function is to indicate the presence of a vehicle or to give information about the intentions of a driver or the future movement of the vehicle to others. Such signal lighting includes front position lamps, rear position lamps, side marker lamps, turn lamps, hazard flashers, stop lamps, rear fog lamps, reversing lamps, daytime running lamps, as well as retro-reflectors. Some signal lamps, such as front and rear position lamps, are used only at night or in conditions of poor daytime visibility while others, such as daytime running lamps, are, as their name implies, used by day, and yet others, such as turn lamps and stop lamps, have to be visible at all times, both day and night. To fulfil its function of transmitting information from the vehicle to other people, the signal lamp has to be visible and its meaning has to be clear. It also helps if it is conspicuous. The second function is to contribute to the branding of a vehicle by arranging the signal lamps in a form that supports the easy identification of the vehicle's manufacturer and what type of vehicle it is. In other words, to give the vehicle a signature. This element of signal lighting may not contribute much to traffic safety, but it is an important element in the vehicle's styling and offers vehicle designers an opportunity to display their skill (Styliois et al., 2020).

Today, signal lighting is regulated, but this has not always been the case (Moore and Rumar, 1999). Early motor vehicles had little by way of signal lighting using carbide lamps for forward lighting and oil lamps for rear lighting. It was not until the time of the First World War that motor vehicles started to have electric signal lighting, this taking the form of one lamp showing a red light to the rear and a white light to the side to illuminate the licence plate. In the 1930s signal lamps that could indicate braking and turning began to appear. These were usually red but sometimes green and yellow. By the 1950s most cars had two front position lamps, two rear position lamps, two stop lamps, and a license plate lamp. Vehicles in the USA also had flashing turn signals, but Europe was still using the trafficator, an internally illuminated yellow semaphore that swung out from the side of the vehicle to indicate the intention to turn (Figure 7.1).

It was in the 1960s that vehicle signal lighting really became regulated, a development brought on by the increased density of traffic and the need to ensure consistency between vehicles so that the meaning of the signal was clear. Since then, the main developments have been the introduction of the centre high-mounted stop lamp, rear fog lamps, reversing lamps, and daytime running lamps, at various times, in various combinations, in different countries. Today, thanks to the global nature of the vehicle market, all these forms of vehicle signal lighting are to be found all over the world.

DOI: 10.1201/9781003388906-7

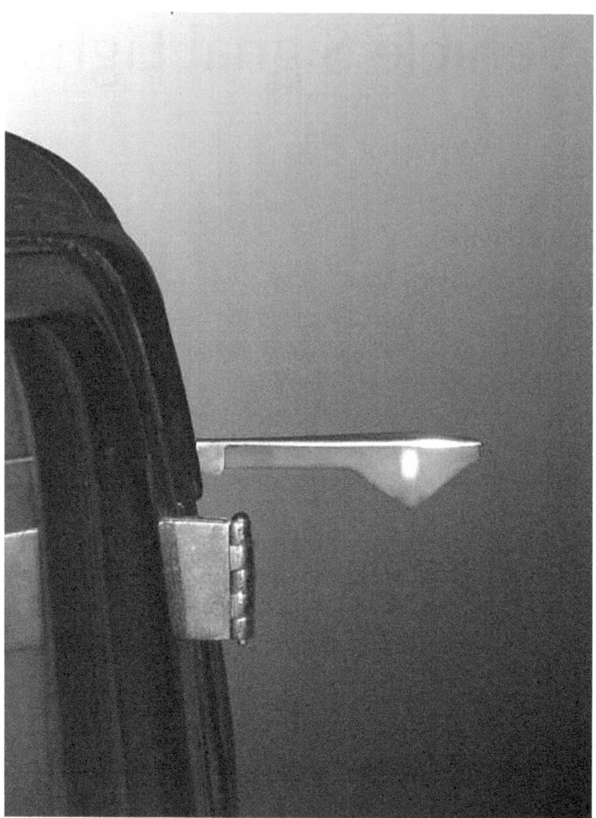

FIGURE 7.1 An early form of turn indicator, the trafficator. This is an internally illuminated semaphore that swings out from the upper part of one or the other side of the vehicle body to indicate the intention to turn.

7.2 THE TECHNOLOGY OF VEHICLE SIGNAL LIGHTING

For many years the technology of vehicle signal lighting hardly changed, consisting of little more than an incandescent light source covered by a clear or coloured lens. However, over the last two decades this situation has been transformed. Today, signal lighting primarily uses the LED light source using different methods of optical control and is an integral part of the styling of the vehicle.

7.2.1 LIGHT SOURCES

Since the 1920s there have been two dominant light sources in the field of vehicle signal lighting, incandescent and LEDs. Until the 1990s, the most widely used light source for vehicle signal lighting was the incandescent but this has now been replaced by the LED. The incandescent has the advantage of being simple and inexpensive but the disadvantage of requiring regular replacement throughout the life of the vehicle. For example, a double filament incandescent light source for use in a combined stop/rear position lamp has a life of approximately 100 h for the stop function and 1,500 h for the rear position function. Based on typical patterns of use, after 150,000 km (93,200 miles) the stop lamp will have been used for about 600 h, which suggests that the light source will have had to been replaced six times. This problem is overcome by the LED. LEDs have much longer lives than incandescent light sources, so much so that they should not need to be replaced during the life of the vehicle. LEDs have other advantages over the incandescent. They have a much higher luminous efficacy and so require less power to produce the same light output. Further, to create a coloured signal, the incandescent has to be filtered while the LED, if judiciously chosen, emits light of the desired colour, without filtering. Taken together, this means that LED signal lamps have smaller power demands than incandescent signal lamps fulfilling the same function. LEDs, being solid-state devices, they are also less sensitive to vibration and have a faster rise time than light sources that rely on a heated filament. Also, because they are smaller, they offer the designer a wider range of possibilities. LEDs are now the light source of choice for vehicle signal lighting.

7.2.2 OPTICAL CONTROL

The regulations applicable to signal lighting specify different luminous intensity magnitudes and distributions for each signal function. To meet these regulations, signal lamps require some form of optical control. There are several systems used in signal lamps based on reflection, refraction, and total internal reflection (Wordenweber et al., 2007) as well as multiple LED arrays in which each LED is individually addressable. For the reflector system, the light distribution is determined by the position of the light source relative to the reflector, the shape of the reflector and any optical patterning of the front cover glass. Where a smooth cover glass is preferred, some combination of smooth, segmented, or faceted reflectors can be designed to produce the desired luminous intensity distribution.

Rear position lamps and turn lamps commonly use refractor optics. In this system, a Fresnel lens cover glass is used. Where a smooth clear cover glass is preferred, a Fresnel lens system behind the cover glass is used. With this approach, no reflector is needed.

Total internal reflection is the physical principle used in light guides. A light guide consists of a transparent material with a refractive index higher than the surroundings, which for motor vehicles is air. The light source is closely coupled to the light guide. Light is transmitted down the light guide by repeated reflections at the surface of the guide. Light is extracted from the light guide by having a prismatic element. Such an element increases the angle of incidence of the light in the guide and thereby defeats total internal reflection. LEDs are the preferred light source for light guides, their low temperature and small size making it easy to couple the LED closely to the light guide. The light distribution is largely determined by the prismatic element. Light guides offer the stylist a great deal of freedom.

7.2.3 Structure

Although the different types of signal lamps are considered separately, it is necessary to appreciate that although they can be packaged individually, they are much more usually combined in a common structure called a cluster (Figure 7.2a and b).

In a cluster there will be multiple light sources for the different signal lamps but even when individual signal lamps are used it is possible to have multiple light sources in a number of different compartments. Until the advent of LEDs, the main reason for using multiple compartments for the same lamp was to increase the visible surface area without having to make the cluster too deep, although there was also a benefit in introducing redundancy in that the signal did not disappear if one light source failed. Where LEDs are used as the light source, it is usually necessary to have multiple LEDs in the same lamp because the light output of a single LED is insufficient. Whether or not such a use constitutes a multi-compartment lamp has caused some confusion in the interpretation of regulations.

The structure of signal lamps can be considered in mechanical, electrical, environmental, and colour terms. The mechanical elements involve making sure that the light source and optical control elements are consistently related to each other and the whole is mounted securely in the vehicle body. The electrical elements call for the connection of the light source to the vehicle's electrical system, either directly or through any necessary power supply, as well as the means to change the lamp when it fails. The environmental elements require careful consideration of the materials used so that the lamp can cope with the extremes of heat, cold, moisture, and vibration. Finally, it is necessary to consider the means used to generate the colour of light. One approach is to use a light source that emits light of the required colour,

FIGURE 7.2 Two clusters of lamps from a Volkswagen Up! a) a front cluster containing low-beam and high-beam headlamps, a front position lamp, a daytime running lamp and a turn indicator, b) a rear cluster containing a rear position lamp, a stop lamp, a turn indicator, and a reversing lamp.

such as an LED or a painted incandescent. This light source can then be used with a clear cover glass or a coloured cover glass, both having a high spectral transmittance in the relevant wavelengths. The advantage of the coloured cover glass is that the signal looks the expected colour by day, even when the lamp is not lit.

7.3   THE REGULATION OF VEHICLE SIGNAL LIGHTING

The visibility of a signal lamp depends on its luminance, size, shape, and colour, the background against which it is seen, and the state of adaptation of the driver. Much effort has been put into measuring the minimum values of some of these variables necessary to make the signal lamp visible under different conditions (Dunbar, 1938; de Boer, 1951; Moore, 1952; Hills, 1975a, 1975b; Sivak et al., 1998). Figure 7.3 shows the relationship between luminance and size for red rear position lamps, disc obstacles, and pedestrian dummies to be "just visible", i.e., at threshold, under no road lighting and no glare conditions (Hills, 1976). It can be seen that log luminance plotted against log visual area gives a nearly straight line; the smaller the signal area, the greater its luminance has to be before it is just visible. This is just what would be expected from the spatial summation of the visual system for small targets (see Section 3.4.3).

Using data similar to that shown in Figure 7.3, Hills (1976) produced a predictive model of the relationship between luminance increment and area for small objects to be just visible for a wide range of background luminances (Figure 7.4). A small object in this model is one for which spatial summation occurs in the visual system. For foveal vision, spatial summation is complete within a circle of diameter subtending about 6 min arc. For targets that occur five degrees off-axis, spatial summation

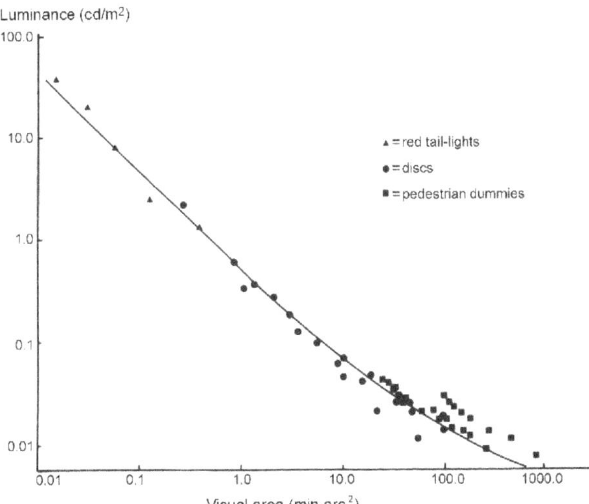

FIGURE 7.3   The relationship between luminance (cd/m²) and visual area ((min arc)²) for rear position lamps (red tail lights), discs, and pedestrian dummies to be "just visible" under a no road lighting/no glare conditions (after Hills, 1976).

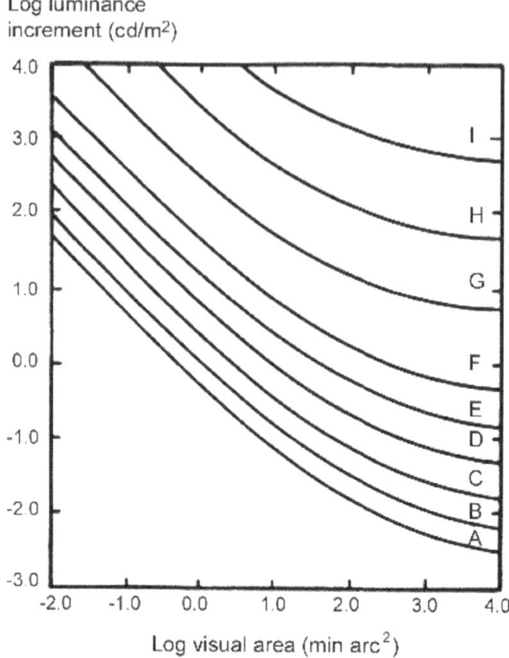

FIGURE 7.4   Relationships between luminance increment (cd/m²) and visual area ((min arc)²) at different background luminances, for small targets to be just visible. Each curve is for one background luminance as follows: A = 0.01, B = 0.1, C = 0.32, D = 1.0, E = 3.2, F = 10, G = 100, H = 1000, I = 10,000 cd/m² (after Hills, 1976).

occurs over a circle of diameter about 0.5 degrees (Boff and Lincoln, 1988). Given the usual size of signal lamps and the distance from which they need to be seen, spatial summation should apply. The ordinate in Figure 7.4 is the logarithm of the increment of the object luminance necessary for it to be just visible against the background luminance. Different values of background luminance enable the effects of different lighting conditions to be estimated, from starlight, through road lighting, to daylight. Hills (1976) also shows that by using such curves he can plausibly predict the field results of Dunbar (1938) and Moore (1952). Such information forms the background to the regulations governing signal lighting.

The two most widely used sources of standards relevant to vehicle signal lighting are the US Federal Motor Vehicle Safety Standard 108 (FMVSS) and the Economic Commission for Europe (ECE) recommendations. Most countries follow one or the other of these standards though with modifications to suit local circumstances. The FMVSS regulations are confined to North America and the ECE regulations are adopted by most of the rest of the world. These standards are used to form a legal framework for vehicle lighting but most of the actual work is done by two other bodies, the Society of Automotive Engineers (SAE) for the FMVSS regulations and the Groupe De Travail-Bruxelles (GTB) for the ECE recommendations. The tendency is for the recommendations of the SAE or the GTB eventually to be incorporated into their respective standards. FMVSS regulations are implemented by a self-certification process in which manufacturers have to test any proposed lighting systems to ensure they meet the requirements of the regulations. ECE recommendations are implemented through a type approval process in which manufacturers have to submit products to an independent testing laboratory. The type approval process is usually more stringent than self-certification. Over the last three decades, there has been convergence between the FMVSS and ECE standards, a process called harmonization. For many types of signal lamps, harmonization, although not complete, has resulted in sufficient overlap between standards for a manufacturer to be able to make one signal lamp that can meet both sets of requirements, something that is desirable for a global industry that is conscious of costs.

Both the FMVSS and ECE standards cover the number, colour, position, angles from which the signals should be visible, electrical connections and tell-tale warning of operation to the driver as well as the minimum and maximum luminous intensities and, if flashing or sweeping is called for, what the frequency of flashing or sweeping should be. Although the standards cover many of the same topics, they differ in detail. To expound all the similarities and differences would require a large tome made even larger by the fact that other regulations apply to commercial vehicles and trailers. Fortunately, compliance with these regulations are the concern of manufacturers rather than the public. Therefore if the lighting on your vehicle is fitted as original equipment you can be confident that it meets the relevant regulations but if it has been retrofitted then you should check compliance with the applicable regulations.

The regulated aspects of vehicle signal lighting are measured when the lamps are new and clean, and it is always salutary to be reminded of reality. Schmidt-Clausen (1985) measured the luminous intensities of rear position lamps and stop lamps on new cars and cars in use. He found that the luminous intensities of rear signal lamps in use were half the minimum values required by regulations. Similarly, Cobb (1990) carried out a roadside survey of vehicle lighting in the UK, including rear signal lighting, and found that dirt typically reduced the luminous intensity of vehicle lighting by 30–50%.

7.4 FRONT POSITION LAMPS

In different parts of the world, front position lamps are known as parking lamps, sidelights, position lamps, standing lamps, or city lights. For most vehicles, two front position lamps are required at the same height above the road and at an equal distance from the centre line of the vehicle. Figure 7.2a shows a front position lamp in a cluster containing a headlamp, a turn indicator, and a daytime running lamp, as well as the front position lamp. The signal it is intended to convey is that of presence when stopped or parked. In America, front position lamps can be either white or amber, but in countries that follow the ECE regulations, white is the only colour allowed, except for motorcycles which are also allowed to be amber. Front position lamps are required to remain illuminated once the vehicle forward lighting is lit. In many countries, it is illegal to drive a vehicle at night using front position lamps alone, but in others it is allowed in urban areas.

7.5 REAR POSITION LAMPS

Rear position lamps, also known as tail lights or rear lights, are always red in colour. Their function is to indicate presence when moving or stopped. Figure 7.2b shows a rear position lamp in a cluster with a stop lamp, a turn indicator, and a reversing lamp. The different information conveyed by the rear position lamp and the stop lamp is revealed by brightness, the stop lamp looking much brighter than the rear position lamp. To ensure this perception, regulations specify a much higher minimum luminous intensity for the stop lamp than for the rear position lamp (see Section 7.11). The different information between the rear position lamp and the others are revealed by colour. The rear position lamp is red, the reversing lamp is white, and the turn indicator is amber and flashing or sweeping, at least in vehicles following ECE regulations. In America, the turn indicator can be red but still has to flash or sweep. Rear position lamps have to be lit when the front position lamps are lit. Depending on the country, rear position lamps are required, permitted, or forbidden to be lit when daytime running lights are in use.

7.6 SIDE MARKER LAMPS

The basic function of side marker lamps is to indicate the presence of the vehicle to drivers at oblique angles to the direction of movement, a situation of importance at intersections. In America, both side marker lamps and side retro-reflectors are required, both being amber at the front and red at the rear of the vehicle. The minimum luminous intensities are 0.62 cd for amber and 0.25 cd for red. The side marker lamps are connected so that they are lit whenever the front and rear position lamps are lit. The front amber side marker lamps may also be wired so that the lamp on the relevant side flashes when a turn signal is activated, thereby adding another function.

In countries following the ECE regulations, side marker lamps are permitted rather than required. If they are installed, the side marker lamps are required to be visible over a wider range of angles than those in America. Further, the side marker lamps must be continuously lit when the front and rear position lamps or headlamps are lit, which means they cannot be connected to the turn signal. They must be amber at the front and rear, unless incorporated into a rear lamp cluster in which case the lamp can be red or amber.

Large vehicles are often required to carry additional marker lights. In North America, large vehicles have to have amber side marker lamps or retro-reflectors midway between the front and rear marker lights. Also, vehicles over 80 inches (2.03 m) wide have to carry identification lighting. This consists of three lamps at 6–12 inches (15–30 cm) separation, mounted centrally on the front and rear of the vehicle, amber at the front and red at the rear, and as high up as possible. These are intended to indicate the presence of a large vehicle. In countries that follow the ECE regulations, vehicles wider than 2.1 m have to carry marker lamps, usually mounted on stalks, at the front and rear of the vehicle, white at the front and red at the rear. These are to indicate the maximum width of the vehicle.

7.7 RETRO-REFLECTORS

Retro-reflectors are sometimes referred to as passive lighting, not an unreasonable description given the fact that such elements reflect incident light predominantly towards the light source illuminating them (see Section 2.8). They are regulated as vehicle lighting devices. The function of retro-reflectors is to signal presence, even when there is no power on the vehicle. Both FMVSS and ECE-based regulations require vehicles to have red retro-reflectors at the rear. American regulations also require amber, front-facing retro-reflectors and red, side-facing retro-reflectors at the rear. Some other countries require white, front-facing retro-reflectors.

In addition to side marker lamps, in North America heavy truck trailers are required to have alternating red and white, retro-reflector tape outlining the bottom of the sides and the bottom and underguard of the rear of the trailer. White retro-reflecting tape is used to define the top corners of the rear of the trailer. The use of retro-reflectors on the rear of heavy truck trailers is common in many other countries (Figure 7.5). Such marking is useful because heavy truck trailers can be difficult to see at night, particularly when emerging from an intersection on an unlit road. Morgan (2001) used crash data from Florida and Pennsylvania to examine the effectiveness of the American marking. He found that trucks so equipped were involved in 29% fewer side and rear crashes in the dark than truck trailers without the tape.

Another situation where retro-reflectors are sometimes used is on the inside edge of a door. An open door is a hazard to passing traffic, and passing traffic is a hazard to people exiting the vehicle. The door open warning retro-reflector is red in colour and is mounted so that it is visible to drivers approaching from the rear. Sometimes a red warning lamp is used instead of a retro-reflector. If so, this is activated by opening the door and extinguished when the door is closed.

7.8 LICENSE PLATE LAMPS

All countries require that the rear licence plate should be illuminated by white light using one or two hidden lamps after dark. This is for identification purposes, not for signaling presence, although the illumination of the licence plate does add some variation in luminance contrast within the vehicle, which should help maintain visibility. Moore and Rumar (1999) have argued that now that licence plates are almost always retro-reflective and the level of illumination from the licence plate lamps is low, the requirement for licence plate lamps should be abolished. However, given that authorities are increasingly using automatic reading of licence plates as a means of identification for catching toll dodgers and for revenue generation through fines for speeding offences, this seems unlikely to happen.

7.9 TURN LAMPS

So far, all the lamps considered are intended to make the vehicle more conspicuous and are permanently lit when the vehicle is moving after dark. Now, attention will be switched to signal lamps with a wider range of messages that are intermittently lit

FIGURE 7.5 Two rectangular retro-reflecting markers and a retro-reflective tape outlining the rear of a truck trailer.

when the message is sent. The first to be considered are turn lamps, also known as direction indicators, flashers, or blinkers. These are lamps mounted at the front and rear corners and on the sides of the vehicle and are used to indicate to other drivers that the vehicle is about to turn or change lane. Regulations for turn lamps specify minimum and maximum luminous intensities, angles from which the signal should be visible, colour and flash rate, as well as feedback signals for the driver. Minimum luminous intensities on the optical axis of front turn lamps vary according to the separation of the turn lamp from the nearest headlamp. In America, the minimum such luminous intensity is increased from 200 cd to 500 cd if the separation is less than 100 mm, the separation being measured as the distance between the optical centre of the turn lamp to the edge of the light-emitting area of the headlamp. In countries following ECE recommendations, the minimum luminous intensity on the optical axis is increased from 175 cd to 400 cd when the separation of the edge of the turn lamp and the edge of the headlamp is not more than 20 mm. This increase in minimum turn signal luminous intensity when in close proximity to a headlamp is intended to overcome the masking effect of disability glare from the headlamp. Turn lamps flash at a rate of 1–2 Hz, with all the turn lamps on one side operating in phase. In countries following the ECE recommendations, front, rear, and side turn lamps are amber in colour but the FMVSS regulations also allow the rear turn signal to be red, if desired. The side-mounted turn signal can take several different forms. In countries following the ECE recommendations, it takes the form of a dedicated turn signal. In America, the side marker lamp may also be used as a turn signal by making it flash in phase or anti-phase when the front and rear turn lamps are activated. Both ECE and FMVSS standards require that audiovisual feedback of the operation of turn lamps be given to the driver. This usually takes the form of a flashing light on the instrument panel together with an audible click at the same frequency.

Current practice in turn lamps has given rise to four questions about their effectiveness. The first question concerns the use of red rather than amber as a colour for the rear turn lamp in the USA. Current styling tends to group all the rear signal lamps into two clusters, one on each side of the vehicle and usually wrapped around the corner of the vehicle although some have links between the two clusters extending across the back of the vehicle. Stop lamps and rear position lamps are both red in colour. It might be thought that a rear turn lamp of the same colour would be less conspicuous than one of a different colour, specifically, amber, although the flashing of the signal may be sufficiently potent to negate the difference in colour. However, Luoma et al. (1997) have shown a faster reaction time to red brake signals when the turn signal is amber. Further, Allen (2009) has shown, by examining crash data for vehicle models that have changed turn indicators from red to amber or amber to red, that, relative to red, amber turn indicators reduce the probability of being rear-ended when the leading vehicle is turning or changing lanes.

Another trend in vehicle design is the insertion of side turn lamps in the mirror housings (Figure 7.6).

An analysis of the geometry between vehicles when one is about to pass the other in adjacent lanes indicates that a mirror-mounted turn lamp can be seen over a wider range of relative positions by the driver of the overtaking vehicle than conventional side turn lamps because the latter are often obscured by the bodywork of the passing vehicle (Reed and Flannagan, 2003). Further, the mirror-mounted turn lamp is closer to the line of sight of the passing driver and hence more likely to be detected, a probability confirmed by Schumann et al. (2003). This is an important finding for one particular type of accident, lane change/merging. In this, the driver wishing to change lanes is unaware of the vehicle in the adjacent lane because it is the blind spot between the driver's peripheral vision and what can be seen in the side mirror (Wang and Knipling, 1994). As long as the lane-changing vehicle is using the turn lamp, a greater ability to see it by the driver in the blind spot may allow that driver to take evasive action before contact. So far, attempts to test this hypothesis using crash data have failed to show a statistically significant effect (Sivak et al., 2006b) but this may

FIGURE 7.6 A mirror-mounted turn signal on an MG M3.

be due to a lack of sensitivity caused by a small sample size rather than the absence of a real effect. However, this question may soon become moot as advanced driver assistance systems are now available indicating to the driver when there is another vehicle in the blind spot (see Section 15.4).

Yet another trend in signal lamp design is the use of a clear lens and a coloured light source rather than a coloured lens and a white light source. This difference is trivial in darkness but in daylight it is not because then the greater amount of daylight that is reflected from the inside of the clear-lens turn lamp will increase the luminance on which the luminance produced by the operation of the light source is superimposed. The effect of this is to reduce the luminance increment when the turn lamp is lit. In addition, daylight entering and then exiting through a clear lens will desaturate the colour of the turn lamp when lit although the colour difference between the lamp on and off will still be greater for the clear lens than for the coloured lens. Sullivan and Flannagan (2001) measured the reaction time to the onset of turn lamps using coloured and clear lenses in bright sunlight. The turn lamp with the coloured lens had a slightly shorter mean reaction time than those with clear lenses. Sivak et al. (2006c) measured the luminance contrasts for turn lamps on, with and without the sun, for a wide range of commercially available turn lamps. On average, the clear-lens turn lamps provided lower luminance contrasts for turn lamps on and off in sunlight than did the coloured lens turn lamps. However, examination of the results shows that this is not inevitable. Clear-lens turn lamps can be designed that are resistant to strong sunlight but it does require an awareness of the potential problem and attention to detail. Sivak et al. (2006c) also point out that the higher luminous intensities required for front turn lamps in close proximity to headlamps mean that clear-lens turn lamps of this type will be more resistant to confusion in bright sunlight.

Recently, some turn signals have appeared with a sweeping function. This is not a new invention as such signals were used in the 1960s, but the adoption of multiple LEDs has made the implementation much easier. Specifically, the LEDs are arranged in a line, and each LED is powered sequentially rather than simultaneously. The visual effect is a perception of movement in the intended direction. Whether or not sequential switching, i.e, sweeping, is more effective than simultaneous switching is open to question. After all, the function of turn signal is first to attract attention and then to indicate the direction of the turn. Both temporal and spatial changes in the peripheral visual field will attract attention. The intended direction of turn is given by the location of the signal and by the apparent motion so the question becomes what does the apparent movement created by sequential switching add to what is already provided by simultaneous switching? One answer is provided by Bullough and Skinner (2016). They conducted a field experiment where observers stood 30 m (100 ft) from vehicles with turn signals operating, one flashing and one sweeping turn signal. The observers initially stood with their backs to the signal and then turned slowly round so that the turn signal first appeared in the far periphery of their visual field. They kept turning until they could correctly identify the direction indicated by the turn signal. It was established that the sweeping turn signal could be correctly identified further out into the periphery than the flashing turn signal, about 70 degrees rather than about 50 degrees. How useful such an increase in off-axis identification is remains to be determined.

7.10   HAZARD FLASHERS

Since the 1960s, turn lamps have been adapted to give a warning to other drivers that a vehicle that is stopped in or near moving traffic, is disabled, or is moving very slowly. This signal is given by making all front, side, and rear turn lamps flash in phase. The photometric conditions produced and the flash rate of the turn lamps are the same as when they are used to signal a turn.

7.11   STOP LAMPS

Stop lamps, also known as brake lights, are mounted at the rear of the vehicle. They are red in colour and are lit when the driver applies pressure to the brake pedal. The function of the stop lamps is to inform drivers behind that the vehicle is decelerating, although it may not stop. For many years, there were only two stop lamps mounted symmetrically on the left and right rear corners of the vehicle. However, over the last 20 years, many countries have adopted a third stop lamp for cars and light trucks.

This is called the centre high-mounted stop lamp (CHMSL) or centre brake lamp, third brake lamp, eye-level brake lamp, safety brake lamp, or high-level brake lamp. The CHMSL is mounted on the centre-line of the vehicle at a level above the left and right stop lamps although either one lamp with off-centre mounting or two lamps mounted symmetrically about the centre line are allowed if doors at the rear of the vehicle are divided centrally. Different vehicle forms mean that the CHMSL is sometimes mounted at the roofline, sometimes at the top of the boot and at all points between.

An increase in luminous intensity is used to differentiate the stop lamp from the rear position lamp. Exactly how big a difference is required is open to question. Rockwell and Safford (1968) found that up to a luminous intensity ratio of 5.3 (stop lamp/rear position lamp), reaction time to the onset of the stop lamp was reduced, as was the likelihood of confusion between stop lamp and rear position lamp. In the USA, the allowed luminous intensity for stop lamps is in the range 80–300 cd while in those countries that follow the ECE recommendations, the allowed luminous intensity range is 60–185 cd.

Visually, the problems associated with detecting the onset of a stop lamp are different by day and night. By day, detection can be difficult because daylight reflected from the stop lamp increases the luminance on which the luminance of the stop lamp is superimposed and hence reduces its luminance increment. In addition, some designers like to use rear lamp clusters that are the same colour as the vehicle bodywork when the lamp is unlit. Such a body-coloured lamp may reduce or enhance the colour difference between the stop lamp and the surrounding area of the vehicle depending on the colour of the vehicle body. Chandra et al. (1992) measured reaction times to the onset of body-coloured stop lamps in simulated sunshine. Provided the vehicle body is not red, the onset of the stop lamp produces a change in both luminance and colour. Four chromatically neutral lamps varying from black to white when off and four body-coloured lamps when off were examined. Figure 7.7 shows the mean reaction times for the stop lamps divided into two groups. One group had equal lightnesses (Munsell Value = 4) but different colours when the stop lamp was off, so they had equal luminance shift but different chromaticity shifts as the stop lamp changed from off to on. The other group had similar chromaticity shifts but different luminance shifts when the stop lamp changed from off to on because they were all chromatically neutral but differed in lightness when the stop lamp was off (Munsell Values of 2, 4, 7, and 9 for black, dark grey, light grey, and white, respectively). The smallest chromaticity shift occurred for the red body-coloured stop lamp. The smallest luminance shift occurred for the white body-coloured stop lamp. An examination of Figure 7.7 shows that increasing the magnitudes of both luminance shifts and chromatic shifts is effective in decreasing reaction time although the effects are small over the range examined. This suggests that the colour appearance of the stop lamp when off could be used to enhance its effectiveness.

Another aspect is the form of the stop lamp. The two stop lamps at the ends of the vehicle are usually part of a rear signal lamp cluster and the CHMSL is separate (Figure 7.8). For both types, the stop lamp can vary in area and that area can have different aspect ratios, particularly CHMSLs. Sayer et al. (1996) found that both

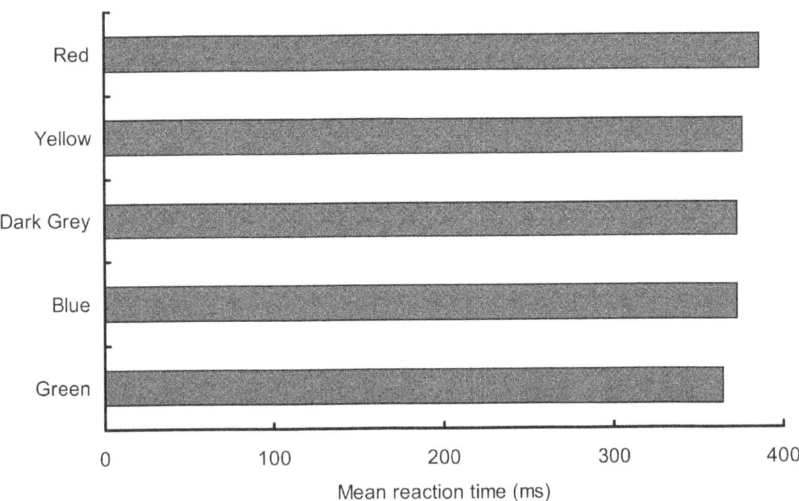

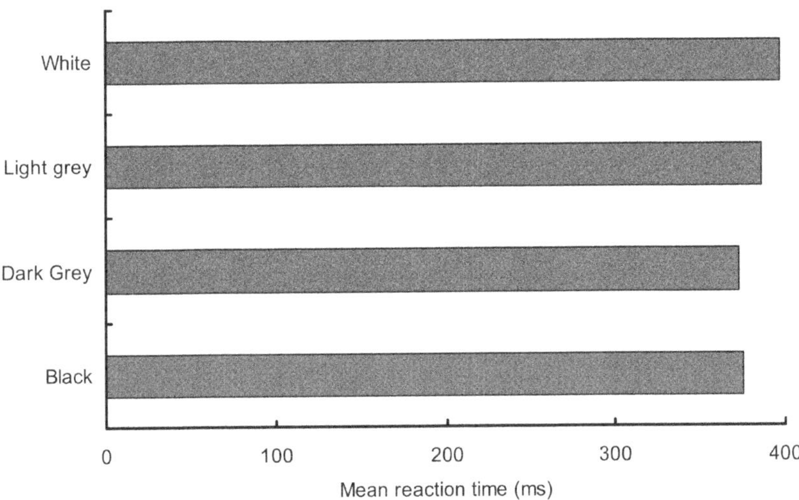

FIGURE 7.7 Mean reaction times (ms) to body-coloured stop lamps sorted into two groups. One group has equal lightnesses (Munsell Value = 4) but different colours when off and hence equal luminance shifts but different chromaticity shifts when the red stop lamp comes on. The other has equal chromaticity shifts to red when the stop lamp comes on but varies in luminance shift when changing from off to on because the stop lamps have different lightnesses when off (Munsell Values of 2, 4, 7, and 9 for black, dark grey, light grey, and white, respectively) (after Chandra et al., 1992).

FIGURE 7.8 The rear of a Skoda Karoq showing two light clusters each containing a rear position lamp, stop lamp, turn lamp, reversing lamp, hazard flasher fog lamp, and daytime running lamp as well as a centre high-mounted stop lamp (CHMSL). There is also a number plate lamp but this is not visible.

luminous intensity and aspect ratio influenced reaction time to the onset of a stop lamp. While luminous intensity was the major factor in determining reaction time, the impact of a low luminous intensity was particularly marked when the stop lamp had a high aspect ratio.

By night, the problem facing anyone wishing to detect the onset of a stop lamp is one of confusion because the rear position lamps will be lit, as may rear fog lamps (see Section 7.12) and, on occasion, the rear turn signal. Rear position lamps, rear fog lamps, and stop lamps are all red in colour and rear turn lamps may be in the USA. Further, rear fog lamps, and rear turn lamps may have similar luminous intensities to stop lamps. Luoma et al. (2006) examined the influence of having rear stop lamps in a cluster or separate on rear-end crashes at night using data from Florida and North Carolina. They concluded that separate stop lamps tended to make rear-end collisions less likely although the effect was complex and deserved further investigation. This finding is consistent with the work of Helliar-Symons and Irving (1981) who found that greater separation between stop lamps and rear fog lamps resulted in more accurate recognition of the onset of stop lamps and recommended a separation of at least 100 mm. Another approach to making stop lamps more noticeable is to use different colours for the rear turn lamps. Louma et al. (1997) found that having amber rear turn signals rather than red, produced shorter reaction times to the onset of stop lamps, the reduction being about 110 ms.

While both separation and colour are used to enhance the difference between stop lamps, rear position lamps, rear fog lamps, and rear turn lamps in some vehicles, there is one difference that is consistent across all cars and light trucks, namely, the number of lamps lit simultaneously. Rear position lamps and rear fog lamps occur in pairs. Rear turn lamps occur on one side only, unless used as hazard lamps, and flash rhythmically at 1–2 Hz. Stop lamps occur in a set of three and while they may flash on and off according to the driver's pressure on the brake, the flashing is rarely rhythmic. The third stop lamp is the CHMSL. The CHMSL was introduced with the twin objectives of making the signal that the vehicle ahead is braking more obvious to the vehicle immediately behind and of having a stop lamp that would be visible to vehicles separated from the braking vehicle by one or more vehicles between. Whether this second objective is met will depend on the nature of the vehicles between and the position of the CHMSL on the braking vehicle. A vehicle without windows at the rear, as is the case for most vans, does not allow a driver to see through the vehicle to the braking vehicle ahead. Even if a view through the vehicle ahead is possible, where the braking vehicle has a low-mounted CHMSL, e.g., at boot level on some sports cars, the CHMSL may not be visible. When the CHMSL is visible, there is some evidence that vehicles separated from the braking vehicle by another had shorter reaction times to the application of the brakes than when there was no CHMSL on the braking vehicle (Crosley and Allen, 1966). If the use of CHMSLs leads to shorter reaction times to the onset of stop lamps, it seems reasonable to suppose that this would have an effect on the number of rear-end collisions. Rear-end collisions are one of the most frequent types of crashes, and although they are rarely fatal, they often cause injury, particularly to the neck. Kahane and Hertz (1998) have used accident data to estimate that the widespread use of CHMSLs is responsible for a long-term 4.3% reduction in rear-end collisions.

One aspect of stop lamp design that is applicable to stop lamps in all positions, by day and night, is rise time. The rise time of the light source in a stop lamp is determined by the technology used to generate the light and the voltage applied. Incandescent light sources have a slow rise time, of the order of 200 ms, but LEDs can have a very fast rise time, of the order of nanoseconds. The effect of rise time in light output on reaction time to the onset of the stop lamp can be understood by considering the reaction time as the sum of two components, a visual reaction time and a non-visual reaction time. The visual reaction time is the time it takes for the light received from the stop lamp to be transformed to an electrical signal in the retina and transmitted up the optic nerve to the visual cortex. The non-visual component includes the time required for information perceived at the visual cortex to be processed and for neural signals to be sent to the muscles that are used to make the response. Differences in rise time of light output for the stop lamp can be expected to influence the visual reaction time but not the non-visual reaction time. The effect of different rise times in light output can be estimated from the fact that to see the signal, a constant level of energy, i.e., luminance integrated over time, is required to reach the retina (Teichner and Krebs, 1972). In other words, the visual reaction time is determined by the temporal summation properties of the visual system (see Section 3.4.4). Given a fixed maximum luminous intensity of the stop lamp, the shorter the rise time of the light output, the shorter the reaction time should be. This suggests that a fast rise stop lamp, such as those using LEDs, would lead to shorter reaction times to onset and this, in turn, might help diminish the number and severity of rear-end collisions (Sivak and Flannagan, 1993). However, despite a suggested 3.6% reduction in rear-end crashes for vehicles equipped with LED stop lamps and CHMSL, Kahane (2015) found that deficiencies in the real-world crash data relating to such collisions did not conclusively support this reduction.

7.12   REAR FOG LAMPS

In countries that follow the ECE regulations, rear fog lamps are required. Rear fog lamps are the same as rear position lamps in colour but have a much higher luminous intensity in the range 150–300 cd. They may be included in a cluster or mounted separately (Figure 7.8). Rear fog lamps are not required in the USA although they may be used, following the relevant SAE standard, which is the same as for the one compartment stop lamp. The function of rear fog lamps is, as the name implies, to increase the conspicuity of the vehicle in poor atmospheric conditions. Rear fog lamps are controlled manually by the driver, an action that sometimes leads to complaints of glare from drivers behind when the rear fog lamps are lit in only slightly degraded atmospheric conditions. When used, rear fog lamps can be installed as a single lamp on the driver's side of the vehicle or as a pair mounted symmetrically about the centre line. The case for the single rear fog lamp is that it is easily distinguished from stop lamps while a pair of rear fog lights is not, even when a CHMSL is present (Akerboom et al., 1993). The disadvantage of the single rear fog lamp is that it does not provide a cue to distance for an approaching driver. The need to differentiate rear fog lamps from stop lamps is the reason why the ECE regulations require a minimum separation, edge to edge, of 100 mm between adjacent stop lamps and rear fog lamps.

7.13   REVERSING LAMPS

The reversing lamp, also known as the back-up lamp, is unusual in rear lighting in that it is white in colour. The reversing lamp has two functions. One is to illuminate the road behind the vehicle so that the driver can see any obstructions directly by turning round, or indirectly via a rear view mirror or via a rear view camera. The other is to alert other drivers and pedestrians that the vehicle is about to reverse. The reversing lamp is automatically lit by placing the vehicle in reverse gear. One or two reversing lamps can be installed on a vehicle, but, in the USA, if only one reversing lamp is used it has to provide twice the minimum luminous intensity.

The problem with reversing lamps is not discriminating them from other rear signals; the different colour is enough to ensure that. The problem is in providing enough light, suitably distributed, to allow the driver of the reversing vehicle to see clearly in vehicles without a reversing camera. This problem is often exacerbated by the use of low transmittance glazing in rear windows. Passenger cars in the USA are required to have glazing with a transmittance of at least 0.70. However, some common vehicle types, such as minivans and sports utility vehicles (SUVs), are classified as light trucks for which there are no limits on transmittance of the rear window. An option available for many minivans and SUVs is privacy glazing for the windows behind the driver. The transmittance of privacy glazing is of the order of 0.18. The installation of privacy glazing has consequences. Freedman et al. (1993) found a decreased probability of detection of children and debris with lower transmittance glazing when preparing to reverse. Sayer et al. (2001) used a US database to examine the impact of lighting conditions, driver age, and vehicle type on backing crashes as a proportion of all crashes. The database, the General Estimates System, contains a nationally representative sample of police-reported crashes. Table 7.1 shows the percentage of reversing crashes and all crashes for cars, minivans, and SUVs involving drivers less than 66 years of age, who had not been drinking. Minivans and SUVs are involved in a higher percentage of backing crashes than would be expected from all crashes. Table 7.2 shows the percentage of reversing crashes and all crashes for minivans and SUVs in different ambient lighting conditions, involving drivers less than 66 years of age who had not been drinking.

The percentage of reversing crashes is what would be expected from the overall accident distribution for daylight, dark but lit, and after dark conditions but for dawn and dusk conditions, the percentage of reversing crashes is statistically significantly

---

TABLE 7.1

Percentages of Reversing Crashes and Total Crashes for Cars, Minivans, and SUVs for Drivers Less Than 66 years of Age Who Had Not Been Drinking. Statistically Significant Differences Are Marked with an Asterisk (from Sayer et al., 2001)

Vehicle Type	Reversing Crashes (%)	All Crashes (%)
Cars	82.1	88.3
Minivans/SUVs*	17.9	11.7

TABLE 7.2

Percentages of Reversing Crashes and Total Crashes for Minivans and SUVs by Ambient Illumination, for Drivers Less Than 66 Years of Age Who Had Not Been Drinking. Statistically Significant Differences Are Marked with an Asterisk (from Sayer et al., 2001)

Ambient Illumination	Reversing Crashes (%)	All Crashes (%)
Daylight	66.6	76.1
Dark	7.2	7.6
Dark/Lit	15.4	11.5
Dawn/Dusk*	9.3	3.7
Unknown	1.5	1.2

greater than would be expected. Taken together, these results suggest that in dark conditions, drivers reversing are careful because they are aware that they cannot see well. In daylight, there is no problem in seeing well but in dawn and dusk conditions, and possibly in dark but lit conditions, drivers of minivans and SUVs overestimate how well they can see when reversing. There are a number of possible solutions to this problem, among them being the banning of low transmittance glazing, an increase in luminous intensity for reversing lamps where low transmittance glazing is used, the use of sensors to detect obstacles behind the vehicle or the compulsory installation of rearview cameras. Keall et al. (2017) provide some support for the benefits of rearview sensors and cameras. They used records of crashes in which pedestrians were injured by a reversing vehicle in Australia and New Zealand, for vehicles equipped with rear sensors alone, a rearview camera alone, both sensors and a camera or neither. They found that the odds of a pedestrian being injured by a reversing vehicle were reduced by 33% for sensors alone, by 41% for a rearview camera alone, and by 30% for both sensors and a camera relative to vehicles not equipped with either sensors or a camera. Such technology would be particularly helpful for older drivers who are over-involved in reversing crashes (see Section 13.2).

7.14 DAYTIME RUNNING LAMPS

Daytime running lamps are pairs of lamps positioned at the front and sometimes at the rear of the vehicle, used during the day to increase the conspicuity of the vehicle. Daytime running lamps are provided by dedicated lamps or by the use of low-beam headlamps at the front and rear position lamps at the rear, although exactly which are required or just allowed varies from country to country. ECE regulations call for dedicated daytime running lamps to emit white light with a minimum luminous intensity of 400 cd on axis and no more than 1200 cd in any direction. In the USA, daytime running lamps are allowed but not required and can be white or amber with a minimum luminous intensity for dedicated daytime running lamps of 500 cd on axis with a maximum luminous intensity in any part of the beam of 7,000 cd. However, if a low-beam headlamp is being used as a daytime running lamp there is no maximum luminous intensity set.

But why should a vehicle need to have its conspicuity increased in daytime when the visibility of everything on the road should be high? There are two answers to this question. The first is the sad fact that about half of fatal vehicle crashes occur in daytime, so there is certainly a problem (Bergkvist, 2001). The second is that the most basic error of drivers is late detection (Rumar, 1990). Increasing conspicuity should enable a driver to detect other vehicles earlier. But why does late detection occur so frequently? The answer to this question involves both cognitive and visual factors (Hughes and Cole, 1984; Rumar, 1990). The cognitive factor is a matter of expectation and hence the allocation of limited attention. If attention is given, externally, to the wrong part of the visual world or, internally, to something other than driving, late detection of another vehicle is likely. Conspicuity is essentially a measure of the ability to attract attention. The visual factor is a matter of a weak stimulus, particularly in the peripheral visual field. A vehicle approaching or being approached head-on at speed causes little change in the retinal image which may make it difficult to detect until too late. Daytime running lamps are an attempt to attract attention to the other vehicle and hence to get drivers to use foveal vision to examine the other vehicle's movement.

From the above, it would seem that daytime running lamps are an obvious means to reduce daytime crashes, but that may not be. Daytime running lamps have potential drawbacks as well. Concerns have been expressed about the possibility that daytime running lamps may cause glare; may mask turn signals and may reduce the conspicuity of vehicles who already use them, such as motorcycles, or who do not have them, such as bicycles (Rumar, 2003). The extent to which daytime running lamps might cause glare will depend on the ambient illuminance. Studies of discomfort glare received through rearview mirrors show that for a low ambient illuminance of 700 lx, a luminous intensity of 1,000 cd is just permissible but for a high ambient illuminance of 90,000 lx, a luminous intensity of 5,000 cd is acceptable (Kirkpatrick et al., 1987; SAE, 1990). The FMVSS regulations applicable to dedicated daytime running lamps allow luminous intensities that seem likely to cause discomfort. As for masking of turn signals, SAE (1990) found masking when the daytime running lamps had a luminous intensity of 5,000 cd and higher and were observed from a short distance but at longer distances masking could occur for luminous intensities of 1,000 cd, especially if the separation between the turn signal and the daytime running lamp was small, findings that are consistent with what is known about disability glare (see Section 6.6.2). Again, the FMVSS regulations applicable to daytime running lamps allow luminous intensities that are capable of masking turn signals.

In response to these concerns it would seem to be a simple matter to reduce the maximum luminous intensity allowed, but there is a problem with this. It is simply that the higher the ambient illuminance, the greater the luminous intensity required for daytime running lamps to increase the conspicuity of the vehicle (Rumar, 2003). This observation suggests that daytime running lamps, as currently regulated, should have a greater effect on conspicuity and hence crashes in countries at high latitudes where the ambient illuminance is low more frequently and the sun is low in the sky for longer (Koornstra, 1993). Elvik (1996) has examined this possibility and concluded that it is correct, which may explain why many of the countries that first required daytime running lamps are at high latitudes.

There is also concern about the relative conspicuity of different vehicle types. This concern arises because if having daytime running lamps makes one vehicle more conspicuous and the available attention is limited, then another vehicle without daytime running lamps should become less conspicuous. Attwood (1979) has shown that it is more difficult to detect a car without daytime running lamps when it is between two cars that have them than when none of the cars has them. The inverse of this situation is a particular concern for motorcyclists who, of all road users, are the most likely to be killed or injured for the same mileage covered (see Section 10.2). Even where daytime running lamps are not required, motorcyclists are encouraged to drive with headlamps on during the day to increase their conspicuity, something that is very desirable given the small frontal area of a motorcycle and the consequent difficulty in detecting presence and estimating distance and speed. Wells et al.

(2004) have shown that motorcyclists who use headlamps by day have a 27% lower risk of being killed or injured than those who do not. With regard to daytime running lamps, the motorcyclists' concern is that if every vehicle were to have daytime running lamps, the conspicuity of motorcycles would be reduced. Whether or not such a reduction would matter is almost certainly related to traffic density and attentional capacity. Where there are only a few vehicles on the road, there should be enough attentional capacity for a driver to examine all of them, starting with those that are using headlamps or daytime running lamps. Where there are many vehicles on the road, there may not be enough attentional capacity to examine all the vehicles. If daytime running lamps are widespread and motorcycles use their headlamps by day, the only advantage motorcycles have is the greater luminous intensity of headlamps over daytime running lamps. This argument implies that motorcyclists' concerns about reduced conspicuity following the widespread introduction of daytime running lamps are justified for heavy traffic. One possibility would be to maintain the conspicuity advantage of motorcycles above other vehicles by making their headlamps flash or pulse by day.

As for other road users without daytime running lamps such as cyclists, Cobb (1992) examined the conspicuity of bicycles near cars equipped with daytime running lamps of different luminous intensities. He found that daytime running lamps increased the conspicuity of cars but did not reduce the conspicuity of bicycles until the luminous intensity of the daytime running lamps was very high, presumably high enough to produce masking by disability glare. This unexpected finding may simply indicate a halo effect around daytime running lamps. If the daytime running lamps attract attention to the car and the bicycle is close to the car, the retinal image of the bicycle is closer to the fovea and is more likely to be detected. However, if the bicycle is some way away from the car, directing attention to the car may reduce the chances of the bicycle being detected. A similar argument may apply to pedestrians in the road. Daytime running lamps do not make pedestrians more visible to drivers so the presence of daytime running lamps on other vehicles may attract drivers' attention away from pedestrians, particularly in heavy traffic. Fortunately, this disadvantage may be more than offset by making a vehicle with daytime running lamps easier to detect by the pedestrian. Thompson (2003) found that the largest accident reduction associated with the use of daytime running lamps concerned collisions with pedestrians, particularly children.

These observations indicate that daytime running lamps are a matter of balance between the positive effect of enhanced conspicuity for some and the negative effects of glare and reduced conspicuity for others. A legal requirement for daytime running lamps has been introduced in a number of countries and the consequences studied (Andersson and Nilsson, 1981; Elvik, 1993; Arora et al., 1994). These studies indicate that daytime running lamps are effective in reducing multi-vehicle crashes. However, Theeuwes and Riemersma (1995) have criticized the reliability of the statistical method used in these evaluations and argued that the effects found in Sweden are due to an unexplained increase in single vehicle daytime crashes. Fortunately, Elvik (1996) has carried out a meta-analysis of 17 such studies and concludes that the beneficial effects of daytime running lamps are robust. He further concludes that the use of daytime running lamps on cars reduces the number of multi-party daytime crashes by about 10% to 15% for vehicles having daytime running lamps and reduces the total number of multi-party daytime crashes by about 3% to 12%. He also states that there is no evidence that the use of daytime running lamps affects any type of accident other than multi-party crashes. Such reductions are somewhat greater than indicated by Farmer and Williams (2002) who found that, in the USA, vehicles with automatic daytime running lamps were involved in 3.2% fewer multi-party crashes than those without. As for the claim that daytime running lamps only reduce multi-party crashes, this is partly supported by Wang (2008), This study used crash data from nine US states to examine the benefit of daytime running lamps on three crash types: two passenger vehicle crashes excluding rear-end crashes, single passenger vehicle crashes involving pedestrians or cyclists, and single passenger vehicle crashes with motorcyclists. Each crash type and all three crash types together were examined at two severity levels, fatal and injury, and with and without crashes occurring at dawn and dusk. Out of the 24 possible comparisons made, only one achieved statistical significance at the $p<0.05$ level. The statistically significant effect was for light trucks/vans. For these, the presence of daytime running lights reduced these vehicles' involvement in two-vehicle crashes by 5.7%. Taken together, these results suggest that daytime running lamps are of limited value to traffic safety, apart from in high-latitude countries.

All the above has been concerned with the effectiveness of daytime running lights in reducing crashes. However, since the advent of LEDs, they have become an essential element in the styling of the vehicle. Once upon a time, all signal lamps were round. Today, daytime running lamps can be almost any shape from annular to linear to convoluted (Figure 7.9). Whether there is any difference in effectiveness for these different shapes is an open question. There have also been attempts at simplification. For example, on some vehicles, individual daytime running lights have doubled up as turn lamps, with the change in function being indicated by a change in colour from white to amber and from stable to flashing/sweeping light output.

7.15 DEVELOPMENTS

In many ways the development of vehicle signal lighting has been characterized by the piecemeal addition and removal of signals as the need arises. For example, in the UK there was, for a number of years, a requirement for a dim-dip lamp. This was a low-beam headlamp operated at about 20% of normal luminous intensity. It was intended for use in towns where road lighting was present and where front position lamps were considered inadequate but low-beam headlamps too glaring. In the USA, sequential turn signals were used for a number of years in some vehicles. In these, the turn signal did not flash as one unit, rather a number of small light sources were energized in sequence so that there appeared to be a sweeping movement in the direction of the turn or lane change. Both were eventually prohibited for many years but the sweeping turn signal has recently been approved by the FHTSA showing that innovation has not been completely crushed by bureaucracy. Indeed, there have been a number of suggestions made for improving the effectiveness of vehicle

FIGURE 7.9 An array of daytime running lamps showing the influence of vehicle styling on the layout of vehicle signal lighting: (a) Toyota Corolla, (b) Tesla Y, (c) Volvo XC40, (d) Mini Countryman.

signal lamps. These suggestions have varied from those attempting to improve existing signal lamps by increasing visibility and removing ambiguity, through those that aim to increase the amount and type of information conveyed by the signal lamp to those that combine signal lamps with sensors to make the signal more responsive to prevailing conditions.

In the first group comes the suggestion by Mortimer (1981) that rear signal lamps should be colour-coded for function rather than the present situation in USA where rear position lamps, stop lamps, and rear turn signals can all be the same colour. What was suggested was that rear position lamps should be greenish-blue, turn signals should be amber, and stop lamps red. The idea was that by colour-coding, the speed of the response to the signal would be increased. Another suggestion was to change the location and number of all signal lamps so as to minimize the probability of them being hidden by parts of other vehicles. For heavy trucks, this is already common, with many having additional high-level rear position and stop lamps as well as multiple side marker lamps. A similar approach could readily be implemented in small trucks and vans. Even if this were unacceptable, it would be a good idea for the CHMSL in cars really to be mounted high up on the vehicle and not somewhere convenient. Yet another proposal made by several authors is for turn and stop signal lamps to have two levels of light output, one for use by day and one by night (Moore and Rumar, 1999). The idea behind this proposal is that the ambient lighting is very different by day and night, yet current turn and stop lamps have a fixed luminous intensity distribution which is inevitably a compromise between providing a high enough luminous intensity for the lamp to be conspicuous by day without causing glare at night. By having different luminous intensities by day and night, the luminous intensity could be increased by day and decreased by night so that conspicuity is increased by day and glare is reduced at night. Finally, Huhn et al. (1997) have suggested that hazard flashers would be more easily discriminated from turn signals by having these two signals flash at different frequencies, unlike the present situation where they flash at the same frequency. Despite the logic of these proposals and the ease with which they could be implemented most of them have fallen on stony ground.

The second group reflects the desire to provide more information by signal lamps. One example that has already attracted attention is the stop lamp. At the moment, the activation of the stop lamp simply tells drivers behind that the brakes have been applied in the vehicle ahead but nothing about how strongly they have been applied. Horowitz (1994) suggests the use of a combination of flashing and colours to discriminate between deceleration without breaking, sudden accelerator release, antilock braking system activated, and braking at low speeds or stopped. However, early studies of the use of a signal indicating sharp deceleration have given only limited support to its value for enhancing traffic safety (Rutley and Mace, 1969; Voevodsky, 1974; Mortimer, 1981). Since 2013, several attempts have been made by manufacturers to provide what are called emergency stop systems. These modify the existing stop lamps by making them brighter, larger, or flashing in phase at 3–5 Hz when the vehicle's deceleration exceeds a threshold. It remains to be determined if this information is helpful in reducing rear-end crashes when the response is left to the driver behind. It may be that technology has already gone beyond reliance on the driver's response (see Section 15.4). Some vehicles have been fitted with automatic emergency braking in which a camera and sensors detect the rapid deceleration of the vehicle ahead and automatically apply the brakes of the vehicle following. Studies of insurance claims in the USA and the UK have shown that the vehicles equipped with automatic emergency braking are involved in fewer claims for property damage than vehicles without such systems (Highway Loss Data Institute, 2015; Doyle et al., 2015). Further, Tan et al. (2020) have concluded that, depending on market penetration, automatic emergency braking systems have the potential to reduce fatalities and injuries significantly.

Another suggestion is to arrange signal lamps so as to make it easier to identify the type of vehicle at night and to estimate the rate of closure. Identifying the type of vehicle is useful because different vehicles have different dynamics. Estimating the rate of closure is valuable for avoiding rear-end and head-on collisions. One approach to identifying the type of vehicle is to use retro-reflective material to outline the vehicle. Support for this approach comes from the finding that contour lighting of heavy trucks is effective in reducing collisions at night (Schmidt-Clausen and Finsterer, 1989). As for estimating the rate of closure at night, the primary cue used is the angular separation of the rear position lamps of the vehicle ahead, so much so that Janssen et al. (1976) suggested that the separation between rear position lamps should be standardized and set as wide as possible. Estimating relative speed is a particular problem for vehicles with a single rear signal lamp or headlamp, a fact that has led to the idea of fitting motorcycles with a specific pattern of multiple daytime running lamps (CIE, 1993).

The third group, integration with sensors, is already evident in modern vehicles. For example, many cars now have a sensor to activate and deactivate headlamps and position lamps according to the ambient illuminance. It is not too difficult to see a similar sensor being used to adjust the luminous intensity of rear fog lamps and of daytime running lamps so as to maintain a constant level of conspicuity in different ambient conditions. Another role for sensors is to ensure that the correct signal is sent every time it is required. This would overcome the problem of drivers turning or changing lanes without signaling or carrying on straight ahead while signaling a turn (Ponziani, 2006). It should be possible to develop a system whereby any attempt to change lane or turn without signaling would trigger the relevant turn signal, although this would still give little notice to nearby vehicles. More useful would be a more sensitive system to automatically cancel a turn signal after completion of the manoeuvre.

Clearly, there is no shortage of ideas for improving the rather ambiguous and confusing system that currently constitutes regulated vehicle signal lighting (Bullough et al., 2007b). What is required to get some of these proposals implemented is evidence that the perceived problems of current practice are real, that the proposed changes are effective in changing driver behaviour in a desirable direction, that the new equipment is reliable in use and, when installed on a large scale, that the new equipment does indeed reduce crashes, fatalities, and injuries. However, using signal lights to tell other drivers where you are and what you are doing or are about to do may soon be made redundant by technology. Vehicle-to-vehicle communication is developing rapidly and may provide a more immediate and comprehensive means to alert drivers about hazards ahead (see Chapter 15). Even more revolutionary is the move to intervene in the driver's control of the vehicle. The automatic emergency braking system may only be the start of this process. Despite these trends, it is likely that vehicle signal lights will be around for some time yet. This is because for several years there are likely to be a mixture of vehicles with different combinations of basic signal lighting as well as advisory and interventionist automated systems together on the road. There will also be pedestrians and other road users who will always need to be made aware of the presence of a vehicle and its intended movements.

Another reason why vehicle signal lighting is likely to persist is that, today, it has another purpose – to provide an identity for the vehicle. The possibilities for designers created by LEDs has already led to additional lighting being applied to the front and rear of vehicles to create a signature. This is in addition to or combined with the signal lighting required by regulations and is there simply to enhance the customer appeal of the vehicle. It would be interesting to know if such signature lighting, which can be extensive and convoluted, enhances or damages the essential information transfer that vehicle signal lighting is intended to achieve. This question needs to be addressed because the use of lighting to give a vehicle a signature is likely to be encouraged with the introduction of electric vehicles as these do not require a front grill to provide air flow for engine cooling. For more than a century, the front grill has been the signature of many vehicles. If it is no longer required, lighting provides an inexpensive and simple way to reveal identity.

7.16 SUMMARY

Today, signal lighting has two purposes. One is functional, to indicate the presence of or give information about the movement of a vehicle. The other is decorative, to be an important part of the branding of a vehicle by giving it a visual signature. Signal lighting includes front position lamps, rear position lamps, side marker lamps, licence plate lamps, turn lamps, hazard flashers, stop lamps, rear fog lamps, reversing lamps, daytime running lamps, as well as retro-reflectors. Some signal lamps are used only at night or in conditions of poor daytime visibility while others have to be visible at all times, both day and night.

For many years, vehicle signal lighting was based on incandescent light sources but over the last two decades LEDs have taken over. The optical control of vehicle signal lamps is usually done through reflection, refraction, total internal reflection, or by individually addressing arrays of LEDs. Vehicle signal lights are usually contained in a cluster, the structure of which has to meet mechanical, electrical, environmental, and colour objectives.

Vehicle signal lighting is closely regulated. The two most widely recognized bases for regulations are the US Federal Motor Vehicle Safety Standard 108 (FMVSS) and the recommendations of the Economic Commission for Europe (ECE). Most countries follow one or the other of these standards though with modifications to suit local circumstances. Both the FMVSS and ECE recommendations cover the allowed colour of the light, the minimum and maximum luminous intensities that should be provided in different directions, the lit area of the lamp, the allowed positions of the lamp on the vehicle and, if flashing or sweeping is provided, what the frequency of flashing or sweeping should be. Although these recommendations cover many of the same topics, they differ in detail.

Vehicle signal lamps can be divided into two classes: those lamps intended to make the vehicle more conspicuous and which are permanently lit when the vehicle is moving after dark, and those with a wider range of messages that are intermittently lit when the message is sent. The first group includes front and rear position lamps, side marker lamps, rear fog lamps, and retro-reflective markers. The second group includes turn lamps, stop lamps, hazard flashers, reversing lamps, and daytime running lamps. Discriminating the second group from the first is usually done by using a different colour, e.g., reversing lamps, by increasing luminous intensity, e.g., stop lamps, by flashing or sweeping, e.g., turn lamps, or by changing the number of lamps lit, e.g., hazard flashers.

Despite these differences, there is still concern over how rapidly the information presented by a signal lamp can be accessed and how limited the information is, even when it is accessible. There have been a number of suggestions made for improving the effectiveness of vehicle signal lamps. These suggestions have varied from those attempting to improve existing signal lamps by increasing visibility and removing ambiguity, through those that aim to increase the amount and type of information conveyed by the signal lamp to those that combine signal lamps with sensors to make the signal more responsive to prevailing conditions. It would be worthwhile examining some of these suggestions because past innovations in signal lighting have been beneficial. The introduction of the centre high-mounted stop lamp has been estimated to have reduced the number of rear-end crashes by about 4%. Similarly, the widespread adoption of daytime running lamps has been estimated to have reduced multi-party crashes. The history of vehicle signal lighting is one of ad hoc development. The potential for more information to be made available through a combination of sensors, software, and LEDs suggests that a more systematic approach to vehicle signal lighting will be possible in the future.

8 Vehicle Interior Lighting

8.1 INTRODUCTION

From the start, it should be made clear that interior lighting, as discussed here, refers to the lighting of the internal compartments of the vehicle and does not include the lighting of the instrument panel and displays. The lighting of instrument panels and displays is a specialist field that is dealt with using different techniques and technologies (Wada et al., 2007; Chang and Wang, 2007; Maciej and Vollrath, 2009). Having said that, it is as well to recognize that the amount of information now being delivered to drivers via in-vehicle technology using screens has increased enormously, some of it helping with immediate questions facing the driver such as navigation, speed, and lane-changing, some of it being useful for selecting options on how the vehicle will perform and some of it for the purpose of entertainment. The greater the amount of information delivered in this way, the greater is the risk of the driver being distracted from the task of driving with possible consequences for traffic safety (Ziakopoulos et al., 2019; Starkey and Charlton, 2020; Zulkefli et al., 2021). This is an interesting area of study, but it is not vehicle interior lighting.

Until the last two decades, interior lighting was the Cinderella of vehicle lighting. This was for a number of reasons. First, there were no detailed legal requirements for interior lighting. Second, the interior lighting was switched off once the vehicle was moving, so it was considered of little practical significance. Third, there were many other aspects of design that had bigger impacts on vehicle reliability and attractiveness to potential purchasers than interior lighting. However, since the 2000s the situation has changed. There are still no detailed legal requirements for interior lighting, but as vehicle reliability has improved, competition between vehicle manufacturers has increased, the need to find other ways to differentiate one product from another. The number and size of cup holders was one such differentiator; interior lighting is now another. Further, great care goes into the selection of the materials used to furnish the vehicle. The appearance of these materials in daylight and in the showroom has always been considered important, but the impact of interior lighting on the appearance of these materials when entering the vehicle after dark has now begun to be appreciated.

8.2 PURPOSES OF INTERIOR LIGHTING

For many years, the only form of interior lighting was the dome lamp, a simple translucent diffuser mounted on the centre of the roof and operated either automatically, by switches on the doors, or manually, by a switch on the lamp itself. This provided some ambient lighting so that the driver and any passengers were not getting

DOI: 10.1201/9781003388906-8

"

into a black box. The simple dome lamp is now restricted to smaller, less-expensive vehicles (Figure 8.1). The next step up consists of a pair of individually controlled light sources aimed at the individual seats, often with one pair for the front seats and another for the rear seats.

Up-market vehicles have multiple interior lighting systems for a range of purposes. Table 8.1 lists the various types of vehicle interior lighting and the circumstances when they are likely to be used. Given the tendency for innovations to first occur in up-market vehicles and then, if found to be attractive to buyers, to percolate down to less expensive vehicles it is not surprising that dedicated interior lighting for some of these functions has now become commonplace.

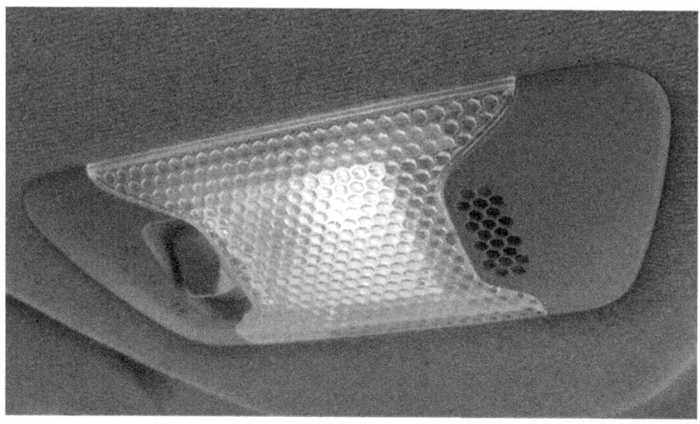

FIGURE 8.1   A simple dome lamp mounted on the centre line just above the windscreen of a Toyota Aygo.

TABLE 8.1

Types of Vehicle Interior Lighting and the Circumstances When They Are Likely to Be Used

Circumstances for Use	Lighting Types
Entering or exiting the vehicle	Door threshold lighting
	Puddle lighting
	Foot well lighting
	Ambient lighting
	Vanity mirror lighting
Getting ready to start driving	Marker lighting of all controls
	Ambient lighting
	Storage lighting
Driving	Reading lighting
Maintenance	Boot lighting
	Engine compartment lighting

The factors that are important in selecting the equipment used to provide the various types of interior lighting are not only light output, power demand, light colour properties, and cost but also the heat gain and depth of the product. The reasons for the importance of the latter two factors is that interior lighting has to be installed within the body shell so minimizing the space taken up by lighting equipment is important for maximizing interior space. Whatever the equipment needed to produce the various types of interior lighting, it has to be controlled. Two forms of control are used, switching and dimming. Where the lighting is confined to a closed space such as the boot, engine compartment, or glove box, or the lighting is unwanted after a specific event, such as the puddle lighting after the door is closed, switching is used. For all others, dimming is used to reduce the light output slowly, usually to zero.

8.3 TYPES OF FUNCTIONAL INTERIOR LIGHTING

8.3.1 DOOR LOCK AND HANDLE LIGHTING

The purpose of door lock and handle lighting is simply to show the driver and passengers where the locks and handles are. Externally this can be done very simply by using LEDs to mark the locations (Figure 8.2). Such door lock and handle lighting is usually activated by the remote key system and will be switched off when the door is closed. Internally, marking the door locks and handles forms part of the marking of controls and facilities (see Section 8.3.5).

8.3.2 DOOR THRESHOLD LIGHTING

Tripping over the threshold of the door will spoil any attempt at elegance. Specific lighting of the threshold can be done by having a wide-beam LED light source mounted in the top or side of the door frame although ambient lighting can also do the job. Such light sources may also be used for puddle lighting or foot well lighting. An alternative is to use OLEDs, electroluminescent strips or light guides mounted actually on the threshold. Specific door threshold lighting is switched on by opening the door and switched off when the door is closed.

8.3.3 PUDDLE LIGHTING

The purpose of puddle lighting is to reveal what is on the ground immediately adjacent to the open door. It gets its name from the risk that there might be a deep puddle just outside the door. Puddle lighting can be provided by a wide-beam LED light source, OLED, or electroluminescent panel mounted either in the door frame, below the storage bin on the door (Figure 8.3), or in the bottom of the door. Puddle lighting is switched on by the opening of the door and switched off when the door is closed.

8.3.4 FOOT WELL LIGHTING

Foot well lighting has two functions, to reveal the interior of the vehicle so that the whole of the cabin can be seen before entering and to help occupants pick up items

FIGURE 8.2 Door handle lighting on a Volvo XC40.

that have been dropped at any time. For the former, all the foot wells have to be lit, while for the latter, only the foot well of interest has to be illuminated. Foot well lighting can be provided by using OLED, LED, or electroluminescent light sources with a diffuse light distribution mounted under the bulkhead for the front foot wells (Figure 8.4) and under the front seats for the rear foot wells. Foot well lighting can be switched on by the operation of the remote key system and dimmed to off with a time delay after the closing of the doors. Ad hoc operation of the foot well lighting during driving requires a separate control.

8.3.5 Marker Lighting of Controls and Facilities

Marker lighting of controls can take two forms, either simple marking which indicates where the control is but relies on some other form of lighting to illuminate how

FIGURE 8.3 Puddle lighting on the passenger door of a Volvo XC40. The light source is mounted under the door storage bin.

to use the control, or marking and illuminating combined. Marking can be done by internally illuminating the controls. Marking and illuminating can be done by having high contrast icons on the controls and then lighting the area around them. The controls in question extend beyond the obvious stalks on the steering column and audio controls to such items as window switches, door handles and locks, seat adjustments, and fuel tank cap release. Anyone who has got into an unfamiliar rental car after dark is aware of the value of being able to easily see, identify, and understand all the things that need to be adjusted before and during a journey. Internal illumination of controls can be done using individual LEDs to deliver the light where it is needed. Marking and illuminating can be done by either LEDs mounted close to the controls or general ambient lighting. In addition to the controls, the interior will contain other facilities such as cup holders and storage compartments. These can be more easily located if they are outlined using LED light pipes. Lighting for all the

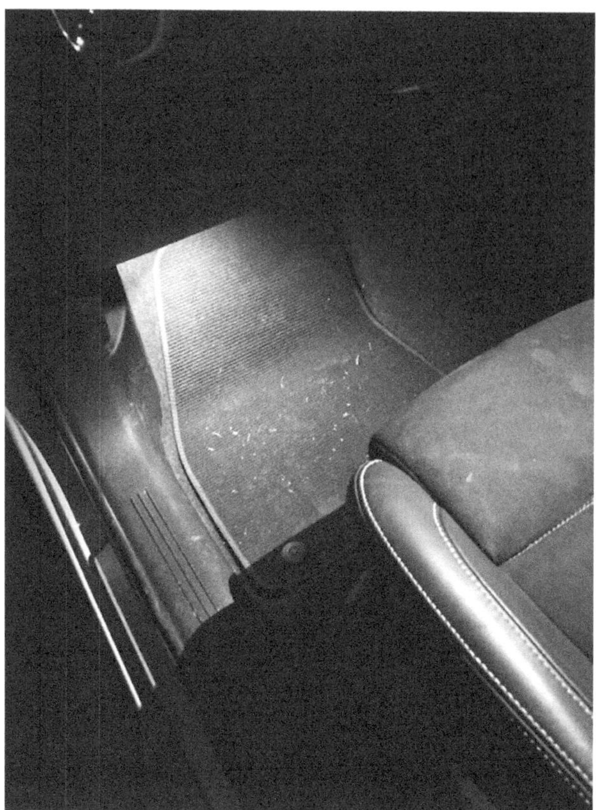

FIGURE 8.4 Foot well lighting ahead of the front passenger seat in a Volvo XC40.

controls and facilities is activated by the insertion of the key into the ignition system before starting the engine.

8.3.6 AMBIENT LIGHTING

Ambient lighting is designed to illuminate the whole of the vehicle cabin. In small, low-cost vehicles it is the only form of interior lighting provided (Figure 8.1). In larger vehicles, the pairs of lights aimed at the individual seats are used for ambient lighting (Figures 8.5 and 8.6).

Usually, ambient lighting is mounted on the centre line of the roof. The light distribution is diffuse. Ambient lighting is switched on when any door is unlocked and remains on until all the doors are closed after which there is a slow decline in light output to extinction, typically over a period of 5–15 s. This allows the people to check for any unwanted passengers before entering the vehicle and to orientate themselves after entering the vehicle. The ambient lighting is usually extinguished while driving unless an override switch is used. On stopping the vehicle and removing the ignition key, the ambient lighting is switched on and stays on for a few seconds to

FIGURE 8.5   A view of the interior lighting of the front seats of a Volvo XC40. The lighting provides ambient lighting from a pair of individually controlled light sources mounted on the centre line, each light source being aimed at a different seat. There is another pair of light sources providing ambient lighting for the rear seats (see Figure 8.6).

allow time for the doors to be opened. Opening the doors returns the ambient lighting to full-light output where it stays until all the doors are shut after which there is a slow dimming to extinction.

8.3.7   STORAGE LIGHTING

Most vehicles now contain a number of separate boxes and bins for storing items, such as glove boxes, door bins, and parts of the central console (Figure 8.7). Finding an item in one of these boxes or bins without some form of illumination can be difficult. Glove box lighting is designed to illuminate the inside of the box and is switched on and off when the lid is opened and closed. Storage compartments in the console may use the same system or may use an overhead lamp with a narrow beam and a manual switch. The problem with storage lighting is usually that items inside the box or bin may obstruct the lighting. Where the lighting is built into the box the solution to this problem is to use an area light source such as OLEDs or electroluminescent panels. Where the lighting is remote, there is little that can be done.

8.3.8   VANITY MIRROR LIGHTING

Vanity mirrors are commonly mounted on the back of sun visors for use by the driver and front seat passenger and, sometimes, in the back of the front seats for use by rear seat passengers. When the visor is lowered, the mirror becomes available.

FIGURE 8.6 A view of the interior lighting of the rear seats of a Volvo XC40. The lighting is provided by a pair of individually controlled light sources mounted on the centre line, each light source being aimed at a different seat. There is another pair of light sources for the front seats (see Figure 8.5).

The lighting for use with vanity mirrors is intended to illuminate the face of the person using the mirror and is usually done by LED light sources mounted around the mirror or close by in the headliner (Figure 8.8). These are activated by opening the cover over the mirror and extinguished by the closing of the cover. Sometimes they have a dimming facility. However, such lighting is often unsatisfactory at night because the small size of the mirror and the position of the lamps close to the mirror mean that the lamps are directly in front of the person looking into the mirror. In this situation, the lighting may produce glare to the viewer and may not illuminate the sides of the face well. A better solution would be to light the face of the viewer by more widely spaced lamps, as would be the case in bathroom vanity mirror lighting.

FIGURE 8.7 Lighting of a storage bin on the front passenger door of a Volvo XC40. The light source is mounted under the armrest on the door and cannot be seen by the driver or passenger.

FIGURE 8.8 Vanity mirror lighting in a Skoda Karoq. Note the proximity of the lighting to the mirror.

8.3.9 Reading Lighting

Today, more and more vehicles make provision for the entertainment of passengers during a journey through a touch screen providing access to radio, mobile phones, streaming services, and social media. For others who simply want peace and quiet and prefer the printed page to a self-luminous screen, a reading lamp can be provided for each seat. The essence of the reading lamp is that it should provide a tightly controlled beam of light that produces a sharp-edged area of relatively high illuminance. A tightly controlled beam is necessary to limit the amount of light that will reach the eyes of the driver directly. Some additional light at the driver's eyes is inevitable when reading lamps are used because light will be diffusely reflected from most printed publications, but unless the interior of the vehicle is finished in very light materials, this should not be enough to cause disturbance to the driver. Reading lamps can use LED light sources. They are usually controlled by manual switching by the user, but there is also a lot to be said for having a dimming control so that the amount of light can be adjusted by the user.

8.3.10 Boot Lighting

The boot or trunk is often filled with loot or junk. Lighting is necessary to find what is being sought after dark. Like the storage boxes and bins in the interior of a vehicle, the problem facing boot lighting is obstruction. One solution is to use large area light sources such as OLEDs, or electroluminescent panels. These lamps have a diffuse light distribution. The boot lighting is switched on and off by the opening and closing of the boot lid. In the USA, it is now a requirement for there to be an emergency release latch that can be operated by someone trapped inside the boot. A simple marking system for this latch, employing an LED, would be of value.

8.3.11 Engine Compartment Lighting

Occasionally, someone may need to access the engine compartment after dark. Lighting that is switched on and off by the opening and closing of the bonnet is useful. As is the case for the boot, the main problem for lighting is obstruction, most engine compartments being very full. Again, the solution is to use a large area light source with a diffuse light distribution rather than a single small lamp. It also helps if the underside of the bonnet has a high reflectance finish.

8.4 DECORATION AND CUSTOMIZING

So far, this discussion of interior lighting has been confined to functional factors. However, vehicle interior lighting has increasingly become one of the means used to give added value to a vehicle and to establish a brand image. This implies that interior lighting has not only to fulfil its functions but also to be a coherent part of the design. The concept behind this approach to vehicle interior lighting is that every lighting installation does two things: it makes whatever it illuminates visible and it sends a message about the people who bought or use the installation. Probably the most obvious example of this approach in buildings is the lighting of shops. In retail premises, the lighting is designed to display the merchandise to advantage and to deliver a message about the nature of the shop, the type of service that can be expected, and the clientele the owners are trying to attract. For example, anyone viewing a shop lit by a uniform array of bare lamps is likely to conclude that what is on offer is going to be cheap and of moderate quality, the service will be self-service, and the clientele will be those who are mainly interested in value for money. For the interior lighting of a car, the functional requirements discussed above should be met but consideration also needs to be given to the message sent. It could be argued that meeting the functional requirements alone sends a message, and so it does, but there are other messages that might be sent. For example, it is possible to use lighting to emphasize the style of the interior, the quality of the materials, and the status of the occupants. If any of these possibilities are of interest, consideration has to be given to the lighting of the whole interior space and the people in it rather than just the locations where tasks have to be performed.

There are a number of different technologies available for this purpose. In addition to the incandescent and LED light sources, there is the possibility of using organic LEDs which are flat area light sources that can be formed into complex shapes (Kraus et al., 2007). Another possibility is to use electroluminescent foils. Again, these are flat area light sources available in a range of colours. There are also different light distribution systems (Jalink, 2002). Conventionally, electricity is distributed to the point where light is required by wire and there it is converted into light by the light source. An alternative is to generate light remotely by grouping LEDs into what is called a light engine and distributing the light emitted to the desired locations using a fibre optic network. The advantage of such remote lighting is that fibre optic cable is very fine and so requires very little depth, the light engine being located where space is available. The disadvantage is that a light source failure extinguishes a part or the whole of the network. Such systems can be used to produce starlight patterns over the headliner that are purely decorative and can be customized. Further, by using monochromatic LEDs, specific coloured light can be generated giving the occupants the ability to choose the lighting colour they find most pleasant (Blankenbach et al., 2020). This ability to customize the interior lighting can be enhanced by using mixtures of red, green, and blue LEDs in the light engine so that by adjusting the light output of each element, a wide range of light colours can be generated. In addition to light colour, customizing requires the ability to adjust the amount of light by dimming and ideally the ability to adjust the shape of any beam. All these options may offer the occupants of the vehicle too much choice, in which case preselected scenes can be offered making switching between them a simple matter of touching a screen. Individual control of lighting has been available in buildings for many years. If what usually happens in buildings is followed in vehicles, it is likely that a new owner will play around to identify lighting conditions they find attractive after which few changes will be made. Systems that allow the driver or passengers to customize the interior lighting are available on up-market vehicles.

8.5 INTERIOR LIGHTING RECOMMENDATIONS

While there are no detailed legal requirements for vehicle interior lighting, there are some quantitative recommendations for functional lighting derived from preferences (Lighting Research Center, 1996) or based on current practice (Wordenweber et al., 2007). Table 8.2 is an amalgamation of these recommendations.

The mean illuminances are consistent with what is known about the interior lighting of buildings, specifically, the illuminance requirements for safe movement under emergency lighting (Boyce, 2014) and for ensuring adequate visual performance for high-contrast printing (Rea and Ouellette, 1991). Where ambient lighting is the only form of interior lighting, a reasonable mean illuminance is 20 lx measured on a horizontal plane at the height of the seat cushions. There are no illuminance uniformity criteria for the door lock and handle lighting and for the marking of the controls.

TABLE 8.2

Vehicle Interior Lighting Recommendations

Function	Mean Illuminance (lx)	Illuminance Uniformity (Maximum/ Minimum)	CIE General Colour Rendering Index	Coverage
Ambient lighting only	20	15:1	>80	Horizontal plane over whole cabin at seat level
Door lock and handle	1	--	–	Local area of lock and door handle
Door threshold	5	15:1	>50	Whole threshold
Puddle light	10	15:1	>50	1 m out from door and full width of door
Foot well lighting	10	15:1	>50	Whole foot well without person present
Marking and illuminating controls	1	–	–	Local area of controls
Storage lighting	20	15:1	>50	Floor of storage box when empty
Vanity mirror lighting	20	5:1	>80	Area of face when using mirror
Reading lighting	80	3:1	>80	300 mm diameter disc on lap of reader
Boot lighting	20	15:1	>50	Floor of boot when empty
Engine compartment lighting	20	15:1	>50	Top of engine compartment with bonnet raised

This is because such lighting is designed to mark where the lock, handle, and the controls are, not to illuminate these objects. Higher CIE colour rendering index limits are given for applications where the appearance of skin tones and the interior finishes of the vehicle are important, e.g., vanity mirror lighting, than for applications where exposure is brief, e.g., puddle lighting.

8.6 PERCEPTIONS

There are clearly plenty of opportunities to provide different forms of interior lighting in vehicles but how are they perceived by the driver? Caberletti et al. (2010) carried out an evaluation of 12 different forms of ambient interior lighting using a BMW 3 Series car connected to a driving simulator providing a 135-degree field of view. The room was blacked out so that the driver appeared to be following another car along a road with a surface luminance in the range 0.1–1.5 cd.m^{-2}. The car was connected to the simulator in such a way that the driver could steer but could not change the speed which was fixed at 100 km.h^{-1} (62 mph). In successive 3-min periods, 31 drivers experienced 12 different ambient lighting conditions. The ambient lighting was focused on the central console, the foot wells, and the doors, with the door handles and pulls sometimes accented. The ambient lighting was varied in brightness providing three luminances of the illuminated areas (high = > 0.04, middle = > 0.01 and < 0.02, low = 0.007 cd.m^{-2}). The car's instrument panel and touch screen were lit at a fixed, low-level representative of current practice while driving. The 12 ambient lighting conditions examined are listed in Table 8.3. During each minute of the 3-min drive, the driver was given a secondary task to do such as finding a particular control button. During the 3-min drive the driver's performance was

---

TABLE 8.3

Lighting Scenarios Used to Examine the Perception of Ambient Interior Lighting of a Car while Driving (from Caberletti et al., 2010)

Scenario Number	Ambient Lighting	Brightness	Colour
1	Doors with accents on handles, centre console, and foot wells	High	Orange
2	Centre console and upper part of doors	Middle	Orange
3	Doors	High	Orange
4	Doors	Low	Orange
5	Without any ambient lighting	–	–
6	Doors, centre console, and foot wells	High	Orange
7	Doors, centre console, and foot wells	Low	Orange
8	Doors, centre console, and foot wells	Middle	Orange
9	Foot wells	High	Orange
10	Foot wells	Low	Orange
11	Centre console	Middle	Orange
12	Doors, centre console, and foot wells	Low	Blue

measured by the standard deviation of the distance from the road's edge. After three minutes, the drive was stopped, the ambient lighting was turned off, and the driver was asked to complete a questionnaire using only a reading lamp delivering 10 lx on the form which is enough to ensure it could be read. The questionnaire consisted of 18 semantic differential pairs addressing perceptions of space, comfort, interior quality, interior attractiveness, safety, alertness, and functionality. An assessment of the driver's emotional state was also made using a self-assessment manikin.

The results showed no statistically significant effects of the ambient lighting conditions on the driver's emotional state or their driving performance. This is not surprising given the limited duration of each drive. However, there were statistically significant differences found in the questionnaire. For brightness, it was found that the presence of ambient lighting at a low level rather than no ambient lighting (condition 7 versus 5) made the car seem more spacious, more attractive, more comfortable, of better quality, and having better perceived safety. But there was a limit to these positive results. Compared to no ambient lighting, increasing the brightness to a high level with accents (condition 1 versus 5) reduced the number of statistically significant questions from 16 to 4 showing decreases in pleasantness, comfort, and perceived safety as well as more distraction to the driver. This suggests that only a low level of ambient lighting is required to enhance the perceptions of the driver when driving. This conclusion is supported by another comparison. Comparing ambient lighting of the central console alone with no ambient lighting (conditions 11 and 5) showed statistically significant improvements in interior attractiveness, quality, spaciousness, and functionality.

As for the different positions, lighting of the foot wells alone to a high level relative to the lighting of the doors alone to a low level (condition 9 versus 4) was considered cheap, less comfortable, and less pleasant. There was also a comparison between the colour of the ambient lighting. For conditions 1 to 11, orange LED lighting (peak wavelength 605 nm) was used, but for condition 12, blue LED lighting (peak wavelength 471 nm) was used. For the same brightness setting and positions (conditions 7 versus 12), the blue light was seen as brighter and helped in locating control elements but was also considered uncomfortable. Orange light looked more luxurious and gave a perception of better quality, although this may be influenced by the colour of the materials used in the interior.

Weirich et al. (2022) carried out a more extensive study of colour preference for vehicle interior lighting using an open on-line survey. In this study, 238 drivers from Europe and China responded to schematic drawings of the interior of a vehicle viewed looking forward and by light pipes emitting light of ten different colours and in nine different layouts. Seven of the colours used were monochromatic and arranged around the hue circle (red, orange, yellow, green, cyan, blue, and purple). The other three were white with different correlated colour temperatures (5,000 K, 6,500 K, and 10,000 K). The least-preferred monochromatic light colours were red and purple and the most-preferred were cyan and blue. As for the white colours, the most-preferred correlated colour temperature was 5,000 K.

The obvious conclusion from these studies is that care is required in the design of ambient lighting if negative perceptions are to be avoided. When driving, the overall appearance of the car is improved with a low level of ambient lighting but too much brightness, aimed at unimportant surfaces, will not be appreciated. Higher brightness ambient lighting might well be welcome when entering or leaving the vehicle. Light colour also needs careful consideration, the degree of satisfaction likely being linked to the appearance of the occupants of the vehicle and the materials used in the interior as well as any cultural significance the colour may have.

8.7 DISTURBANCE TO THE DRIVER

Vehicle interior lighting used when entering or exiting a car has no impact on the driver. However, interior lighting used when the vehicle is in motion may be disturbing to the driver. This can occur even when the ambient lighting is not directly visible to the driver, being outside the driver's field of view when looking ahead. The reason such interior lighting may cause disturbance is that the light produced is diffusely reflected around the interior of the vehicle including from the windscreen. Light reflected from the windscreen has two potentially disturbing effects. First, it produces a veil of light over the view of the road ahead which reduces the luminance contrast of the scene. The higher the luminance reflected from the windscreen, the more difficult it will be to see the road ahead and anything on it (Devonshire and Flannagan, 2007). Second, the reflections from the windscreen form an image of the interior of the vehicle so the driver has two alternative views of the world, the road ahead and the interior of the vehicle. These effects imply that the reflectances of the interior finishes are important, particularly those of the dashboard and bulkhead. If these reflectances are high, the risk of disturbance to the driver's ability to see down the road at night is increased. To accommodate the different interior finishes used in the same vehicle, it is essential to provide a dimming system for all the interior lighting, similar to that provided to allow the driver to adjust the brightness of the instrument display.

Another way to consider the problem of disturbance is to assess the illuminance received at the driver's eyes. The higher the illuminance received at the driver's eye from the interior lighting, the more likely it is that disturbance will occur. Table 8.4 shows the percentage of drivers disturbed by various forms of interior lighting and the illuminance received at the eye from the interior lighting installed in a Cadillac (Lighting Research Centre, 1996). It is clear that the illuminance received at the eye is an important factor in determining the disturbance felt by the driver, but the variation in the percentage of disturbance at similar illuminances suggests it is not the whole story. The other factor is most likely the distraction produced when the lamp providing the illuminance is itself visible to the driver, e.g., the glove box lamp when the glove box is open. In passing, it is worth pointing out that a similar pattern of results probably exists for stray light from information and entertainment displays in use during driving.

Another important factor is the location of the areas lit. In the Weirich et al. (2022) study described above (see Section 8.6), the preference for the areas lit followed the closeness to the driver's line of sight. The most preferred lit areas were the lower half of the doors, foot wells, and seats. The least preferred lit areas were

TABLE 8.4

Percentage of Drivers Disturbed and the Illuminance (lx) Received at Their Eyes from Various Forms of Interior Lighting in a Left-hand Driver Car (from LRC, 1996)

Source	Illuminance at the Eye (lx)	Percentage Disturbed
No interior lighting	0.11	0
Rear foot well lighting	0.11	5
Glove box lighting	0.12	50
Front foot well lighting	0.14	15
Left rear door ambient lighting	0.16	39
Right rear door ambient lighting	0.18	55
Rear reading lighting	0.25	50
Right front door ambient lighting	0.30	49
Front reading lighting	0.33	40
Dome light ambient lighting	0.60	90
Left front door ambient lighting	1.05	90

around the steering wheel and on the A-pillars, with the absolute worst being when all nine different positions were lit. These results support the opinion that it is easy to disturb the driver with vehicle interior lighting. Careful design is required, and a distinction needs to be made between any lighting intended to welcome the driver and passengers to the vehicle and the lighting intended for use when the vehicle is in motion.

8.8 SUMMARY

For many decades, interior lighting was the Cinderella of vehicle lighting, but since 2000 this has changed. Since then, it has become increasingly important to differentiate one product from another. Interior lighting is one way to do this. For many years, the only form of interior lighting was the dome light, a simple translucent diffuser mounted in the centre of the roof and operated either automatically, by switches on the doors, or manually, by a switch on the lamp itself. These have now become restricted to small, low-cost vehicles. The next step up is to provide individual lighting to individual seats, both front and rear. Up-market vehicles have multiple interior lighting systems designed to fulfil a range of functions.

The more sophisticated interior lighting systems provide different lighting for entering and exiting the vehicle, preparing to drive, actually driving, and vehicle loading and maintenance. These systems consist of different combinations of door lock and handle lighting, door threshold lighting, puddle lighting, foot well lighting, marking of controls, storage lighting, vanity mirror lighting, ambient lighting, reading lighting, boot lighting, and engine compartment lighting. The technologies used to provide this lighting commonly involve the use of mainly LED light sources, the light being distributed through a conventional electric network or through a fibre optic network.

There are no detailed legal requirements for interior lighting but there are lighting recommendations. The main restriction on interior lighting is the need to avoid interfering with the vision of the driver when driving. This can be done by limiting the amount of light reaching the driver's eyes directly and by reflection from the windscreen. Achieving such limits requires attention to the distribution of light within the vehicle and the ability to dim the interior lighting as desired, particularly where the interior is finished in high reflectance materials.

The role of lighting in many buildings today is expanding from the visual performance of tasks alone to the appearance of the space as well. The interior lighting of vehicles is following the same path. This means that the interior lighting of vehicles is becoming concerned with revealing the style of the interior, the quality of the materials, and the status of the occupants, as well as fulfilling its functions. Cinderella has gone to the ball.

9 The Interaction between Road and Vehicle Forward Lighting

9.1 INTRODUCTION

Up to this point, road and vehicle lighting have been considered separately. This is a reflection of current practice. Road lighting is designed to promote visibility without reference to vehicle lighting. Vehicle lighting is designed to provide visibility in the absence of road lighting and, with the exception of the town beam proposed as part of the adaptive forward lighting system (see Section 6.7.3), ignores the existence of road lighting. The justification for this apartheid is that some road lighting has to provide visibility for road users who have little by way of lighting, such as cyclists and pedestrians, and vehicle lighting has to be adequate where there is no road lighting. While there is some truth in these explanations, with today's technology there is less and less excuse for failing to consider how current road and vehicle lighting interact and how one or the other might be adjusted to enhance their combined effect.

9.2 EFFECTS ON VISIBILITY

The usual way that the interaction between road lighting and vehicle lighting is considered is as the impact of introducing vehicle forward lighting into a road lighting installation (Gallagher et al., 1974; Janoff, 1992b; Bacelar, 2004). This is reasonable for anyone primarily interested in road lighting but unreasonable from the point of view of the driver. The driver always has vehicle lighting but does not always have road lighting. The approach used here will be to consider the integration of road lighting with existing vehicle lighting.

Both vehicle forward lighting and road lighting are designed to make what is ahead visible to the driver. For objects ahead to be visible they have to have a visual size and a luminance contrast or a colour difference above threshold. Lighting can do little to change the visual size, and the colour difference is only of importance when luminance contrast is low. Further, several of the light sources used for road lighting in the past have had non-existent or poor colour-rendering properties. Therefore, the most fitting way to examine the effect of adding road lighting to existing vehicle forward lighting is to estimate the consequences for luminance contrast. The first step in this process is to look at the illuminances received by a target from both vehicle forward lighting and road lighting at different distances from the vehicle. Figure 9.1

DOI: 10.1201/9781003388906-9

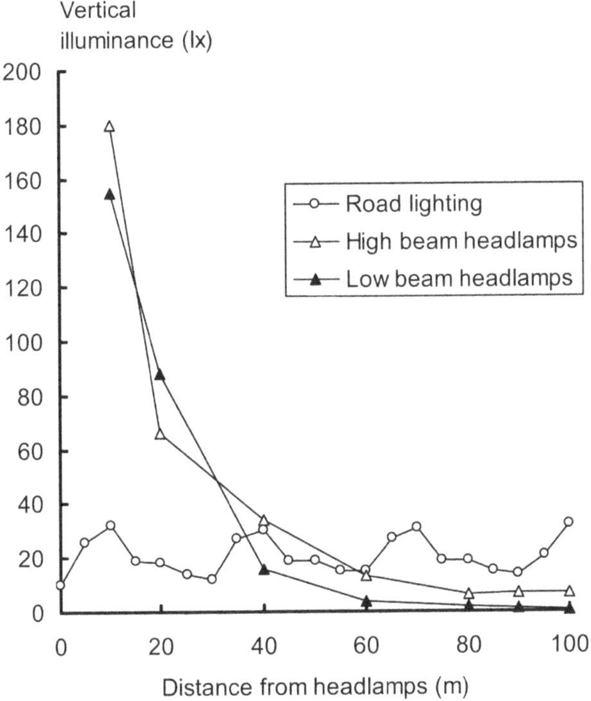

FIGURE 9.1 Vertical illuminance (lx) at road level plotted against the distance from the headlamps (m) for road lighting alone, low beam headlamps alone and high beam headlamps alone. The spacing between the road lighting columns was thirty metres (after Bacelar, 2004).

shows the illuminances on a square target of side 20 cm with a reflectance of 0.2, placed vertically on the road at different distances from the vehicle and oriented so that the normal to the plane of the target is along the axis of the road (Bacelar, 2004). The vehicle forward lighting used was that fitted to a Renault Clio, this being halogen headlamps conforming to the ECE recommendations. The road lighting consisted of a single-sided layout of five luminaires at 30-m spacing. Each luminaire contained a 150 W high-pressure sodium light source and was mounted 8 m above the road surface. The road surface had reflection characteristics where $Q_0 = 0.08$ and $S1 = 0.42$ meaning it lies at the boundary of road classes R1 and R2 (see Table 4.11) and is representative of many pavement materials. The resulting photometric characteristics of the road lighting were average road surface luminance = 2.45 cd.m^{-2}, overall luminance uniformity ratio = 0.6, and longitudinal luminance uniformity ratio = 0.7.

From Figure 9.1, it can be seen that the distance from the vehicle can be divided into three zones. From 10 m to 40 m from the vehicle, the illuminance on the vertical target is largely due to the vehicle forward lighting. Between 40 m and 60 m, the road lighting and vehicle forward lighting make similar contributions to the vertical illuminance. Beyond 60 m from the vehicle, road lighting makes the major contribution to the illumination of the vertical target, particularly when low beam headlamps are used. Of course, these boundaries are somewhat movable, depending on the forms of the vehicle forward lighting and the road lighting. The road lighting used by Bacelar (2004) produces a higher average road surface luminance than is normally recommended (see Section 4.4). For road lighting producing lower average road surface luminances but with the same light distribution, it can be assumed that the boundaries of the three zones will be shifted further away from the vehicle. The same is true for vehicles equipped with HID or LED headlamps. Nonetheless, there will still be three zones, one where vehicle forward lighting is dominant, one where road lighting is dominant, and one where the two forms of lighting are approximately equal.

For the distant zone, where very little light from the vehicle forward lighting reaches the target, the presence of road lighting will usually increase the target's visibility. This is measured as the visibility level, which is the ratio of the actual luminance contrast of the target to the threshold luminance contrast of the target. Increasing the adaptation luminance by increasing the road surface luminance using road lighting will increase the contrast sensitivity of the visual system (see Section 3.4.3), thereby reducing the threshold luminance contrast and tending to increase the visibility levels of all targets. While this is generally true, there are some targets for which the visibility level will be reduced. This is because the actual luminance contrast of the target may be reduced by the use of road lighting. One factor that determines whether or not this happens is the relative reflection characteristics of the target and the road surface. Targets that are seen in negative luminance contrast against the road surface, i.e., darker than the road surface, will show an increased luminance contrast when road lighting is introduced unless the luminance of the target is increased proportionally more than the road. Targets that are seen in positive luminance contrast, i.e., brighter than the road, may show a decreased luminance contrast when the road surface luminance is increased, unless the luminance of the target is increased proportionally. Another important factor is the luminance uniformity of the road lighting. Guler and Onaygil (2003) have shown that road lighting with overall and longitudinal luminance uniform ratios below the minima recommended tend to have larger areas where visibility levels are close to zero.

What this means is that, for the distant zone, introducing road lighting meeting the recommendations will generally increase visibility but may reduce it for specific targets. Within this zone, the range over which targets will remain visible will depend on their visual size. Threshold luminance contrast increases with decreasing visual size (see Figure 3.7) so the visibility level of a target will decrease as the distance between the observer and the target increases until the threshold luminance contrast approaches the actual luminance contrast and the target disappears.

For the near zone, the illuminance on the target is dominated by the vehicle forward lighting, as is the road surface luminance. The increase in adaptation luminance produced by introducing road lighting will again reduce threshold luminance contrasts although because of the dominance of the vehicle forward lighting, this effect will be small. As for the actual luminance contrast, the impact of introducing road lighting will depend on the relative increases in luminance produced for the target and the road surface. Given that road lighting is designed primarily to light the horizontal road surface, it is likely that the increase in road surface luminance will be greater than the luminance of the vertical target. This implies that for targets seen in positive luminance contrast against the road surface when lit by vehicle forward lighting alone, the actual luminance contrast will be reduced. Whether this reduction leads to a decreased visibility level will depend on the extent to which the decrease in actual luminance contrast is compensated by the reduction in threshold luminance contrast. For targets seen in negative luminance contrast against the road surface when lit by vehicle forward lighting alone, the introduction of road lighting will most likely lead to an increase in actual luminance contrast which, together with the reduction in threshold luminance contrast, will always produce an increase in visibility level.

It is in the intermediate zone that things get really interesting. In this zone both road lighting and vehicle forward lighting make similar contributions although the road lighting emphasizes the horizontal road surface while vehicle forward lighting emphasizes the vertical target. Bacelar (2004) has calculated visibility level from measurements of target and background luminance for the conditions described above and using the model of target visibility developed by Adrian (1989). The target was placed at a constant distance of 40 m from the vehicle for low beam headlamps and 90 m for high-beam headlamps. Forty metres is assumed to be the stopping distance for inner-city areas where vehicle speeds are of the order of 50 km.h^{-1} (31 mph) and 90 m is the stopping distance for suburban areas where vehicle speeds are in the range 75–110 km.h^{-1} (47–68 mph). The target was moved in 5-m steps along the road between the second and third road lighting columns, successive columns being separated by 30 m. Figure 9.2 shows the variation in visibility level for headlamps alone, road lighting alone, and headlamps and road lighting together.

For headlamps alone, the visibility levels are constant because the target is at a constant distance from the vehicle. Visibility levels are lower at 90 m using high-beam headlamps alone than at 40 m using low beam headlamps alone because of the lower illuminance on and smaller angular size of the target at the greater distance. A lower illuminance implies a lower adaptation luminance and consequently a higher threshold luminance contrast, as does the smaller angular size of the target.

For road lighting alone, there is some variation in visibility level because of the variations in illuminance on the road and target at different positions relative to the road lighting luminaires. The visibility levels are lower at 90 m than at 40 m because of the smaller visual size of the target. When the target is between 5 m and 20 m from the column and 40 m from the vehicle, low beam headlamps and road lighting together produce lower visibility levels than either system alone. When the target is 90 m from the vehicle and high-beam headlamps are used, high-beam headlamps and road lighting together produce lower visibility levels than either system alone, at all positions. A similar pattern of visibility levels for different combinations of vehicle forward lighting, road lighting, and target reflectances has been found by others (Akashi et al., 2003; Guler et al., 2005; Bullough and Rea, 2010; Erkan et al., 2023, Vogel et al., 2024).

So far, this discussion of visibility has concentrated on the effect of introducing road lighting on the luminance contrast of the target and the adaptation luminance

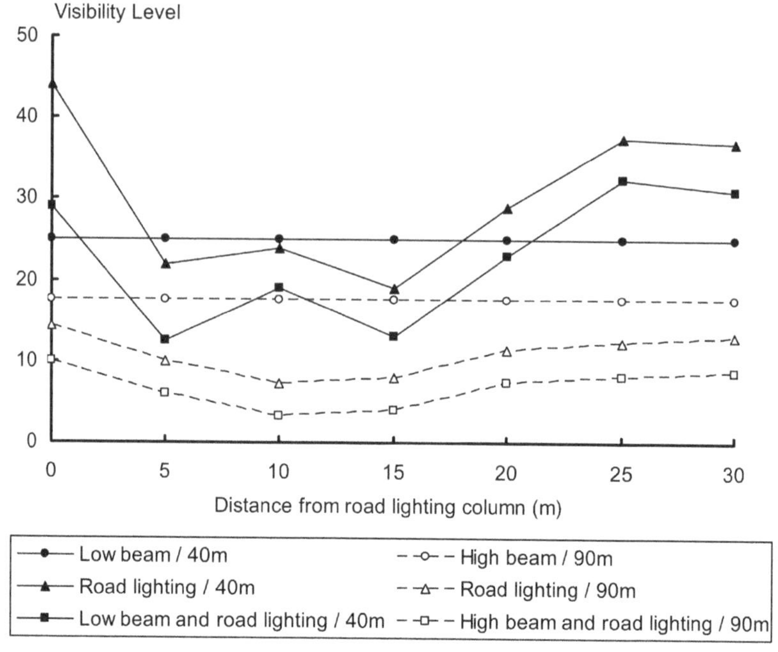

FIGURE 9.2 Visibility levels calculated from luminance measurements taken for a vertical target of reflectance 0.2 at a distance from the driver of 40 m for low-beam headlamps and 90 m for high-beam headlamps, plotted against the position of the target relative to a road lighting column (m). Successive road lighting columns were separated by 30 m (after Bacelar, 2004).

of the visual system, but there is another effect that needs to be considered. This is the effect on disability glare from the road lighting (see Section 4.3). Fortunately, Bacelar (2004) also calculated the visibility level of the target in a fixed position relative to the road lighting and 40 m ahead of the vehicle, which was using low beam headlamps. Figure 9.3 shows that, for this position, introducing road lighting results in an increase in the visibility level from 25 m to 33 m, suggesting that, with respect to visibility, the changes in road surface luminance and the actual luminance contrast of the target caused by adding road lighting are more than enough to offset any additional scattered light in the eye.

Of course, this may not always be true. Another interaction of road lighting and vehicle lighting involves the change in the effects of the disability glare caused by the headlamps of opposing vehicles. Bacelar (2004) also reports changes in visibility level for the target positioned at a fixed point 40 m ahead of a vehicle using low beam headlamps, with and without road lighting, in the presence of one or three opposing vehicles using low beam headlamps. Figure 9.3 shows the calculated visibility levels, for different distances between the vehicles, with and without the road lighting. It is clear that disability glare from opposing vehicles reduces visibility levels, that three opposing vehicles produce greater reductions in visibility levels than one

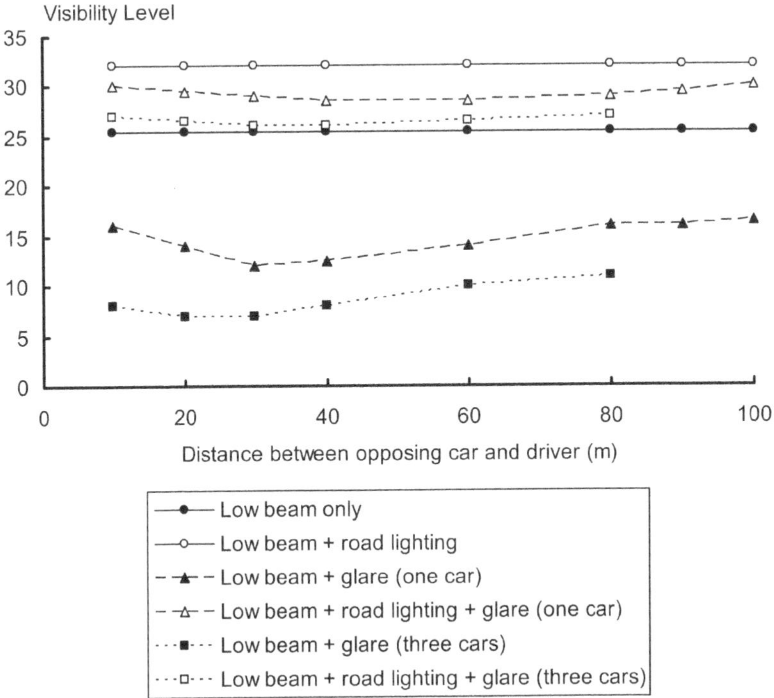

FIGURE 9.3 Visibility levels calculated from luminance measurements taken for a vertical target of reflectance 0.2 at a fixed position relative to the road lighting and 40 m from the driver of a vehicle using low beam headlamps, with and without road lighting, and with none, one, or three opposing vehicles, plotted against the distance (m) between the opposing vehicles and the driver (after Bacelar, 2004).

opposing vehicle and that the reduction in visibility level caused by disability glare from opposing vehicles is less when road lighting is present.

These results suggest three conclusions. The first is that introducing road lighting is likely to improve the visibility of most targets, particularly when they are in the distant zone. The second is that there can be no guarantees that visibility will improve for all targets. There are some targets for which the combination of light distributions from the vehicle forward lighting and road lighting and the reflection properties of the target and the road surface may lead to reduced visibility. The third is that introducing road lighting alleviates the effects of disability glare from opposing vehicles on visibility.

The question that now arises is what is the visibility level that is enough for driving. If the changes in visibility level discussed above leave the driver with sufficient visibility, these changes are of little significance. If the changes in visibility level are enough to make the target difficult to detect, they are important. Gallagher and Meguire (1975) placed a cone as an obstacle on a lit public road and observed the distance at which unwarned drivers took action to avoid colliding with the cone, as well as the speed of the vehicle at the time. The combination of distance and speed was used to calculate the time to target. Blackwell and Blackwell (1977) calculated the visibility levels for the cone target from luminance measurements taken at the site of the experiment. Their results showed that time to target increased with increasing visibility level (Figure 9.4) following a law of diminishing returns with little improvement in time to target occurring above a visibility level of 20.

In another study, Brusque et al. (1999) showed observers a series of street scenes, some of which had a pedestrian at the side of the road or on the carriageway 65 m away. The presentation time was 100 m. The street scenes varied in the complexity of the background, quantified as the uniformity of background luminance (minimum/maximum). Figure 9.5 shows the percentage of presentations in which the pedestrian was detected and correctly located plotted against the visibility level of the pedestrian in a low complexity urban environment. Figure 9.6 shows the rating of ease of detection of the pedestrian plotted against the visibility level of the pedestrian.

Again, it seems that a visibility level of around 25 is required for a high level of correct detection that is visually easy. It should be noted that this is somewhat higher than the visibility levels of 10–20 suggested as necessary for safe driving under road lighting by Adrian (1987a) but less than the visibility level of 30 recommended for driving at night by Hills (1976).

Given that a visibility level of about 25 is desirable for easy and accurate detection of targets on and near the road and that the combination of vehicle forward lighting and road lighting can reduce the visibility level of some targets to considerably less than this level (see Figure 9.2), why is difficulty to detect targets on the road when road lighting is present not reported by drivers? One answer is that an important class of targets, other vehicles on the road, have very high, positive visibility levels, guaranteed by the presence of lit headlamps and rear position lamps. Introducing road lighting will reduce the visibility levels of such vehicles but not by enough to

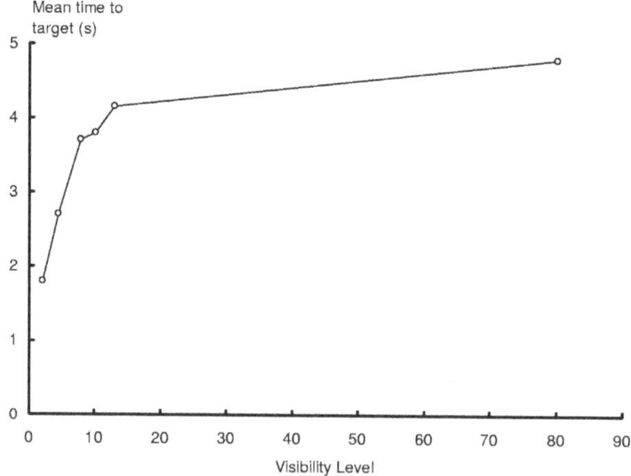

FIGURE 9.4 Mean time to target (s) when the first response to a cone obstacle on the carriageway occurred plotted against the visibility level of the obstacle (after Blackwell and Blackwell, 1977).

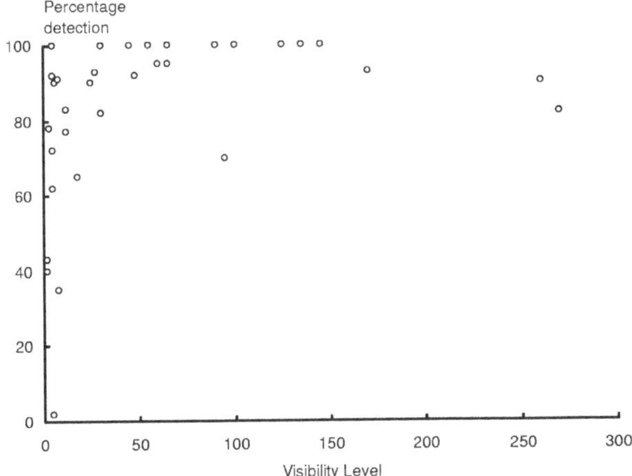

FIGURE 9.5 Percentage detection of a pedestrian at the edge of the road or on the carriageway plotted against the visibility level of the pedestrian for a background with a luminance uniformity (minimum/maximum) around the pedestrian of greater than 0.25 (after Brusque et al., 1999).

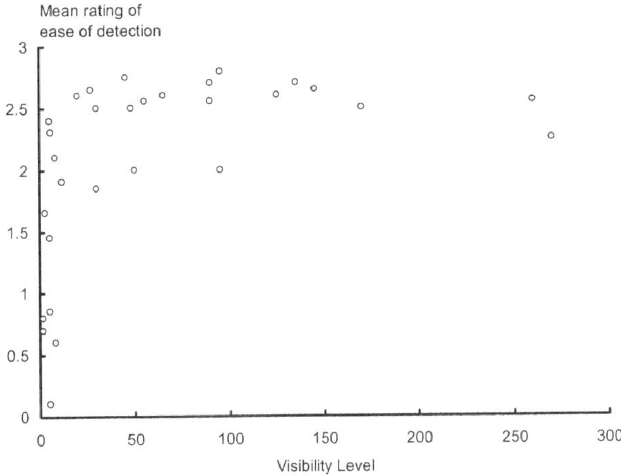

FIGURE 9.6 Mean rating of ease of detection of a pedestrian at the edge of the road or on the carriageway plotted against the visibility level of the pedestrian for a background with a luminance uniformity (minimum/maximum) around the pedestrian of greater than 0.25. The rating scale is 0 = not seen, 1 = difficult task, 2 = average difficulty task, and 3 = easy task (after Brusque et al., 1999).

make any difference to the ease with which they can be detected, i.e., the reduced visibility level will still be on the saturated part of the responses shown in Figures 9.4, 9.5, and 9.6. It is targets that rely on reflected light for their visibility, such as pedestrians, unlit vehicles, and cyclists without lights that are sensitive to the introduction of road lighting.

Another reason for the lack of complaint about difficulties in detecting targets under road lighting may be that unlit targets with the reflectance characteristics necessary for a significant reduction in visibility level are rare, and even when they do occur the conditions necessary for difficulty exist only temporarily. As the target and the driver change their relative positions, the balance between road lighting and vehicle forward lighting changes, and, in consequence, so does the visibility level of the target. These changes mean that a target is unlikely to remain difficult to detect for long.

Yet another reason, and one of more general relevance, is that many of the targets used in studies of visibility level represent a very simplified version of reality. Targets on the road rarely consist of a single diffusely reflecting surface of one reflectance in a single vertical plane. Often, they are three-dimensional, composed of surfaces on many different planes with reflection properties that differ in both kind and degree, i.e., they can be both diffusely and specularly reflecting to different extents. Because these surfaces are in different orientations to the observer, the illumination falling on each surface will be different. All this means that many targets found on roads will have a wide range of luminances and hence will have multiple luminance contrasts. Further, targets of significance to drivers are not always seen against the road surface. A pedestrian in the middle of the road will be seen against the road surface but one at the edge of the road may not. This variability in target and background luminances implies that most three-dimensional targets have multiple visibility levels. Introducing road lighting will change the balance of visibilities between different parts of the target. If this happens, it is unlikely that the target will disappear completely. It is much more likely that the appearance of the target will change as the balance between vehicle forward lighting and road lighting changes. Provided drivers are familiar with such changes and hence know what to expect, such changes are unlikely to cause any problems in perception and hence produce few complaints.

Attempts have been made to extend the visibility approach to deal with multiple luminance contrasts within the same target (Hentschel, 1971; Frederiksen and Rotne, 1978) but with little impact. This is probably because the resulting complexity is too much to digest. Even if it were to be developed, there are two difficulties lying in wait. The first is that while a more sophisticated visibility approach would be useful for specific targets in specific locations under specific conditions, for determining the benefits of introducing road lighting in general it raises the additional complexity of having to choose the nature of the targets and the conditions under which they should be measured from a multitude of possibilities. The second concerns the concentration on visibility level and hence on detection. Hall and Fisher (1977) argue that by focussing the evaluation of road lighting on detection alone, there is a tendency to ignore other important requirements for driving that are necessary after detection has occurred.

Despite these concerns, there can be no doubt that introducing road lighting does have advantages for visibility. This conclusion was confirmed in an experiment examining the changes in detection rate and reaction time following the transition from an unlit section of road to a lit section and vice versa (Fotios et al., 2019). This experiment used a 10:1 scale model of a three-lane road. An observer sat in a position that simulated a driver of a vehicle in the middle lane. The headlamps of the vehicle were fixed in low beam mode while the lighting of the road was either off or on. When on, the luminances of the road surface were representative of current practice. The observer was asked to follow a dynamic fixation task and at the same time detect the appearance of a matt black obstacle in the lane ahead or the movement of one or the other of two unlit grey model cars from the adjacent lanes into the middle lane. When the transition between the unlit road to the lit road occurred, the detection rates improved little for the car lane change (0.97–0.99), but much more for the obstacle in the lane ahead (0.37–0.89). Mean reaction times also improved when entering the lit road, from 2,728 ms to 2,156 ms for the obstacle and from 2,722 ms to 2,034 ms for the car lane change. An opposite pattern of results was found when going from the lit road to the unlit road. Detection rates and reaction times did not change in the 20 min following the transition in either direction.

While it is clear that even a low level of road lighting delivers greater visibility and greater visibility promotes a higher probability of detection and faster reaction times, this process is not without limits. Specifically, Fotios et al. (2019) found that even a road surface luminance of 0.1 cd.m^{-2} produced an improvement in detection rate, an improvement that continued up to 1 cd.m^{-2}, but a further increase to 2 cd.m^{-2} produced no further improvement.

Detection of a change in the road ahead is useful, but it is only the start of what the driver has to do. Once a target has been detected, it can be fixated and recognized. Yerrell (1976) quotes measured distances for recognizing a target, with and without road lighting, for vehicles with different headlamp beam patterns. In the absence of road lighting, recognition distances on straight, curved, and hilly roads range from 16 m to 51 m while on a lit road, the recognition distances ranged from 145 m to 225 m depending on the position of the target. Further, detecting and recognizing targets at greater distances allows more time in which to examine other aspects of the road ahead and more time in which to select an appropriate response. In addition, lighting the road far beyond the range of low- and high-beam headlamps provides a strong cue for visual guidance, easier estimates of changes in speed and distance, and a larger optical flow field, all these being important for driving performance (Hills, 1980).

9.3 EFFECTS ON DISCOMFORT

Anyone who has driven for some distance on an unlit motorway will be aware of the feeling of relaxation experienced when a section of lit road is reached. This feeling is primarily due to the greater distances over which targets can be detected and recognized and the consequent increase in time available in which to make decisions. But there are other factors that can affect discomfort. The first is the possible increase in discomfort produced by the presence of road lighting luminaires themselves. To what extent road lighting may cause discomfort glare is open to question, but one fact that is not is that road lighting standards do not include criteria for discomfort, the belief apparently being that limiting the disability glare caused by road lighting to less than a 10–15% increase in threshold increment results in road lighting that is acceptable as far as discomfort glare is concerned.

This should not be taken to mean that road lighting can never cause discomfort glare. The variables that affect discomfort glare are the luminaire luminance, the solid angle subtended at the eye by the luminaire, the angular deviation of the luminaire from the line of sight, and the adaptation luminance (see Section 4.3). LED road lighting luminaires are usually smaller than those using earlier light sources such as high-pressure sodium, but they also have a higher luminaire luminance, characteristics that have raised concerns about the level of discomfort glare likely to be produced by LED road lighting. Lin et al. (2014) have constructed a model of the discomfort glare produced by LED road lighting luminaires, expressed on the de Boer rating scale, and based on 72 different combinations of luminaire luminance, the solid angle subtended at the driver's eyes, background luminance, and deviation from the line of sight, all at values representative of those experienced by drivers. The model was later verified in the laboratory with a new dataset and in the real world. Their model took the form

$$R = 3.45 - \mathrm{Log}\, ((L_{\mathrm{g}} \cdot \omega)^{2.21} / L_{\mathrm{b}}^{1.02} \cdot \theta^{1.62})$$

where R = the de Boer scale rating, L_{g} is the luminance of the glare source in cd.m^{-2}, ω is the solid angle subtended by the luminaire at the driver's eyes in steradians, L_{b} is the luminance of the background in cd.m^{-2}, and θ is the deviation from the line of sight in degrees. Lin et al. (2014) found that the de Boer rating at which 50% of observers found the lighting comfortable was 4.7, for which the nearest named point on the scale is "just acceptable".

Interestingly, this equation has a very similar form to that developed by Schmidt-Clausen and Bindels (1974) for HID headlamps although the exponents are different. One additional factor shown to be important but not considered in either equation is the correlated colour temperature (CCT) of the light emitted by the glare source. Lin et al. (2014) found that a light source with a higher CCT was considered to produce greater discomfort. This is probably because light sources with a higher CCT will have greater power at the short wavelength end of the visible spectrum and this has been shown to lead to a greater perception of brightness (Fotios et al., 2005b). Increasing the CCT of the road lighting from 3,000 K to 6,500 K had the effect of decreasing the de Boer rating by 2.1. This suggests that care should be taken when selecting LED road lighting to avoid high CCTs.

Another factor to be considered is that when driving along a lit road, luminaire luminance, the solid angle and deviation from the line of sight all vary rhythmically, the rhythm depending on the spacing of the road lighting luminaires and the speed of the vehicle. The result is a steady oscillation of the discomfort glare that may itself enhance the sense of discomfort (Raynham, 2004). Girard et al. (2022) examined this possibility but did not find any effect of such oscillation on discomfort glare except for luminaire luminance at a low frequency (0.65 Hz). Fortunately, such oscillations may be more of a theoretical than a real problem because most road lighting on traffic routes is mounted so high that the top of the windscreen cuts off the driver's view of the luminaires before the luminaire luminance becomes high enough to cause discomfort, although the recent vehicle design trend to extend the windscreen further along the roof may cause problems in the future. The most common situation where road lighting causes discomfort is the use of "historic" luminaires in urban centres with the luminaire mounted at low height, closer to the driver's line of sight (Figure 9.7).

While road lighting itself has the potential to cause discomfort glare, it will certainly reduce discomfort glare from headlamps. It does this because the introduction of road lighting increases the background luminance against which the headlamps are seen. The models of Schmidt-Clausen and Bindels (1974) and Lin et al. (2014) both have the background luminance in the denominator of the equation, the effect being to increase the de Boer rating indicating a reduction in the magnitude of discomfort glare. Figure 9.8 shows the magnitude of the effect of introducing road lighting (Schmidt-Clausen and Bindels, 1974). Mean glare ratings on the de Boer scale are plotted against the logarithm of the illuminance at the eye for two different levels of adaptation luminance corresponding to two levels of road surface luminance. The deviation of the glare source from the line of sight was fixed at one degree.

From Figure 9.8, it can be seen that changing the average road surface luminance from 0.015 cd.m^{-2} to 1.9 cd.m^{-2}, i.e., from unlit to the upper limits of what is recommended for road lighting, increases the glare rating by about 1.4 units on the de Boer glare scale. This nine-point scale is unusual in that higher values indicate less discomfort glare. For example, if the headlamps on an unlit road were causing a discomfort glare rating of three, corresponding to disturbing glare, introducing road lighting giving a road surface luminance of 1.9 cd.m^{-2} would move the rating much closer to a rating of just admissible glare.

FIGURE 9.7 A "historic" urban area lighting installation in Albany, New York. Note the low mounting height relative to conventional road lighting. Such an installation is likely to cause discomfort glare to drivers (by Naomi Miller).

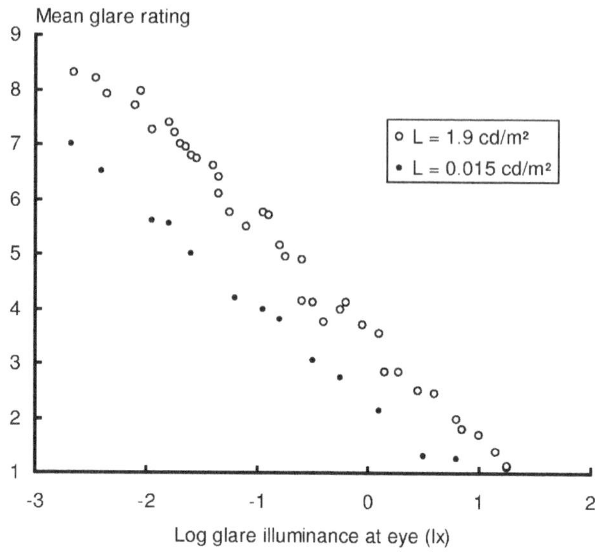

FIGURE 9.8   Mean discomfort glare (de Boer) ratings plotted against the illuminance (lx) at the eye for two different levels of average road surface luminance, one corresponding to an unlit road (0.015 cd/m²) and one to a lit road (1.9 cd/m²) (after Schmidt-Clausen and Bindels, 1974).

There can be no doubt that introducing road lighting will reduce the level of discomfort caused by glare from headlamps. Whether or not this is enough to offset the increase in discomfort glare produced by the road lighting will depend on how the road lighting is done. Experience of current road lighting practice and the absence of criteria for discomfort glare in current road lighting standards suggest that it usually is.

9.4   ADAPTIVE ROAD LIGHTING

Given that the introduction of road lighting is primarily associated with better visibility at greater distances and a reduction in disability and discomfort glare from headlamps, why is its use limited to specific parts of the road network? Why not simply use road lighting everywhere? The most serious answer to these questions is that road lighting costs money, consumes energy, contributes to light pollution and, depending on the source of electricity, can generate carbon emissions. This means there is always interest in minimizing its use and thereby placing more reliance on vehicle forward lighting (Boyce et al., 2009).

For most road lighting, the simplest way to reduce use is by switching it off or to a default low level. Sometimes this switching takes the form of extinguishing every second or third luminaire, but this destroys the luminance uniformity. A better and more common approach is to switch the lighting off when the traffic density is low, typically from midnight to 05:00. Such schemes are often implemented at times of financial stress but tend to cause disquiet among the public. However, an extensive study by Steinbach et al. (2015) of reduced road lighting in England and Wales found little evidence of any detrimental effects of late-night switch-offs on crash rates over the years 2000–2013. This suggests that when traffic density is low and pedestrians and cyclists are rare, the better visibility provided by road lighting has less influence on crashes, and other factors, such as driver fatigue and drunkenness, become more significant. It should also be noted that this finding does not contradict the evidence for the effectiveness of road lighting on crashes as described in Section 1.5 because that evidence is based on the changes in lighting conditions around daylight saving time, changes that occur in the morning and evening. What it does suggest is that road lighting can be reduced when traffic densities are low with only a limited impact on traffic safety. This opens the door to another possibility for reducing energy consumption: dimming. LED luminaires are easily dimmed. Dimming can reduce road lighting energy consumption in two ways. The first is purely technical and applies at all times. The light output of LEDs, and all other electric light sources, declines slowly over life. Without dimming, road lighting installations are designed to meet the relevant standard at the end of their life. This means that at the start of their life, such installations will exceed what is required, i.e., over-lighting will be evident. By dimming the light output at the start of life and reducing the level of dimming through life, the required standard can be met at all times with reduced energy consumption.

The other benefit of dimming is the ability to match the light output of the road lighting to the prevailing traffic pattern. This is called adaptive road lighting. There are two approaches used in adaptive road lighting. One is to use controls integrated into each luminaire. Typically, an occupancy sensor is used to detect the presence of a vehicle, cyclist, or pedestrian and the luminaire is constrained to operate at a high- or low-light output. When no vehicle, cyclist, or pedestrian is detected, the luminaire operates at the low light output. When a vehicle, cyclist, or pedestrian is detected, the luminaire changes to the high light output which meets the standard appropriate for the road.

The other approach uses a network system. In this, wireless control modules on each luminaire are connected to a gateway, a central control point for the network. Using network control software and internet access, staff can adjust the light output of individual luminaires or groups of luminaires as desired. Input to the gateway can also be accepted from cameras and weather stations and the appropriate response selected using artificial intelligence (Gagliardi et al., 2020). Such technology is impressive and has the potential to reduce energy consumption by significant amounts but it does leave one question open – what factors are important in determining when and how far road lighting should be dimmed? Gibbons et al. (2014) have developed a set of design criteria for such adaptive road lighting to be used in the USA. The approach adopted is based on the Commission Internationale de l'Eclairage (CIE) approach (CIE, 2010b) and linked to road lighting standards in the USA (see Section 4.4.1). In these recommendations, three broad classes of road lighting are identified; highway lighting for freeways and expressways where pedestrians and cyclists are not usually present; street lighting for major, collector, and local roads where pedestrians and cyclists are generally present; and residential/ pedestrian area lighting where the lighting is provided for the benefit of pedestrians rather than drivers. Each of these road classes has a number of categories, each with different photometric recommendations. For adaptive lighting a number of weighting factors are identified and their sum is subtracted from the base level to identify the appropriate category to be used. The factors considered are vehicle speeds, traffic volume, the presence of a median between traffic moving in opposite directions, the density of intersections, the ambient luminance, i.e., light provided by sources other than the road lighting, the guidance given to the driver by the layout of the road lighting, pedestrian–cyclist interaction, the presence of parked vehicles, and the importance of facial recognition. The weighting factors for each road class are different. Table 9.1 shows the values considered when deriving the weighting factors for street lighting where the base level is 6. For obvious reasons, the pedestrian–cyclist interaction, the presence of parked vehicles, and the importance of facial recognition

TABLE 9.1

Selection Criteria for Adaptive Street Lighting (ADT) and the Associated Weighting Values

Parameter	Options	Criteria	Weighting Value
Speed	High	>45 mph (70 km.h^{-1})	1
	Moderate	35–45 mph (55–70 km.h^{-1})	0.5
	Low	<35 mph (55 km.h^{-1})	0
Traffic volume	High	>15,000 ADT	1
	Moderate	5,000–15,000 ADT	0
	Low	<5,000 ADT	−1
Median	No	No median	1.
	Yes or one-way road	Must block glare	0
Intersection/ Interchange density	High	>5 per mile (1.6 km)	1
	Moderate	1-5 per mile (1.6 km)	0
	Low	<1 per mile (1.6 km)	−1
Ambient luminance	High	LZ3 and LZ4	1
	Moderate	LZ2	0
	Low	LZ1	−1
Guidance	Good	>100 mcd.m^{-2}.lx^{-1}	0
	Poor	<100 mcd.m^{-2}.lx^{-1}	0
PedestrianBicycle interaction	High	>100 pedestrians per hour	2
	Moderate	10–100 pedestrians per hour	1
	Low	<10 pedestrians per hour	0
Parked vehicles	Yes	Parked vehicles present	1
	No	No parked vehicles present	0

Note. ADT is the average daily traffic. LZ is lighting zone (from Gibbons et al., 2014)

TABLE 9.2
The Five Categories of Street Lighting that Can Be Used with Adaptive Road Lighting. The Category to Be Used Is Selected by Subtracting the Sum of the Weighting Factors Given in Table 9.1 from a Base Level of 6. For example, if the Sum of the Weighting Factors at a Specific Time Is 4 Then the Class to Be Used at that Time Is S2 (from Gibbons et al., 2014)

Category	Average Road Surface Luminance (cd.m^{-2})	Maximum Luminance Uniformity Ratio (Average/ Minimum)	Maximum Luminance Uniformity Ratio (Maximum/ Minimum)	Veiling Luminance Ratio
S1	1.2	3.0	5.0	0.3
S2	0.9	3.5	6.0	0.4
S3	0.6	4.0	6.0	0.4
S4	0.4	6.0	8.0	0.4
S5	0.3	6.0	10.0	0.4

are ignored for highway lighting. For equally obvious reasons, the presence of a median and the guidance given to drivers by the layout of the luminaires are ignored but facial recognition is included for residential/pedestrian area lighting.

Table 9.2 gives the five categories of recommended photometric conditions for adaptive street lighting (Gibbons et al. 2014). Subtracting the sum of the weighting factors for a street indicates which of these categories should be used as conditions change over time. If the result of the subtraction is not an integer, the next lower integer category should be used. If the subtraction leads to a negative number, the lowest category should be used. For this system to be used, it is necessary to have information on the pattern of use of the street at different times. This can be obtained by sample observations or road monitoring. Clearly, the advent of LED luminaires and adaptive road lighting has the potential to change the whole basis of how roads are lit at different times as well as to significantly reduce the energy consumption used and the light pollution produced by road lighting. Advice on when and how to use adaptive road lighting is given in IESNA (2021b).

9.5 SUMMARY

As a general rule, road lighting is designed to promote visibility without reference to vehicle lighting, and vehicle lighting is designed to provide visibility in the absence of road lighting. The approach used here is to assume that the driver will always have vehicle lighting, so the chapter is devoted to the effects of introducing road lighting in terms of its effects on visibility and comfort and how it might be adjusted to match prevailing traffic conditions.

Both vehicle forward lighting and road lighting are designed to make what is ahead visible to the driver. The most suitable measure to determine whether or not an object ahead of the driver will be visible is its luminance contrast. The luminance contrast of an object will depend on the illuminances on and reflection properties of the object and the surfaces against which it is seen. Measurements of the illuminance ahead of a vehicle on a lit road show there are three distinct zones: a near zone where vehicle forward lighting is dominant, a distant zone where road lighting is dominant, and an intermediate zone where the two forms of lighting are approximately equal. Examination of the impact on visibility of introducing road lighting for the three zones leads to three conclusions. The first is that introducing road lighting is likely to improve the visibility of most targets when they are in the distant zone. The second is that there can be no guarantees that visibility will improve for all targets, particularly in the intermediate zone. There are some objects for which the combination of light distributions from the vehicle forward lighting and road lighting and the reflection properties of the object and the road surface may lead to reduced visibility. The third is that introducing road lighting alleviates the effects of disability glare from opposing vehicle's headlamps on visibility. There can be no doubt that introducing road lighting has advantages for visibility and for other aspects of driving. Lighting the road far beyond the range of low- or high-beam headlamps alone increases the probability of detection of most targets, provides a strong cue for visual guidance, makes estimates of changes in speed and distance easier, and ensures a larger optical flow field, all these being important for driving.

The introduction of road lighting also changes the discomfort experienced by the driver at night. This feeling is primarily due to the greater distances over which targets can be detected and recognized and the consequent increase in time available in which to make decisions. On narrow roads, there is an additional reason why the introduction of road lighting reduces discomfort: the reduction in discomfort glare from the headlamps of opposing vehicles.

Given that the introduction of road lighting is primarily associated with better visibility at greater distances and a reduction in glare from headlamps, why not use road lighting everywhere? The basic answer to this question is that road lighting costs money, consumes energy, produces light pollution, and, depending on the source of electricity, can generate carbon emissions. This means there is always interest in reducing the energy consumption of road lighting. One way to do this is by switching it off late at night when traffic densities are low and pedestrians and cyclists are rare. An extensive study of the consequences of such late-night switch-off on crashes did not find any evidence that such an action was detrimental to traffic safety. This is just as well because one of the benefits of using LED luminaires is that they can be easily dimmed. This introduces the possibility of adaptive road lighting in which the light level is adjusted to match the prevailing traffic conditions, either by a luminaire sensor approach or a more sophisticated networking system. The factors involved in determining how the lighting should be adjusted are the traffic speed, the traffic density, the presence of a median, the intersection density, the ambient luminance, the guidance given by the layout of the luminaires, the presence of pedestrians, the presence of parked vehicles, and the need for facial recognition. Adaptive road lighting has the potential to change the way in which road lighting is designed and used at different times as well as reducing the energy consumption of and light pollution produced by road lighting.

10 Different Vehicle Types

10.1 INTRODUCTION

Whenever the topic of lighting for driving is considered, the mind automatically turns to the familiar, which usually means cars and trucks and other vehicles with four or more wheels. However, there are other vehicles likely to be found on the road that have only two wheels, such as motorcycles and bicycles, or even no wheels, just legs. Although the latter are not strictly vehicles, they are included here because drivers need to be able to see them if they are to avoid causing death and injury. Further and regardless of the number of wheels, there are other vehicles that have additional lighting requirements, such as heavy goods vehicles, emergency vehicles, slow-moving vehicles, and personal mobility vehicles. This chapter is concerned with how lighting, colour, and retro-reflection can improve traffic safety for these different vehicle types by making them easier to see.

10.2 FATALITIES AND SERIOUS INJURIES FOR DIFFERENT VEHICLE TYPES

Different vehicle types have different levels of risk associated with them. Table 10.1 shows the total number of fatalities and serious injuries suffered by drivers or occupants of different vehicle types in the UK in 2019. As would be expected the highest number of fatalities and seriously injured occurs for cars because the number of cars on the road is greater than the any other vehicle type. Nonetheless, it is interesting to note that pedestrians, cyclists, and motorcyclists all have a higher number of fatalities and serious injuries than do the drivers and occupants of light and heavy goods vehicles (HGV) as well as those using buses and coaches.

As explained in Section 1.4, the best way to quantify the risk involved in using different vehicle types is to examine the number of events occurring for a fixed level of exposure. Table 10.2 does this by showing the number of fatalities and serious injuries per billion miles travelled. Looked at this way a different picture emerges.

Riding a motorcycle, as either rider or passenger, has the highest risk by far, followed by pedestrians and cyclists for fatalities and cyclist and pedestrians for serious injuries. Compared to these vehicle types, the drivers and occupants of cars, light and heavy goods vehicles, and buses are exposed to much less risk. This pattern of risk is related to a combination of speed and level of protection. Motorcyclists travel at high speeds with little protection other than a crash helmet. Cyclists and pedestrians travel more slowly, very slowly in the case of some pedestrians, but they too have little by way of protection. Cars certainly do travel fast, but the protection given to drivers and occupants has been improved greatly by the compulsory wearing of

DOI: 10.1201/9781003388906-10

TABLE 10.1

Number of Fatalities and Serious Injuries Suffered by Drivers and Occupants of Different Vehicle Types in the UK in 2019 (from Department for Transport, 2022b)

Vehicle Type	Fatalities	Seriously Injured
Pedestrian	470	6,528
Cyclist	100	4,247
Motorcyclist	658	11,295
Car	1,244	18,993
Light goods vehicle	79	963
Bus or coach	16	354
Heavy goods vehicle	36	335

TABLE 10.2

Number of Fatalities and Serious Injuries per Billion Miles Travelled for Drivers and Passengers of Different Vehicle Types in the UK in 2019 (from Department for Transport, 2022b)

Vehicle Type	Fatalities	Seriously Injured
Pedestrian	35	492
Cyclist	28	1,170
Motorcyclist	247	4,207
Car	5	72
Light goods vehicle	2	18
Bus or coach	2	150
Heavy goods vehicle	2	20

seat belts and the introduction of vehicle design features such as crumple zones and airbags. Even less risk is associated with heavy goods vehicles which tend to be slower than cars and heavier so the level of protection for the occupants is enhanced. The only anomaly in this analysis is the relatively high level of serious injury to the occupants of buses and coaches although the level of fatalities is about what would be expected. A plausible explanation for this anomaly is that passengers on buses and coaches rarely wear seat belts, even if they are provided, so that in the event of a crash, the occupants will be thrown about which can cause serious injury but is less likely to be fatal.

Table 10.2 gives the number of fatalities and serious injuries per billion miles travelled for the driver/riders and passengers of different vehicle types. Another aspect to consider is what risk these different vehicle types pose to other road users. Table 10.3 gives the estimated number of fatalities and serious injuries per billion miles

TABLE 10.3

Number of Fatalities and Serious Injuries per Billion Miles Travelled by Other Road Users Caused by Drivers of Specific Vehicle Types in England Over the Years 2005–2015 (from Aldred et al., 2021)

Vehicle Type Causing Crash	Fatalities	Seriously Injured
Cyclist	1.1	32.8
Motorcyclist	7.6	100.1
Car	3.3	49.7
Light goods vehicle	2.6	25.7
Bus or coach	19.2	144.7
Heavy goods vehicle	17.1	62.5

travelled among other road users that can be ascribed to different vehicle types, in England, over the years 2005 to 2015 (Aldred et al., 2021).

Viewed this way, it can be seen that the most dangerous vehicles to other road users are buses with heavy goods vehicles in second place for fatalities but third for serious injuries while motorcycles are third for fatalities but second for serious injuries. Compared to these, bicycles pose a minimal risk to other road users. The obvious explanation for this pattern is the momentum of these vehicles. Buses and trucks have a modest velocity but high mass giving them considerable momentum. Motorcycles have a smaller mass but a higher velocity also resulting in a high momentum. Bicycles have a relatively low mass and modest velocity giving them limited momentum. Any road user with limited protection, such as a pedestrian or cyclist, struck by another vehicle with high momentum is likely to be seriously injured if not killed.

Given these variations in the level of risk associated with these different vehicle types it is appropriate to consider what contribution lighting can make to reducing the risk of death and serious injury, particularly for motorcyclists, cyclists, and pedestrians.

10.3 MOTORCYCLES

Motorcycles come in various forms ranging from mopeds through scooters to high-power machines designed for a rider, with or without a pillion passenger. The different types of motorcycle are not separated in the crash statistics. The lighting requirements for motorcycles used on the roads in the UK are for one or more headlamps providing both a high beam and a low beam, rear position lamps, one or two stop lamps, two turn indicators, a rear registration plate lamp, and a red rear reflector. There is no requirement for a daytime running lamp or a front position lamp if there is a headlamp, although one headlamp is often used as a daytime running lamp by being automatically turned on when the engine is started. The headlamp and any front position lamp should emit white light. The headlamps should have high-beam and low-beam light distributions similar to those for the forward lighting of cars. If there is one headlamp, that should provide both high-beam and low-beam lighting. If there are two headlamps either both can provide high- and low-beam lighting or one can provide high beam and one low beam provided the high beam headlamp is turned off when low beam is selected. The rear position lamps have to emit red light and can be incorporated into the turn indicator housing. The turn indicators have to emit amber light and be separated by at least 240 mm and flash at a frequency between 1 Hz and 2 Hz. A tell-tale for the turn indicators does not need to be fitted if they are visible from the rider's position. The stop lamp has to emit red light and be mounted on the longitudinal axis of the machine. If there are two stop lamps, they can be mounted one above the other on the longitudinal axis or symmetrically on either side of it.

Given this extensive list of lighting requirements and their regular testing by the state, there has to be some other explanation for the high mortality and serious injury associated with riding a motorcycle. An explanation might be sought by examining the type of crashes in which motorcyclists are involved and where and when they occur. Table 10.4 shows the number of motorcyclists killed or seriously injured in crashes in the UK in 2019 and the other vehicles involved in the crash, if any (Department for Transport, 2021a). From Table 10.4 it is clear that the main situation in which motorcyclists are killed or seriously injured involves a collision with another vehicle although a significant minority of crashes do not. This minority is probably due to a failure to match speed to the layout of the road and/or the road surface conditions.

Table 10.5 shows the percentage of motorcycle fatalities and serious injuries occurring at different road locations over the years 2015–2020 (Department for Transport, 2021a). From Table 10.5 it is clear that although the majority of fatalities occur away from a junction or roundabout, there are a significant number of fatalities that do, particularly for T, Y, and staggered junctions. The pattern of percentages for serious injuries is similar but now the percentage for crashes at junctions

TABLE 10.4

Number of Motorcycle Fatalities and Serious Injuries in the UK in 2019 with the Involvement of Other Vehicle Types (from Department for Transport, 2021a)

Other Vehicle Involved	Fatalities	Seriously Injured
None	445	8,128
Bicycle	4	152
Motorcycle	35	673
Car	804	20,737
Light goods vehicle	118	4,147
Bus or coach	21	169
Heavy goods vehicle	106	476

TABLE 10.5

Percentages of Motorcycle Fatalities and Serious Injuries Occurring at Different Road Locations Over the Years 2015–2020 (from Department for Transport 2021a)

Road Location	Percentage Fatalities	Percentage Serious Injuries
Crossroads	5.8	8.0
T, Y, or staggered junction	24.2	33.9
Junction with more than four arms, not a roundabout	0.3	0.8
Other junction	3.7	3.8
Roundabout	2.5	7.3
Mini-roundabout	0.2	0.9
Private drive or entrance	4.6	4.6
Slip road	1.1	1.2
Not at junction or within 20 m of one	57.6	39.4
Unknown	0.0	0.1

and roundabouts exceeds the percentage away from junctions and roundabouts. The reason why junctions and roundabouts are so hazardous for motorcyclists is that they are places where other vehicles are likely to cross the path of a motorcycle even when the motorcycle has the right-of-way. Although Tables 10.1 to 10.5 are taken from the UK data, similar conclusions can be drawn from data in the US (National Highway Traffic Safety Administration, 2021b).

The next question to consider is when do these crashes occur? Interestingly, there is clear difference between weekdays and weekends (Department for Transport, 2021a). During weekdays there are two peaks in the distribution of the number of motorcyclists killed or seriously injured by the hour, one between 07:00 and 10:00 and a bigger peak between 16:00 and 19:00. These peaks align with travel to and from work when traffic will be heavy and drivers may be in a hurry. At the weekend there is only one peak, comparable in size to the evening peak during the working week, but located between 12:00 and 16:00 suggesting recreational riding. As regards the visual conditions, it is worth noting that the peak in fatalities and serious injuries that occur at the weekend will all have occurred in daylight for the whole year. This will only be true for the weekday peaks in summer. The fact that many fatalities and serious injuries occur in daylight indicates that it is not the visibility of the road ahead that is the problem but rather the ability of others to detect an approaching motorcycle and judge its speed and distance. This supposition is supported by the factors alleged to have contributed to crashes resulting in fatalities or serious injury to motorcyclists in the UK between 2015 and 2020. These contributory factors are subjective and reflect the opinion of police officers attending the crash. The most frequent contributory factor was "Driver or rider failed to look properly". The next most frequent contributory factor was "Driver or rider failed to judge other person's path or speed" (Department for Transport, 2021a). To what extent these two contributory factors are really different is unclear but the chances of a driver looking but failing to see a motorcycle should be reduced by making the motorcycle more conspicuous.

Motorcycle conspicuity has been a topic of interest for some time. Olson et al. (1989b) examined 30 different motorcycle conspicuity treatments and evaluated the most promising ones by day and night on public roads. The evaluation sought to determine if greater conspicuity deterred drivers from accepting shorter times before the motorcycle would be expected to pass in front of them. The results showed that the daytime conspicuity of a motorcycle is significantly improved if the low-beam headlamp is on or the high-beam headlamp is on and modulated at 3 Hz, or the rider is wearing a high-visibility vest and helmet cover. Nighttime conspicuity is enhanced when the motorcycle is equipped with additional running lights or the rider is wearing a retro-reflective vest and helmet cover. The conclusion was that motorcyclists should be required to drive during the day with their low-beam headlamp on. This conclusion has been implemented in many countries either by advice or by motorcycle technology or legal activity. The result has been a reduction in motorcycle crash risk estimated to be between 4% and 20% (Davoodi and Hossayni, 2015).

However, many cars and trucks are now equipped with daytime running lights. This means the difference between motorcycles and other traffic is lessened, so motorcycles using headlamps to improve conspicuity may be more exposed when in traffic. Cavallo and Pinto (2012) tested this possibility using photographs of a complex urban traffic scene displayed for 250 ms. Observers were asked to detect vulnerable road users, specifically motorcyclists, cyclists, and pedestrians who appeared at different locations and distances. The presence of cars with daytime running lights hampered the detection of all three types of vulnerable road user. This demonstrates that increasing conspicuity is not just a matter of increasing contrast, it is also necessary to make sure that the object to be made conspicuous looks clearly different from other objects that surround it. Fortunately, there are many different ways to increase the conspicuity of motorcycles and riders including headlamp colour, number and locations of running lamps, flashing or pulsing lamps, motorcycle colour, rider's clothing colour and reflectance, and helmet colour and reflectance (Abdul Khalid et al., 2021). Many of these can be effective, but many are also matters of personal choice. Probably, the easiest to implement generally that would certainly improve motorcycle conspicuity would be to make any installed running lights flash or pulse at a low frequency, always assuming that other vehicle types do not do the same.

While improved conspicuity is important for detection that is just the start of what a driver waiting to enter or cross a road on which a motorcycle is approaching has to do. Having detected a motorcycle, the driver has to decide how far away it is and its speed and hence decide whether or not there is enough time to enter or cross the road in front of the motorcycle. This introduces the size/arrival effect (DeLucia, 2013). This effect describes the experimental observation that if two objects of different size are approaching from the same distance at the same speed, the larger one will be perceived to reach the observer before the smaller one. If the size/arrival effect applies to drivers waiting at a junction it implies that the actual time before the arrival of a motorcycle approaching from the same distance and travelling at the same speed as a car will be shorter than expected for the motorcycle because the small frontal area of the motorcycle means it will be perceived as further away or moving more slowly. Robbins et al. (2018) tested this prediction in driving simulators. Participants in this study were asked to drive towards a four-way intersection on the minor road and cross the major road. As they reached the intersection they were forced to stop to let a car pass coming from the left. After this had passed, they were then faced with either a car or a motorcycle approaching from the right at a constant speed of 30 mph (48 km.h^{-1}) from different distances. The scene was set in daylight but neither car nor motorcycle were using headlamps nor daytime running lights. The driver had to decide whether to move across the junction or to wait for the vehicle approaching from the right to pass before driving across. Three times were measured. Two were measured only when the driver decided to cross; the gap accepted lag and the time difference. One, the cross time, was measured regardless of the decision made. The gap accepted lag is the time taken for the approaching vehicle to reach the centre of the junction starting from when the driver's vehicle first moved to cross the road. The time difference is the time after which the approaching vehicle would reach the centre of the junction starting from when the rear of the driver's car had moved far enough across the junction that the approaching vehicle could pass behind it. The cross time was simply the time the driver took to cross the junction when the vehicle was approaching and after it had passed. Statistically significant differences were found for the gap acceptance lag and the time difference for the car and the motorcycle. For the car the mean gap acceptance lag was 4 s and for the motorcycle it was 3.5 s. For the car, the mean time difference was 3.2 s and for the motorcycle it was 2.8 s. As for the cross time, there was a statistically significant difference in cross time when the decision had been made to cross in front of the approaching vehicle (0.9 s) and after it had passed (1.2 s), but there was no difference between a car approaching and a motorcycle approaching. These results suggest that drivers accept smaller gaps for motorcycles than cars and do not try to clear the junction faster for motorcycles. It would be instructive to replicate this experiment but with all the vehicles involved using daytime running lights or headlamps, in mixed traffic, and at night.

Despite such limitations, these results do indicate that the size/arrival effect does apply to vehicles so it is necessary to consider why size matters. There are two competing theories for the effect. One is attitude bias which says that larger vehicles are considered to be more dangerous as a collision with them is more likely to injure the driver. The other is a perceptual bias which involves the visual difficulty of judging the apparent distance and speed of small vehicles. One reason for this difficulty is that changes in angle subtended at the driver's eye as the motorcycle approaches are less than for a car or truck. This makes it difficult to detect how far away the motorcycle is and how fast it is moving, particularly at low light levels when the background against which vehicles are moving will be less clear (Gould et al., 2012). Whether attitude bias or perceptual bias is dominant or whether both are involved in a decision to cross the path of an approaching vehicle remains to be determined (DeLucia, 2013), but the remedy for the reduced gap times allocated to motorcycles seems clear. It is to increase the apparent size of the motorcycle. Various attempts have been made to do this by adding more lamps on the front forks, mirror mounts, or around the headlamp (Gould et al., 2012; Helman et al., 2014; Robger et al., 2015). The effects found on detection and minimum gap decision are generally positive, particularly at night, but such lighting is rarely seen on the road because a lit headlamp is considered adequate.

While lighting clearly has a role to play in making motorcycles easier to detect and distances and speeds easier to estimate, there is another aspect of human behaviour that has nothing to do with lighting. This is inattentional blindness. This is what happens when an observer looks at a scene but fails to see something in the scene despite it being clearly visible. Many readers will have seen a demonstration of this effect using a video of a group of people passing a basketball between them. The observer is asked to count how many times the ball is passed. In the middle of the video, a person in a gorilla suit appears and waves. Significant number of observers fail to see the gorilla at all. This effect has to do with the allocation of attention which is inherently time-limited. Pammer et al. (2018) examined the possibility that drivers detected motorcycles less frequently than other vehicles because they are lower down on the attentional hierarchy. Using photographs displayed for a short time, they found that the observers were twice as likely to miss a motorcycle compared with a taxi. Of course, inattentional blindness only occurs when attention is directed elsewhere, either externally or internally. Conspicuity may have the potential to override inattentional blindness which is why making motorcycles highly conspicuous is necessary if the number of fatalities and serious injuries of riders and pillion passengers is to be reduced. It is also worth noting that when evaluating the relevance of experimental results to real life it is important to know if the participants were expecting to see a motorcycle or if their attention had been directed elsewhere.

But this is not the end of the story. While crashes involving other vehicles are important, Tables 10.4 and 10.5 show that there are a large number of fatality and serious injury crashes where no other vehicle is involved and that occur away from a junction. This opens up the cause to a wide range of factors that have little to do with lighting. Among them are excessive speed, failure to anticipate changes in road direction, road surface conditions, rider's alcohol or drug use, rider distraction and inattention as well as adverse weather (National Highway Traffic Safety Administration, 2021b). Many of these causal factors apply to many situations apart from traffic safety which may explain why research has largely focused on the well-defined topic of conspicuity. The fact is that riding a motorcycle is inherently dangerous. The most that lighting can do is to reduce the risks involved by making the way ahead clear, the intentions of the rider and others obvious and both machine and rider more conspicuous and larger to others, by day and night. Advice on how to improve motorcycle conspicuity is readily available (Robger and Lennie, 2017; Abdul Khalid et al., 2021).

10.4 BICYCLES

Like motorcycles, bicycles come in several different forms ranging from those designed for trips to the shops to those intended for long-distance touring, with and without battery-powered assistance, as well as oddities such as recumbent and prone bicycles, tandems and tricycles. These different types are not separated in the accident statistics so attention here will be focused on the most common, the upright two-wheeled bicycle for a single person.

The lighting requirements for bicycles used on the roads in the UK are somewhat limited and confusing (Bever, 2022). For a start, lamps are only required on bicycles between sunset and sunrise. No lamps are required for use during the day, even in conditions of seriously reduced visibility. When the bicycle is being used at night, the minimum lighting requirements are a white forward lamp and a red rear lamp as well as a red rear reflector and amber reflectors on the front and rear edges of both pedals. The front lamp should be mounted no more than 1,500 mm above the road, either on the longitudinal axis of the bicycle or on the right (the UK drives on the left) and should be visible from ahead. The rear lamp should be mounted between 350 mm and 1,500 mm centrally or to the right. If the front and rear lamps emit a steady light output they should conform with BS 6102 Part 3 (BSI, 1986) or a European equivalent such as the German StVZO K-mark system. Flashing is allowed for both front and rear lamps provided they flash at 1–4 Hz and have a luminous intensity of at least 4 cd. The red rear reflector should conform with BS 6102 Part 2 (BSI, 1982) and be mounted between 250 mm and 900 mm, centrally or to the right. The pedal reflectors should be amber, conform with BS6102 Part 2, and be visible from both the front and the rear. Lamps can be powered by a battery or by a dynamo although the latter means that the lamp will cease operating when the bicycle stops unless the circuit is fitted with a delay system to keep the lamp on for a time after stopping. Although these lighting regulations are extensive they are rarely enforced. Many cyclists add additional lamps to their bicycles and use them during the day. There is also a trend for cyclists to use helmet-mounted lamps which can cause glare to other traffic since their aiming is determined by where the rider is looking.

It is now interesting to examine the crash statistics of cyclists. In 2019, 100 cyclist were killed and 4,121 were seriously injured on roads in the UK. Table 10.2 shows that, for the same distance travelled, cyclists are much less at risk of being killed or seriously injured on the road than motorcyclists, but they are still much more at risk than the occupants of motor vehicles. This is because most cyclists travel slower than motorcyclists although they suffer from a similar lack of protection. The generally slower speeds of bicycles is both an advantage and a disadvantage. The advantage is that should the rider fall the momentum is much less and so the consequential damage to the rider is likely to be less. The disadvantage is that travelling at a slower speed than other vehicles in traffic means that bicycles are continually being overtaken. This increases the risk of being hit by the overtaking vehicle. Motorcycles and bicycles share a common problem of having a small area when viewed from in front or behind compared to other vehicles on the road which makes them more difficult to detect.

Table 10.6 shows the number of cyclists killed or seriously injured in crashes in the UK in the years 2015–2020 and the other vehicles involved in the crash, if any (Department for Transport, 2021b). From Table 10.6 it is clear that one of the main situations in which cyclists are killed or seriously injured involves collisions with other vehicles, particularly cars.

TABLE 10.6

Number of Cyclist Fatalities and Serious Injuries in the UK Over the Years 2015–2020 with the Involvement of Other Vehicle Types (from Department for Transport, 2021b)

Other Vehicle Involved	Fatalities	Seriously Injured
None	98	2,028
Bicycle	5	337
Motorcycle	6	406
Car	298	18,836
Light goods vehicle	47	1,842
Bus or coach	15	358
Heavy goods vehicle	92	487

TABLE 10.7

Percentages of Cyclist Fatalities and Serious Injuries Occurring at Different Road Locations Over the Years 2015–2020 (from Department for Transport 2021b)

Road Location	Percentage Fatalities	Percentage Serious Injuries
Crossroads	9.5	9.7
T, Y, or staggered junction	21.6	36.4
Junction with more than four arms, not a roundabout	1.1	1.1
Other junction	2.3	3.4
Roundabout	4.5	11.1
Mini-roundabout	0.3	1.9
Private drive or entrance	1.1	3.4
Slip road	0.9	0.6
Not at junction or within 20 m of one	58.6	32.1
Unknown	0.0	0.4

Table 10.7 shows the percentage of cyclist fatalities and serious injuries occurring at different road locations over the years 2015–2020 (Department for Transport, 2021b). From Table 10.7 it is clear that although the majority of fatalities occur away from a junction or roundabout, there are many that occur at T, Y, and staggered junctions. The pattern of percentages for serious injuries is similar but not identical because then more happen at junctions and roundabouts than away from these locations. This is because junctions and roundabouts are places where other vehicles are likely to cross a cyclist's path. Although Tables 10.2, 10,6, and 10.7 are taken from UK data, similar conclusions can be found in the data collected in the US (National Highway Traffic

Safety Administration, 2022b). As for the lighting conditions when cyclist fatalities occur, 50% occur in daylight and 50% occur after dark, or at dawn and dusk (National Highway Traffic Safety Administration, 2022b), despite fewer cyclists riding after dark particularly when there is no road lighting (Uttley et al., 2020).

What role does lighting have to play in delivering traffic safety for cyclists? Given that no lighting is required for riding a bicycle on the roads during the day in the UK the first question to consider is, what percentage of cyclists use the lamps they do have. There is some evidence to indicate that this is unlikely to be 100%. Observations in Oxford on a lit road at dusk found that only 42% of cyclists had both front and rear lamps lit (McGuire and Smith, 2000). O'Boyle (2010) made observations of cyclists on a minor road in London at dusk on a weekend evening and again on a major road at dusk in the rush hour of a workday. Only 28% of cyclists were using both front and rear lamps on the minor road and 58% on the major road. Similar results were found in Sweden (Setiawan, 2009). Of course, these observations were all made on lit roads. It would be interesting to know how many cyclists indulge a death wish by riding without any lights on unlit roads at night (Figure 10.1). In the past, many did.

Using data from 1981 to 1984 Hoque (1990) found that 61% of cyclists involved in fatal crashes at night did not have any lights on their bicycles. Hoque (1990) also found that unlit streets were the scene of more nighttime fatal cycle crashes than lit streets. More recently, Uttley et al. (2023) have used data on cyclist fatalities occurring in the UK between 2004 and 2019 to examine to what extent the absence of road lighting increases the risk of a cyclist being fatally injured when involved in a crash after dark. An odds ratio was calculated based on the relative numbers of fatal and

FIGURE 10.1   A group of cyclists on the road at night. How many are there? (iStock by Getty Images / tornal).

non-fatal crashes involving cyclists occurring in darkness in the absence and presence of lighting. Precisely,

$$\text{Odds ratio} = (FDN / NFDN) / (FDL / NFDL)$$

where

FDN = Fatal crashes occurring after dark when lighting was absent
NFDN = Non-fatal crashes occurring after dark when lighting was absent
FDL = Fatal crashes occurring after dark when lighting was present
NFDL = Non-fatal crashes occurring after dark when lighting was present

This odds ratio was calculated separately for crashes involving other vehicles and for cyclist-only crashes. For other vehicle crashes the odds ratio was 8.50. For cyclist-only crashes the odds ratio was 2.45. Both odds ratios are well above 1.0 implying that cyclists are at much greater risk of being killed on unlit roads, particularly where other vehicles are involved. Subsequent calculations suggest that adjusting for the speed limits on unlit roads reduces the odds ratio for cyclist fatalities involving other vehicles but still leaves it well above 1.0. Such statistics emphasise the need to persuade cyclists to use lamps at night. There have been and there still are many attempts to do this, not all of them are successful (Ferguson and Blampied, 1991), but it is certainly worth trying.

Given that cyclists can be persuaded to use both front and rear lamps, night and day, it could be useful to identify where a bicycle is likely to be struck. Unfortunately, the answer is everywhere. A number of investigations of types of crashes involving cyclists and other vehicles, usually cars, have shown that cyclists can be struck from in front (National Highway Traffic Safety Administration, 2022b); in the rear by a car approaching from behind and not seeing the cyclist until too late, particularly at night (Hutchinson and Lindsay, 2009), or on the side by a vehicle changing lanes or turning at a junction without adequate warning to a cyclist (Johnson et al., 2010). Evidently, cyclists are also victims of the "Looked but failed to see" phenomenon discussed in relation to motorcycles (see Section 10.3), so making the cyclist more conspicuous should make a contribution to safety (Bil et al., 2010; Johnson et al., 2010; Filipovic et al., 2022). There are many different ways to increase the conspicuity of cyclists, ranging from making the lamps flash or pulse, to adding more lamps and reflectors, to encouraging cyclists to wear higher reflectance clothing and retro-reflectors. They are not all equally effective, some being better by day and others by night. Further, aids to conspicuity need to do more than simply enhance the ability to detect a cyclist; to be effective they also have to ensure that the cyclist is recognized as a cyclist.

One of the easiest conspicuity aids to implement would be to ensure that all cyclists use their front and rear lamps during the day. In other words, cyclist should have daytime running lights as is often required for other vehicles. Madsen et al. (2013) carried out a study in Odense, Denmark, in which the number of crashes per person month of 1,845 cyclists with permanent running lights were compared with

2,000 other cyclists without daytime running lights, over a year. The number of bicycle crashes per person month, including those that led to injury to the cyclist, was 19% less for those with permanent running lights.

Another possibility suitable for use during the day is to make the front and rear lights flash. This is allowed in the UK but not in other countries such as Germany and The Netherlands. Flashing front lights may be useful for conspicuity during the day but at night they may not be because then the front light is also the means for seeing the road ahead. As for rear lights, Edewaard et al. (2019), using video images and eye tracking, examined the distance at which a cyclist was first glanced at and the distance at which the observer was certain there was a cyclist present, for various types of rear lamps, in daylight. There was no statistically significant difference in the distance at which the cyclist was first glanced at with or without a seat post-mounted rear lamp, but there was for the mean distance at which the cyclist was recognized (140 m without a rear lamp; 204 m with steady rear lamp, and 226 m with a flashing rear lamp). The difference between the distances for the steady and flashing seat post-mounted rear lamp was not statistically different. At night the situation is rather different. Edewaard et al. (2017) drove participants along a test route where they encountered two cyclists, one on a straight part of the road and one just after a 90-degree curve. The participants were asked to press a button when they were confident there was a cyclist present. The timing of this was used to calculate the distance at which the cyclist was recognized. For the cyclist straight ahead, there was a statistically significant difference in the recognition distance between a steady and a flashing seat post-mounted rear lamp (flashing 123 m, steady 41 m) but for the cyclist after the curve there was not (flashing 43 m, steady 45 m) either because the brightness of the rear light would be much reduced until the driver had rounded the curve and was directly behind the cyclist or the driver was paying attention to steering round the curve. Nonetheless, these results do demonstrate the benefits for cyclists of using a flashing rear lamp at night.

Another method for increasing the cyclist's conspicuity is clothing. Lahrmann et al. (2018) conducted a randomized controlled trial in Denmark. More than 3,000 participants were enrolled in a test group and a control group. Both groups cycled regularly during summer and winter. The test group was given a fluorescent yellow jacket with a large retro-reflective cross on it. The trial was conducted for one year. At the beginning of each month the participants received a web-based questionnaire asking if they had had a crash in the previous month; if they had, the questionnaire requested details of the crash. In total 694 crashes were reported, 274 by the test group and 420 by the control group, indicating the benefits of increased conspicuity. Table 10.8 shows the number of multi-party crashes that resulted in personal injury for different seasons, times of day, and different parties. Wearing the fluorescent jacket reduced the number of crashes for the test group relative to the control group for all the variables listed.

The results of Lahrmann et al. (2018) provide good evidence that wearing high-visibility clothing is an effective means to enhance cyclist safety, but is that all that is necessary? Wood et al. (2012) used a very different method to examine the effect of cyclist's clothing on conspicuity at night. Twenty-four drivers, 12 young and 12 old,

TABLE 10.8

Number of Multi-party, Personal Injury Accidents Occurring in Different Seasons, at Different Times, with Different Vehicles, for a Test Group of Cyclists with a Yellow Fluorescent Jacket Available and a Control Group without Such a Jacket (from Lahrmann et al., 2018)

Variable	Test Group	Control Group
Winter	890	1,730
Summer	1,400	1,960
Day	850	1,490
Night	130	220
Collision with pedestrian or cyclist	610	820
Collision with motor vehicle	530	1,020
High jacket use	860	1,840
Low jacket use	1,450	1,840

were driven 11 times round a closed test track at night. The test track had no ambient lighting and no traffic. What it did have were two cyclists pedaling in place, i.e., the bicycle was mounted on a support frame so that the rider was cycling but the bicycle was not moving. The test cyclist was positioned at the end of a 400-m long, three lane, straight part of the test track and was next to a pair of car headlights intended to represent the glare of an oncoming vehicle. The other bicycle and assorted traffic cones, bollards and road signs were distributed around the test track to provide some clutter but were not near the test cyclist. The test cyclist wore three different sets of clothing, a black sweatshirt and sweatpants and black shoe covers; the same but with a yellow vest with silver retro-reflective material on the shoulders back and front; and all this plus bands of silver retro-reflective material around the ankles and knees. The bicycle itself presented three different lighting conditions, no lighting, as well as white front and red rear lamps, either steady or flashing. The bicycle also had front white and red rear retro-reflectors as well as amber pedal and spoke reflectors as is required by the state authorities. The drivers were instructed to drive around the circuit at a comfortable speed and to press a touch pad when they were confident that what they saw was a person. To increase the workload of the driver, they were also asked to read aloud all the road signs they encountered. Two performance measures were used; the percentage of times the driver confidently and correctly identified the presence of the test cyclist and the distance at which this identification occurred. The percentage of trials in which the test cyclist was recognized averaged 70% ranging from 27% for black clothing, no bicycle lights, and older drivers, to 100% for the vest, ankle, and knee retro-reflectors, bicycle lights and young drivers. Table 10.9 shows the mean distances at which the drivers recognized the test cyclist for the different clothing and lighting conditions.

The first conclusion to draw from Table 10.9 is that riding at night on unlit roads, in black clothing and without lights is very foolish, the mean recognition distance

TABLE 10.9

Mean Distance (m) at which a Pedaling Cyclist Was Confidently Recognized by a Driver, at Night, on an Unlit Road When Wearing Different Clothing and With Different Lighting on the Bicycle (from Wood et al., 2012)

Lighting on Bicycle	Black Clothing	Black Clothing and Yellow Vest	Black Clothing, Yellow Vest, and Retro-reflective Bands on Ankles and Knees
None	18	44	165
Static front light	20	33	92
Flashing front light	20	38	98

being only 18 m. The second is that having a bicycle light makes little difference to the mean recognition distance when black clothing alone is worn. This shows the difference between detection and recognition. There is no doubt that having a light on a bicycle increases the chances of detection, but does little for the ability to recognize what has been detected. The third is that wearing the yellow vest increases the mean distance at which the cyclist is confidently recognized, but what really makes a difference is adding the retro-reflective bands on ankles and knees. Interestingly, for this condition when the bicycle is equipped with lights the mean recognition is less than when there are no lights on the bicycle. The explanation given for this unexpected result is that the front light may have acted as a glare source masking the ankle and knee retro-reflectors to some extent.

These results demonstrate that clothing can have a beneficial effect on a driver's ability to recognize a cyclist in the presence of glare, at night, but it is the addition of retro-reflectors on ankles and knees that produces a major improvement in recognition distance. One explanation for this effect is that when pedaling at night the retro-reflectors add movement to the luminance contrast. Threshold contrast in the peripheral visual field when movement is present is much lower than without movement (see Figure 3.8). Another aspect to consider is the results of Johansson (1973) who showed that when people wearing dark clothing are fitted with small lights attached to the major moveable joints, in the dark observers could easily identify people walking, cycling, and dancing. This implies that by attaching retro-reflective bands to the ankles and knees of a pedaling cyclist not just any movement but meaningful movement is being provided at night. The benefits to conspicuity of providing what is called biomotion has been demonstrated for both cyclist and pedestrians (Owens et al., 1994; Wood et al., 2022). Sadly, it is rarely seen on the roads at night, possibly because, for cyclists at least, it is assumed that the retro-reflectors mounted on the pedals are sufficient to provide a biomotion signal.

So far this discussion has concentrated on conspicuity of cyclists to other road users because that is where vision has a major part to play but it should not be thought that conspicuity is the only factor involved in the fatal and serious injury to cyclists.

Lahrmann et al. (2018) also report a number of personal injury crashes in which no other road user was involved. Specifically, the test group had 80 personal injury crashes in total while the control group had 96, a small difference that suggests that wearing the yellow jacket had little influence. Possible explanations for the personal injury crashes without the involvement of other road users include poor forward lighting making it difficult to see hazards ahead, reckless cycling in the weather conditions, loss of control after hitting a pothole, swerving to avoid an obstruction in the road, and rider fatigue or drunkenness.

So what should the cyclist do to minimize the risk of death or serious injury? The first step would be to stop wearing low reflectance clothing and riding without lights both day and night. The second step would be to use bicycle lanes where these are available to increase separation from other road users as these have been shown to minimize risk (Reynolds et al., 2009). The third is to wear high reflectance clothing with both fluorescent and retro-reflective elements, both day and night. Both fluorescent and retro-reflective elements are necessary because fluorescence will only occur when there is ultra-violet radiation present, as in daylight, but will not occur at night or under road lighting. Conversely, retro-reflective elements will not be effective in daylight when the daytime running lights of other vehicles will not be bright enough to produce a high luminance reflection but they will at night when other vehicles are using forward lighting. As for lighting, during the day use both front and rear lamps, preferably flashing. At night, use a static front lamp that illuminates the road well and a flashing rear lamp. Further, and most significantly, if you want to be really conspicuous at night, on unlit roads, get a set of retro-reflectance or self-luminous bands to wear round your ankles and knees (Wood et al., 2010). All these actions are possible but they would be more likely to be implemented if regulations specifying what cyclists should wear were introduced and the current legal requirements for the lighting of bicycles were regularly enforced and then updated (Fotios and Castleton, 2017). Hopefully, this latter action would improve the design and testing of lighting to ensure that cyclists could be both more easily seen by other road users and could see the road ahead clearly (Wu et al., 2016; Kircher and Niska, 2021). It would also be a good idea, given the uncontrolled direction of the light emitted, to ban the use of helmet-mounted lamps by cyclists.

10.5 PEDESTRIANS

Pedestrians come in different forms. Some are young and nimble, others are old and doddery. Some stroll, others march, yet others run. Some walk alone, some are aided by walking frames and some are encumbered with pushchairs or strollers. It is also fair to point out that, strictly, pedestrians are not vehicles as they are not driven. Nonetheless, pedestrians are frequently killed or injured because, in urban areas they are often found in close proximity to a busy road on a sidewalk. They also often have to cross roads. In rural areas there may be no sidewalk so pedestrians are forced to walk in the road. Therefore, in both urban and rural areas pedestrians often come into conflict with vehicles.

There are no lighting requirements for pedestrians, although in rural areas without road lighting it is common for pedestrians to carry a torch. In 2019, 470 pedestrians were killed and 6,528 were seriously injured on roads in the UK. Table 10.2 shows that, for the same distance travelled, pedestrians are much less at risk of being killed or seriously injured on the road than motorcyclists but they are comparable with cyclists and are much more at risk than the occupants of motor vehicles. Further, there is rarely a barrier to stop traffic leaving the road and entering the sidewalk. When this happens the pedestrian is very vulnerable because they have no protection, not even a helmet.

Table 10.10 shows the number of pedestrians killed or seriously injured in the UK in the years 2016–2021 and the other vehicles involved in the crash. From Table 10.10 it is clear that the main situation in which pedestrians are killed or seriously injured involves collisions with cars although being struck by a heavy goods vehicle is more likely to result in death than injury. Table 10.11 shows the percentage of pedestrian fatalities and serious injuries occurring at different road locations over the years 2016–2021 (Department for Transport, 2022c). Table 10.11 shows that a significant minority of fatalities and serious injuries occur at T, Y, and staggered junctions.

One site that is missing from Table 10.11 is the pedestrian crossing where pedestrians are given priority over approaching traffic (see Section 11.3). This is important because some 77% of pedestrian fatalities and serious injuries occur when the pedestrian is attempting to cross the road (Department of Transport, 2015c). Uttley and Fotios (2017) have examined the risk to pedestrians of using such crossings by day and after dark by applying the odds ratio method to daylight saving time changes (see Section 1.4) for UK data of crashes in which pedestrians were killed or injured over the years 2005-2015. They found that there was an increased risk of death or injury when using a pedestrian crossing compared to traversing at a point more than 50 m from a crossing (odds ratio = 1.23). They also found that the risk of death or injury when using a crossing increased after dark (odds ratio = 1.70) despite virtually all such crossings being lit by road lighting. This finding casts doubts on the adequacy of present lighting practice for pedestrian crossing, a subject that will be addressed in Section 11.3. Malin et al. (2020) examined the factors associated with

TABLE 10.10

Number of Pedestrian Fatalities and Serious Injuries in the UK over the Years 2016–2021 with the Involvement of Other Vehicle Types (from Department for Transport, 2022c)

Other Vehicle Involved	Fatalities	Seriously Injured
Bicycle	15	771
Motorcycle	66	1,472
Car	1,469	26,404
Light goods vehicle	166	2,276
Bus or coach	119	1,288
Heavy goods vehicle	277	578

TABLE 10.11

Percentages of Pedestrian Fatalities and Serious Injuries Occurring at Different Road Locations over the Years 2016–2021 (from Department for Transport 2022c)

Road Location	Percentage Fatalities	Percentage Serious Injuries
Crossroads	7.9	8.6
T, Y, or staggered junction	25.6	32.1
Junction with more than four arms, not a roundabout	0.9	1.4
Other junction	3.9	5.2
Roundabout	2.5	2.8
Mini-roundabout	0.5	0.9
Private drive or entrance	1.5	1.9
Slip road	1.6	0.6
Not at junction or within 20 m of one	55.5	46.3
Unknown	0.0	0.3

pedestrian fatalities and serious injuries in Finland over the years 2014–2017. They found that the main factors involved in such crashes were speed limits, municipality type, and lighting. Speed limits matter because the speed of the vehicle that hits a pedestrian influences the severity of injury (Tefft, 2013). Municipality type presumably relates to the likely traffic density and road infrastructure. As for lighting, there is a greater risk of a pedestrian fatality in daylight than after dark when road lighting is available. This is different to the situation in the US where the majority of pedestrian fatalities occur after dark (National Highway Traffic Safety Administration, 2021c). What this suggests is that pedestrians suffer from poor visibility after dark and poor conspicuity during daylight.

Fortunately, both visibility and conspicuity of pedestrians can be improved by increasing their contrast against the immediate background. For pedestrians, this is usually a matter of clothing. Edwards and Gibbons (2007) had people drive a vehicle equipped with halogen headlamps over a closed test track and report when they detected a pedestrian on a crossing. The test track was lit by road lighting set to provide four different vertical illuminances on the crossings. The pedestrians were clothed in white, denim, or black hospital scrubs. Figure 10.2 shows the mean detection distance plotted against vertical illuminance.

In Figure 10.2 the very limited effect of illuminance over the range examined indicates that the reflectance of the clothing is vital for seeing the pedestrian. Of course, this is at night. By day, given the variety of other coloured stimuli visible, wearing high-reflectance clothing may not be enough but wearing high-visibility fluorescent clothing will be, as it is different from most clothing and has been shown to be beneficial for both detection and recognition of pedestrians (Kwan and Mapstone, 2006). Such clothing covers large areas of the body but there are additional benefits

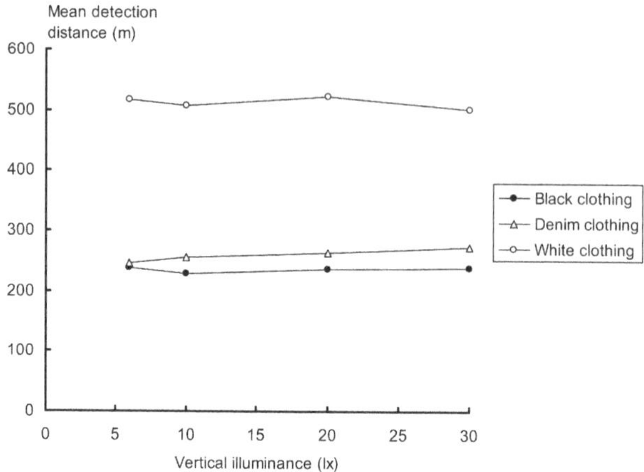

FIGURE 10.2 Mean detection distance (m) for pedestrians wearing black, denim, or white clothing on pedestrian crossings platted against vertical illuminance (lx) at the crossing (after Edwards and Gibbons, 2007).

to adding retro-reflective patches to the clothing at night. Wood et al. (2005) had young and old drivers drive round a closed test track using either low- or high-beam headlamps and looking for pedestrians, with and without glare from another set of headlamps. The pedestrians were wearing either black or white clothing, black clothing with a retro-reflective vest or black clothing with retro-reflective bands on leg and arm joints so as to provide a perception of the movements associated with walking (biomotion). Drivers recognized only 5% of pedestrians in the most difficult conditions (low-beam headlamps, black clothing, glare), but 100% of pedestrians in good conditions (high-beam headlamps, black clothing with retro-reflective bands configured to give biomotion, no glare). In the absence of glare, the mean recognition distances in the best conditions (young drivers, high-beam headlamps, black clothing with retro-reflective bands giving biomotion) was 220 m, but it was 0.0 m in the worst conditions (low-beam headlamps, old drivers, black clothing).

There can be little doubt that retro-reflective bands configured to display biomotion are of great benefit to making pedestrians conspicuous at night. Even in the presence of road lighting, the greater light output from the headlamps of approaching vehicles will produce high contrast stimuli representing biomotion. Unfortunately, this will not occur by day as daytime running lights do not produce much by way of a beam. In order to produce a biomotion stimulus by day it is necessary to use some light-emitting source instead of retro-reflective material. Battery-powered LEDs or electroluminescent patches are a possibility. Fekety et al. (2006) have demonstrated that electroluminescent patches further enhance the conspicuity of pedestrians at night especially where the pedestrians are outside the forward lighting beams of an approaching vehicle. It would be interesting to know if they were also beneficial when attached to fluorescent clothing in daylight.

Thus, there are several ways to improve the conspicuity of pedestrians by day and night, but it should not be thought that improving conspicuity is enough to ensure absolute pedestrian safety. It is not. This is because there are many factors involved in crashes that lead to pedestrian fatalities and injuries. Limited conspicuity is just one of them. Others include driver and/or pedestrian distraction, either external or internal, driver and/or pedestrian drunkenness, driver and/or pedestrian cognitive overload, and driver and/or pedestrian stupidity.

10.6 HEAVY GOODS VEHICLES

In many countries, heavy goods vehicles are the means by which food, fuel, materials, and manufactured goods are distributed. Heavy goods vehicles (HGVs) come in two forms: rigid and articulated. Rigid vehicles have the tractor unit permanently attached to the trailer. Articulated vehicles have a tractor unit that can be separated from the trailer. HGVs do not represent a high risk of death or serious injury to their drivers or passengers (see Table 10.2), but they do pose a serious risk of harm to other road users, particularly those with little protection (see Table 10.3). The lighting requirements for HGVs are rather more extensive than for vans and cars. In addition to the forward lighting and signal lighting required for all vehicles, HGVs are required to have amber side lamps as well as red marker lamps at the rear and white markers lamps at the front of the vehicle to indicate its width. They also have additional high-mounted red rear position and stop lamps. In many countries, heavy goods vehicles are also required to have conspicuity marking at the front, rear, and sides conforming to ECE regulation 48. Specifically, in Europe and the UK conspicuity marking is required on all HGVs with a gross weight of 7,500 kg, a width greater than 2.1 m, and a length more than 6 m. For articulated HGVs, such conspicuity marking is required for the trailers with a gross weight of more than 3,500 kg and exceeding the same width and length limits. The tractor unit of an articulated HGV is exempt from these requirements. The Freight Transport Association (2013) illustrates what form these conspicuity marking should take for different HGV types. The colours of the retro-reflective markings are limited to red or amber at the rear of the vehicle and white or amber on the side. These conspicuity markings replace the previous requirement for retro-reflective plates mounted on the rear of the HGV although these can still be used (see Figure 7.5).

Other countries may have different definitions of what constitutes a HGV and different types of conspicuity marking but they all have the same objective, to make the HGV more conspicuous at night from the rear and the side. A number of studies on the effectiveness of retro-reflective contour marking on HGVs have been undertaken. Several have indicated that the reaction time to the presence of a HGV is shortened and, consequently, the distance at which it can be detected is increased by retro-reflective contour marking (Schmidt-Clausen, 2000; Sullivan and Flannagan, 2004; TUV Rheinland Group, 2004). Morgan (2001) examined the effect of retro-reflective contour marking on crashes into HGVs in Florida and Pennsylvania. She found that such marking reduced the number of fatalities and injuries caused by a crash into the side and rear of trailers by 44%, but only in the dark. By day and even at night but when road lighting was present, the contour marking made no difference.

One practical aspect that determines how effective conspicuity marking will be is how clean the retro-reflective tapes are. Morgan (2001) found that clean tape reduced the number of rear collisions by 51%, but when the tape was dirty, the reduction was only 27%. Everyday observation suggests that dirty contour marking is common.

10.7 EMERGENCY VEHICLES

Emergency vehicles are used by the emergency services, most commonly the fire and rescue service and the police. Emergency vehicles are often painted in high contrast chevron or chequerboard (also called Battenburg) patterns using fluorescing and retro-reflective material. Chevrons are often used on the rear of the vehicle and chequerboard patterns on the side where space is more limited (Figure 10.3).

These vehicles are also equipped with special lighting in addition to the conventional forward and signal lighting to indicate their status to other traffic and pedestrians. This lighting is only used when responding to an emergency or, in the case of the police, when making a traffic stop. The lighting is often combined with a siren to provide an auditory signal of the presence of an emergency vehicle in a hurry. The siren is certainly alerting, but it is often difficult for other drivers to determine where the sound is coming from. The additional lighting of the emergency vehicle enables the source of the sound to be located. The response of the drivers of other vehicles should be to move over so as to clear the way for the emergency vehicle to pass. At

FIGURE 10.3 A UK police car with a light bar on the roof, chevron markings on the rear and chequerboard markings on the side. The vehicle in front of the police car is a fire engine, also with chevron markings on the rear (iStock by Getty Images / Alan Lagado).

junctions, even those controlled by traffic lights, attempts should be made to clear the junction so as not to impede the emergency vehicle.

The lighting provided on emergency vehicles can take several different forms from a single flashing beacon to a light bar containing several light sources that provide flashing or spatially alternating signals (Figure 10.3). The beacon can consist of a single lamp around which a reflector is spun. This produces a flash of light when viewed from a fixed direction which can be anywhere in a 360-degree circle around the beacon. If a halogen light source is used, the transparent cover of the beacon will be of a solid colour. If LEDs are used, this is unnecessary as the LED itself produces light of the required colour. LEDs also have the advantage that they can be programmed to produce a wide variety of flash patterns. Light bars can contain a variety of lamps of different colours, light distributions, steady or flashing. Both rotating beacons and light bars can be permanently mounted, usually on the roof, but rotating beacons can also be detachable using a magnet to fix them to the vehicle's roof when required. Other equipment in the form of LED variable message signs are sometimes installed in the vehicle so as to be visible to other drivers. These are used to tell other drivers what they should do. The conventional front and rear vehicle lighting can also be modified electronically or mechanically to provide a flashing signal or, where there are two lamps such as headlamps, a spatially alternating signal. Which of these options is used varies greatly in different countries and even within countries depending on the authority responsible for emergency services. The same can be said of the colour of the light sources used. In the UK, the colour for emergency vehicle lighting is blue although there is also widespread use of flashing headlamps (white) and rear lamps (red). Doctors on an emergency call may use green flashing lamps in addition to blue. In the USA, emergency vehicle lighting colour is determined at state level. Many states use red lights facing forward or blue lights. These are often combined with white light, steady or flashing. Some federal government agencies use green lights. In Australia and China, red and blue lights are used by most emergency vehicles. In Japan, red lights are used by emergency vehicles, but spatially alternating headlamps are not, and so on around the world. All this variation should cause confusion amongst road users, but it rarely does. The simple rule, wherever you are, is if you hear a siren and see a vehicle lit up like a Christmas tree with a lot of flashing lights, you can assume it is an emergency vehicle in a hurry so it would be wise to get out of its way and follow any instructions it issues. Unfortunately, this is not always successful. Missikpode et al. (2018) examined the incidence of crashes involving emergency vehicles when using lights and sirens. The data used was taken from the Iowa Department of Transportation over the years 2005–2013. Over those years, police vehicles were involved in 2,406 crashes while ambulances and fire engines were involved in 528 crashes. Unsurprisingly, when police vehicles were being driven with lights and sirens, they were 1.8 times more likely to crash than when being driven in a normal manner, but there was no increase in crash risk for ambulances and fire engines when using lights and sirens. However, Watanabe et al. (2019), using a US national database for a single year, 2016, found that the crash rate for ambulances responding to a call increased from 4.6 per 100,000 calls to 5.4 per 100,000 calls when lights and siren were used. When transporting a patient from an incident to hospital the crash rate increased from 7.0 per 100,000 calls to 17.1 per 100,000 calls when lights and siren were used. The National Safety Council (2023) report that in 2020, in the US, 50% of people killed in crashes involving emergency vehicles were the drivers or occupants of other vehicles, 25% were pedestrians, 11% were the drivers of the emergency vehicles, and 5% were passengers in the emergency vehicle. It seems reasonable to assume that such crashes have little to do with the use of lights and siren as such and more to do with the manner in which the emergency vehicle is being driven given the urgency of the journey. Nonetheless, in an age where vehicle-to-vehicle communication is becoming a real possibility it must be possible to devise a better system to give vehicles ahead a more timely and accurate warning about the approach of an emergency vehicle than is possible with just lights and siren.

The lighting described above is valuable to draw the attention of other drivers and pedestrians to the emergency vehicle and to tell them to get out of the way. This is useful when travelling to the emergency and when returning from it with casualties. However, when at the emergency a wider range of requirements occur. It is still necessary to warn approaching drivers that there is something unusual ahead but the emergency workers have to be able to see what they need to do and the approaching drivers need to be able to see the emergency workers. This is important because there is a steady stream of media reports of emergency workers being struck by passing drivers when outside their vehicles, particularly at night where there is no road lighting. This is because the bulk of the scene will be in darkness and any flashing lamps will attract attention and may cause disability glare making any emergency worker more difficult to detect. This is becoming increasingly common as there is no maximum luminous intensity set for flashing lamps, only a minimum of 300 cd (Society of Automotive Engineers, 2014). Also, two common assumptions made about emergency vehicle lighting are that brighter is better and more lamps are better than fewer. Neither of these assumptions is true. The beneficial effects of both brightness and number of flashing lamps on conspicuity soon saturate after which further increases simply cause confusion to approaching drivers with their attention being divided between the many stimuli. Bullough et al. (2019) conducted an experiment in which observers were asked to look at a model police car fitted with flashing blue or red lights, in an otherwise unlighted black room, from a distance scaled to be 75 m. Sometimes a model police officer was placed beside the car. The observer's task was to determine if the police officer was present or absent. The percentage of correct response was 57% when there were no flashing lights, but 45% when the blue or red lights were flashing. Sadly, neither of these percentages can be considered acceptable. What really made a difference was having white LED lights mounted on the police car so as to provide 1 lx on the ground close to the car. When these were lit, the percentage of correct identification increased to 81%. This suggests that what is needed at an incident is some form of area lighting, either mounted on the emergency vehicle or carried as portable equipment. A telescopic floodlight would be an example of the former, while a balloon light would be an example of the latter. The essential characteristics of this lighting would be that it provides diffuse illumination over a large area without causing disability or discomfort glare to emergency workers or approaching drivers. Such area lighting ensures that the immediate background brightness to any flashing lights is increased so that any disability or discomfort glare they cause will be reduced. This could be further diminished by dimming to reduce the maximum luminous intensity of the flashing light at night or by increasing the minimum luminous intensity to make the light source pulse rather than flash. It has been found that lights that pulse rather than flash are less glaring and easier to navigate past as well as giving faster detection of closure (Rea and Bullough, 2016). Essentially, the objective when lighting the scene of an incident should be to create conditions similar to those that occur in daylight, not in terms of the absolute amount of light, rather in terms of the relative illumination of the scene and the emergency vehicles attending. Until due consideration is given to lighting the scene uniformly, without glare, it is likely that emergency workers will continue to be killed at night while doing their duty.

10.8 SLOW AND FREQUENTLY STATIONARY VEHICLES

Amongst all the cars, vans, buses, and trucks found on many roads are some vehicles likely to be travelling at lower speeds. Examples are agricultural tractors, wide loads, snowploughs, and towed caravans and horseboxes. There may also be vehicles that stop frequently, such as postal delivery vans and dustcarts. In addition, there may be vehicles stationary in unusual places such as utility repair vans and breakdown recovery vehicles. The presence of any of these vehicles is likely to disrupt the smooth flow of traffic which means other drivers need to be made aware of their presence.

All motorized vehicles are equipped with hazard flashers (see Section 7.10). Drivers of any vehicle that has partially or totally broken down should use them. They are also useful when an unexpected hazard is encountered. Indeed, it is quite common to see hazard flashers activated by vehicles arriving at the back of a queue on a motorway to warn approaching drivers about the need for a dramatic reduction in speed. For many drivers a breakdown or sudden blocked road is a rare event, but where the slow speed or regular stopping or being stopped in unexpected locations is a common aspect of the vehicle's use, the vehicle is often fitted with additional well-separated, amber, flashing beacons. Unlike emergency vehicles, the installation of amber lights does not confer any rights on the drivers of these vehicles, the amber lights are there simply to provide a warning to other drivers.

Rea and Bullough (2016) conducted two laboratory experiments to devise a better performance specification for flashing warning beacons than was then available. One experiment displayed an amber beacon subtending 1 min arc at the observer's eye and appearing at the edge of a straight road at background luminances simulating the road surface by day (300 cd.m^{-2}) and at night (1 cd.m^{-2}), in simulated urban and rural conditions, the urban condition having visual clutter in the form of 120 flashing lights randomly located along the sides of the road, the rural condition having no other light sources. The beacon simulated four peak luminous intensities ranging from 80 cd to 3100 cd, flashed at 1 Hz from full light output to off with a 50% duty cycle. The observers fixated on a low contrast Landolt ring target (contrast

= 0.2) located where the beacon would appear or 5 degrees away from it and were instructed to press a button as soon as they detected the appearance of the beacon. The observers' reaction times to the onset of the beacon were measured as were the subjective visibilities of the beacon itself and of the low contrast Landolt ring. The visibility ratings of the beacon was made relative to a beacon of peak luminous intensity of 850 cd which was given a value of 10. The rating scale is open-ended so a rating of 5 indicated a beacon half as visible as the 850 cd beacon and a rating of 20 indicated a beacon twice as visible as the 850 cd beacon. The same method was used for the visibility rating of the low contrast Landolt ring, but for this the reference value of 10 was associated with the Landolt ring when there was no beacon present. Figure 10.4 shows the mean reaction times for on- and off-axis detection, by day and at night. The mean reaction times show little change beyond a peak luminous intensity of 190 cd, for day or night conditions, seen on- or off-axis. Below a peak luminous intensity of 190 cd, the on-axis reaction times showed little change but the off-axis reaction times increased, particularly for the day condition. Figure 10.5 shows the mean visibility ratings of the beacon plotted against peak luminous intensity for on- and off-axis viewing. Unlike the mean reaction times, the rated visibility continues to increase up to the maximum peak luminous intensity of 3,100 cd, the differences between the four viewing conditions being slight. Figure 10.6 shows the mean visibility rating of the low-contrast target when the beacon is viewed on- and off-axis, by day and night. There is little change in mean visibility rating until the peak luminous intensity of the beacon reaches 850 cd and then mainly when seeing the low-contrast target on-axis, at night. This deterioration of mean rating is due to disability glare from the beacon.

The other experiment used a video animation of a straight road, at night, in a rural setting (background luminance 1 cd.m^{-2}). At the edge of the road were one or two amber lamps producing a steady peak luminous intensity of 300 cd or alternating from 300 cd to 0 cd (flashing) or from 300 cd to 30 cd (pulsing), at a frequency of 1Hz with a 50% duty cycle. These lamps simulated a stationary utility vehicle with

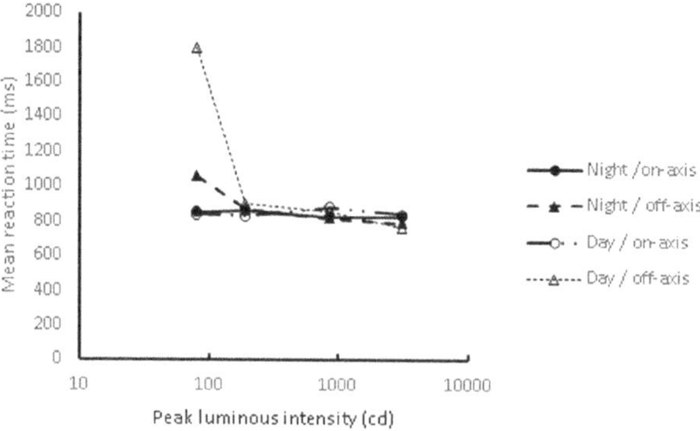

FIGURE 10.4 Mean reaction times (ms) to the onset of warning beacons as a function of peak luminous intensity for simulated day and night conditions when viewed on- and off-axis (from Rea and Bullough, 2016).

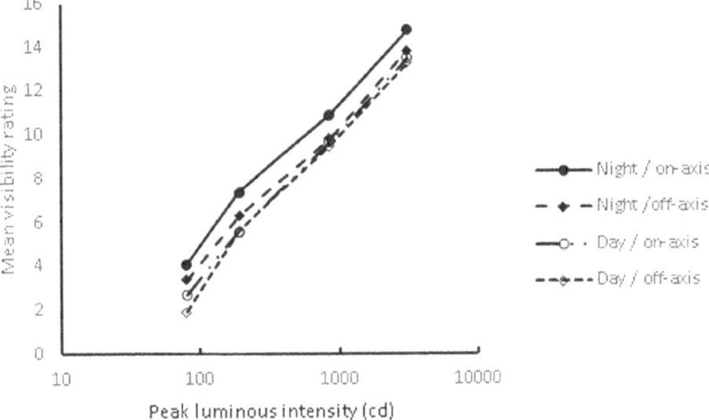

FIGURE 10.5 Mean visibility rating for the warning beacons as a function of peak luminous intensity for simulated day and night conditions when viewed on- and off-axis (from Rea and Bullough, 2016).

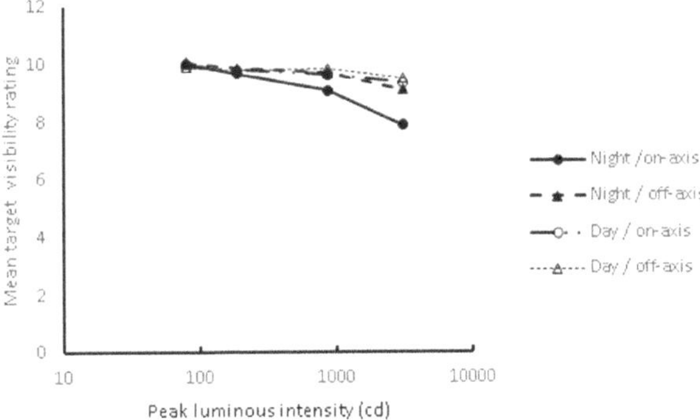

FIGURE 10.6 Mean visibility rating of the low contrast Landolt Ring target as a function of peak luminous intensity of the warning beacons for simulated day and night conditions when viewed on- and off-axis (from Rea and Bullough, 2016).

amber warning beacons operating. After a random delay, the lamps began to move towards the observer at a speed of 10 mph (16 km.h^{-1}) from an apparent distance of 100 m. The observers were instructed to view the lamps on-axis or to fixate on an LED target 5 degrees off-axis and to respond as soon as they realized the lamps were getting closer. The mean closure detection times were obtained for 1 or 2 lamps, for on- or off-axis viewing and for the steady, flashing and pulsing beacons. These data showed that two lamps give statistically significant shorter mean closure detection times than one lamp (one lamp = 6.52s, two lamps 4.81s), as does on-axis viewing rather than off-axis viewing (on-axis = 4.56s, off-axis 6.75s). Most interestingly, the pulsing lamps give statistically significantly shorter mean closure detection times than the flashing lamps, times that are not statistically significantly different from the times for the steady lamp (flashing lamps = 6.37s, pulsing lamps = 5.29s, steady lamps = 5.33s).

From these experiments Rea and Bullough (2016) developed a set of recommendations for the flashing warning beacons. Specifically, the vehicle should be equipped with two beacons, each with a peak luminous intensity in the range of 750 cd to 2000 cd pulsing down to a minimum luminous intensity of 10% of the peak. The justifications given for these recommendations were that two beacons are better for closure detection than one; peak luminous intensities in the range 750 cd to 2000 cd are enough to reach an asymptotic reaction time to the onset of a beacon, both day and night, without causing disability glare; and pulsing the luminous intensity down to 10% of the peak supports closure detection.

In addition to beacons, vehicles that are likely to be parked on the road frequently are also often painted in high contrast chevron patterns on the rear using fluorescing and retro-reflective material (Figures 3.18). The pairs of colours used vary from authority to authority. Common combinations are blue and yellow, green and yellow, and red and yellow. The idea of such markings is to make the vehicle conspicuous and easy to identify although quick identification requires a consistent application of such markings, something that is difficult to achieve in such an unregulated market.

10.9 PERSONAL MOBILITY VEHICLES

Personal mobility vehicles come in a number of different forms and are increasingly likely to be seen on or near the road. Examples are mobility scooters, powered wheelchairs, e-scooters, e-skateboards, and Segways. They tend to be electrically powered and are designed for use by one person. The personal mobility vehicles most likely to be seen by drivers are the mobility scooter and, increasingly, the e-scooter, the former being used primarily by the elderly and the latter by the young.

In many wealthy countries there is an ageing population and increasing levels of obesity. Both these afflictions can make walking any distance difficult. The mobility scooter gives such people greater independence and helps them maintain social contact. The mobility scooter is effectively a battery-powered wheelchair (Figure 10.7a). In the UK, there are two classes of mobility scooter: one (class 2) being limited to use on sidewalks and the other (class 3) being allowed on both sidewalks and roads. A class 2 mobility scooter is limited to a top speed of 4 mph (6.4 km.h^{-1}), but a class 3 can travel at up to 8 mph (12.9 km.h^{-1}) when on a road. For use on roads, a class 3 mobility scooter has to be equipped with a forward headlamp, rear position lamps, flashing turn lamps that can also be used as hazard flashers, as well as retro-reflectors. In the unlikely event that a mobility scooter is used on a dual carriageway, then it must also carry a flashing amber beacon to identify it as a slow-moving vehicle. With the exception of the amber beacon, these lights will be installed by the manufacturer, although there is a flourishing aftermarket for adding lights to class 2 and for enhancing the lighting of class 3 mobility scooters. Given what has been said above about the benefits of improved conspicuity for relatively unprotected drivers and walkers, it would seem that people using a mobility scooter would be well advised to employ such lighting, both night and day, to wear fluorescent and retro-reflective clothing, and to mark the scooter with retro-reflective tape.

FIGURE 10.7 a) A mobility scooter crossing a road (iStock by Getty Images/Eileen Groome) and b) an e-scooter also crossing a road (iStock by Getty Images/LukaTBD).

Mobility scooters are not without risk to their users and to pedestrians. The Australian Institute for Health and Welfare has made a study of mobility-related injuries and deaths, using data taken from the National Hospital Morbidity Database for the five years, 2011–2016 (AIHW, 2019). In Australia, a mobility scooter is legally considered to be a pedestrian and can go anywhere that a pedestrian can go. They can only be used on a road where there is no pavement and then they are required to stay close to the edge of the road and travel in the opposite direction to traffic. Over the five years, 4,500 people were admitted to hospital following a fall involving other and unspecified pedestrian conveyances. Exactly what these conveyances were is not known but some will have been mobility scooters, the fall happening when the user attempted to mount or dismount from the machine or when it toppled over after a collision or an unwise manoeuvre. As for pedestrians, 121 pedestrians were admitted to hospital for injuries caused by collisions with mobility scooters, only 19% occurring in a traffic setting such as crossing the road or walking on a sidewalk. Most of those injured were aged 65 years and older and were most likely to suffer fracture injuries to their legs and hips. Unfortunately, there is a paucity of crash statistics for mobility scooters. Amongst the missing information is data on when the crash occurred and hence whether it occurred in daylight or after dark on a lit or unlit road. Until such information becomes available it is not possible to determine if the lighting of mobility scooters should be improved to enhance the visibility provided to the rider and make them more conspicuous to other users of sidewalks and roads.

Another battery-powered means of transport that is increasingly popular in cities is the e-scooter (Figure 10.7b). These have a higher range of speeds than mobility scooters, typically 15–30 mph (24–48 km.h^{-1}). The rules about where an e-scooter can be used and what the legal requirements are for riding one vary widely between cities in different countries and are changing frequently. In the UK, at present, privately owned e-scooters cannot be used on public roads although there are some pilot trials going on where rented e-scooters can be used on roads but not on sidewalks. In these trial areas the maximum speed limit for e-scooters is 15.5 mph (25 km.h^{-1}). However, a quick search on the internet will reveal copious advice on how to make your e-scooter faster. In the US, the situation varies from state to state and sometimes between cities in the same state. Many states allow e-scooters to be used on public roads, but some do not. Some allow their use on sidewalks, but others do not. The situation is similar in Europe were some cities permit their use while others have changed their mind after initially allowing e-scooters to be used but have now banned them. This confused situation is typical of a new technology that does not fit into existing legal frameworks for vehicles. No doubt in time a common basis for legal requirements and a pattern of use will develop.

Most e-scooters are fitted with white forward lighting and red rear signal lighting, but these are often inadequate because they are mounted so low down that they only illuminate the road surface for a short distance ahead and are difficult for other drivers to discern. Additional lighting equipment is recommended for the rider to be able to see the road far enough ahead to see any obstacle or potholes and to take evasive action if necessary and for the drivers of other vehicles to be able to see the scooter. Figure 10.7b shows how unprotected riders of e-scooters are. To ensure their conspicuity, riders of e-scooters would be well advised to employ such lighting, both night and day, to wear fluorescent and retro-reflective clothing, and to mark the scooter with retro-reflective tape.

E-scooters are inherently less stable than mobility scooters. A mobility scooter typically has four wheels on the corners and the rider sits on a seat between them. On an e-scooter, the rider stands on a narrow platform with two or three wheels, one or two at the front and one behind. Of course, an e-scooter is easier to mount onto and dismount from than a mobility scooter, but that also means it is easier to fall off when moving. The lighting of e-scooters is often rather primitive. It usually consists of a white LED headlamp mounted just above the front wheel(s) capable of being used in continuous or flashing mode; a red rear position/stop lamp also mounted low and amber turn indicators. It is recommended that the headlamp and rear lamps should have a light output of at least 300 lm. For better visibility and conspicuity it is also recommended that the headlamp should be mounted at the level of the handlebars.

Injury statistics for e-scooters are limited. In the UK, the Department for Transport has published some casualty data for 2020, based on reports from police forces of slight and serious injuries arising from interactions with e-scooters (DfT, 2021c). In 2020, it was reported that there were 484 casualties associated with e-scooters. Of these 384 were injuries to e-scooter riders, 57 were injuries to pedestrians, and 21 were injuries to cyclists. Only one e-scooter rider was killed, but it is estimated that 128 people suffered serious injuries and 355 slight injuries. The majority of these crashes happened in daylight.

With an ageing population in many countries, it is likely that the use of mobility scooters will increase. A similar prediction can be made for e-scooters because they provide an economic, efficient, and environmentally friendly means of personal transport in urban areas. This makes it essential that clearer regulations on where and when they can be used and how they should be lit are developed. It is also essential that the means of systematically recording crashes and injuries are revised. At present, in the STATS19 database, mobility scooters and e-scooters come under the heading of "other vehicle". With the expected rapid increase in their presence on sidewalks and roads, there is clear need to collect accurate data on their crash involvement.

10.10 SUMMARY

All vehicles are equal, but some are more equal than others, at least as far as the risk of death or serious injury is concerned. When expressed in terms of equal level of exposure, riding a motorcycle, as either rider or passenger, has the highest risk by far, followed by cyclists and pedestrians. Compared to these, the drivers and occupants of cars, light and heavy goods vehicles and buses are exposed to much less risk. This pattern of risk is related to a combination of speed and level of protection. Motorcyclists travel at high speeds with little protection other than a crash helmet. Cyclists and pedestrians travel more slowly, but they too have little protection. Cars certainly do travel fast, but the protection given to drivers and occupants has been improved greatly by making the wearing of seat belts compulsory, and the introduction of vehicle design features such as crumple zones and airbags. Even less risk is associated with goods vehicles which tend to be slow and more substantial so the driver is well-protected. However, when the question is reversed and the risk that different vehicle types pose to other vehicles, it turns out that heavy goods vehicles and buses pose the most risk to other road users. This is a matter of the momentum each vehicle brings to a crash. The subject of this chapter is what lighting can do to reduce the risk of death or serious injury for the users of different vehicle types.

Motorcyclists, and any pillion passengers, are most at risk, often, but not entirely, from collision with other vehicles. Collisions occur because the small frontal area of the motorcycle and the use of one or two headlamps close together make it difficult for a driver waiting at a junction to judge how far away the motorcycle is and how fast it is moving, particularly at night. During the day, there is an additional problem. This is the allocation of attention that drivers have to give to many different stimuli. The way to alleviate this is to increase the conspicuity of the motorcycle. This is why motorcyclists are encouraged to use low-beam headlamps during the day. Unfortunately, now that many other vehicles have daytime running lights, the conspicuity benefits have been reduced but they could be restored by making the headlamp flash or pulse.

Cyclists are not required to use lights during the day, but only at night. Nonetheless, they would be well advised to use front and rear lighting during the day to increase their conspicuity, preferably lighting that flashes or pulses. They should also consider wearing high reflectance clothing that fluoresces in daylight with integrated retro-reflectors that provide a high contrast stimulus when illuminated by the headlamps of an approaching car or truck at night. Even better would be the use of retro-reflector bands on the ankles and knees of the cyclist. When pedaling these provide a biomotion stimulus which makes it clear to the approaching driver that it is a cyclist ahead.

Pedestrians are not strictly a vehicle type but are included here because they have a high risk of death or serious injury when struck by a vehicle so drivers need to see them. Pedestrians do not carry lighting unless they are in a rural area without road lighting but they do have a choice about what sort of clothing to wear. Like cyclists, they can increase their visibility by wearing high reflectance clothing that fluoresces during the day integrated with retro-reflective material for use at night. Again, like cyclists, retro-reflective strips positioned on moving joints make it easier for an approaching driver to identify a pedestrian from the biomotion.

Heavy goods vehicles (HGVs) do not pose much risk of death or injury to their drivers but they do pose a high risk to other road users because of their momentum. HGVs are required to have additional lighting to what is required for smaller vehicles to make them conspicuous when viewed from behind or the side. This applies to both rigid and articulated HGVs although for the latter the additional lighting applies to the trailer but not the tractor unit. This lighting consists of side marking lamps evenly spaced along the edge of the vehicle. Further, retro-reflective strip is applied to define the contour of the HGV on the rear and sides of the vehicle. Such markings have been shown to reduce the number of crashes in which another vehicle crashes into the rear or side of a HGV at night.

Emergency vehicles are used by the emergency services, most commonly the fire and rescue services and the police. These vehicles are equipped with lighting in addition to the conventional forward and signal lighting of the vehicle to indicate their status to other traffic and pedestrians. This lighting can take various forms from a simple rotating beacon to a light bar containing several different lamps, combined with modifications to the conventional lighting to provide flashing and spatially alternating stimuli, all in different colours. This lighting is only to be used when responding to an emergency or, in the case of the police, when making a traffic stop. Lighting is often combined with a siren to warn drivers of the approach of an emergency vehicle, this is because sound is often difficult to locate, lighting is easier. When a driver detects the approach of an emergency vehicle the correct action is to get out of the way to let it pass. Needless to say, emergency vehicles are more likely to be involved in a crash when they responding to an emergency than when being driven normally. This has more to do with the way the vehicle is being driven than the lighting. Once emergency vehicles have arrived at the incident they are responding to then another form of lighting becomes important. There have been a steady stream of media reports of emergency workers being killed or injured when outside their vehicles. What is required to avoid this fate is some diffuse area lighting to increase the visibility of the people in the area of the incident. Reducing the luminous intensity of the flashing lamps of the emergency vehicles at the scene or making them pulse rather than flash by not switching completely off would also help.

Another vehicle type that has special lighting or marking is the slow-moving or frequently stationary vehicle. These are often fitted with well-separated amber flashing lamps. Like emergency vehicles, frequently stationary vehicles are often marked with large chevrons or chequerboard patterns in retro-reflective material to warn drivers of their presence. Unlike emergency vehicles, this lighting and marking does not confer any freedom from the established rules of the road. It is simply there for identification.

In the future, personal mobility vehicles in the form of mobility scooters and e-scooters are increasingly likely to be found on roads and sidewalks. At present, where, when and how these vehicles can be used varies widely between and within countries. Both mobility scooters and e-scooters usually have some form of LED forward and rear signal lighting fitted but it is often rather primitive. The involvement of mobility scooters and e-scooters in crashes causing death or injury to their riders or others is not well documented. There is a clear need for lighting standards and the collection of crash data to be updated to take these vehicles into account.

These considerations of what lighting and marking can do to reduce the risk of death or serious injury to relatively unprotected motorcyclists, cyclists, riders of mobility scooters and e-scooters, and pedestrians, and to minimize the risk to other road uses posed by heavy goods vehicles, emergency vehicles, and slow-moving or frequently stationery vehicles have a number of common features. Nearly all use lighting and marking to enhance conspicuity, day and night. The problem with current practice is twofold. The first is that too often attention is focused on conspicuity alone without considering the consequences for other visual capabilities and cognitive functions. The second is that different countries and different authorities within countries have their own preferences for lighting and marking, particularly as regards colour. A more consistent and holistic assessment of how different vehicle types interact on the roads and on pavements would be beneficial to everyone.

11 Special Locations

11.1 INTRODUCTION

The lighting discussed in Chapter 4 is applicable to the vast majority of roads but there are still a number of locations that require special treatment, for a number of reasons. Some locations such as tunnels pose visual problems during the day. Other locations such as pedestrian crossings, railway crossings, and car parks are places where vulnerable individuals are exposed to hazards. Intersections are high-risk areas because traffic streams cross and merge. Road works may temporarily restrict traffic flow in an unexpected manner and contain workers on foot in the carriageway. Locations such as roads with traffic-calming measures may pose a risk to drivers unfamiliar with such obstructions. Roads near docks and airports may cause confusion to other forms of transport. This chapter is devoted to how conventional road lighting should be adapted to cope with these special locations.

11.2 TUNNELS

The first question that needs to be addressed when considering the lighting of tunnels is what constitutes a tunnel? One answer would be a section of road that is not exposed to the sky but that definition would also apply to an underpass. The difference between these two structures is a matter of their length and a driver's ability to see what is inside. A tunnel can be defined as a section of road that is not exposed to the sky and that requires lighting during the day for drivers to be able to see vehicles and obstructions within it. An underpass is usually short in length and may not require daytime lighting for drivers to be able to see vehicles and obstructions inside it. CIE (2004) simplifies the decision on whether a structure is a tunnel or an underpass by recommending that a structure less than 25 m in length does not need lighting by day while a structure more than 125 m in length does. For structures between these limits, the look-through percentage should be considered. The look-through percentage is the ratio of the area of the exit aperture visible to the approaching driver to the area of the entrance aperture up to a line formed by the top of the exit aperture, times 100 (Schreuder, 1998; van Bommel, 2015). Look-through percentages greater than 50% are believed not to need lighting during daylight hours because vehicles inside can be seen in silhouette against the exit aperture. Look-through percentages in the range 20% to 50% may need lighting during daytime hours depending on how much daylight enters the tunnel, the reflectance of the road surface and tunnel walls, and the type of vehicles likely to be found in the tunnel. Look-through percentages of less than 20% certainly do need lighting during daytime. Such structures should be classed as tunnels regardless of their length.

DOI: 10.1201/9781003388906-11

The lighting of tunnels has to address two different problems. The first is the black-hole effect experienced by a driver approaching a tunnel. The second is the black-out effect caused by a lag in visual adaptation on entering the tunnel. Neither of these problems occurs at night because then the average road surface luminance inside the tunnel is recommended to be a minimum of 2.5 cd.m^{-2}, a value likely to be similar to, if not more than, the road surface luminances outside the tunnel (IESNA, 2021b). By day, this is not the case. The luminances around the tunnel portal will be much higher than those inside the tunnel so both the black-hole effect and the black-out effect may be experienced, and driver safety may suffer (Ueki et al., 1992).

The black-hole effect refers to the perception that from the distance at which a driver needs to be able to see vehicles and obstructions in the entrance to the tunnel, that entrance is seen as a black hole. The original studies of tunnel lighting assumed that misadaptation was the major cause of the black-hole effect (Schreuder, 1964). However, Narisada and Yoshikawa (1974) showed that drivers typically start to fixate on the tunnel entrance at about 150–200 m from the entrance. At a speed of 100 km.h^{-1} (62 mph), this means the driver is fixated on the tunnel entrance for about 5–7 seconds before entering, which is sufficient for some foveal adaptation to occur. This finding, together with an awareness of the fast neural process of adaptation, led Adrian (1982) and Vos and Padmos (1983) to conclude that the major cause of the black-hole effect is the reduction in luminance contrasts of the retinal images of vehicles and obstructions in the tunnel entrance caused by light scattered in the eye. This is certainly the case in a clear atmosphere with a clean windscreen but when the atmosphere is misty and the windscreen is dirty, light scattered in the atmosphere and at the windscreen also make significant contributions to the reductions in luminance contrasts (Padmos and Alferdinck, 1983a, 1983b).

Given that the low retinal image luminance contrasts of vehicles and objects in the tunnel entrance are the problem drivers experience when approaching a tunnel in daytime, there are two possible solutions. The first is to reduce the luminance of the surroundings to the tunnel because this will decrease the amount of scattered light in the eyes. The luminance of the surroundings can be reduced given appropriate construction e.g., by ensuring that the tunnel portal is of low reflectance, by shading the tunnel portal and the road close to the tunnel entrance with louvres designed to exclude sunlight, by using low-reflectance road surface materials outside the tunnel, and by landscaping to shield the view of high-luminance sources, such as the sun and sky. This last possibility is particularly valuable where the tunnel is orientated East–West so that the sun can be seen immediately above the tunnel entrance at some times of the day and year. The second is to increase the luminance contrast of vehicles and obstacles inside the tunnel entrance. This can be done by a careful choice of the materials used in the tunnel entrance. The road surface inside the tunnel entrance should be of higher reflectance than that immediately outside and the walls of the tunnel against which vehicles and objects in the tunnel are usually seen should have a reflectance of at least 0.5. There are a number of ways to determine the luminance required in the tunnel entrance. One is by calculating the equivalent veiling luminance corresponding to the scattered light at the fovea, given the luminances that occur in different parts of the visual field around the tunnel entrance, and then estimating the tunnel entrance luminance required to keep targets of different sizes and luminance contrasts above visual threshold (Adrian, 1982, 1987b; Blazer and Dudli, 1993, van Bommel, 2015). Figure 11.1 shows the luminance at the entrance of a tunnel necessary to ensure a target subtending 10 min arc at the approaching driver's eye and glimpsed for 100 ms is above threshold for a given target contrast and different levels of equivalent veiling luminance. It is apparent that the higher the equivalent veiling luminance and the lower the target luminance contrast, the higher is the required luminance at the tunnel entrance. For the calculation of the threshold road surface luminance, the CIE (2004) recommends the use of a luminance contrast of 0.28 and an object reflectance of 0.2. This usually leads to road surface luminances in the range 200–500 cd.m^{-2} at the entrance to the tunnel, the actual value depending on the specific equivalent veiling luminance.

As the driver approaches the tunnel, the proportion of the visual field taken up by the tunnel entrance increases and the equivalent veiling luminance caused by the scattered light is reduced so that, at the tunnel entrance, the equivalent veiling luminance is much diminished (Hartmann et al., 1986). This is evident in Table 11.1 where, for all the tunnels measured, the veiling luminance is less at 20 m from the tunnel entrance than at 200 m. The magnitude of the equivalent veiling luminance will depend on the surroundings of the tunnel. The Schipohl tunnel in Table 11.1 has an extensive view of the sky immediately above the tunnel entrance which is likely to demand a high road surface luminance at the entrance but the Schonegg tunnel is a mountain tunnel where the surround to the tunnel entrance is rock. In the absence of snow, most rock has a low luminance so the entrance road surface luminance is reduced.

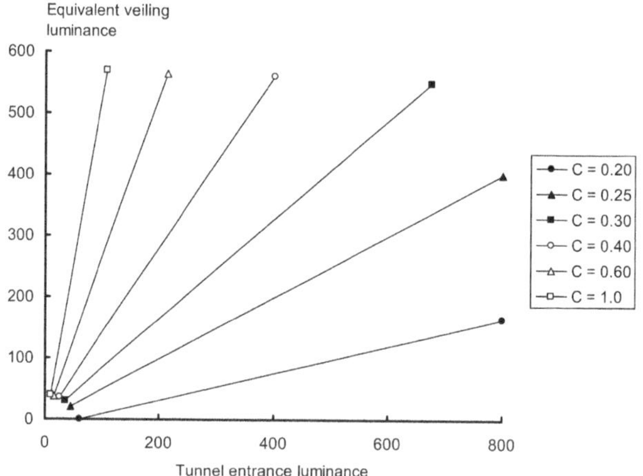

FIGURE 11.1   The luminance of the tunnel entrance (cd/m^2) necessary for a target subtending 10 min arc and glimpsed for 100 ms to be visible, plotted against the equivalent veiling luminance from the tunnel surroundings (cd/m^2) for different target luminance contrasts (after Adrian, 1982).

TABLE 11.1

Equivalent Veiling Luminances (cd.m^{-2}) from Distances of 200 m and 20 m from the Tunnel Entrance for 13 Tunnels in 7 Different Countries (from Adrian, 1982)

Tunnel	Equivalent Veiling Luminance (cd.m^{-2}) from 200 m	Equivalent Veiling Luminance (cd.m^{-2}) from 20 m
Tennozen, Japan	125	50
Shin-Kobe, Japan	200	55
Hami, Japan	295	135
Plaza de Fernando, Spain	315	130
Vlake, The Netherlands	200	35
Heinenoord, The Netherlands	270	90
Schipohl, The Netherlands	560	190
Loen, The Netherlands	190	55
Castellar, France	130	30
Wiltener, Austria	200	35
Arlberg, Austria	340	150
Schonegg, Switzerland	70	40
Schloßplatz, Germany	100	45

Although the approach to the tunnel starts the process of adaptation across the retina, there is no guarantee that adaptation will be complete by the time the driver enters the tunnel (Bourdy et al., 1988). If adaptation to the luminances in the tunnel is incomplete, i.e., the driver is misadapted, the driver may not be able to see what is inside the tunnel after entering. Whether or not this black-out effect occurs depends on the driver's speed and the distance allowed for the change of luminance from the outside to inside the tunnel. Given a slow enough speed and a long enough distance, the visual system is well able to look after itself so there would be no need for the tunnel lighting to be designed to allow for adaptation. However, reducing speed on entering a tunnel is not beneficial in terms of traffic flow and may be dangerous in heavy traffic, so such a behavioural solution is not often adopted. Rather, the lighting is designed to overcome the black-out problem. The approach used is to gradually reduce the road surface luminance in the tunnel, from a threshold zone starting at the entrance, through a transition zone, to the interior zone. The length of these zones is determined by what is termed the safe stopping sight distance at a given speed (SSSD) (American Association of State Transportation Officials, 2011). This is the estimated distance in which a vehicle can stop on a straight and level but wet tunnel approach when travelling at or near the speed limit. The faster the speed of the vehicle approaching the tunnel, the longer the SSSD. Figure 11.2 shows the three zones into which the lighting of a tunnel is divided: the threshold zone, the transition zone, and the interior zone.

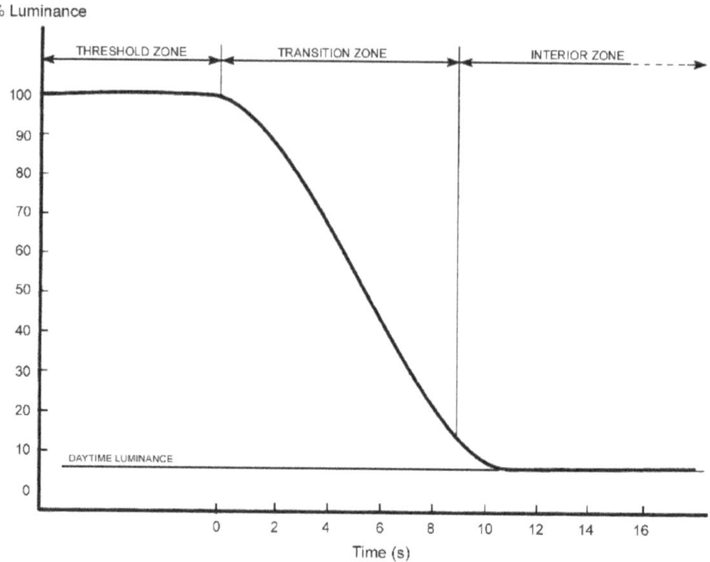

FIGURE 11.2 The three zones into which the lighting of a tunnel is divided: the threshold zone, the transition zone, and the interior zone, plotted against the time from entering the transition zone. The length of the transition zone will depend on the assumed speed of the vehicles entering the tunnel (after IESNA, 2005).

Different points on Figure 11.2 are determined by different conditions. The actual threshold zone entry luminance is determined by the tunnel orientation, the surroundings of the tunnel portal, and the speed limit, and adjusted for different traffic volumes, traffic mixes, tunnel lengths, tunnel road and wall surface reflectances, and daylight penetration. The length of the threshold zone is one SSSD less the distance from the tunnel portal at which the entrance fills a significant proportion of the visual field and adaptation is assumed to be occurring. This distance will usually correspond to the top of the tunnel entrance being 22–25 degrees above the horizontal at the driver's eye height, this being a typical cutoff imposed by the top of the windscreen. Where screening is used to filter out direct sunlight, the start of the threshold zone is taken to be the start of the screening. For the first half of the threshold zone the road surface luminance is held constant at what is provided at the entrance but over the second half it can be linearly reduced to reach 40% of the entry luminance at the start of the transition zone (CIE, 2004). The length of the transition zone is determined by the assumed vehicle speed, the distance being set so as to allow at least 10 s for adaptation. The transition can be smooth where dimming is possible or achieved by a series of steps where it is not (CIE, 2004). The maximum road surface luminance ratio between adjacent steps should be less than three and the luminance of the last step before the interior zone should be not more than twice the luminance of the interior zone. The road surface luminance of the interior zone in daytime depends on the speed and density of traffic in the tunnel and covers a range of 3–10 cd.m^{-2}, the faster the speed and the higher the traffic density, the higher the average road surface luminance recommended in the interior zone. The minimum overall luminance uniformity in the interior zone should be 0.4 and the minimum longitudinal uniformity along each lane in the interior zone should be at least 0.6 (CIE, 2004). The luminance of the walls of the tunnel up to 2 m should be at least 0.6 of the average road surface luminance. The interior zone luminance is usually continued up to the exit from the tunnel, special treatment rarely being needed at the exit because of the high luminance of the tunnel exit and the faster speed of adaptation from lower luminances to higher luminances than from higher luminances to lower luminances. However, if hazards are expected at the exit or the tunnel is very long, the luminance of the road over the stopping distance up to 20 m from the exit should be linearly increased to five times the luminance of the interior zone.

This approach of grading the luminance from the threshold zone through the transition zone to the interior zone is only used where the tunnel is longer than the SSSD or where the tunnel is curved. If the tunnel is shorter than the SSSD but only a small part of the tunnel exit can be seen as the tunnel is approached, i.e., if the tunnel has a low look-through percentage, the same road surface luminance is used throughout the tunnel, the luminance being that recommended for the threshold zone in longer tunnels.

At night, the form of tunnel lighting depends on the lighting of the roads approaching and leaving the tunnel. If these roads are lit, then the lighting of the tunnel should match that of these roads in terms of average road surface luminance, overall uniformity, longitudinal uniformity, and glare control. If the road approaching the tunnel is unlit, the road surface luminance of the tunnel should be at least 1 cd.m^{-2}, overall luminance uniformity should be at least 0.4, and longitudinal luminance uniformity at least 0.6 (CIE, 2004).

In the event of a power failure or a crash in a tunnel, some form of emergency lighting is needed. The purpose of this emergency lighting is to enable people in the tunnel to reach the exit, either in their vehicles or, if this is not possible because a crash has blocked the tunnel, by walking to the exit or a refuge. Some form of signage at the entrance to the tunnel to stop approaching drivers from entering the tunnel is also essential. In the event of a mains power failure, an uninterruptible power supply for a limited number of luminaires delivering an average road surface illuminance of 10 lx and a minimum of 2 lx is recommended (van Bommel, 2015). Guidance on tunnel emergency lighting is given in CIE Publication 193 (2010c).

As for the type of lighting used in tunnels, the light source of choice is now the LED because of its high luminous efficacy, long life, good colour properties, and ease of dimming although there are still many tunnels lit by one of the discharge light sources. The luminaires used in tunnels have to be of rugged construction to deal with vibration, dirt, chemical corrosion, and washing with pressure jets. Three types of light distribution are used, symmetrical, counter-beam, and pro-beam lighting. Symmetrical light distributions produce uniform luminance lighting throughout the tunnel so vehicles of different reflectances will have either positive or negative luminance contrasts with the road. Counter-beam light distributions are those where the light is directed predominantly against the traffic flow. This gives a high pavement luminance so that vehicles tend to be seen in negative contrast, but there is some risk of the driver experiencing discomfort and disability glare. Pro-beam light distributions are those where the light is directed predominantly in the direction of the traffic flow. This gives a low road surface luminance but high luminances for vehicles so the vehicles tend to be seen in positive contrast. Various claims have been made about the benefits of these different systems (Novellas and Perrier, 1985; Schreuder, 1993; van Bommel, 2015) but no consensus about which is best has developed.

Finally, it is necessary to consider the potential the tunnel lighting has for producing perceptions of glare or flicker which may cause the driver discomfort and distraction. For glare, the threshold increment metric used for road lighting (see Section 4.3) should be applied throughout the tunnel. The threshold increment should be less than 15% in all parts of the tunnel, day and night (CIE, 2004). As for flicker, when tunnel lighting is provided by a series of regularly spaced, discrete luminaires, there is always a possibility of flicker being perceived. It is recommended that care be taken to avoid spacing individual luminaires so that drivers moving at representative speeds in the tunnel are not exposed to flicker in the range 4–11 Hz for more than 20 s (CIE, 2004). Of course, flicker is only a consideration if the lighting is provided by discrete luminaires. An alternative system based on a continuous linear luminaires through the tunnel avoids any flicker problem and provides good visual guidance for the tunnel, a feature that is particularly valuable where the tunnel curves.

11.3 PEDESTRIAN CROSSINGS

Pedestrians crossing the road are exposed to a high risk of death or injury by collision with high-momentum vehicles. In the USA, in 2019, 6,205 pedestrians were killed, this number representing 17% of the total road fatalities. This is why special crossing points are identified where pedestrians have priority over vehicles. Pedestrians who do not use these crossing points are at much greater risk of death or injury than those who do (Martin, 2006). Nonetheless, about one-quarter to one-third of pedestrian fatalities occur on or close to a pedestrian crossing (Hunter et al., 1996; OECD, 1998). This is almost certainly a reflection of the level of pedestrian exposure rather than the inherent hazard of the crossing, many more pedestrians using crossings than other locations. Visibility of pedestrians on a crossing is also likely to be a problem. This is shown by the fact that 80% of pedestrian fatalities in the USA in 2019 occurred after dark or at dawn and dusk (NHTSA, 2021c).

There are two broad types of pedestrian crossings, those associated with traffic signals where the signals are arranged so as to give priority to vehicles and pedestrians at different times, and those without traffic signals where a pedestrian waiting to cross or actually on the crossing has priority over vehicles at all times. The signal-controlled crossings are primarily a feature of urban areas where traffic volumes, both vehicular and on foot, are high. The crossings without traffic signals are most frequently found in suburban areas where traffic of both types is lighter. To minimize the danger of using either type of pedestrian crossing, it is first necessary to make the driver aware of the existence of the crossing. The first step to meet this objective is careful site selection, the aims being to make sure that the driver's view of the crossing and people waiting to use it as well as the pedestrian's view of traffic are not obstructed by buildings, parked vehicles, landscaping, etc. (DfT, 2005). The second is to ensure consistency of application so that both drivers and pedestrians can identify a pedestrian crossing and know what is expected of them. The third step is to place the appropriate warning signs, signals, and road markings. Where traffic signals are used, these are of the standard type for traffic control although the amber element sometimes flashes before the signal turns green. A flashing amber element allows a driver to move over the crossing if there is no pedestrian using or waiting to use the crossing. The signals facing drivers are supplemented by signals directed towards the pedestrians waiting to cross. These can be in the form of text or pictograms. Sometimes there is also a countdown signal telling the pedestrian how long a time is available to complete the crossing before the traffic signal changes. Where traffic signals are not used, the crossing itself is marked by some combination of side or overhead mounted signs that are externally or internally illuminated; large, high-contrast road markings, and, sometimes, side- or overhead-mounted flashing beacons (Huang et al., 2000; DfT, 2005) (Figure 11.3).

In addition to these measures, special lighting is sometimes used to emphasize the presence of the crossing and to increase the visibility of a pedestrian on the crossing at night. In both the UK and the USA, pedestrian crossings are considered as a conflict area (see Section 4.4). In the UK, the recommendations for such areas are given in terms of a maintained average horizontal illuminance and cover a range of 7.5 -50 lx, with a minimum overall illuminance uniformity of 0.4 (BSI, 2015). In the USA, for crossings associated with intersections or roundabouts it is recommended that the maintained average horizontal illuminance on the crossing should

FIGURE 11.3 A pedestrian crossing made conspicuous by high-contrast road marking and lit beacons on striped poles with lit elements. The whole crossing is positioned on a speed table (iStock by Getty Images / Francesca Leslie).

match or preferably exceed that of the road nearby with a minimum overall illuminance uniformity of 0.33. A maintained minimum vertical illuminance at a height of 1.5 m matching the horizontal illuminance on the crossing should also be provided (IESNA, 2021a).

In both countries, where a pedestrian crossing is close to an intersection or roundabout the lighting is designed as part of the wider conflict area but where it occurs in isolation as, for example, halfway along one side of a city block but where people wish to cross the road, there are two possibilities for lighting. One is to use the normal lighting of the traffic route but with the road lighting luminaires arranged so that the crossing is positioned at the midpoint between luminaires. The other is to supplement the road lighting with additional lighting. The supplementary lighting approach is recommended when the average road surface luminance is less than 1 cd.m^{-2} or the crossing is located on a bend or on the brow of a hill. The supplementary lighting should illuminate the crossing to a higher illuminance than that used to produce the average road surface luminance of the road approaching the crossing. The supplementary lighting should also have strong vertical component to ensure that pedestrians are seen in positive contrast, which is why it is recommended that where conventional road lighting is used, the crossing should be at the midpoint between the luminaires. The value of seeing the pedestrian in positive contrast is that the light from the headlamps of an approaching vehicle will tend to increase the contrast of the pedestrian.

The outcome of the supplementary lighting approach is a bright stripe of light over the crossing and a higher vertical illuminance on pedestrians using the crossing. The benefits of this approach are evident in a study by Hasson et al. (2002). In this study at two mid-block crossings in an American city, the ability of observers sitting in a car 82 m away to detect the correct number of pedestrian-sized cutouts near or on the crossing was measured, the cutouts having a diffuse reflectance of 0.18. The crossing was lit using either conventional road lighting giving a road surface luminance of less than 2 cd.m^{-2} and producing vertical illuminances in the range 8 - 11 lx, or with supplementary lighting resulting in a vertical illuminance at the crossing of 40 lx. The car in which the observers sat used low-beam headlamps. Table 11.2

TABLE 11.2

Percentage of Detection of Fewer Than, More Than and the Correct Number of Pedestrian Cutouts for a 2-s Observation Period, for Two Pedestrian Crossing Sites Lit by Conventional Road Lighting with and without Supplementary Lighting (from Hasson et al., 2002)

Site	Type of Lighting	Percentage (Fewer Than)	Percentage (More Than)	Percentage (Correct)
1	Conventional	50	17	33
1	Conventional + Supplementary	10	10	80
2	Conventional	20	7	73
2	Conventional + Supplementary	13	0	87

shows the percentage of presentations in which the drivers were able to detect fewer than, more than, or the correct number of pedestrian cutouts in 2 s.

It is evident that the supplementary lighting improves the ability to quickly detect the correct number of pedestrians, although much more at one site than the other, probably because Site 2 was more well-lit by the conventional road lighting alone than was Site 1. Gibbons and Hankey (2006) carried out a study on the effect of vertical illuminance on pedestrian visibility, covering a range of vertical illuminances from 5 - 60 lx. They found that a vertical illuminance of 20 lx was enough to ensure adequate pedestrian visibility when on the crossing. Bullough and Skinner (2015) report a series of studies on a rather different way to light a pedestrian crossing, using pairs of bollards placed at either end of the crossing, the bollards being designed to deliver a narrow beam of light across the crossing so as to maintain positive contrast of people on the crossing. They recommend that a vertical illuminance of at least 10 lx at a height of 0.9 m is enough to enhance pedestrian contrast. Given this disagreement, it is fortunate that Bhagavathula and Gibbons (2023) have evaluated the effectiveness of several different methods of lighting a mid-block crosswalk. They constructed a pedestrian crossing on a rural, closed road system. The crossing was illuminated and marked in a number of different ways: A single, conventional, overhead, road lighting luminaire positioned at the edge of the road at a height of 9.1 m offset by about 5 m in front and behind the crossing to produce positive and negative contrast of people on the crossing depending on the direction of approach; two conventional, overhead, road lighting luminaires matched to the direction of approach so that positive contrast was produced for both directions (staggered); two commercially available crosswalk illuminators mounted at the ends of the crossing to provide a bright stripe of light across the crossing; and two warning signs, one being a rectangular, rapid flashing beacon and the other a flashing pedestrian sign. The two warning signs were used alone and in combination with one conventional, overhead, road lighting luminaire producing positive contrast. The overhead lighting was adjusted to produce average vertical illuminances on the crossing of 2 lx, 10 lx, and 20 lx at a height of 1.5 m. The crosswalk illuminators produced average vertical illuminances of 8 lx and 10 lx at the same height. A 1.2-m high mannequin clad in grey clothing was sometimes placed at the entrance, exit, or middle of the crossing as a driver approached the crossing at 35 mph (56 km.h^{-1}). Through practice, the drivers were aware of where the crossing was and the nature of the mannequin so they knew what to look for. The effectiveness of the different crosswalk treatments was quantified by the distance at which the drivers reported seeing the pedestrian. At the low vertical illuminance (2 lx), the crosswalk illuminators produced much longer detection distances (175 m) than any of the other treatments (80–110 m). At the medium vertical illuminances (10 lx), the staggered overhead lighting, the single overhead lighting providing positive contrast, and the crosswalk illuminators gave the longest detection distances (165–190 m). Interestingly, adding the warning signs to the single overhead lighting producing positive contrast made the detection distances shorter, possibly because the flashing interfered with the visibility of the mannequin or attracted attention away from whatever was on the crossing. At the high vertical illuminance (20 lx), the staggered overhead lighting gave a longer detection distance

(205 m) than any of the other crossing treatments. The two flashing warning signs gave the shortest detection distance (80 m). From these results, it is clear that providing enough light on the crossing in a way that delivers a positive contrast for whatever is on the crossing is the key to enhancing pedestrian safety at night. Bhagavathula and Gibbons (2023) concluded that where overhead lighting is available, a vertical illuminance of 10 lx should be provided in a way that produces positive contrast for a pedestrian when viewed by an approaching driver. Where overhead lighting is not available, crosswalk illuminators should be used.

While the visibility of the pedestrian on the crossing is important, lighting can also be used to prepare the approaching driver for action. This is important both day and night. There are a number of treatments that can be used for this purpose, their common characteristic being that they produce a change in the road ahead as seen by the approaching driver. At night, a smart lighting system that raises the illuminance on a crossing when a pedestrian is crossing or wishing to cross can be used (Nambisan et al., 2009). For both night and day, widely used treatments are specific, high-contrast road markings and flashing beacons (Figure 11.3). At night, supplementary lighting of the crossing of any type increases the brightness of the crossing relative to the rest of the road and thereby makes it more conspicuous, but using a light source of a different colour to illuminate the crossing is even better (Figure 11.4). This increases the conspicuity of the crossing further because it adds another dimension on which the crossing differs from its surroundings. Janoff et al. (1977) report a study in which low-pressure sodium lighting was installed over pedestrian

FIGURE 11.4 A pedestrian crossing with high-level supplementary lighting designed to make the crossing conspicuous. The supplementary lighting uses low-pressure sodium light sources so it differs in colour from the rest of the road lighting which uses metal halide light sources.

crossings on roads that were lit by other light sources. As would be expected, the increased illuminance on the crossing increased the distance at which a target on the crossing could be detected by an approaching driver, but observations also suggested safer behaviour by both drivers and pedestrians. This use of a different colour of light is part of the recipe for better pedestrian crossing lighting developed by Freedman et al. (1975).

Providing more light of the same or a different colour will make the crossing more conspicuous at night, but does not discriminate between an empty crossing and one in use and does nothing to increase the conspicuity of the crossing during the day. The rectangular rapid flashing beacons and the flashing pedestrian sign evaluated by Bhagavathula and Gibbons (2023) can address both these limitations, but from their results it is clear that they should not be used at night without supplementary lighting. Another approach is to place lights into the road surface at intervals across the road at the edge of the crossing (Figure 11.5). These in-road lights, which are similar in many ways to airport runway marking lights, provide a narrow beam of light towards the driver. They can be made to flash, either by a simple manual control available to the pedestrian or automatically by the use of a system for detecting the presence of a pedestrian entering or on the crossing (Whitlock and Weinberger Transportation, 1998; Van Derlofske et al., 2003; Arnold, 2004). Such a system can signal a crossing in use by day or night, although the light output needs to be reduced at night if glare is to be avoided.

The question that needs to be addressed now is how effective are these different forms of lighting for pedestrian crossings in reducing accidents? Polus and Katz (1978) report a study undertaken in Israel on the impact of installing an internally illuminated sign mounted above a crossing that also provided an average horizontal illuminance on the crossing of 30 lx. As a result of installing the special lighting, crashes involving pedestrians using the crossings at night decreased by 39%. By

FIGURE 11.5   A pedestrian using a crossing with conventional road markings and in-road flashing lights.

comparison, daytime crashes on the crossings and crashes at any time on nearby crossings without special lighting but with similar pedestrian and vehicle traffic profiles showed no statistically significant changes. There can be little doubt that special lighting of crossings is beneficial at night because it increases both visibility and conspicuity, but that is not the complete answer. Light can also be used to provide useful information to the driver by day. To be really useful, what drivers need to know is not only that they are approaching a pedestrian crossing but also that it is in use. Van Houten et al. (1998) examined the effectiveness of a combination of signs on driver behaviour. He found that simple signs indicating the presence of a pedestrian crossing had little effect on drivers but a combination of a sign indicating "stop when flashing" and an overhead flashing beacon activated by the pedestrian was effective in reducing pedestrian/vehicle conflicts. In-road warning lights triggered either manually or automatically by the pedestrian have similar effects. A series of before and after studies have shown a limited number of effects on pedestrian behaviour but much more significant effects on drivers (Arnold, 2004). Specifically, the percentage of drivers who gave way to pedestrians increased, as did the distance from the crossing at which drivers braked, while the speeds at which vehicles approached the crossing were decreased. Consequently, pedestrian waiting time before crossing and the percentage of times a pedestrian on the crossing had to run to avoid approaching traffic were both reduced.

The message from such findings is clear – lighting has an essential role to play in making pedestrian crossing safer places (Mitran et al., 2020, Tomczuk et al., 2023). Enhancing the conspicuity of the crossing and the visibility of pedestrians on and near a crossing by means of special lighting is useful at night but still relies on the driver searching the crossing and its environs before deciding if any action is necessary. To improve the safety of pedestrians using crossings significantly, by day and night, it is also necessary to signal when there are pedestrians on the crossing or waiting to use the crossing. This can be done by signals triggered manually by the pedestrians or automatically by sensors. Such a system should be more effective because it would warn approaching drivers about an actual conflict rather than a potential conflict.

11.4 RAILWAY CROSSINGS

Like pedestrian crossings, railway crossings involve the movement of one type of transport across the path of another and are therefore considered a conflict area. Unlike pedestrian crossings, at railway crossings it is the driver of the motor vehicle who is more likely to suffer death or injury unless the train is derailed in which case both the train driver and the passengers are likely to be hurt. Railway crossings are marked in various ways depending on the amount of traffic on the road and track. For remote rural crossings with low levels of traffic on both, all that is usually provided is a standard warning sign and possibly a pair of manually operated gates. These arrangements are inherently hazardous because they provide no warning of an approaching train. Even when a driver does look to see if a train is approaching, there remains the difficulty of accurately estimating the distance and approach speed of the train, particularly at night when the engine lamps will be seen as bright spots in an otherwise dark landscape showing very little angular movement. Even where gates are fitted, they are subject to abuse, the temptation being to open one gate and drive onto the track and wait there while opening the other gate rather than opening both gates before driving across the track. From this description, it might be thought that such crossings are very dangerous, and they are. Fortunately, the low level of exposure due to the low-traffic levels means that deaths at such crossings are rare although stupidity ensures they are also regular.

For railway crossings where there is a high level of traffic on either the road or track, much more elaborate systems are used. Where the roads approaching the railway crossing are lit, there is little need for additional lighting. Where they are unlit, the railway crossing should be treated as an isolated intersection (IESNA, 2021b). In addition to any such conflict area lighting, the crossing is usually marked with road signs, two pairs of alternately lit lamps, a bell or siren, and either full- or half-width barriers (Figure 11.6).

As a train approaches the crossing, the bell or siren sounds, the alternately lit lamps flash, and the barriers come down. The bell or siren is silenced once the barriers are down. The lights continue to flash until the train has passed and no other train is approaching, at which time the alternately lit lamps are extinguished and the barriers are raised. Full barriers cover the whole road and prevent drivers moving

FIGURE 11.6   A car stopped at level crossing in the UK as a train approaches. The crossing is lit by two high-pressure sodium floodlights. On each side, the crossing has a full-width barrier and two pairs of alternating flashing red lamps. Both lamps of each pair appear to be lit because of the long exposure time needed for the photograph. Surrounding each pair of flashing lamps is a black shield with a red and white rectangular border made from retro-reflective material. This track carries a lot of traffic (by Mick Stevens).

over the crossing but anyone foolish enough to stop on the crossing may be trapped if the barriers come down before they can move off. Half-width barriers cover only half the road and therefore avoid trapping the driver, but they do allow drivers with a death wish to traverse the crossing when a train is approaching.

The systems used at railway crossings are slightly different in different countries but most follow the approach described above. In all countries, lighting, as such, has a very limited role to play, most of the information provided to drivers is through signs and signals. However, it is worth noting that the approach advocated for pedestrian crossings, namely to tell the driver when the crossing is actually in use, is already widely used for railway crossings.

11.5   CAR PARKS

Another location where pedestrians appear to be at risk because they are mixed with vehicles is in car parks. Car parks can be divided into parking lots and parking garages, the former being completely open to the sky while the latter are enclosed to some extent, sometimes totally and sometimes only with a roof. Fortunately for pedestrians, the risk of being injured by a vehicle in these locations is more theoretical than actual, probably because of the much slower speeds used by vehicles in parking lots and parking garages. Box (1981) examined the nature of vehicle crashes in parking lots in the USA and found that about two-thirds involved a moving vehicle hitting a parked vehicle, less than one-third involved a moving vehicle striking another moving vehicle, about 6% involved a moving vehicle striking a fixed object but only 1% involved a vehicle hitting a pedestrian. Accidents to pedestrians are much more likely to involve tripping, slipping, and falling while walking in the parking lot (Monahan, 1995). Other research suggests a higher risk to pedestrians in parking lots although to what extent vehicles are involved is unclear. According to Fayard (2008), who examined the Census of Fatal Occupational Injury to determine the percentage of worker fatalities in parking lots in the USA over the years 1993–2002, the most common reason for fatalities in parking lots was homicides (36%), followed by contact with objects (15%), pedestrian fatalities (13%), and suicides (11%). Such a high percentage for homicides seems unlikely to be found in other countries where gun ownership is more restricted. Nevertheless, pedestrian worker fatalities are high enough to suggest that the lighting of car parks deserves consideration.

The lighting approaches used in parking lots and parking garages are markedly different from the lighting of roads and from each other. For parking lots, luminaires mounted on columns are conventionally used but the luminaires have a much wider range of luminous intensity distributions than those used for lighting roads. The specific luminous intensity distribution used depends on the shape of the parking lot and the positioning of the columns in and around it. As for the light sources used, these are typically some form of LEDs or high-intensity discharge, usually high-pressure sodium or metal halide (see Section 2.5). In the UK, the maintained mean horizontal illuminance covers a range of 5 - 20 lx depending on the level of activity with an overall uniformity (minimum to average) of 0.25 (BSI 2020). In the USA, a maintained minimum horizontal illuminance of 2 lx, a maximum horizontal illuminance uniformity ratio (maximum/minimum) of 20:1, and a maintained minimum vertical illuminance of 1 lx are recommended (IESNA, 2021b). The minimum horizontal illuminance recommendation is supported by the results of Bhagavathula and Gibbons (2020). They carried out an evaluation of parking lots with asphalt and concrete pavements, lit to horizontal illuminances in the range 1 - 12.5 lx by high-pressure sodium light sources or two LED light sources with different correlated colour temperatures (3,000 K and 5,000 K). The evaluation was made by observers representing both pedestrians and drivers in two age ranges (18–35 years and >65 years). The evaluation involved both visual performance measures and subjective assessments of visibility, comfort, and safety. The visual performance tasks for pedestrians were detecting if glasses were being worn by a pedestrian in dark clothing facing the observer at distances of 3.0 m (10 ft) and 15.2 m (50 ft), if the hands of the front-facing pedestrian at the same distances were by his sides or behind his back, if a side-on pedestrian emerging from behind a parked car at a distance of 15.2 m (50 ft) could be seen, if a car parked side on but with rear position lamps and reversing lamps lit at a distance of 15.2 m (50 ft) could be seen, and if a wheel stop of one of several different reflectances at a distance of 15.2 m (50 ft) could be seen. The observers viewed the scene for 1 s, one object at a time. After each 1-s exposure, the observer reported if they had seen the object. After all the objects had been presented at the same light level, the observers answered a Likert scale questionnaire about the visibility, comfort, and safety of the scene. Observers acting as a driver sat in the driver's seat of a car with low-beam headlamps operating. With one exception the tasks were the same as those undertaken by the simulated pedestrian but some were at different distances. For the driver, the wheel stop was placed at 7.6 m (25 ft), and the front-facing pedestrians were adjacent to the parked cars at 15.2 m (50 ft). The exception was a pedestrian placed at a stopping sight distance of 47.2 m (155 ft).

The results of this evaluation indicated that an increase in the pavement illuminance produced an increase in the ability to detect all the various objects. However, a plateau in the percentage of detections in both parking lots occurred around a horizontal illuminance of 2 lx. This is consistent with the results of Fotios and Cheal (2009). As for the perceptions of visibility, safety, and comfort, these too tended to saturate around a horizontal pavement illuminance of 2 lx. The effects of the light source were inconsistent because of a possible learning effect, but the effects of the observer's age were what would be expected. The older observers had lower detection probabilities than the young observers in the asphalt parking lot but not in the concrete parking lot, suggesting that older pedestrians and drivers would benefit from higher illuminances where surface reflectances are low.

While these lighting criteria are clear and supported by research, they are often ignored in the direction of over-lighting. It is not uncommon to find parking lots lit to much higher illuminances, not because higher illuminances are required for driver or pedestrian visibility, but to give an impression of brightness and hence safety. This is particularly the case for parking lots attached to retail premises where the penalty in terms of lost business is particularly severe if the parking lot is perceived as dim and unevenly lit and hence unsafe. Boyce et al. (2000a) collected perceptions of safety when walking alone in a number of urban and suburban parking lots, by day and night, on a seven-point rating scale. The effectiveness of the nighttime lighting is given by the difference between the safety ratings collected from the same people in the same parking lots by day and night. Figure 11.7 shows this difference plotted against the median parking lot illuminance at night. There is an asymptotic approach to zero difference.

At a median illuminance of 10 lx the difference is less than one scale unit and above 30 lx it is less than half a scale unit. This study has the advantage of using real parking lots but is limited by the fact that the parking lots differed not only in horizontal illuminance but also in the illuminance uniformity and light source colour properties. Narendran et al. (2016) carried out a similar evaluation in a single parking lot where the mean surface illuminance and illuminance uniformity could be systematically varied. When the illuminance uniformity (maximum to minimum) was 10:1 the results were very similar to those found by Boyce et al. (2000a) but when illuminance uniformity was improved to 3:1 a higher level of perceived safety was achieved at a lower illuminance. These results imply that if you wish to ensure that your parking lot is perceived to be safe to use at night, you will need to provide a higher illuminance than the minimum recommended. This will inevitably increase energy costs, but this increase can be reduced if you aim for a very uniform illuminance distribution (Lighting Energy Alliance, 2020).

For parking garages, the low ceiling heights and high level of obstruction caused by the structure and parked vehicles require a different approach, except on the top floor when it is open to the sky. The top floor is usually treated as a parking lot where crime and vandalism are likely. For the other floors, luminaires are mounted directly

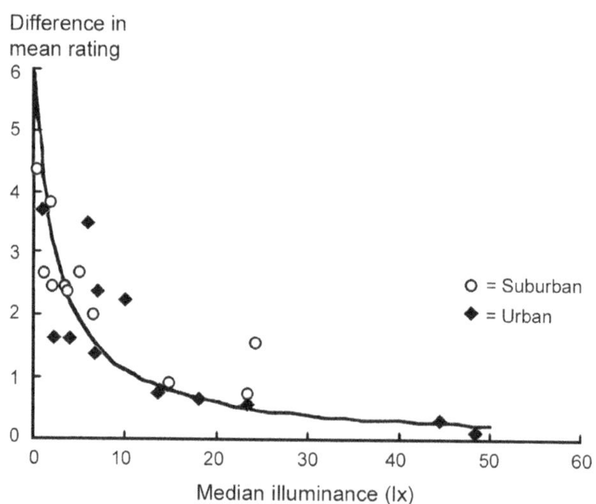

FIGURE 11.7   The difference in the mean ratings of perceived safety when walking alone in a parking lot by day and night, for a number of urban and suburban parking lots, plotted against the median illuminance (lx) on the parking lot surface at night (from Boyce et al., 2000a).

on a ceiling that is often formed from concrete. Where the ceiling is flat, a full cutoff luminaire with a flat lens is used to avoid glare (see Section 4.2). Where the ceiling is constructed of coffers, a non-cutoff luminaire can be used, the glare shielding being provided by the coffer. In both cases, care has to be taken with luminaire spacing to ensure a high level of illuminance uniformity although this is easier to achieve if the surface reflectances of the garage are high. The light sources used in parking garages are typically LED or high-pressure sodium or metal halide discharge (see Section 2.5). Regardless of the light source used, the luminaire has to be constructed to deal with a corrosive, humid, and dirty atmosphere. In the USA, a maintained minimum horizontal illuminance of 10 lx, a maximum horizontal illuminance ratio (maximum/minimum) of 10:1, and a maintained minimum vertical illuminance of 5 lx are recommended (IESNA, 2021b). These recommendations for parking garages are consistent with the results that Bhagavathula and Gibbons obtained in a concrete parking garage using the evaluation method described above for parking lots. However, the horizontal illuminance at which saturation occurred (10 lx) was determined with the parking garage virtually empty. To achieve this minimum when the parking garage is filled with vehicles, a higher average horizontal illuminance will be required.

These illuminances are recommended for use by day and night with one exception; during the day, the maintained minimum horizontal illuminance in the entrance should be increased to 500 lx and the maintained minimum vertical illuminance increased to 250 lx so as to form a transition zone from daylight to the interior of the parking garage. One final point worth mentioning is that the above lighting recommendations are based on the visual performance and perceptions of people. If a CCTV surveillance system is in use, it may be necessary to enhance the illuminance, the illuminance uniformity, and the glare control of the lighting to enable the surveillance system to operate effectively.

11.6 RURAL INTERSECTIONS

Intersections and roundabouts are inherently dangerous locations because they are conflict areas where traffic streams interact. There is copious advice available for designing the lighting of intersections and roundabouts (BSI, 2015; IESNA, 2021b). Where the intersection forms part of a network of lit roads, it too will be lit. When the intersection is isolated in a network of unlit roads, a decision on whether or not to light the intersection has to be taken. This is not an easy decision to make as it involves a trade-off between benefits and costs where the costs are certain but the benefits are not (Bullough et al., 2013). Fortunately, Isebrands et al. (2004, 2006) present an analysis of the effect of providing lighting at rural intersections in Minnesota. Data was collected on crashes occurring over a period of two years at 3,622 rural intersections, 223 with lighting and 3,399 without. The crash rate based on the number of crashes per million vehicles entering the intersection was calculated for each intersection. Table 11.3 shows the mean crash rates for the lit and unlit intersections, by day and night. This metric takes traffic flow through the intersection into account, this being much higher for the lit than the unlit intersections. This

TABLE 11.3

Mean Crash Rates Based on Number of Crashes Per Million Vehicles Entering an Intersection, for Lit and Unlit Rural Intersections, by Day and Night (after Isebrands et al., 2004)

Condition	Lit Intersections	Unlit Intersections
Mean day crash rate	0.40	0.29
Mean night crash rate	0.57	0.59
Night/Day ratio of crash rates	1.43	2.03

TABLE 11.4

Mean Crash Rates Based on Number of Crashes Per Million Vehicles Entering a Rural Intersection, by Day and Night, before and after the Installation of Lighting (after Isebrands et al., 2004)

Condition	Before Installation of Lighting	After Installation of Lighting
Mean day crash rate	0.30	0.39
Mean night crash rate	1.12	0.73
Night/Day ratio of crash rates	3.73	1.87

higher level of exposure at the lit junctions may explain why the average crash rate is higher for the lit than unlit junctions during the day. What is interesting is that the night/day ratio of crash rates is much higher for unlit than for lit intersections. This suggests that providing lighting at rural intersections leads to a reduction in the number of crashes at night.

Isebrands et al. (2004, 2006) also carried out a before-and-after study on the effect of installing lighting at 33 rural intersections. Accident data was examined for three years before and after installation of the lighting. The crash rate, i.e., the number of crashes per million vehicles entering the intersection, was calculated for day and night conditions, before and after the installation of the lighting. Table 11.4 shows the mean crash rates before and after the installation of the lighting, by day and night.

There are two interesting points about this table. The first is that the mean crash rates are much higher by night than in the wider survey shown in Table 11.3 suggesting that the intersections where lighting had been installed recently had been identified as hazardous. The second is that after the installation of the lighting at these intersections the mean crash rate by day increased but the mean crash rate by night decreased. The result of this is that the night/day crash rate ratio decreased from 3.73 before the installation of the lighting to 1.87 after. Again, the installation of lighting at these rural intersections has led to a reduction in crashes at night.

Given that lighting is an effective crash countermeasure for isolated rural intersections, does the form of the lighting matter? Bruneau and Morin (2005) describe a comparison of the effects of standard and non-standard road lighting at rural intersections in the Province of Quebec, Canada. Standard lighting consists of specific lighting columns and luminaires located around the intersection so as to meet the recommendations of the Ministere des Transports du Quebec. Non-standard lighting is lighting installed by municipalities that makes use of existing power utility poles to mount a luminaire. While standard lighting is designed to provide specified photometric conditions, the same is not true of non-standard lighting. The lighting conditions provided by non-standard lighting can vary widely depending on where the power utility poles are in relation to the intersection and what type of luminaire and light source are used. Crash data for 376 rural intersections were examined and expressed as crash rates. Table 11.5 shows the mean crash rates for unlit intersections and intersections with standard and non-standard lighting, by night and day. Also shown are the resulting night/day ratios of crash rates. As would be expected, Table 11.5 shows little difference in the mean daytime crash rates for the three lighting types. However, there is a clear increase in the mean crash rate at night at all types of intersection, but the increase is less at the intersections with either standard or non-standard lighting. As a result, the night/day ratios of crash rates are less for intersections with lighting than intersections without lighting.

All types of crashes are considered in Table 11.5. Table 11.6 shows the night/day ratios of crash rates for all crashes, crashes involving personal injury, and those producing property damage only. The interesting point about these results is that the only statistically significant changes in night/day crash rate ratios occur for property damage only crashes. For property damage only crashes, both the night/day ratios for crash rates for intersections with standard lighting and non-standard lighting are statistically significantly less than for intersections with no lighting. For crashes causing personal injury, there is little difference in the night/day ratios for any of the lighting conditions which suggests that factors other than visibility are involved in these types of crashes. Also interesting is the fact that, regardless of the crash category, there are no statistically significant differences between the night/day ratios for intersections with the two forms of lighting. This suggests that almost any form of lighting has a beneficial effect on accidents at rural intersections where the approach roads are unlit.

TABLE 11.5

Mean Crash Rates Based on Number of Crashes Per Million Vehicles Entering a Rural Intersection, by Day and Night, for Intersections without Lighting and with Standard or Non-standard Lighting (after Bruneau and Morin, 2005)

Condition	No Lighting	Standard Lighting	Non-standard Lighting
Mean day crash rate	0.9	0.7	0.8
Mean night crash rate	2.8	1.3	1.7
Night/Day ratio of crash rates	3.2	2.0	2.3

TABLE 11.6

Night/Day Ratios of Crash Rates Based on Number of Crashes Per Million Vehicles Entering Rural Intersections, for Intersections without Lighting and with Standard or Non-standard Lighting, for All Crashes, those Causing Personal Injury, and those Involving Property Damage Only (after Bruneau and Morin, 2005)

Condition	No Lighting	Standard Lighting	Non-standard Lighting
Night/Day ratio of crash rates for all crashes	3.24	1.96	2.30
Night/Day ratio of crash rates for crashes causing personal injury	1.94	1.74	1.87
Night/Day ratio of crash rates for crashes causing property damage only	3.87	2.02	2.45

Given that non-standard lighting is effective in reducing accidents at many rural intersections, the question that arises is why this should be? This question can be answered by considering what lighting can do. One thing that almost any form of road lighting at an intersection of otherwise unlit roads will do is capture the driver's attention. This means that drivers approaching a lit intersection are more likely to be prepared to take action. To determine if action is necessary, drivers may carry out a visual search of the surrounding traffic and of the intersection. How important lighting is to the effectiveness of these visual searches will depend on the nature of the traffic. In heavy traffic, the main concern will be what other vehicles nearby are doing, in which case the role of the intersection lighting will be limited. In light traffic, the main concern will be anything that is moving into the intersection, particularly anything on a collision course. In this situation, the form of lighting of the intersection is likely to matter. Lighting that provides good visibility over the whole intersection and all the roads that enter it will be more effective than something that only covers part of the intersection; i.e., standard lighting should be more effective than non-standard lighting. This is because to avoid a collision it is necessary for a driver to be able to judge the velocity and direction of movement of other vehicles on intersecting roads. Wertheim (1981) has shown that velocity and direction of movement can be finely judged if the object of interest is seen against a fixed frame of reference. If there is no frame of reference, it is still possible to judge velocity and direction of movement but with much less precision. At night, good quality lighting of the intersection will provide a frame of reference against which the movement of other vehicles can be judged. This is likely to be most important at intersections where the time for response is limited, and such intersections are likely to have high crash rates. This may be why Bruneau and Morin (2005) found that standard lighting had significantly lower nighttime crash rates than non-standard lighting for intersections with high crash rates.

However, Bhagavathula and Gibbons (2022) conducted an experiment on the visual performance of drivers approaching an isolated intersection. The task required was the detection of an 18-cm square, grey target of reflectance 0.5 placed, one at a time, at five different locations; on the four corners of the intersection and in the middle of the approach road. Three different lighting configurations were used: lighting the approach road alone, lighting the central box of the intersection that was common to the two intersecting roads, and lighting both the approach road and the box. Five different road surface illuminances were used: 0 lx, 8 lx, 12 lx, 16 lx, and 21 lx. The measure used to quantify visual performance was the distance at which the target was detected. Drivers drove towards the intersection at 35 mph (56 km.h^{-1}) using low beam headlamps and told the experimenter when they saw the target. The results showed that lighting the box alone had a longer mean detection distance (108 m) than lighting the approach road and the box (76 m) or lighting the approach road alone (52 m). As for the effect of illuminance, the detection distances were always less for all target locations when there was no lighting, but when the different lighting configurations were used, there was no consistency across target locations, presumably because of the differences in positive and negative target contrast produced by the different lighting configurations. However, when only the box was illuminated, all detection distances were longer than the safe stopping distance when the illuminance was 12 lx or more. For the other lighting configurations, higher illuminances were necessary to achieve saturation of visual performance. These results are attractive from the point of view of cost because the box alone can often be illuminated by a single luminaire. However, this experiment does have some limitations. The target is essentially that used for the small target visibility metric of road lighting (see Section 4.3). It bears little relation to the vehicles, cyclists, or pedestrians that are likely to be involved in crashes at intersections. It is stationary, not moving, and is located around the actual intersection. Thus, detecting this target tells us little about how well an approaching driver can evaluate the movements of other traffic approaching the intersection. More research is needed to determine if lighting the box alone is sufficient to reduce crashes at isolated intersections at night.

Taken together, these studies demonstrate that providing lighting at rural intersections is a useful crash countermeasure. But what form the lighting should take is a more open question. Almost any lighting will be useful in attracting attention to the intersection. Beyond that, the form of the lighting needed will depend on the nature of the intersection and the traffic flow after dark. The essential points are that the lighting should provide enough light and a wide enough coverage area to allow fast and accurate judgements of presence, velocity, and direction of movement. Current guidance for lighting isolated intersections in the USA recommends an average road surface illuminance in the range of 4 - 9 lx depending on the road surface type. This is less than that suggested as necessary by Bhagavathula and Gibbons (2022) and is much less than the illuminance suggested by the results of Oya et al. (2002) who carried out a statistical analysis of night crashes resulting in death or injury before and after the installation of lighting at intersections on major roads in Japan. Their results suggest that an average road surface illuminance of 30 lx or more is necessary to ensure a statistically significant reduction in nighttime accidents. Clearly, there is some way to go before the form necessary for lighting isolated rural intersections can be said to be established.

11.7　WORK ZONES

Work zones are inherently hazardous to those working in them and to drivers moving through them. The hazard for workers comes from the proximity of traffic while for drivers it is the change in road layout. In 2019, in the USA, 845 people died in work zones, 690 of them being drivers or passengers and 141 people on foot or cyclists (Federal Highways Administration, 2022b). The most common reason why those on foot were killed is that they were struck by a vehicle. Road works can take many different forms, varying in size, duration, mobility, and timing. The methods used to inform drivers about road works can also take many forms, from variable message signs to traffic signals, manually controlled stop/go signs, cone or barrel arrays defining the road layout, flashing beacons on cones, and large flashing arrows mounted on trucks. Road workers are sometimes protected from traffic by crash barriers but they almost always have their visibility enhanced by wearing fluorescent, retro-reflective vests, jackets, or suits. Attempts to influence driver behaviour are made by setting lower speed limits, sometimes enforced through average speed cameras and higher fines for speeding in work zones.

What combination of safety measures is used depends on the form of the road works. Among the simplest are such mobile off-road activities as clearing roadside litter where the only measure is usually the wearing of high-visibility clothing and the presence of a truck indicating a lane change away from the edge of the road is required. Small temporary obstructions such a digging a hole in the road to repair an electricity cable are usually dealt with by marking off the work area with cones and barriers and controlling traffic with manual stop/go signs or portable traffic signals. Large-scale road works that are planned to take several days or weeks use advance warning signs and extensive coning to define new traffic lanes with flashing lights to make the road layout clear at night as well as concrete crash barriers to protect the workers.

Fixed road works planned to last more than a few hours will have some form of lighting and signage. The exact form will depend on the nature of the work and when it will occur. When work is only planned to take place during daylight hours, temporary signage indicating that there are road works ahead will be installed but lighting will be limited to identifying the road works at night and advising drivers of what they need to do. When work is planned to take place at night, which is often the case on major roads to avoid causing traffic congestion during the day, the lighting has two purposes. The first is to make drivers aware of the changed road layout and the movements of people and vehicles around the road works. The second is to enable the workers to see what they are doing.

The approach used for identifying the work zone and directing drivers round it, by day and night, is usually a series of coloured cones or drums arranged to direct the driver away from and then around the working area (Figure 11.8). A proportion of these cones or drums have amber-flashing warning lamps mounted on top. One

FIGURE 11.8 An array of high-contrast cones with amber-flashing lights on the top of every second cone. This arrangement is designed to ensure the array is visible from a distance both day and night (iStock by Getty Images / Norrie3699).

standard for such lamps is given by the Institute of Transportation Engineers (2001). This specifies a minimum effective luminous intensity of 4 cd for night operation only or 35 cd for operation both day and night. These recommendations are for effective luminous intensities because they apply to flashing lamps so the luminous intensity integrated over time depends on the duty cycle used. There is no maximum set for the effective luminous intensity. The flash frequency is limited to the range 0.9 - 1.25 Hz.

Rea et al. (2018) conducted an experiment to develop an improved standard for the amber-flashing lamps used in work zones. A series of traffic drums with amber lamps on top were arranged on an unlit two-lane road. The drums were arranged to gradually taper in and out so at one point drivers were limited to one lane. This meant that drivers approaching from one direction had to change lane but those coming from the other direction did not. A pick-up truck was parked at the widest part of the restricted part of the road. The first amber-flashing lamp, positioned at the edge of the road, had a fully-on luminous intensity of 0 cd or 750 cd. The subsequent lamps had fully-on luminous intensities of 7.5 cd, 25 cd, 75 cd, or 250 cd during the day and 2.5 cd, 7.5 cd, 25 cd, or 75 cd at night. Both day and night the amber lamps changed at a frequency of 1 Hz from fully on to 10% on so they pulsed rather than flashed. Rea and Bullough (2016) found that pulsing in this manner provided drivers with better information about movement than flashing, i.e., fully-on to fully-off. These lamps, including the first lamp when lit, were arranged to pulse in a random manner, synchronously so they all pulsed together, or sequentially in which the timing of successive lamps was arranged to give a perception of forward movement in the direction the driver had to go. Twelve people rode in a vehicle as either driver or passenger along the road by day and at night, keeping to the speed limit of 30 mph (48 km.h⁻¹) and navigating the work zone. After completing the manoeuvre, the participants were asked two questions about the level of discomfort glare from the first lamp seen and what was the level of discomfort glare from the entire array of lamps? The answers were given on a slightly modified, nine-point, de Boer discomfort glare scale, the ratings being labelled 1 = unbearable, 3 = disturbing, 5 = just permissible, 7 = satisfactory, and 9 = just noticeable glare. Participants were also asked how easy or difficult it was to drive through the work zone, how quickly or slowly could the driver navigate through the work zone, and how accurately could the driver navigate through the work zone. Responses to these questions were given on five-point scales labelled very easy/quickly/accurately to very difficult/slowly/inaccurately. In daytime, neither the full-on luminous intensity nor mode of pulsing had an effect on discomfort glare. However, the mode of pulsing did have a small but statistically significant effect on the speed with which the vehicle passed through the work zone. Specifically, the synchronized pulsing lamps produced a greater reduction in speed when navigating the work zone (2 mph) than the random (1.5mph) or sequential (1.4 mph) modes. At night, when the first lamp was lit to a 750 cd full-on luminous intensity the discomfort glare rating was between "Satisfactory" and "Just permissible". For all the lamps together, there was a two-way interaction between the full-on luminous intensity and the pulsing mode. For the sequential pulsing mode there was little change in discomfort glare as the full-on luminous intensity increased up to 75 cd, probably because only two of the lamps were lit simultaneously. Random and synchronised pulsing both produced a worsening of discomfort glare with increasing luminous intensity although none of the conditions made the rating of discomfort glare worse than "Just permissible". There was a similar two-way interaction when participants were asked how easy it was to navigate through the work zone. When the full-on luminous intensity of the array of lamps was 2.5 cd there was no effect of pulsing mode but as the full-on luminous intensity increased, the sequential mode was assessed as making navigation through the work zone easier than the random mode with the synchronized mode in between. From these findings, Rea et al. (2018) conclude that both night and day, the array of warning lamps should be operated in a sequential mode and from earlier work (Rea and Bullough, 2016) that they should operate from fully-on to 10% on, i.e., should pulse rather than flash. Where day and night differ it is in the fully-on luminous intensity of the first lamp. For daytime, the fully-on luminous intensity should be 750 cd. At night, the fully-on luminous intensity of the first lamp in the array should be reduced to 75 cd or 250 cd and the fully-on luminous intensity of the lamps in the rest of the array should be 25 cd.

Delivering information to approaching drivers is only one part of what work-zone lighting has to do. The other part is to allow the workers at the work zone to see what they are doing and to be seen while they are doing it. A brightly lit work site attracts attention and that, in itself, will warn drivers approaching the site that something requiring their attention is near, but the lighting technology used will depend on the work being done. For relaying the road, vehicle lighting attached to the paver is often all that is used. For constructing new lanes or rebuilding bridges, a much more extensive range of work is needed, so lighting covering a much wider area is necessary. This is usually provided either by a cluster of luminaires mounted on a telescopic mast or by a balloon light, both being powered by a diesel generator. The light sources used can vary from LEDs to high-pressure sodium or metal halide discharge. Where the lights are positioned and where the luminaires are aimed is mainly determined by the needs of the work, although one or more are usually positioned wherever there is a change in direction, narrowing, or merging of lanes for passing traffic. With the cluster of luminaires, this flexibility can result in severe discomfort and disability glare to approaching drivers. This is not inevitable. Rather, it is a matter of poor luminaire design, in the sense that the light distribution is too wide for the limited mounting height, and poor luminaire aiming. This is less likely to be a problem with balloon lighting because the light distribution is completely diffuse and the large surface area of the balloon ensures a low surface luminance so that discomfort or disability glare is less likely (Hassan et al., 2010). The minimum road surface illuminances recommended for work zones range from 54 lx throughout the work zone, to 108 lx around equipment to 215 lx for road surface repairs (Ellis et al., 2003). Further advice on the general lighting of work zones can be found in the publications of the American Traffic Safety Services Association (2013) in the USA and the Department for Transport (2013) in the UK.

All the above is concerned with temporary but fixed work zones; however, some activities, such as line painting, pothole mending, and setting out cone arrays, involve continuous but slow movement or intermittent, slow movement with brief stops. These are known as mobile work zones. There is no special lighting of mobile work zones, the safety of drivers and workers being enhanced either by a convoy of vehicles bearing illuminated messages and signals warning an approaching driver of the slow-moving vehicles forming the work zone or even a mobile barrier sheltering the workers from the passing traffic. The usual arrangement for the convoy system is for the truck and crew carrying out the work to be shadowed by another truck about 40 m behind painted with high contrast fluorescing and retro-reflective chevrons and carrying an array of both amber-flashing beacons or light bars and self-luminous direction indicators (Figure 11.9). These are intended to warn the approaching driver that there are slow-moving vehicles in the lane and that they should slow down and prepare to change lane to avoid them. On high-speed roads, this shadowing vehicle is sometimes fitted with a crash attenuator. These are devices designed to absorb the energy released by a rear-end crash into the shadowing truck. Sometimes there is another vehicle, 40 m behind the shadowing truck carrying amber-flashing beacons and a variable message sign telling drivers that they are approaching a mobile work zone. A comprehensive review of equipment and layouts for mobile work zones is given by Aroke et al. (2022).

What needs to be considered now is how effective all these measures are. Freeman et al. (2003) monitored personal injury accidents at 29 major road work sites on motorways in the UK for 20 months. The road works covered 730 km (453 miles) of road having an exposure of 4,178 million vehicle kilometres; 423 personal injury accidents occurred in the road works in this time. Interestingly, this rate of personal injury accidents is not statistically significantly different from the personal injury accident rate for the same sites when there were no road works and is very close to the national average personal injury accident rate for motorways. Further, factors such as weather conditions and ambient lighting did not affect the personal injury accident rate in the road works. What this suggests is that the safety measures used at

FIGURE 11.9 A truck carrying flashing warning lights and a retro-reflective lane change indicator as well as a large lane change arrow formed from individual lights. Such trucks are used to warn drivers they are approaching a work zone and a lane change is required (iStock by Getty Images / Wirestock).

such major road works; advance warning signs, cones marking lane changes, truck-mounted lane change warnings, special lighting etc., have been effective in overcoming the inherent hazard of work zones although given the lower speed limits used, personal injury accidents should have been less than the average for all motorways if the hazard had been completely nullified. It should also be noted that this conclusion refers only to personal injury accidents. Given that the most common forms of accidents at road works are multiple rear-end collisions, collisions when overtaking, and collisions with objects other than vehicles, it would have been interesting to know how the safety measures influenced property damage only accidents, but these data were not collected. Despite this limitation, there can be little doubt that the safety treatment of major road work sites, which includes special lighting, is on the right track and that remaining incidents of crashes at work zones have more to do with driver inattention, fatigue, recklessness, and inebriation than visibility. Whether this is also true for lesser road works remains to be determined. What is certain is that with some care over glare control, the lighting of fixed road works can be improved to the benefit of driver and worker comfort.

11.8 TRAFFIC-CALMING MEASURES

Traffic calming is a euphemism for making the driver's task more difficult and hazardous by placing obstacles in the road to produce a reduction in speed. Such measures usually take one of four forms. The most common is a series of large bumps placed across the road. The sudden change in road height jolts the driver and passengers of the vehicle making the ride uncomfortable unless the bump is traversed slowly. It may also cause damage to the vehicle's suspension and tyres. A variation on this is the speed table, where there is a wide flat section between two slopes giving a vertical deflection to the road (Figure 11.10). This flat section sometimes carries a pedestrian crossing. Speed bumps and speed tables are typically used in residential streets where excess speed has become a problem or near schools where children are exposed to traffic. They are usually preceded by a traffic sign warning of their presence. In addition, each bump or table is marked in some way either with a high-contrast white marking or a different road surface treatment. The only lighting is usually whatever is provided by the street lighting, but this is not necessarily aligned with the speed bumps. For speed tables, there may be special lighting where the flat part is marked as a pedestrian crossing (see Figure 11.3).

Another traffic-calming measure is road narrowing by promontories projecting into the road (Figure 11.11). These are typically placed at the entrance to a town or village to slow drivers before they enter. These promontories usually contain some sort of structure fitted with retro-reflectors, a speed limit traffic sign, and a notice welcoming visitors to the village. These promontories are either positioned so as to be illuminated by a nearby road lighting luminaire or have a specific luminaire located close to them. A variation on this is the chicane. In this, two or more promontories extending into the road from opposite sides, but offset from each other, turn what was a straight road into a zigzag requiring the driver to slow down. Both road narrowing and speed bumps or tables are sometimes applied together. Advice on

FIGURE 11.10 A speed table installed outside a primary school to achieve traffic calming. The speed table is finished in a different colour material to the rest of the road.

FIGURE 11.11   An island projecting into the road so as to reduce a two-lane road into a one-lane road, thereby achieving traffic calming at the entrance to a village. Note the structure warning drivers of the obstruction and the speed limit. There is a road lighting luminaire illuminating the island, but it is hidden by the surrounding vegetation.

TABLE 11.7

Number of Crashes and the Number Involving Fatalities or Injuries over Five Years, before and after the Installation of Different Traffic-Calming Measures at Three Locations in Italy (from Distefano and Leonardi, 2019)

Traffic-Calming Measure	Timing	Number of Crashes	Number Killed	Number Injured
Speed table	Before	18	2	29
	After	10	0	18
Road narrowing	Before	15	0	28
	After	10	0	19
Chicane	Before	14	1	30
	After	9	0	15

the lighting of traffic-calming measures is available from the Institution of Lighting Professionals (2002).

While traffic-calming measures are generally disliked by drivers, it is undeniable that they do produce a reduction in average speeds. Consequently, there are likely to be benefits in terms of fatalities and injuries occurring in traffic-calmed areas, particularly among pedestrians (Huang and Cynecki, 2001; Distephano and Leonardi, 2019). Table 11.7 shows the number of crashes and the number of fatalities and injuries in the five years before and after traffic-calming measures were applied to three roads in Italy. It is clear that all three traffic-calming measures examined, speed table, road narrowing, and chicane, produce a reduction in the number of crashes with an associated reduction in the number of injuries. Others have found similar benefits in other countries (Zein et al. 1997). Advice on the application of traffic-calming measures can be found in the Department for Transport (2007) and Ewing (2019).

11.9   ROADS NEAR DOCKS AND AIRPORTS

Roads near docks and airports do not need special lighting to enhance the visibility and comfort of drivers but it is required to avoid confusion to ships and aircraft. Ships entering the harbour and aircraft attempting to land rely on signal lamps for guidance. Often, these signal lamps have to be identified from a great distance. The risk is that road lighting luminaires might be confused with the signal lamps. This risk is minimized by restricting the type of road lighting luminaires to those that do not have high luminous intensity in the relevant direction. For docks, the relevant directions are well known so this can be done by careful mounting or shielding of the luminaires. For airports, there are many possible directions as aircraft circle around before attempting to land so the simplest approach is to use only luminaires without any direct upward light emission, i.e., full cut-off luminaires mounted horizontally. Airports impose another constraint on the form of road lighting where the road crosses the flight path, a low mounting height. This restriction leads to the use of many closely spaced columns each carrying a lower wattage light source than would usually be the case. It is important to appreciate that the presence of docks and airports place restrictions on the technology that can be used to provide road lighting not the photometric conditions that should be created. Photometrically, the lighting of roads near docks and airports is the same as any other road of the same class.

11.10   SUMMARY

There are a number of road locations that require special lighting or marking, such as tunnels, pedestrian and railway crossings, car parks, rural intersections, road works, traffic-calming areas, and roads near docks and airports. Road tunnels require special lighting during the day because of two visual effects: the black-hole effect in which the tunnel entrance is seen as a black-hole by approaching drivers and the black-out effect in which drivers entering the tunnel find themselves misadapted. The black-hole effect is due to light scattered in the eye from the surroundings of the tunnel while the black-out effect is due to the driver's visual system having insufficient time to adapt to the much lower luminances in the tunnel. Both these problems can be overcome by appropriate construction of the tunnel entrance and by following the recommendations for grading the luminances of the road surface in the tunnel through a threshold zone followed by a transition zone to the interior of the tunnel.

Pedestrians crossing the road are exposed to a high risk of death or injury. There are two types of pedestrian crossings, those associated with traffic signals where the signals are arranged so as to give priority to vehicles and pedestrians at different times, and those without traffic signals where a pedestrian waiting to cross or actually on the crossing has priority over vehicles at all times. Special supplementary lighting is sometimes used on crossings without traffic signals. The supplementary lighting is designed to provide a higher illuminance than that used to produce the average road surface luminance of the road approaching the crossing with a strong vertical component to ensure that pedestrians are seen in positive luminance contrast. Such lighting will increase the visibility of pedestrians on the crossing and increase the conspicuity of the crossing itself. The conspicuity of the crossing can be further increased by using supplementary lighting of a different colour to that used for the adjacent road lighting. Of course, supplementary lighting does not discriminate between an empty crossing and one in use and does nothing to increase the conspicuity of the crossing during the day. An approach that addresses both these limitations is the insertion of lamps into the road surface at the edge of the crossing that flash when a pedestrian enters and is on the crossing. This has the advantage of warning the driver that there is an actual conflict rather than just a potential conflict.

Like pedestrian crossings, railway crossings involve the movement of one type of transport across the path of another. Unlike pedestrian crossings, at railway crossings it is the driver of the motor vehicle who is more likely to suffer death or injury. Railway crossings are marked in various ways depending on the amount of traffic on the road and track. For remote rural crossings with low levels of traffic on both, all that is usually provided is a standard warning sign and possibly a pair of manually operated gates. For railway crossings where there is a high level of traffic on either the road or track, much more elaborate systems are used. Specifically, the crossing is marked with road signs as well as two pairs of alternately lit lamps, a bell or siren, and either full or half-width barriers. The crossing is also lit at night by conventional road lighting luminaires or floodlights. Lighting has a very limited role to play in railway crossings, most of the information provided to drivers being through signs and signals.

Another location where pedestrians appear to be at risk because they are mixed with vehicles is in car parks. Car parks can be divided into two types: parking lots and parking garages. The former are completely open to the weather while the latter are enclosed to some extent, sometimes totally and sometimes with only a roof. The risk of a pedestrian being injured by a vehicle in these locations is more theoretical than actual, probably because of the much slower speeds used by vehicles in parking lots and parking garages. Accidents involving pedestrians are much more likely to involve tripping, slipping, and falling. The lighting approaches used in parking lots and parking garages are markedly different from the lighting of roads and from each other. For parking lots, luminaires mounted on columns are conventionally used but the luminaires have a much wider range of luminous intensity distributions than those used for lighting roads. For parking garages, the low ceiling heights and high level of obstruction caused by the structure and parked vehicles require a different approach. In parking garages, luminaires are mounted directly on the ceiling, care being taken with the choice of luminaire to avoid glare and with luminaire spacing to ensure a high level of illuminance uniformity.

One situation where crashes are unexpectedly frequent is rural intersections. Most rural roads are unlit, the driver having to rely on headlamps alone for visibility. Studies of crash data have shown that providing lighting at rural intersections is a useful countermeasure, albeit one that mainly affects property damage crashes. But what form the lighting should take is a more open question. Almost any lighting will be useful in attracting attention to the intersection. Beyond that, the form of the lighting needed will depend on the nature of the intersection and the traffic flow after dark. The essential points are that the lighting should provide a high-enough luminance and a wide-enough coverage area to allow easy detection of objects in the intersection as well as fast and accurate judgements of the velocity and direction of movement of other vehicles approaching the intersection.

Road works are inherently hazardous to those working on them and to drivers moving through them. Fixed road works rarely have special lighting unless work is planned at night, but they always have warning signs and a tapering arrangement of cones to divert traffic away from and past the work site. These cones often have flashing amber beacons mounted on top. The flashing can be arranged to be simultaneous or sequential to give a perception of movement. The actual lighting of work site is usually provided by a cluster of luminaires mounted on a telescopic mast or by balloon lights, both being powered by a diesel generator. Balloon lights provide diffuse lighting but the cluster of luminaires can be much more directional. Where these are positioned and where the luminaires are aimed is mainly determined by the needs of the work, although one or more are usually positioned wherever there is a change in direction, narrowing, or merging of lanes for passing traffic. The problem with this flexibility is that often it results in severe discomfort and disability glare to approaching drivers. Accident records at major road works on motorways in the UK have shown that the rate of personal injury accidents is not statistically significantly different from the personal injury accident rate for the same sites when there were no road works. This suggests that the safety measures used at such major sites, advance warning signs, cones marking lane changes, special lighting, protective barriers, reduced speed limits etc., can be effective in overcoming the inherent hazard of major road works. Whether this is also true for lesser road works remains to be determined. What is certain is that with some care over glare control, the lighting of fixed road works could be improved to the benefit of driver and worker comfort. A special case is the mobile work zone where the work involves continuous or intermittent slow movement. Mobile work zones are protected by having a convoy of vehicles. At the front are the machines and workers doing the actual work, behind these is another truck with an array of flashing or pulsing warning lights and a directional arrow and, sometimes, an energy-absorbing attenuator designed to minimize the consequences of a high-speed rear-end crash. Behind this is another vehicle carrying information about the presence of the mobile road works ahead.

Traffic-calming measures are obstacles placed in the road with the aim of causing a driver to reduce speed. There are four common forms. Two consist of speed bumps or speed tables placed across the road, the speed table having a flat area at the top while a speed bump does not. These make the vehicle's ride uncomfortable unless taken at low speed. They are used in residential areas and are identified by high-contrast road markings or different road surface treatments. Two other traffic-calming measures are road narrowing and chicanes. These involve construction of islands projecting into the road so as to reduce two lanes to one or to convert a straight road into a zigzag. These are used at the entrance to towns and villages where speeding is a problem. The islands contain a high-contrast structure usually fitted with retro-reflectors, a speed limit traffic sign, and a traffic sign indicating the priority direction of travel. These islands are either positioned so as to be illuminated by a nearby road lighting luminaire or have a specific luminaire located close to them. Traffic-calming measures do cause drivers to slow down. Consequently, the number of crashes and the number of fatalities and injuries occurring in traffic-calmed areas are reduced.

Roads near docks and airports do not need special lighting to enhance the visibility and comfort of drivers but to avoid confusion to ships and aircraft. The risk is that road lighting luminaires might be confused with the signal lights used by ships entering harbour or aircraft attempting to land. This risk is minimized by restricting the type of road lighting luminaires to those that do not have high luminous intensity in the relevant direction. Airports impose another constraint on the form of road lighting where the road crosses the flight path, a low mounting height. Docks and airports place restrictions on the technology that can be used to provide road lighting not the photometric conditions that should be created.

From these reviews of the lighting of tunnels, pedestrian crossings, railway crossings, car parks, rural intersections, road works, traffic calming, and roads near docks and airports, it is clear that special lighting is required in some locations. Such lighting can have four different purposes: to allow for the limitations of the visual system, to make a situation that requires the driver's attention conspicuous, to make what is happening easily understood so that drivers can modify their behaviour appropriately, and to avoid confusing other modes of transport.

12 Adverse Weather

12.1 INTRODUCTION

The presence of rain, snow, fog, smoke, or dust in the atmosphere and a low-elevation sun in the sky makes the driver's task more difficult, but in different ways. Rain has a direct effect on visibility by absorbing and scattering light and an indirect effect by changing the reflection and friction properties of the road surface. Snow scatters incident light and in doing so doing creates a lot of visual noise in the driver's visual field. In addition, if the snowfall is heavy enough, the snow will cover the road surface, obscuring lane markings and other information designed to aid the driver as well as making the road more slippery. Fog and heavy spray thrown up by other vehicles have their effect by absorbing and scattering light in the atmosphere, thereby reducing the luminance contrasts of everything ahead of the driver. Smoke and dust also absorb or scatter light and thereby reduce the effectiveness of all forms of lighting. A low-elevation sun directly ahead of the driver is a potent source of disability glare that diminishes the luminance contrasts of objects on and around the road ahead. It is the impacts of such adverse weather conditions on drivers and how lighting may be used to alleviate them that are the subjects of this chapter.

12.2 ADVERSE WEATHER AND CRASHES

The presence of rain, snow, fog, smoke, or dust and low-elevation sun undoubtedly make the driver's task more difficult. Becker et al. (2022) examined the effect of including weather data in the predictive power of models of crash risk. They found that adding meteorological data to the models improved their predictive power by up to 24%. This indicates that adverse weather definitely has an effect on crash risk, but it says nothing about the role of lighting conditions. Fortunately, other data does. Table 12.1 shows the number of road crashes in which fatalities occurred in the USA in 2019, classified according to the weather and the lighting conditions (NHTSA, 2021a). Table 12.2 shows the number of road crashes in which personal injury occurred in the USA in 2019, classified in the same way (NHTSA, 2021a). The specific adverse weather conditions considered in these tables are rain and snow or sleet. The category "other" presumably includes fog, smoke, and dust.

An examination of Tables 12.1 and 12.2 reveals a number of interesting facts. The first is that the largest number of road crashes involving death or personal injury, almost 90%, occur when there is no adverse weather. This is partly a matter of exposure and partly a matter of driver experience and behaviour. For most states in the USA, seriously adverse weather occurs infrequently. In states where adverse weather is common at certain times of the year, e.g., snow in winter in the states bordering Canada, most drivers will be familiar with the conditions and will adjust their

DOI: 10.1201/9781003388906-12

TABLE 12.1

Number of Road Crashes in 2019 Involving Fatalities Classified According to the Weather and the Lighting at the Time (from NHTSA, 2021a)

Light Conditions	No Adverse Weather	Rain	Snow and Sleet	Other
Daylight	13,104	1025	235	129
Dark but lit	5,478	595	53	245
Dark	7,425	834	126	205
Dawn and dusk	1,157	106	25	34

TABLE 12.2

Number of Road Crashes in 2019 Involving Personal Injury Classified According to the Weather and the Lighting at the Time (from NHTSA, 2021a)

Light Conditions	No Adverse Weather	Rain	Snow and Sleet	Other
Daylight	1,176,000	116,000	19,000	4,000
Dark but lit	286,000	45,000	7,000	2,000
Dark	151,000	27,000	6,000	2,000
Dawn and dusk	62,000	10,000	2,000	1,000

behaviour accordingly, a change which usually involves reducing speed and keeping greater distances from other vehicles. Both these changes make fatalities in crashes less likely. The second interesting fact is that in the absence of adverse weather 48% of fatal crashes occur in daylight, 27% in darkness, 20% after dark, but on lit roads 4% at dawn or dusk. The third interesting fact is that fatal accidents that occur in rain or snow and sleet are affected by lighting conditions in a similar manner to those that occur in the absence of adverse weather.

Ashley et al. (2015) examined the records of the US Fatality Analysis Reporting System (FARS) from 1994 to 2011 to identify the role of fog, smoke, and dust in fatal crashes. They found that of the fatalities where adverse road or weather conditions were reported, 46% occurred in rain, 12% occurred when there was frozen precipitation, i.e., ice, and 10% occurred in fog, and blowing smoke or dust. Of this 10%, 90% were linked to fog and only 10% were related to blowing smoke or dust. Unfortunately, Ashley et al. (2015) did not consider the lighting conditions in their analysis. However, earlier FARS records do. In 2005, for fog, 24% of fatal crashes occur by day, but 51% occur after dark on unlit roads (NHTSA, 2006). This is a different pattern to that occurring for fatal crashes occurring in rain and in snow or sleet. A plausible explanation for this difference is that while rain and snow and sleet have some effect on visibility, their major effect is to change the coefficient of friction of the road surface, an effect that is independent of lighting condition and that results in longer stopping distances. Conversely, fog does little to affect the grip of the vehicle on the road but can have dramatic effects on the visibility of everything ahead, particularly at night. Abdel-Aty et al. (2011) examined the crash data from Florida between 2003 and 2007 in which fog or smoke were involved. They found that fog- and smoke-related crashes tended to result in more serious injuries and involve more vehicles. These crashes were most likely to occur at night on roads without lighting.

12.3 RAIN

Rain consists of falling water droplets with diameters in the range of 20 - 200 μm (Kocmond and Perchonok, 1970). A photon of light striking such a droplet may be absorbed or scattered, the scattering being caused by some combination of reflection and refraction. As a result of these processes, the proportion of light emitted by vehicle headlamps or signal lamps that reaches the eyes of drivers in rain is reduced relative to what would be the case in dry conditions. In addition, the scattered light forms a luminous veil that reduces the luminance contrasts seen through the rain. Further, water left on the windscreen will be enough to degrade the quality of the retinal image of the scene outside the area served by the windscreen wipers, a degradation that may occur even within the wiped area in heavy rain. Consequently, the direct effect of rain in the atmosphere is to reduce visibility, which is why drivers tend to reduce speed in the presence of rain, the reduction being greater on roads without lighting than on roads with lighting (Jagerbrand and Sjobergh, 2016). However, unless the rain is heavy or the windscreen wipers are ineffective, this is not the main impact of rain on visibility. Rather, it is the indirect effect of the rain on the reflection characteristics of the road surface that is more important, unless the temperature is low enough to convert the water on the road surface to ice. In this situation, the grip of the vehicle's tyres on the road will become more important for traffic safety than any loss of visibility.

When rain covers a road surface, the visual effect is to make the road surface more of a specular reflector than a diffuse reflector (Figure 12.1). This is not usually a problem during the day because the meteorological conditions that tend to produce heavy rain also tend to produce diffuse illumination from the sky rather than direct illumination from the sun. However, at night, when the road is lit either by vehicle forward lighting alone or by a combination of road lighting and vehicle forward lighting, a wet road surface dramatically alters the effectiveness of that lighting. Where only vehicle forward lighting is available to the driver, the effect of a thin film of water on the road surface is threefold. First, the average road surface luminance will be reduced as more of the light from the headlamps is specularly reflected forward rather than diffusely reflected back to the driver. Second, light from headlamps specularly reflected from the water film on the road fails to reach the retro-reflective road markings underneath. As a consequence, the visibility of the road markings used for visual guidance and accurate lane keeping is much reduced (Rumar and Marsh, 1998). Third, the disability and discomfort glare from approaching vehicles will be increased as more light from the headlamps of the approaching vehicle is

FIGURE 12.1   The headlamps of a car operated on low beam reflected from a wet road. Also shown is the reflected image of a low-pressure sodium road lighting luminaire.

specularly reflected towards the driver. The magnitude of all these effects depends on the thickness of the water film. Smooth road surfaces will be affected by thinner films than rough road surfaces, and retro-reflective elements and road studs that stand proud of the road surface are less likely to be degraded in visibility than those that are flush with the road surface.

As for road lighting, when the lit road is wet rather than dry, the average road surface luminance is increased, but this occurs because some parts of the road surface become much higher in luminance, i.e., the road surface luminance distribution becomes much more non-uniform. This is evident from the measurements of average road surface luminance, overall luminance uniformity ratio, and longitudinal luminance uniformity ratio shown in Table 12.3 (Ekrias et al., 2007). As objects on the road are seen in silhouette against the local road surface luminance, this increase in luminance non-uniformity means that the pattern of visibilities produced by other vehicles and pedestrians will change more rapidly as they are seen against different parts of the road. The movements of objects that change their patterns of visibilities from moment to moment are likely to be more difficult to identify. This is a negative effect, but there are two positive effects of introducing road lighting on wet roads. One is to reveal road marking more clearly, provided the road markings have a high diffuse reflectance. The other is to diminish the impact of disability glare from opposing vehicle headlamps by increasing the background luminance.

The most successful solution to ensure good visibility on wet roads lies in the choice of road materials. A road constructed with a high proportion of aggregates will have a more coarse texture. Such a road will be less specular in wet conditions but will take longer to dry (Sorensen, 1977). The permeability of the road is also important. A road constructed of more permeable materials will reduce the

TABLE 12.3

Average Road Surface Luminance (cd.m^{-2}), Overall Luminance Uniformity Ratio and Longitudinal Luminance Uniformity Ratio for Two Road Lighting Installations on a Two-lane Dual Carriageway in Finland, under Three Different Weather Conditions. The Measurements Were Made of the Same Pieces of Road Using an Imaging Photometer Positioned at the Side of the Road at a Height of 1.5 m above the Road Surface (after Ekrais et al., 2007)

Road conditions	Road Lighting Installation 1			Road Lighting Installation 2		
	Average Road Surface Luminance (cd.m^{-2})	Overall Luminance Uniformity Ratio (Minimum/ Average)	Longitudinal Luminance Uniformity Ratio (Minimum/ Maximum)	Average Road Surface Luminance (cd.m^{-2})	Overall Luminance Uniformity Ratio (Minimum/ Average)	Longitudinal Luminance Uniformity Ratio (Minimum/ Maximum)
Dry	1.11	0.740	0.782	1.76	0.577	0.830
Wet	2.37	0.220	0.398	5.01	0.209	0.310
Ploughed and salted after snow	1.66	0.639	0.756	2.35	0.539	0.648

probability of specular reflections occurring because water is less likely to collect on the surface. How much a road surface will change its reflection properties between dry and wet conditions and how it compares in this respect to roads constructed of other materials can be judged from the wet road reflection classification system of Frederiksen and Sorensen (1976).

The other approach to limiting the problems caused by wet roads is to modify either the vehicle forward lighting or the road lighting. For vehicles, adaptive forward lighting systems (see Section 6.7.3) can have a wet road beam as an option. The wet road beam involves a change in the luminous intensity distribution of the forward lighting to produce a reduction in the illumination just in front and increased light to the sides of the vehicle. This beam is activated when either rain is detected on the road or when the windscreen wipers are switched on. The reason for reducing the illumination just ahead of the vehicle is simple geometry. Given conventional mounting heights for headlamps, this is the part of the light distribution that is specularly reflected towards approaching drivers and hence which makes the biggest contribution to the increased disability and discomfort glare experienced (see Figure 12.1). The reason for the increased light to the sides of the vehicle is to counter the reduction in diffuse lighting of the surroundings caused by the increased specularity of the road surface. In the absence of an adaptive forward lighting system with a wet road beam option, the usual advice is to use low beam rather than high beam headlamps.

With road lighting, it is possible to limit the deterioration in road surface luminance uniformity when the road becomes wet by changing the luminous intensity distribution of the luminaires, the spacing between luminaires, and their position relative to the road surface. Van Bommel (1976) developed a metric to identify the suitability of a road lighting luminaire for use where roads are frequently wet. This metric, the cross factor, is defined by the equation:

$$CF = I_c / I_1$$

where CF is the cross factor, I_c is the luminous intensity (cd) of the luminaire in a direction corresponding to the mounting height of the luminaire across the road (45 degrees), and I_1 is the luminous intensity (cd) of the luminaire in a direction corresponding to twice the mounting height of the luminaire (63.4 degrees) but along the road. The higher is the cross factor, the greater will be the overall uniformity of the road surface luminance in wet conditions. The disadvantage of this approach is that it requires a much closer spacing of the luminaires. For this reason, a catenary layout of luminaires is recommended. Luminaires designed for use with a catenary layout can have cross factors of the order of 10 while luminaires intended for use in other layouts can have cross factors of less than 1 (van Bommel, 2015).

Where roads are likely to be wet for a considerable proportion of the time, as in the Scandinavian countries, the national road lighting recommendations call for a minimum overall luminance uniformity of 0.15 for the most specular of the wet road surface reflectances in their classification system (CIE, 1979). How effective this approach is in maintaining visibility of objects on the road is open to question, the point being that a high luminance uniformity achieved at the cost of a low road surface luminance may not represent an improvement in visibility (Lecocq, 1994). This adverse trade-off can be avoided if the better road surface luminance uniformity is achieved by closer spacing of the luminaires or by including some higher reflectance aggregate, known as brighteners, in the pavement mix to increase the diffuse reflectance of the road surface (Sorensen, 1977).

12.4 SNOW

Snow consists of large, high-reflectance particles falling through the atmosphere. Light from vehicle forward lighting striking a snowflake will be scattered, some of the scattered light reaching the eyes of the driver. The result of this will depend on the density of snowflakes. If the density is high then there will be few gaps between the snowflakes, and the result is a uniform veil of high luminance that restricts visibility in all directions to very short distances. This is known as a whiteout. More usually, each snowflake is seen separately against a low-luminance background. This would not be too much of a problem if the snowflakes were stationary and so fixed against the background. After all, in daylight the detailed structure of the background is always visible and causes no complaint. However, snowflakes are not fixed, and their movement against the background provides a strong signal to the peripheral visual system that is difficult to ignore, the peripheral visual system being designed specifically to detect changes in the visual world.

How distracting snowflakes are will depend on their luminance contrast, the higher the luminance contrast, the more distracting the snowflakes. Maximum luminance contrast will occur when the vehicle forward lighting is the only source of light. This is consistent with the finding that drivers reduce their speed in the presence of snow by a greater amount on unlit roads than on lit roads (Jagerbrand and Sjobergh, 2016). When only vehicle forward lighting is available, the background against which the snowflakes are seen will be at a minimum. During daylight or at night when road lighting is present, the background against which the snowflakes are seen is higher in luminance and thus the snowflakes are at a lower luminance contrast.

There are four approaches that can be used to reduce the distraction caused by snowflakes. The one most readily available to the driver is to use the forward lighting on low beam rather than on high beam. The high-beam condition increases the amount of light emitted by the forward lighting, distributes more light closer to the driver's line of sight, and increases the area covered by the beam, with the result that more snowflakes closer to the driver's line of sight are seen at a higher luminance contrast.

Another approach, but one that is only available to the vehicle designer, is to place the forward lighting as far away from the driver's usual line of sight as possible, although this will reduce the effectiveness of retro-reflectors. Field studies have shown that mounting the forward lighting on a snowplough as far away as possible from the driver's line of sight results in better visibility and reduced perceptions of glare in snow (Bajorski et al., 1996; Bullough and Rea, 1997). Further, observations of the choices of lamps made by snowplough drivers, i.e., people who are experienced at driving in snow for long periods, reveal that the lamps furthest from the driver are used more frequently in snow (Eklund et al., 1997). Displacing the light source from the driver's line of sight is effective because the intensity of back-scattered light in snow decreases with increasing angle from the direction of incidence on the snowflake (Hutt et al., 1992).

Yet another possibility is to narrow the luminous intensity distribution of the headlamp because this increases the distance at which significant amounts of light intersect the driver's line of sight. Bullough and Rea (2001a) carried out an assessment of a headlamp fitted with a mosaic of hexagonal louvres to narrow the beam compared to a conventional headlamp in the same position. The assessment was again made by snowplough drivers. The results were a reduction in the average luminance of the falling snow and a small but statistically significant decrease in the perception of discomfort glare for the louvred headlamp even though the louvres reduced the light output of the headlamp.

Bullough and Rea (1997) have incorporated both the displacement of the light source and the narrowing of the luminous intensity distribution into a simple model for predicting satisfaction with forward visibility in snow and fog. Forward visibility was taken as the product of the maximum luminous intensity of the lamp and the displacement of the lamp from the driver's line of sight. The maximum luminous intensity is a measure that combines both the light output and the luminous intensity distribution of the lamp as, for the same lamp wattage, the narrower the beam, the higher the maximum luminous intensity will be. Eight snowplough drivers made ratings of visibility, glare, and satisfaction on three-point scales, each running from −1 to +1, while driving at night in snow or freezing rain. These ratings were found to be highly correlated ($r^2 = 0.91$) so the individual ratings for each condition were summed into a single metric called lighting quality. Figure 12.2 shows the relationship between the lighting quality ratings and the logarithm of forward visibility. It is clear that both displacement from the driver's line of sight and the luminous intensity distribution have a role to play in providing good forward visibility in snow.

A third approach is to modify the spectrum emitted by the forward lighting. As discussed in Section 4.6, there is evidence that light spectra that more effectively stimulate the rod photoreceptors of the retina ensure better off-axis detection in mesopic conditions, at the same photopic luminance. This is to the advantage of the driver in a clear atmosphere at night but when that atmosphere contains snowflakes, ensuring better off-axis detection may be detrimental because it will make the snowflakes more distracting rather than less. Bullough and Rea (2001b) have demonstrated this effect of light spectrum for the performance of a tracking task viewed through a curtain of air bubbles moving through water; light with a spectrum that provided little stimulation to the rod photoreceptors resulting in better task performance than light that did stimulate the rod photoreceptors. This effect is less potent than displacing the light source and narrowing the luminous intensity distribution but may be the only option if displacement and narrowing of the beam are not practical.

All this suggests an opportunity for the adaptive forward lighting system (see Section 6.7.3). When driving in snow, it should be possible to increase the luminous intensity of the headlamp on the passenger's side of the vehicle while decreasing

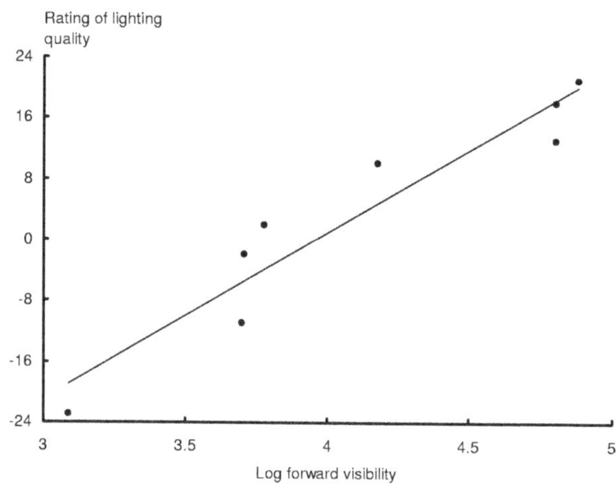

FIGURE 12.2   Ratings of lighting quality made by snowplough operators for halogen lamps with different luminous intensity distributions and displacements from the driver plotted against the logarithm of forward visibility. The rating of lighting quality is the sum of ratings of visibility, glare and satisfaction made on three-point scales by eight drivers. The forward visibility metric is the product of the maximum luminous intensity of the lamp (cd) and the displacement of the lamp from the driver's line of sight (m) (after Bullough and Rea, 1997).

that of the headlamp on the driver's side. If the luminous intensity distribution of the passenger side headlamp could be narrowed at the same time, the forward lighting should provide greater visibility and comfort for the driver in snow.

So far, this discussion has been solely concerned with forward lighting but rear signal lighting also deserves consideration. One type of vehicle certain to be found on the road in snow is the snowplough. Snowploughs are large, heavy trucks moving at slow speeds. Further, when ploughing they are surrounded by a cloud of thrown-up snow that reduce their visibility to other drivers. This may explain why about two-thirds of crashes involving snowploughs in Minnesota are rear-end collisions (Hale et al., 1989). To make their presence conspicuous, snowploughs are commonly fitted with a number of high-mounted, flashing beacons. These lamps are high mounted to place them above the thickest part of the snow cloud, to ensure that they are visible over the top of any closely following vehicles, and to reduce the glare experienced by a driver directly behind the snowplough. The flashing lamps are of a specific colour, usually red or amber, and flashing at a low frequency. Such flashing lamps are undoubtedly effective in enhancing the conspicuity of the snowplough but, if they are all the lighting that is available to an approaching driver, confusion may occur. This is because judgements of position and movement are more difficult under strobe lighting than under stable lighting (Croft, 1971). It would be interesting to measure the ability to judge the relative position and movement of snowploughs under pulsing rather than flashing beacons.

Bullough et al. (2001b) had following drivers evaluate several different forms of supplementary rear lighting on snowploughs at night in clear conditions and in heavy snow. The supplementary conditions ranged from the conventional high-mounted flashing amber lamps (conventional configuration) through floodlights designed to illuminate the rear edges of the truck (indirect edge delineation) and two high-mounted pairs of amber and red flashing lamps that from a distance appear as continuously lit but alternating in colour (alternate colour changes), to continuously lit LED light strips placed vertically at the sides of the rear of the truck (continuous LED strip). Figure 12.3 shows the mean ratings of visibility and feelings of confidence about overtaking made by drivers following a snowplough fitted with the different supplementary rear lighting configurations, at night, in a clear atmosphere, and in heavy snow. There are statistically significant effects of rear lighting configuration and weather conditions. The best rear lighting configuration in a clear atmosphere, and in heavy snow, for both visibility and confidence in overtaking, was the two continuously lit, vertical LED light strips.

Bullough et al. (2001b) also report a field measurement of the ability of three drivers to detect that they were closing on the snowplough fitted with the conventional configuration and the continuously lit LED light strips. The drivers followed a snowplough at a distance of 100 m at a speed of 30 mph (48 km.h^{-1}). Without warning the following driver, the driver of the snowplough was instructed to decelerate without braking, thereby ensuring that the following driver came closer to the snowplough. All three following drivers detected that they were closing on the snowplough sooner with the continuously lit LED light strips operating than with the conventional high-mounted flashing lamps operating.

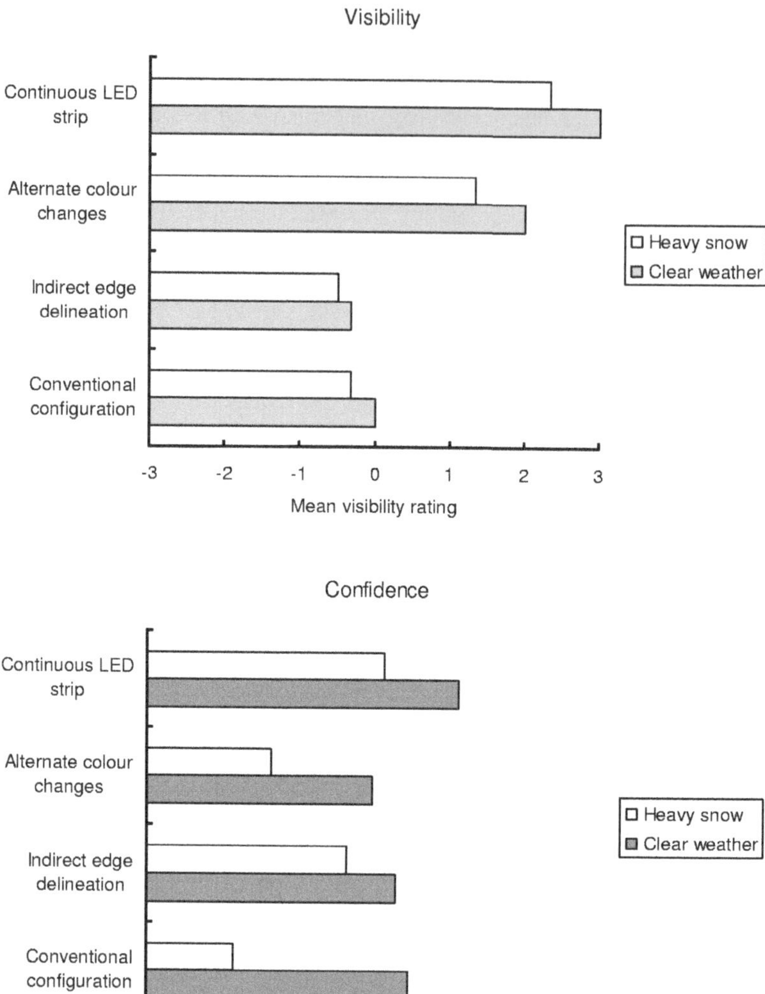

FIGURE 12.3   Mean ratings of visibility and the feelings of confidence about overtaking made by drivers following a snowplough fitted with the different supplementary rear lighting configurations at night in a clear atmosphere, and in heavy snow. The mean ratings were made on a scale running from −3 (worst) to +3 (best) (after Bullough et al., 2001b).

What all this implies is that the rear lighting of snowploughs needs two elements, a temporally varying element to ensure conspicuity and a temporally fixed element to provide a clear and stable cue to movement. This requires some care with balance, as an undue emphasis on one element will be to the detriment of the other. More generally, it should be evident that ensuring conspicuity in adverse weather is a necessary but not sufficient condition for traffic safety. Anyone who has approached the scene of an accident attended by three police cars, a fire engine, and an ambulance, all with multiple flashing lamps of different colours operating on different phases, will recognize the difficulty. Identifying something is there is only part of the problem. It is also necessary to understand what that something is, how it is moving and what is expected of the approaching driver.

After snow has ceased falling, there is a residual effect on visibility due to the snow on the road. At worst, this can lead to the masking of all road markings, making the lanes and edges of the road impossible to identify. Where heavy snow is a regular occurrence, the edges of roads are often lined with posts fitted with retro-reflective markers. Even when most of the snow has been removed by ploughing, the reflectance of the road surface will be increased. The effect of this is evident in Table 12.3, which shows the measured average road surface luminance, overall luminance uniformity, and longitudinal luminance uniformity for two road lighting installations for roads that had been salted and ploughed after a snowfall. The effects of an increase in road surface luminance are to increase the sensitivity of the visual system, to increase the luminance contrast of low-reflectance objects but to reduce the luminance contrast of high-reflectance objects. This implies that once the snowfall has stopped and the road has been ploughed, the net effect of snow on the ground is likely to be beneficial for visibility.

12.5 FOG

Fog consists of small water droplets with diameters in the range 2 - 35 μm suspended in the atmosphere (Kocmond and Perchonok, 1970; Garland, 1971). Photons of light incident on these droplets are absorbed and scattered. The simplest approach to quantifying these effects is to ignore the distinction between scatter and absorption and treat their combined effect on light loss as absorption alone. In mathematical terms, this approach is expressed in Lambert's law which states that the luminous intensity of light propagating through a uniform medium is given by:

$$I = I_0 e^{-\sigma d}$$

where I is the luminous intensity (cd) at distance d (m), I_0 is the unattenuated luminous intensity (cd) at the origin and σ is the extinction coefficient (m^{-1}). The extinction coefficient of fog can range from 0.0015 m^{-1} for thin fog to 0.04 m^{-1} for thick fog. The visual outcome of this absorption and scattering is a reduction in the transmission of light through the atmosphere and the creation of a somewhat uniform veil of luminance covering the driver's visual field. The effect of the uniform luminance veil is to reduce the luminance contrast of all the things in front of the driver and hence to reduce their visibility. This general reduction in visibility has been shown to lead to drivers reducing speed in fog (White and Jeffery, 1980). Hawkins (1988) found that speeds on motorways in the UK started to fall as visibility distance was reduced below 300 m so that when visibility distance had been reduced to 100 m, which corresponds to thick fog, speeds were about 30% less than in a clear atmosphere.

Unfortunately, even these reduced speeds are not slow enough to make the stopping distance of the vehicle less than the distance the driver can see (Sumner et al., 1977). One explanation for this failure to reduce speed sufficiently when there are several vehicles travelling together is the desire to maintain contact with the vehicle immediately in front and thereby to ease the stress of having to find the way ahead. When there is no vehicle immediately ahead, drivers have to find the way ahead themselves. That this is difficult is shown by the fact that the ability to maintain the lateral position the road deteriorates in fog (Tenkink, 1988).

There are a number of ways to help the driver in fog, but none of them is a complete solution. One approach used in locations where fog is common is to inset retro-reflective road studs into the road at regular intervals, as lane and edge markers. When illuminated by the forward lighting of a vehicle, the road studs have a high luminance and hence an increased contrast against the road. This helps the driver to keep the vehicle in the lane and provide visual guidance of the road ahead.

Turning now to the driver, the first choice faced is how to use the vehicle headlamps. It is a common experience that it is often better to use low-beam headlamps than high-beam headlamps in fog. The reason is that high-beam headlamps put more light further down the road and, as a result, project light a greater distance through the fog, thereby producing a higher luminance veil. This means that the luminance contrasts of vehicles and objects on the road are lower when high-beam headlamps are used. Whether or not this is offset by the greater contrast sensitivity induced by the higher adaptation luminance will depend on the fog density. The denser the fog, the more likely it is that the use of high beam headlamps will be detrimental to visibility.

One way that is claimed to help in this situation is to fit the vehicle with supplementary forward lighting intended for use in fog as specified by the ECE or the SAE. Fog lamps are rarely a legal requirement but are widely sold as an optional extra, particularly on more up-market vehicles. Where fog lamps are fitted they are usually mounted low on the vehicle, below the conventional forward lighting, and have a luminous intensity distribution that is both wide and flat, the effect being to put more light on the road immediately in front and to the sides of the vehicle and very little above the horizontal plane through the fog lamps. The low-mounting position is advantageous because the fog lamps are closer to the road and further from the driver's line of sight. Also, fog is usually thinner close to the road. The minimizing of the light distribution above the horizontal is desirable because light directed upwards would intersect the driver's line of sight close to the vehicle and hence increase the veiling luminance seen by the driver.

There has been a long controversy associated with the best light spectrum to be used for forward lighting in fog (Schreuder, 1976). For many years, France required the use of yellow forward lighting on the basis that the yellow light provided better visibility in fog than white light. Whether this is true depends on the size of the water droplets forming the fog. If the water droplet size is less than the wavelength of the incident light, then scattering can occur that is wavelength-dependent. However, many fogs have droplet sizes that are much larger in diameter than the wavelengths that stimulate the human visual system (Middleton, 1952). In these conditions, different degrees of scattering for different visible wavelengths do not occur. It is clear that any advantage of yellow forward lighting in fog has to rely on some factor other than decreased scattering, possibly the removal of wavelengths that provide greater stimulation to the rod photoreceptors.

The visual effect of fog lamps is to provide greater visibility of road edges and nearby lane markings, thereby making lane keeping easier. Fog lamps do little to enhance the visibility of vehicles and objects further along the road. Figure 12.4 shows the calculated luminance contrasts of road markings for fog lamps alone, low-beam headlamps alone, and both fog lamps and low-beam headlamps together, 10 m, 20 m, and 40 m ahead of the vehicle, in a clear atmosphere and in light, medium, and heavy fog (Folks and Kreysar, 2000). Figure 12.4 clearly demonstrates the impact of fog density on visibility by showing a marked reduction in luminance contrast with increasing fog density at all three distances. As for the best form of lighting to use, in a clear atmosphere adding fog lamps to low-beam headlamps increases the luminance contrast at all three distances, although the increase diminishes with distance.

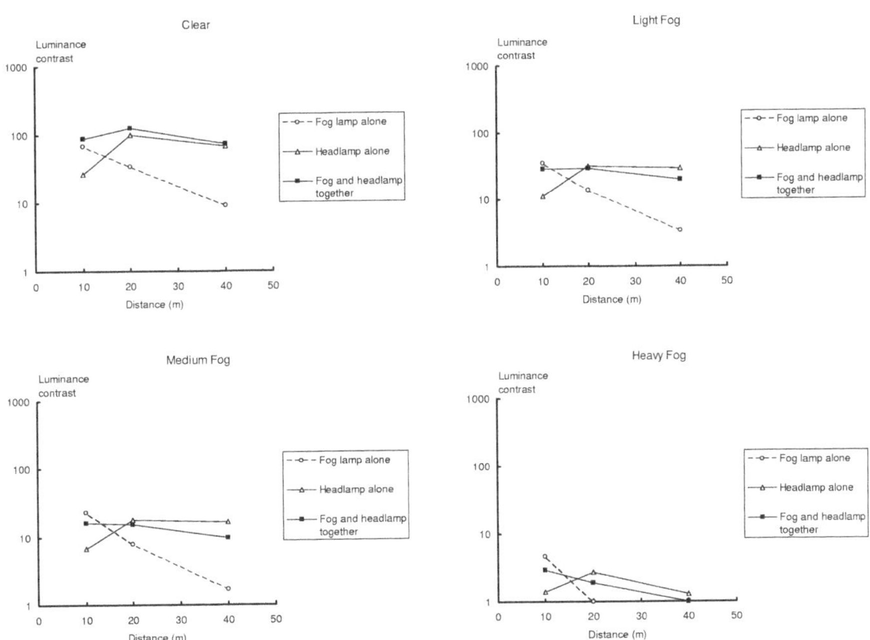

FIGURE 12.4 Calculated luminance contrasts for road markings of reflectance 0.5 in four different atmospheres for fog lamps alone, low-beam headlamps alone and both fog lamps and low-beam headlamps together, the markings being placed at distances of 10 m, 20 m, and 40 m in front of the vehicle. The calculations are made for fog lamps mounted at 0.4 m above the road, eye height at 1.42 m above the road, and a background luminance of 0.017cd m^{-2} in a clear atmosphere. The extinction coefficients for the four atmospheres are: clear = 0.00015 m^{-1}, light fog = 0.003 m^{-1}, medium fog = 0.006 m^{-1}, and heavy fog = 0.03 m^{-1} (after Folks and Kreysar, 2000).

In light, medium, and heavy fogs, fog lamps alone produce the highest luminance contrast at 10 m, but low-beam headlamps alone ensure higher luminance contrasts at 40 m.

The failure of fog lamps to light far down the road may explain why the most common form of crash in a fog is a rear-end collision between vehicles, often several vehicles (Codling, 1971; Koth et al., 1978). It may be that fog lamps contribute to such accidents by making lane-keeping easier without contributing to visibility at a distance, the former giving drivers an unjustified confidence in their ability to see. If so, this is an example of the selective visual degradation hypothesis in action (see Section 6.4) (Leibowitz and Owens, 1977). What is required to increase visibility at a distance and hence diminish the risk of rear-end collisions is a high-intensity lamp with a narrow light distribution aimed down the road and mounted on the vehicle as far away from the driver's line of sight as possible (Bullough and Rea, 2001a).

Given that neither fog lamps nor headlamps as presently designed are effective in avoiding rear-end collisions in fog, another approach is to fit vehicles with high intensity rear lamps for use in fog (see Section 7.12). These are effective because they increase the distance from which the vehicle can be detected. However, if only one rear fog lamp is provided the overestimation of distance common in fog is increased (Cavallo et al., 2001). Two well-separated rear fog lamps enable better estimations of distance to be made. Also, care is required to limit the use of rear fog lamps to foggy conditions because in a clear atmosphere their luminous intensity is so high that they become a source of discomfort and sometimes a source of disability glare to following drivers. This is also a concern for forward lighting fog lamps. Sivak et al. (1996) carried out observations of the use of fog lamps in the State of Michigan. Table 12.4 shows the percentage of vehicles with fog lamps installed actually using them, by day and night, in different weather conditions. The pattern of use during the day is what would be expected, i.e., very few drivers using fog lamps in clear weather but increasing percentages using fog lamps as the weather deteriorates. But at night, there is little difference in the use of fog lamps according to weather conditions, which suggests that, at night, fog lamps are primarily used to supplement conventional forward lighting. Unfortunately, the use of fog lamps with low-beam headlamps at night will certainly increase the level of glare experienced by opposing drivers, particularly when the road surface is wet because then specular reflection of

---

TABLE 12.4

Percentage of Vehicles Fitted with Fog Lamps that Had Them Lit in Different Light and Weather Conditions. No Data Was Available for Light-to-Moderate Fog at Night (from Sivak et al., 1996)

Light Condition	Clear Atmosphere	Moderate Rain	Light-to-Moderate Fog	Moderate-to-Heavy Fog
Day	2.8	10.4	30.8	50.0
Night	64.5	63.0	–	60.6

light from the road immediately in front of the vehicle makes a major contribution to glare and that is where fog lamps deliver their light.

Vehicle lighting alone clearly has its limitations for ensuring traffic safety in fog (Flannagan, 2001), so now it is necessary to consider what road lighting has to offer. Road lighting is sometimes installed in areas prone to fog on otherwise unit roads. How effective such road lighting is in enhancing visibility depends on the density of the fog. This is because adding more light will increase both the luminance of the road and objects on the road and the veiling luminance caused by scatter. Fotios et al. (2018) carried out a study of the interaction of road surface luminance and fog density using a 10:1 model of a three-lane highway. The observer simulated a driver in the middle lane using low-beam forward lighting. Ahead were two cars without rear position lamps, one in the left lane and one in the right lane. In the middle lane was an obstacle simulating a tyre that would briefly appear (300 ms). The observer had three tasks to do. One was to monitor a moving fixation marker that displayed a number between 1 and 9 for 300 ms at random intervals and to call out what the number was. The task was designed to keep the observer's attention fixed on the lane ahead. The second task was to observe when one or the other of the cars started to move across into the middle lane and report it by pressing the appropriate button on the steering wheel. The third was to report the appearance of the obstacle in the middle lane by pressing a foot pedal. Reaction times and detection rates for the car changing lanes and the obstacle appearing were collected for two road surface luminances (0.1 and 1.0 cd.m^{-2}), two spectral power distributions defined by their S/P ratios (0.65 simulating high-pressure sodium lighting and 1.40 simulating metal halide lighting), and three levels of fog density defined by their extinction coefficients (0.000, 0.005, and 0.040 m^{-1}) simulating no fog, thin fog, and thick fog. Two groups of observers of different ages (16 young observers = 18–30 years and 14 older observers = 40–70 years) were used. Statistical analysis of the results revealed no statistically significant main effects of spectral power distribution nor age groups, but there was a consistent pattern of significant main effects of road surface luminance and fog density as well as an interaction between them. The nature of the statistically significant interaction was that for the lane change, increasing the road surface luminance had no effect on the detection rate for no-fog and thin fog conditions but in thick fog increasing the road surface luminance to 1 cd.m^{-2} increased the detection rate to close to that in no fog. For the obstacle, increasing the road surface luminance to 1 cd.m^{-2} increased the detection rate for all levels of fog density but more in thick fog than in no-fog and thin fog conditions. As for the reaction times, for the lane change, increasing the road surface luminance to 1 cd.m^{-2} shortened the mean reaction time for all fog conditions but more for the thick fog condition than for the no-fog and thin fog conditions. A different pattern is apparent for the reaction time for obstacle detection. Increasing the road surface luminance to 1 cd.m^{-2} shortened the mean reaction time for all levels of fog density, but more in thick fog than in no-fog and thin fog conditions. There is also a significant interaction between the fog density and spectral power distribution. In this case, in the no-fog and thin fog conditions, there is no effect of spectral power distribution but in thick fog, the higher S/P ratio produces a shorter mean reaction time, comparable to that achieved in no fog and thin fog. These results imply that up to a road surface luminance of 1 cd.m^{-2}, road lighting can make driving in fog easier and that the spectral power distribution of the light used is a secondary factor.

Of course, this assumes that the road lighting is of the conventional type, mounted 8 m or higher above the road so that light has to pass through a lot of fog before it reaches the road. The adverse effects of such mounting heights can be limited by using catenary lighting fitted with luminaires having a high cross factor (see Section 12.3). This works because the scattering from fog particles is related to the angle of incidence of the light relative to the driver's direction of view, the reflection in the driver's direction being a minimum at 90 degrees. Thus, directing most of the light across the road rather than along it will minimize the amount of scattered light in fog.

An alternative is to mount the luminaires much closer to the road surface. Girasole et al. (1998) carried out a computer simulation, based on a mathematical model of multiple scattering, of the visibility of the lower and upper parts of the back of a truck and a broken tyre on the road, under two different forms of road lighting, in fogs of different density. The two lighting systems were conventional road lighting mounted 9 m above the road and a more closely spaced, low-mounted, pro-beam system (0.9 m above the road) placed on the central reservation, the light being aimed 45 degrees across the road. Figure 12.5 shows the estimated veiling luminance from low-beam headlamps, conventional road lighting, and the low-mounted road lighting, separately, plotted against fog concentration. Figure 12.6 shows the visibility distances for the lower and upper parts of the truck and the tyre, under low-beam headlamps and either the conventional road lighting or the low-mounted system, plotted against fog concentration. Visibility distance is defined as the distance at which the luminance difference between the target and its immediate background reaches threshold.

The low-mounted road lighting system produces much less veiling luminance than the conventional road lighting and hence increases visibility distances dramatically for low fog concentrations, but this benefit is much reduced at high fog concentrations because of the overwhelming effect of the low-beam headlamps on the veiling luminance. These results, and it must be remembered that they are based on a computer simulation rather than field measurements, suggest that new types of road lighting could be designed for use in areas where fog is common and would provide much more effective road lighting than is currently available. In fact, road lighting systems designed for mounting less than one metre above road level are commercially available for use on bridges and other locations where conventional column-mounted road lighting is inappropriate. It would be interesting to see how well such a system would work in areas prone to fog and in high traffic densities.

Finally, it is important to appreciate that there are other ways to improve the flow of visual information to the driver than simply improving the lighting. Nilsson and Alm (1996) examined the impact of a vision-enhancement system on a driver's ability to drive safely in fog, using a driving simulator. The vision-enhancement system produced a clear image of the road ahead as a small window in the scene. With the vision-enhancement system, drivers choose to drive in fog at a speed only slightly less than the speed they used in a clear atmosphere, and their reaction time and

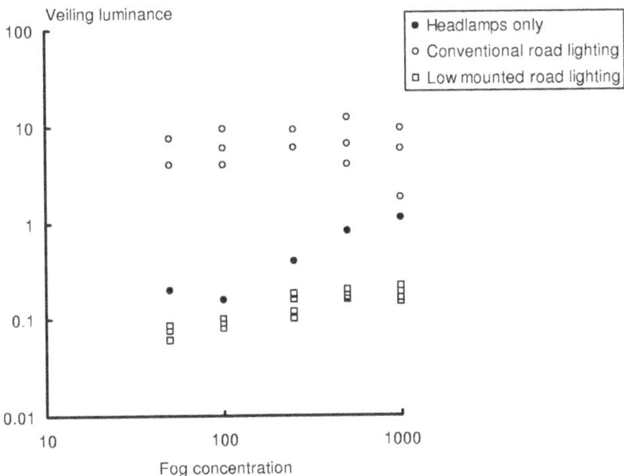

FIGURE 12.5 Veiling luminance (cd/m²) plotted against fog concentration (million droplets.m⁻³) for low-beam headlamps, conventional road lighting and low-mounted road lighting, separately. There are multiple data points for the two road lighting installations because the veiling luminance depends on the position of the driver relative to the road lighting luminaires (after Girasole et al., 1998).

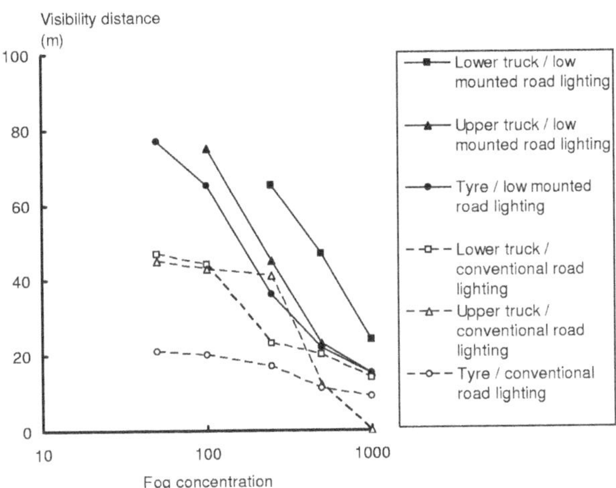

FIGURE 12.6 Visibility distance (m) for the lower and upper parts of a truck and a broken tyre seen under conventional road lighting and under low-mounted road lighting, both with low-beam headlamps, plotted against fog concentration (million droplets.m⁻³) (after Girasole et al., 1998).

the distance they moved after an unexpected hazard appeared were similar to what they were without fog. However, the lateral position of the vehicle varied more with the vision-enhancement system. Technological advances in sensors and computing power are making such a vision-enhancement system a real possibility for everyday use in vehicles. Further, automated driver assistance systems such as autonomous emergency braking are rapidly becoming standard on vehicles and should help reduce rear-end collisions, always assuming that such systems work well in fog.

12.6 SMOKE AND DUST

Smoke and dust consist of particles temporarily airborne, the size of the particles varying widely depending on the source of the smoke or dust. Photons of light incident on these particles are both scattered and absorbed. Which of these processes dominates depends on the reflectance of the particles. In black smoke and dust, absorption dominates, while in white smoke and dust, scattering dominates. Most smoke and dust are somewhere between these two extremes.

For vehicle forward lighting, black smoke and dust simply reduce the amount of light reaching the road, an effect that can be countered by using additional lamps and hence more light. White smoke and dust affect forward lighting by spreading the light distribution and creating a luminous veil that reduces the luminance contrast of everything ahead. Fog lamps will help with lane keeping in this situation, but for visibility at a distance, a high intensity narrow beam lamp mounted far away from the driver's line of sight is required. For vehicle signal lighting, the effect of both black and white dust and smoke is to reduce the maximum luminance of the signal and to spread its apparent area, the amount of spread increasing as the amount of scatter increases. These detrimental effects can be overcome by increasing the luminous intensity of the signal, as in rear fog lamps.

If thick dust is a regular occurrence and of a specific colour, then another approach for increasing visibility is to alter the spectrum of the vehicle forward lighting. For example, a driver following a road train in the outback of Australia will often be enveloped in a cloud of red dust. Lorry drivers in the outback who regularly experience such conditions use green filters on their vehicle's headlamps, the outcome being to reduce the amount of scattered light (Wordenweber et al., 2007).

12.7 LOW-ELEVATION SUN

The sun is a potent source of disability glare and so when it appears at low elevations and is aligned with the road, the driver's task is made much more difficult. Fortunately, this combination of sun elevation and road direction is temporary and occurs infrequently depending on the season of the year, time of day, and the degree of cloud cover, but when it does occur there is an increased risk of a crash. Mitra (2014) examined crash data from signalized intersections in Tucson, Arizona, and found that crashes were more frequent on east-bound roads in the morning and west-bound road in the evening than on north- or south-bound roads at any time, particularly for rear-end collisions. Hagita and Mori (2014) carried out a similar study for crashes in the Chiba prefecture in Japan and found that glare from the sun had a definite effect on crashes involving pedestrians and cyclists and at intersections. This pattern is what would be expected for the reduction in luminance contrasts of objects on the road ahead due to disability glare caused by low-elevation sun. Most vehicles are fitted with a sun visor that can be used to shield a view of the sky thereby cutting out the view of the sun and reducing the disability glare. Visibility can be enhanced even further by ensuring the windscreen is kept clean, both inside and out, so that there is no additional veil of scattered light to be added to the luminance veil caused by light scattered in the eye. The most dangerous situation in which low-elevation sun is present is when it appears unexpectedly as can occur on turning a corner or exiting a tunnel. Then, being able to deploy the sun visor rapidly is essential. .

12.8 SUMMARY

The presence of rain, snow, fog, smoke, or dust in the atmosphere and a low-elevation sun in the sky makes the driver's task more difficult, but in different ways. Rain has a direct effect on visibility by absorbing and scattering light and an indirect effect by changing the reflection properties and friction of the road surface. Snow scatters incident light and in so doing creates a lot of visual noise as well as obscuring road markings and reducing road surface friction. Fog, smoke, and dust have their effects by absorbing and scattering light in the atmosphere, thereby reducing the visibility of everything ahead of the driver. Low-elevation sun is a potent source of disability glare and hence causes a reduction in visibility.

The presence of rain, snow, fog, dust, or smoke undoubtedly makes the driver's task more difficult, but does that lead to more crashes? A study of fatal crashes in the USA shows that almost 90% of crashes involving death occur when there is no adverse weather. This is partly a matter of exposure and partly a matter of driver experience and behaviour. Really adverse weather occurs infrequently and, when it does, drivers tend to reduce speed and keep a greater distance from nearby vehicles, behaviours that make fatalities less likely.

Rain consists of falling water droplets. Unless the rain is heavy or the windscreen wipers are ineffective, the main impact of rain on visibility is due to the change in the reflection characteristics of the road surface. When rain covers a road surface, the visual effect is to make the road surface more specular. For vehicle forward lighting, this change in reflection properties reduces the average road surface luminance, increases the disability and discomfort glare from approaching vehicles and reduces the visibility of the road markings. For road lighting, the effect of a wet road is to make the road surface luminance distribution less uniform.

The problems posed by wet roads can be reduced either by changing the road surface or by modifying the vehicle's forward lighting or the road lighting. The most successful approach to the problem of wet roads lies in the choice of road materials. A road with a coarse texture or a more permeable structure will be less sensitive to rain. For vehicle forward lighting, the proposed adaptive forward lighting system has as one of its options a wet road beam. The wet road beam involves a reduction in the illumination just in front of the vehicle and increased light to the sides of the vehicle, the effect being to diminish discomfort and disability glare to other drivers. For road lighting, the decrease in luminance uniformity can be minimized by changing the luminous intensity distribution of the luminaires, the spacing between luminaires, and their position relative to the road surface.

Snow consists of large, high-reflectance particles falling through the atmosphere. If the density of snowflakes is very high, the result of scattering is a uniform veil of high luminance that restricts visibility in all directions to very short distances. At lower densities, each snowflake is seen separately moving against a lower luminance background. How distracting this is depends on the luminance contrast of the snowflakes, the higher the luminance contrast, the more distracting the snowflakes. Maximum luminance contrast will occur when the vehicle forward lighting is the only source of light. During daylight or at night when road lighting is present, the background against which the snowflakes are seen is higher in luminance and thus the snowflakes are at a lower luminance contrast.

There are four approaches that can be used to reduce the distraction caused by snowflakes. The one most easily available to the driver is to use the forward lighting on low beam rather than on high beam. Another is to place the forward lighting as far away from the driver's usual line of sight as possible. Yet another is to narrow the luminous intensity distribution of the headlamp. The fourth is to modify the spectrum emitted by the forward lighting so as to reduce the stimulus to the rod photoreceptors. Of these, the most potent are to displace the light source as far away from the driver as possible and to narrow the luminous intensity distribution.

Fog consists of small water droplets suspended in the atmosphere. Photons of light incident on a droplet are absorbed and scattered. If the density of the particles is high, the outcome of these processes is a uniform veil of luminance covering the driver's visual field. The effect of this uniform luminance veil is to reduce the luminance contrasts of all the things in front of the driver and hence to reduce their visibility.

There are a number of ways to help the driver in fog but none of them is a complete solution. One approach is to inset retro-reflective road studs into the road at regular intervals, as lane and edge markers. Another is for the driver to use low-beam rather than high-beam headlamps. Yet another is to use dedicated fog lamps. Where fog lamps are fitted they are usually mounted low on the vehicle and have a luminous intensity distribution that is both wide and flat. The visual effect of fog lamps is to provide greater visibility of road edges and nearby lane markings but they do little to enhance the visibility of vehicles and objects further down the road. This may explain why the most common form of crash in a fog is a rear-end collision. One method to reduce such crashes is to fit vehicles with high intensity rear lamps for use in fog. These are effective because they increase the distance from which the vehicle can be detected. However, care is required to limit their use to foggy conditions because in a clear atmosphere their luminous intensity is high enough that they become a source of discomfort glare and sometimes disability glare to following drivers. As for road lighting, how effective this is in enhancing visibility in fog depends on the density of the fog. This is because adding more light will increase both the luminance of the road and objects on the road and the luminance of the luminous veil caused by scatter. There is evidence that up to a mean road surface luminance of 1 cd.m^{-2} road lighting is beneficial, even in thick fog.

Smoke and dust consist of small particles temporarily airborne. Photons of light incident on these particles are both scattered and absorbed. Which of these processes dominates depends on the reflectance of the particles. In black smoke and dust, absorption dominates, while in white smoke and dust, scattering dominates. For vehicle forward lighting, black smoke and dust simply reduce the amount of light reaching the road, an effect that can be countered by using additional lamps. White smoke and dust affect forward lighting by spreading the light distribution and creating a luminous veil that reduces the luminance contrast of everything ahead. For vehicle signal lighting, the effect of both black and white smoke and dust is to reduce the maximum luminance of the signal and to spread its apparent area, the amount of spread increasing as the amount of scatter increases. These detrimental effects can be overcome by increasing the luminous intensity of the signal, as in rear fog lamps.

Low-elevation sun that is aligned with the direction of travel causes disability glare and thereby reduces the visibility of what lies ahead. The rapid deployment of a sun visor to restrict the view of the sky is effective, but it is also useful to ensure that the windscreen is clean, both inside and out. If it is not, another luminance veil will be created to add to that produced by light scatter in the eye.

13 Human Factors

13.1 INTRODUCTION

Driving is a human activity with all the variability that such a statement implies. While driving, humans have to extract information about the world around them, decide on an appropriate course of action, and then implement that course of action. At night or in adverse weather, it may be difficult to extract the relevant information. By day or night, decisions have to be taken quickly. Sometimes, multiple events requiring attention occur simultaneously, so there is competition for cognitive resources. Some people are better equipped to deal with such demands than others. Among the factors that determine how well drivers can respond to such demands are the amount of practice they have had and the capabilities of their visual and cognitive systems. How such individual differences affect driving and how lighting might be used to alleviate any consequent problems are the subjects of this chapter.

13.2 AGE AND CRASHES

One factor that is connected to both driving experience and the deterioration in visual and cognitive abilities is age. Driving on public roads is not allowed until the late teens; the exact age at which driving is allowed varies from country to country. As for deterioration in visual and cognitive abilities, this is continuous from the late teens but accelerates dramatically about the sixth decade of life (Kosnik et al., 1988; Werner et al., 1990). Figure 13.1 shows the crash rate per 100 million miles driven by various age classes for fatal crashes in the USA in 2014–2015 (Tefft, 2017).

From Figure 13.1 it is evident that the numbers killed per 100 million miles travelled show an increase at both extremes of age. The increase at the young age is greater than shown in Figure 13.1 because of the narrow age ranges used for the young in the histogram. The reverse is the case for the old age group because of the wider age range used. Nonetheless, an increase in deaths among young and old drivers is consistent with earlier data collected in the USA (Massie and Campbell, 1993) and later data from the UK (Department for Transport, 2022d). Regev et al. (2018) examined fatal and non-fatal crashes that occurred in Great Britain, over the years 2002–2012, dividing the data into single-vehicle and two-vehicle crashes for different age groups occurring at different times of the day. Table 13.1 shows the number of fatal and non-fatal single-vehicle crashes occurring during the day, evening, and at night, for males and females in different age groups. Table 13.2 shows the same information but for two-vehicle crashes. Daytime was taken as between the hours of 06:00 and 18:00, evening from 18:00 to 21:00, and night from 21:00 to 06:00.

DOI: 10.1201/9781003388906-13

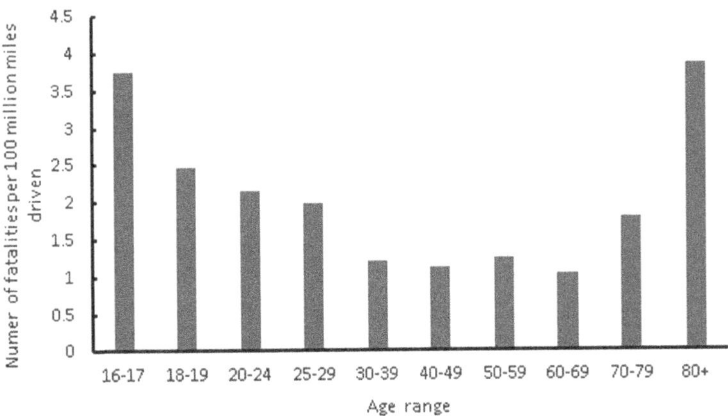

FIGURE 13.1   The number killed in road crashes per 100 million miles driven, classified by age, in the USA, 2014–2015. Note the unequal nature of the age classes (from Tefft, 2017).

There are four features of Tables 13.1 and 13.2 that deserve attention. The first is that females have fewer fatal and non-fatal crashes than males, for all age groups, for the same time of day. The second is that the younger age groups tend to have more fatal and non-fatal crashes than the older age groups. The third is that the younger age groups have more single-vehicle fatal crashes at night than by day, but the reverse is true for the older age groups. For two-vehicle crashes, all age groups have more fatal crashes by day than in the evening or at night. This is reasonable because there are more vehicles about to crash into during the day. The fourth is that there is, at best, a suggestion of an increase in fatal crashes for the oldest age group. However, the oldest age group is limited to 70+ years so this may be too young a cut-off to reveal the effect of really old drivers, as suggested by Figure 13.1.

The data presented in Tables 13.1 and 13.2 are derived from actual crash numbers. In order to determine if such data really justifies the belief that young and very old drivers are a greater risk to themselves and to other road users than those in middle age, these numbers have to be adjusted for exposure. The conventional way to make this adjustment is to divide the number of fatalities or injuries associated with a specified group by the number of miles travelled by the group or the number of licenced drivers in the group or some combination of the two. This is reasonable if the relationship between crash frequency and distance travelled or number of drivers in the group is linear. Unfortunately, there is evidence that the relationship between crash frequency and driving exposure is more likely to be logarithmic than linear (Janke, 1991; Elander et al., 1993). As a result, there is concern that crash rates for groups that differ markedly in their exposure may be biased so that the crash risk for groups with low exposure may be exaggerated and those with high exposure may be underestimated. This is known as the low mileage bias. Both very young and very old drivers are known to have lower driving exposure than other age groups. Regev et al. (2018) examined this possibility by correcting for exposure based on the

TABLE 13.1

The Number of Fatal and Non-fatal Single-vehicle Crashes Occurring during the Day, Evening and Night, for Males and Females in Different Age Groups in Great Britain over the Years 2002–2012. Daytime Is between the Hours of 06:00 and 18:00, Evening from 18:00 to 21:00 and Night from 21:00 to 06:00 (from Regev et al., 2018)

Age Range	Total Fatal Crashes	Fatal Crashes: Day	Fatal Crashes: Evening	Fatal Crashes: Night	Total Non-fatal Crashes	Non-fatal Crashes: Day	Non-fatal Crashes: Evening	Non-fatal Crashes: Night
Male								
17–20	70	16	10	44	5,270	2,081	938	2,251
21–29	117	35	16	66	8,064	4,101	1,347	2,616
30–39	91	40	16	35	7,557	4,702	1,218	1,637
40–49	57	28	10	19	6,311	4,251	948	1,112
50–59	36	17	7	12	4,255	3,086	591	578
60–69	19	12	2	5	2,202	1,682	286	234
70+	23	19	2	2	1,444	1,188	155	101
Female								
17–20	10	3	1	6	1,859	949	342	568
21–29	15	6	2	7	3,218	2,069	539	610
30–39	12	6	2	4	2,978	2,228	426	324
40–49	9	6	1	2	2,379	1,831	333	215
50–59	5	3	1	1	1,369	1,079	182	108
60–69	2	2	0	0	672	553	79	40
70+	6	6	0	0	547	474	50	23

TABLE 13.2

The Number of Fatal and Non-fatal Two-vehicle Crashes Occurring during the Day, Evening and Night, for Males and Females in Different Age Groups in Great Britain over the Years 2002–2012. Daytime Is between the Hours of 06:00 and 18:00, Evening from 18:00 to 21:00 and Night from 21:00 to 06:00 (from Regev et al., 2018)

Gender and Age Range	Total Fatal Crashes	Fatal Crashes: Day	Fatal Crashes: Evening	Fatal Crashes: Night	Total Non-fatal Crashes	Non-fatal Crashes: Day	Non-fatal Crashes: Evening	Non-fatal Crashes: Night
Male								
17–20	68	34	13	21	6,964	4,263	1,462	1,239
21–29	120	66	24	30	12,578	8,449	2,342	1,787
30–39	125	80	21	24	12,571	9,178	2,022	1,371
40–49	112	74	20	18	10,085	7,651	1,501	933
50–59	76	54	10	12	6,603	5,173	913	517
60–69	45	34	5	6	3,682	3,011	452	219
70+	75	64	6	5	3,096	2,671	307	118
Female								
17–20	13	8	2	3	2,741	1,833	551	357
21–29	24	17	3	4	5,694	4,256	952	486
30–39	19	13	3	3	5,506	4,423	778	305
40–49	21	14	4	3	4,487	3,615	635	237
50–59	17	13	2	2	2,742	2,229	373	140
60–69	13	10	2	1	1,401	1,168	172	61
70+	25	22	2	1	1,212	1,084	111	17

assumption that exposure should be high if there are a lot of drivers in a group and the number of trips they make are many and low if the number of drivers is small and the trips are few. When this method is applied to the number of fatal and non-fatal single-vehicle and two-vehicle crashes occurring during the day, evening, and night, for males and females, the risks of crash involvement associated with very young and old drivers are much reduced. The highest risk of crash involvement is associated with drivers in the age range 21–29 years and the lowest with drivers in the age range 60–69 years. Interestingly, this is broadly consistent with the number of fatalities for single-vehicle and two-vehicle crashes given in Tables 13.1 and 13.2, respectively.

Another way to consider the effect of age is to examine the contributory factors identified for crashes in which drivers of different ages are involved. Table 13.3 compares the number of contributory factors identified by police officers attending crashes involving young drivers (17–24 years) and older drivers (60+ years) in Great Britain in 2013 (Department for Transport, 2015d). The same contributory factors are assigned to both age groups but there is an undeniable shift in emphasis between young and old drivers. Both age groups are identified as failing to look properly, failing to judge another person's path, and making a poor turn or manoeuvre, but younger drivers are more likely to be identified as inexperienced, driving in a careless or reckless manner, travelling too fast for the conditions, and indulging in sudden braking or exceeding the speed limit. Conversely, older drivers tend to be involved in more crashes where illness or disability, mental or physical, is considered a contributory factor. There is a hint of this pattern in Table 13.1 where fatalities for young drivers are much higher at night for single-vehicle crashes than older drivers. This is not the case for two-vehicle crashes (Table 13.2) where the vast majority of crashes occur by day, although young drivers still suffer more fatalities than older drivers. Crudely put, what this pattern suggests is that young drivers have no sense and old drivers have no senses. As will become evident later, this is a gross exaggeration, but it does contain a kernel of truth. Crashes in which young drivers are involved tend to be caused by a failure of judgement or an excess of ambition while those involving older people tend to be linked to failures of perception and cognitive processing under time pressure (Gomes-Franco et al., 2020). Why this should be and how lighting might contribute to a reduction in such crashes will now be considered.

13.3 YOUNG DRIVERS

Table 13.1 suggests that it is drivers below 30 years of age who show a marked increase in crash fatalities, particularly at night. Therefore, 30 years of age will be used to define what constitutes a young driver. Before considering how lighting might be used to reduce the higher crash rates of young drivers it is necessary to consider why young drivers are involved in so many crashes. The first thing to say is that it is not likely to be due to visual or cognitive limitations. Young drivers who are in their late teens or twenties will usually have a visual system at the peak of its powers and a cognitive system capable of rapid responses to simple and complex stimuli. One possible reason why young drivers have a higher crash rate is that they may not know how to use these capabilities to allocate attention because they are inexperienced. In most countries, passing a driving test requires the applicant to show knowledge of the rules of the road, the meaning of signs and the expected behaviour in specified situations, as well as a practical demonstration of vehicle handling in a limited number of conditions. What is only sketchily examined in the driving test is the ability to rapidly analyse the situation on the road ahead and to allocate attention appropriately, although the video hazard perception tests now used in some countries are a step in the right direction (Omran et al., 2023). More usually, the ability to allocate attention effectively is gained by experience of driving over the following years. This may be why there is a peak in crashes within one year of drivers passing their driving test (Maycock et al., 1991). Further, there is evidence that some types of crashes become less frequent as drivers gain experience, particularly those involving turning across traffic and rear-end collisions (Clarke et al., 2006).

The effect of experience on the allocation of attention should be evident in the patterns of eye movements made by experienced and inexperienced drivers. Falkmer and Gregersen (2005) report an analysis of the fixations made by experienced and inexperienced drivers while travelling along an urban road in which there were vehicles parked on both sides, a pedestrian crossing, and a speed bump as well as on a rural dual carriageway crossing an intersection. They showed that inexperienced drivers tended to fixate more frequently on in-car objects, keep their fixations within a narrower horizontal band, and fixate slightly more frequently on relevant traffic cues and potential hazards than experienced drivers. Such results hardly suggest that inexperienced drivers are at a disadvantage regarding the use of their visual system. However, Falkmer and Gregersen (2005) also found that experienced drivers altered their pattern of fixations between the urban road and the rural road while the inexperienced drivers did not. Such flexibility may be of advantage given the diversity of conditions that a driver is likely to experience.

Crundall and Underwood (1998) also measured the eye movements of drivers who had recently passed the driving test and young but experienced drivers when driving on a straight rural road, a busy suburban road, and a busy dual carriageway. Again, there were clear differences between the two groups with the older, more experienced drivers showing a wider spread of fixations. This was particularly evident on the dual carriageway where the experienced drivers fixated on traffic in adjacent lanes while the inexperienced tended to concentrate on the lane ahead. There was also a difference in fixation duration. On the rural road where there were few hazards, the experienced drivers made longer fixations than the inexperienced drivers but on the dual carriageway, where there were many hazards, the experienced drivers made a larger number of fixations of shorter duration than the inexperienced drivers. A proposed explanation for such differences is that experienced drivers develop a mental model of the hazards that are likely to occur in different situations (Liu, 1998). This mental model is the endogenous source that generally drives fixation patterns.

So what is the role of lighting, if any, in reducing the crash rate of inexperienced drivers? One answer is to use lighting to speed up the development of a comprehensive mental model. Road lighting could do this by emphasizing locations where hazards are common, e.g., by special lighting for pedestrian crossings and lighting intersections on otherwise unlit roads (see Sections 11.3 and 11.6). Another would be to reduce the mental load on the driver by giving more emphasis to visual guidance in road lighting design. As for vehicle lighting, there is little that can be done that is not also applicable to experienced drivers. Changing both forward lighting and signal lighting to enhance visibility and conspicuity and to more clearly inform other drivers about intentions would be of benefit to all drivers.

While lack of experience is a factor in crash rates for some young drivers, not all young drivers are inexperienced. Anyone who gains a driving licence in their late teens will have several years of driving experience by their mid-twenties but might still be classed as young in crash statistics. The reasons why experienced young drivers have high crash rates are evident from Table 13.3. Basically, the contributory factors where young drivers predominate suggest that they often drive with reduced margins of error, relying on their fast reactions to survive, i.e., are willing to take more risks. Further, young drivers are more likely to be out late at night and more likely to be suffering from the effects of alcohol or drugs, both factors which will tend to slow responses and confuse analysis (Department for Transport, 2015d). In addition, there is social aspect to crashes involving young drivers. Ouimet et al. (2015) conducted a review of the epidemiological literature on fatal crashes involving young drivers, with and without passengers. They found that the presence of passengers increased the likelihood of a fatal crash, particularly when the passengers were male. There is little that lighting can do to overcome such behavioural excesses and social pressures other than to provide more time for decision-making by lighting the road further ahead.

TABLE 13.3

The Number of Contributory Factors Identified by Police Officers Attending Crashes Involving Young Drivers (17–24 years) and Old Drivers (60+ Years) in Great Britain in 2013. Up To Six Contributory Factors Can Be Identified for Each Crash (from Department for Transport 2015d)

Contributory Factor	Young (17–24 Years)	Old (60+ Years)
Failed to look properly	9,402	9,693
Failed to judge another person's path	5,842	5,422
Careless, reckless, or in a hurry	4,397	2,724
Loss of control	4,153	2,310
Poor turn or manoeuvre	3,416	3,153
Travelling too fast for conditions	2,562	988
Learner or inexperienced driver	2,630	412
Slippery road due to weather	3,055	1,289
Sudden braking	2,077	1,297
Exceeding speed limit	1,746	626
Illness or disability, mental or physical	203	1,144
Total	26,628	21,994

13.4 OLD DRIVERS

Tables 13.1 and 13.2 take an old driver to be one greater than 70 years of age. This is the definition used here. Old drivers are of concern because ageing affects all the functions required for driving: visual, cognitive, and motor. Because the subject of this book is lighting, the changes in visual capabilities will be emphasized here but the changes in the other functions should not be forgotten. How important the limitations in the visual, cognitive, and motor functions are will depend on the circumstances. When driving at night, deficiencies of the visual system are likely to be the most important. When driving during the day, the cognitive limitations are likely to be dominant.

13.4.1 OPTICAL CHANGES IN THE VISUAL SYSTEM WITH AGE

The human visual system can be considered as an image-processing system. Like all such systems, the visual system is most effective when it is operating at an appropriate sensitivity with a clear retinal image to process. The optical factors determining the amount of light reaching the retina and hence the sensitivity of the visual system are the pupil size and the spectral absorption of the components of the eye. The area of the pupil varies as the amount of light available changes, the pupil opening to admit more light when there is little and closing when there is plenty. The ratio of maximum to minimum pupil area decreases with age, the maximum decreasing much more than the minimum (Weale, 1992). This means the old are much less able to compensate for low light levels by opening their pupils than are the young. Another change relevant to the operating state of the visual system is the amount of light absorbed in passage through the eye. Such absorption increases with age, the majority taking place on passage through the lens (Murata, 1987), but this absorption is not equal at all wavelengths. Rather, the absorption at short wavelengths increases dramatically with age (Weale, 1992). This goes some way to explain the diminished colour discrimination capabilities of old people. Together, the reduction in pupil size and the increased absorption of light reduces the retinal illumination of old people, so much so that it has been estimated that a 70-year-old receives only one-third of the light at the retina that a 30-year-old will receive when both receive the same illuminance at the cornea (Adrian, 1995).

The optical factors determining the clarity of the retinal image are the ability to focus the image on the retina and the amounts of scattered light and straylight in the eye. In simple optical terms, the eye has a fixed image distance and a variable object distance. To bring objects at different distances to focus on the retina, the optical power of the eye has to change. The range of object distances that can be brought to focus on the retina decreases with age because of the increasing rigidity of the lens. After about 60 years of age, the eye is virtually a fixed focus optical system. Spectacles or contact lenses are used to modify the optical power of the eye, the prescription of the spectacles or contact lens changing over the years as the lens becomes increasingly rigid. As for light scatter, the amount of light scatter increases with age, mainly due to changes in the lens (Wolf and Gardiner, 1965). Scattered light degrades the retinal image by reducing the difference in luminance on either side of an edge, thereby reducing the magnitude of its higher spatial frequencies. Scattered light also degrades the retinal image in terms of colour by adding wavelengths from one area to another, thereby reducing the colour difference at the edge.

Scatter can be quantified by a point spread function, which typically shows that the amount of scattered light decreases with increasing deviation from the beam of light being scattered (Vos and Boogaard, 1963). Straylight within the eye is caused by light back-reflected from the retina and pigment epithelium, by transmission of light through the iris and the eye-wall, and by lens fluorescence (Boynton and Clarke, 1964; van den Berg et al., 1991; van den Berg, 1993). Virtually all these characteristics worsen with age (Werner et al., 1990; Weale, 1992). Straylight matters because it falls uniformly across the retinal image, thereby reducing the luminance contrast of all edges and the saturation of all colours in the image.

These changes with age are the best that can be expected but with increasing age comes a greater likelihood of pathological changes leading to visual impairment (Silverstone et al., 2000). The four most common causes of visual impairment in developed countries are cataract, macular degeneration, glaucoma, and diabetic retinopathy. Cataract is an opacity developing in the lens. The effect of cataract is to absorb and scatter more light as the light passes through the lens. This increased absorption results in reduced visual acuity and reduced contrast sensitivity over the entire visual field, as well as greater sensitivity to glare. Macular degeneration occurs when the macula, which covers the fovea, becomes opaque. An opacity immediately in front of the fovea implies a serious reduction in visual acuity and in contrast sensitivity at high spatial frequencies. Typically, these changes make seeing detail difficult if not impossible. However, peripheral vision is unaffected. Glaucoma is shown by a progressive narrowing of the visual field. Glaucoma is due to an increase in intraocular pressure which damages the blood vessels supplying the retina. Glaucoma will continue until complete blindness occurs unless the intraocular pressure is reduced. Diabetic retinopathy is a consequence of chronic diabetes mellitus and effectively destroys parts of the retina through the changes it produces in the vascular system that supplies the retina. The effect these changes have on visual capabilities depends on where on the retina the damage occurs and the rate at which it progresses.

13.4.2 Neural Changes of the Visual System with Age

The optical changes that occur with age, both normal and pathological, affect the retinal image, but for the visual system to be effective, the retinal image has to be processed by the retina and the visual cortex of the brain. There is no reason to suppose that ageing is limited to the optical elements of the visual system alone. Indeed, morphological changes have been reported in rod and cone photoreceptors in older people (Marshall et al., 1979); the density of rods and cone photoreceptors has been shown to decrease in extreme old age (Feeny-Burns et al., 1990); and the number of ganglion cells in the retina and the number of neurones in the visual cortex both decrease with increasing age (Devaney and Johnson, 1980; Balazsi et al., 1984). Weale (1992) provides a useful review of these neural changes and their possible causes. The fact that the neural elements of the visual system also show changes with age is important because it implies that the compensation for visual system ageing that can be provided by lighting is inevitably limited.

13.4.3 OTHER CHANGES WITH AGE

In addition to the changes in the visual system, ageing introduces more general limitations with respect to decision-making and motor functions. For example, reaction times to all stimuli increase with age and the greater the number of alternatives, the greater the increase in reaction time with age (Sivak et al., 1995). This implies that old drivers need more time to make a safe decision. In addition, old people are more easily distracted and have difficulties in directing attention between several different stimuli (Fildes et al., 1994; Sivak et al., 1995). As for motor functions, muscular strength and joint flexibility decrease rapidly in old age and manual dexterity is diminished (Sivak et al., 1995). There are also changes in the circadian timing system. With increasing age, the amplitude of the circadian timing rhythm diminishes (Brock, 1991; Copinschi and van Cauter, 1995), the period shortens, and the phase advances (Renfrew et al., 1987; Czeisler et al., 1988). The overall effect is to diminish the ability to synchronize the circadian rhythm to the external environment with consequences for many basic physiological functions (Turner et al., 2010). In the absence of clear signals from the suprachiasmatic nuclei, various organs of the body can become uncoordinated, resulting in biochemical confusion. This confusion can lead to disruption of the circadian timing rhythm resulting in declines in alertness, cognitive functioning, mood, and sleep problems (Turner et al., 2010). Lighting has a role to play in minimizing circadian disruption by providing bright light exposure during the day, either daylight or electric light, and dim light at night (Boyce, 2014; Foster, 2022). For drivers, the most common situation where light is encountered at night is when driving on lit roads. However, Gibbons et al. (2022) have used a test track to show that road lighting designed to the US standards for expressways does not have any effect on melatonin concentrations of drivers after two hours exposure at night, indicating that as far as the circadian timing system is concerned road lighting is dim lighting. Alshdaifat et al (2024) have reached a similar conclusion in a laboratory study simulating drving and walking for one hour at night under road lighting.

13.4.4 CHANGES IN VISUAL CAPABILITIES

As might be expected, the changes in the optical and neural characteristics of the visual system that occur with increasing age have an impact on what the visual system is capable of doing (Owsley and McGwin, 2010). A consideration of the effects of age on threshold performance reveals that the effect of increasing age is almost always negative, in the sense that the visual system becomes less discriminating, and more sensitive to adverse conditions. Specifically, old people tend to show reduced visual field size, reduced visual acuity, reduced contrast sensitivity, increased sensitivity to glare, slower recovery from exposure to glare, and degraded colour discrimination.

Figure 13.2 shows the mean radius of the visual field for five-year age groups for off-axis detection alone and with counting of the number of times the fixation point goes off as well. The background luminance of the perimeter was 13 cd.m^{-2} (Haegerstrom-Portnoy et al., 1999). The decrease in visual field with age is clear, but the decrease is much greater when attention has to be divided between the fixation point and the periphery, suggesting that driver distraction is likely to be a significant factor in many crashes (Guo et al., 2016).

Figure 13.3 shows the median distance visual acuity for two-year age groups, expressed as the minimum angle of resolution in minutes of arc, for high- and low-luminance contrasts at a background luminance of 150 cd.m^{-2} (Haegerstrom-Portnoy et al., 1999). Distance visual acuity worsens with increasing age for both high- and low-luminance contrast stimuli but the deterioration is much worse for the low contrast stimuli, particularly after 85 years of age.

Figure 13.4 shows the median contrast sensitivity for two-year age groups measured using the Pelli–Robson chart (Pelli et al., 1988) at a luminance of 150 cd.m^{-2} (Haegerstrom-Portnoy et al., 1999). Unlike visual acuity charts that show high contrast letters of different sizes, the Pelli–Robson chart shows large letters of a fixed size that vary in luminance contrast. Figure 13.4 shows a steady decline in contrast sensitivity from 60 years of age.

Figure 13.5 shows the median number of letters lost from the Berkeley Glare Test due to disability glare for two-year age groups (Haegerstrom-Portnoy et al., 1999). The Berkeley Glare Test (Bailey and Bullimore, 1991) consists of a triangular letter chart illuminated from the front to 80 cd.m^{-2} and surrounded by a translucent panel with a luminance of 3300 cd.m^{-2}. The letters on the chart are of low luminance contrast (0.10). The difference in the number of letters read with and without glare is the number of letters lost. Figure 13.5 shows that the number of letters lost increases above 70 years, the number increasing rapidly above 85 years of age.

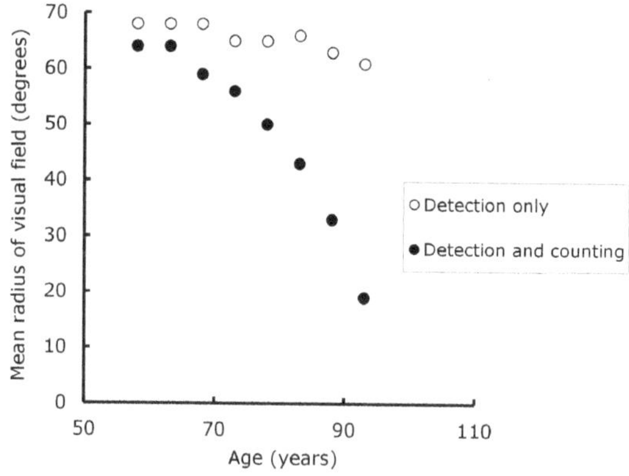

FIGURE 13.2 Mean radius of the visual field, measured in degrees, for five-year age groups for two conditions of off-axis detection alone and with counting the number of times the fixation point goes off as well. The background luminance of the perimeter was 13 cd.m^{-2} (after Haegerstrom-Portnoy et al., 1999).

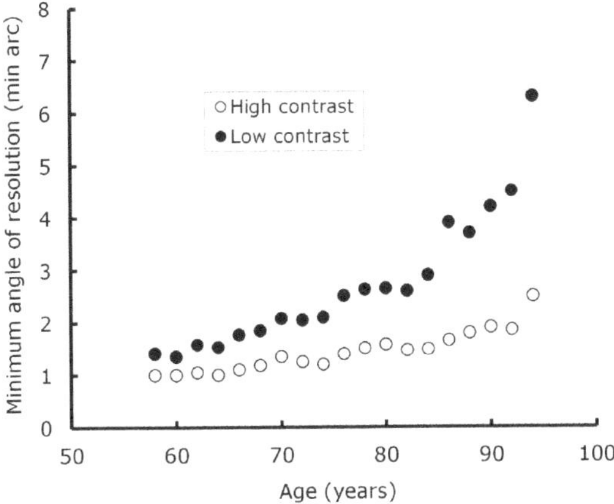

FIGURE 13.3 Median distance visual acuity for two-year age groups using a Bailey–Lovie chart and expressed as minimum angle of resolution in minutes of arc for high- (0.90) and low- (0.17) luminance contrasts at a background luminance of 150 cd.m^{-2} (after Haegerstrom-Portnoy et al., 1999).

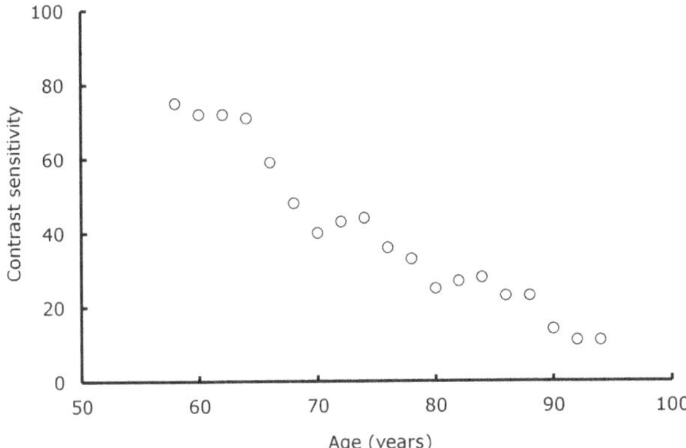

FIGURE 13.4 Median contrast sensitivity for two-year age groups measured using the Pelli–Robson chart at a luminance of 150 cd.m^{-2} (after Haegerstrom-Portnoy et al., 1999).

Figure 13.6 shows the median time taken to recover from exposure to glare for five-year age groups. The SKILL near acuity test (Haegerstrom-Portnoy et al., 1997) was used at a luminance contrast of 0.15 and with a background luminance of 15 cd.m^{-2}. The observer was required to look directly at the Berkeley Glare Test glare source of luminance 3300 cd.m^{-2} for 1 min. The glare source was then turned off and the time taken for the observer to reach a level of visual acuity within two lines of their individual threshold measured without glare was recorded. Figure 13.6 shows that glare recovery time increases dramatically above 85 years of age.

Figure 13.7 shows the mean colour confusion score for five-year age groups, measured on the Farnsworth Panel D-15 colour arrangement test (Farnsworth, 1947)

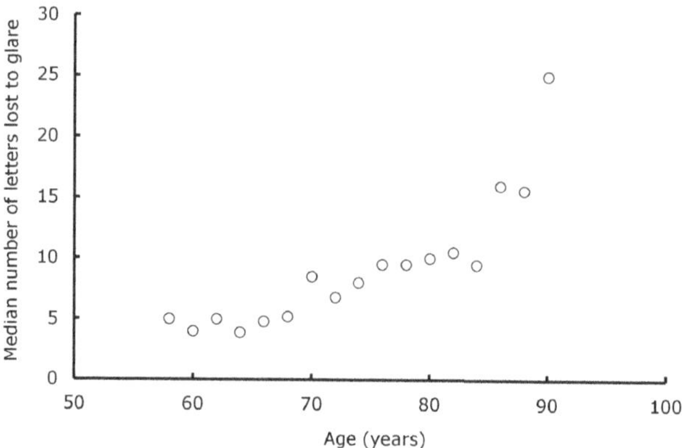

FIGURE 13.5　Median number of letters lost from the Berkeley Glare Test due to disability glare, for two-year age groups (after Haegerstrom-Portnoy et al., 1999).

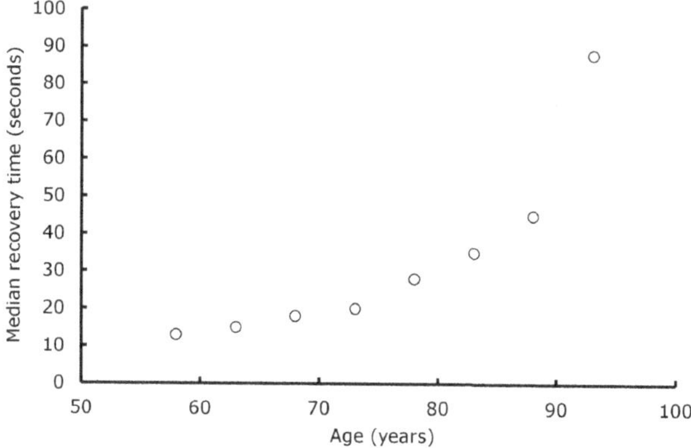

FIGURE 13.6　Median time taken in seconds, to recover visual acuity, following 1-min exposure to a glare source luminance of 3300 $cd.m^{-2}$ for five-year age groups. The SKILL near acuity test was used at a luminance contrast of 0.15 with a background luminance of 15 $cd.m^{-2}$ (after Haegerstrom-Portnoy et al., 1999).

using illuminant C to provide approximately 100 lx (Haegerstrom-Portnoy et al., 1997). For this test, males with defective colour vision were excluded. This test requires people to arrange a series of 15 discs of equal lightness and chroma but different hue into a consistent hue line, i.e., into a line in which the difference in hue between adjacent disks is a minimum. Performance on the test is scored by the distance covered in colour space by a line joining the adjacent disks. The distance for a perfect arrangement scores zero. A distance which is twice that of a perfect arrangement scores 100. From Figure 13.7 it can be seen that the ability to discriminate hues deteriorates rapidly above 75 years of age.

Two features of Figures 13.2 to 13.7 are important. The first is that there is a consistent trend of deteriorating visual abilities as age increases above 60 years of

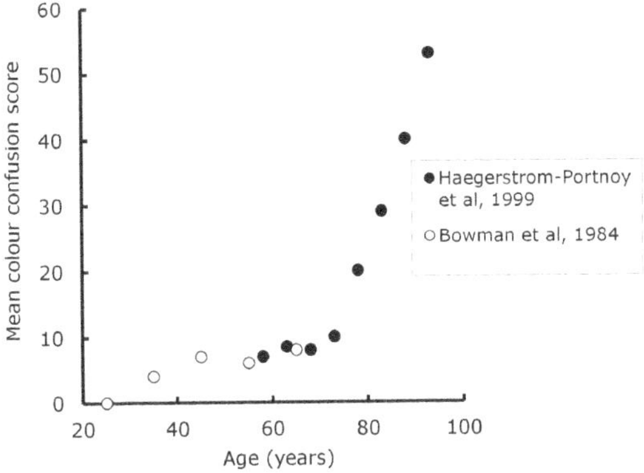

FIGURE 13.7   Mean colour confusion score for five-year age groups, measured on the Farnsworth Panel D-15 colour arrangement test using illuminant C to provide approximately 100 lx (after Haegerstrom-Portnoy et al., 1999. Also shown are mean data for the same test but using younger people (from Bowman et al., 1984).

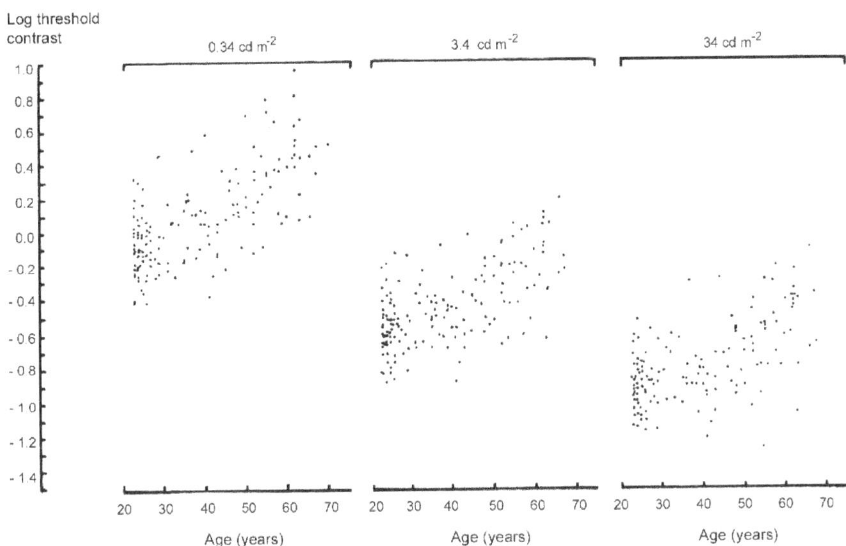

FIGURE 13.8   Log threshold luminance contrasts for individuals of different ages (years) at three different background luminances (cd.m²) (after Blackwell and Blackwell, 1971).

age. The second is that this deterioration accelerates dramatically above 80 years of age. It is also important to appreciate that there are wide individual differences in visual function. This is evident in Figure 13.8, which shows the threshold luminance contrast measured for individuals of different ages at three different adaptation luminances (Blackwell and Blackwell, 1971). While the trend of increasing threshold contrast with increasing age is obvious, it is also clear that the individual differences are large. It should not be assumed that all old drivers, particularly those in their sixties, are visually challenged.

13.4.5 Consequences for Driving

The changes that occur in visual capabilities with age have consequences for driving performance. Wood (2002) measured the driving performance of groups of drivers of different ages and with different degrees of visual impairment. Specifically, 139 drivers in good general health and holding a current Queensland, Australia, driving licence were divided into five groups, labelled young, middle-aged, and old, this last group being subdivided into those with normal vision, mild visual impairment, and moderate or severe visual impairment. The mean ages of the three age groups were 27, 52, and 70 years for the young, middle-aged and old groups, respectively. For the old group with normal, mildly impaired, and moderately impaired vision, the mean ages were 69, 71, and 71 years, respectively. Normal vision was defined as having a static visual acuity of 6/7.5 or better (see Section 3.4.1). Mildly impaired vision was defined as having slight clouding of the lens, early glaucoma, or early macular degeneration in one or both eyes. Moderate to severe visual impairment was defined as having cataract in both eyes or advanced glaucoma or macular degeneration in one or both eyes. All these drivers drove round a 5.1 km (3.2 miles), closed-road circuit, i.e., one closed to the public and hence which was free of other traffic. While driving round the circuit, the drivers were asked to report any road signs they saw; to report any large low-contrast hazards they saw in the road and to avoid them by steering around them; to judge whether or not the gap between a pair of cones was wide enough to get through and, if it was, to drive through it and, if not, to drive around it; and to respond to the onset of one of five LEDs mounted in the car in front of the driver. Having driven round the circuit, the drivers' ability to handle the vehicle was tested by having them manoeuvre in and out of a row of low-contrast cones and reverse into a parking space. Table 13.4 gives the measures of driving performance that showed statistically significant differences between the groups. As would be expected, both age and visual impairment tend to produce worse driving performance but the balance between these two factors changes with the nature of the task. For tasks that involve switching attention, such as detecting the onset of the high-contrast LED stimulus, age is the dominant factor. For tasks where visibility is limited, such as detecting and avoiding low-contrast road hazards, visual impairment is more important. For other tasks, such as seeing road signs and reversing, both age and visual impairment are influential.

Wood (2002) also constructed a composite driving performance score for each individual relative to the performance of all the drivers. The performances included in the composite score were road sign recognition, cone gap perception, cone gap manoeuvring, number of LEDs seen, circuit time, and road hazard detection and avoidance. For each task, a z-score was calculated for each individual. The z-score for the individual is the deviation from the mean of the distribution of the task performance measure for all drivers divided by the associated standard deviation. Then, the mean of the z-scores for each individual over all the components of the composite

TABLE 13.4

Mean Performance Measures for Young, Middle-aged and Old Drivers, the Old Drivers Being Divided into those with Normal Vision, Mild Visual Impairment or Moderate or Severe Visual Impairment. These Measures Were Obtained on a Closed Road Circuit (after Wood, 2002)

Driving Performance Measure	Maximum Possible	Young Drivers	Middle-aged Drivers	Old Drivers with Normal Vision	Old Drivers with Mild Low Vision	Old Drivers with Moderate or Severe Low Vision
Road signs seen	65	51.3	50.0	46.4	46.9	40.7
Road hazards seen	9	8.7	8.7	8.7	8.4	8.0
Road hazards hit	9	0.3	0.3	0.5	0.6	1.8
Number of LEDs seen	15	11.5	10.2	7.3	8.2	7.8
Correct gap manoeuvres	9	8.0	7.7	7.3	7.2	6.5
Cones hit while manoeuvring	9	0.5	0.2	0.3	0.7	0.4
Circuit time (s)	–	428	434	468	482	478
Manoeuvre time (s)	–	38.1	38.7	41.5	49.1	48.8
Reversing time (s)	–	30.9	39.0	48.5	62.6	62.8

score was calculated, meaning that equal weight was given to all the various aspects of performance. Finally, the grand mean composite z-score for each group was calculated from the mean z-scores of the individuals in that group. Figure 13.9 shows the grand mean composite driving performance z-scores for each group. Again, it is evident that both age and visual impairment lead to deterioration in driving performance.

Given the changes in visual and cognitive capabilities that occur with age and the evident impact on driving performance, it would seem remarkable that the increase in crash rates for old drivers shown in Tables 13.1 and 13.2 are not greater, particularly at night, particularly for fatalities when one remembers the greater fragility of old people. One reason for the underwhelming impact of ageing on crash rates is behavioural compensation. What this rather grand term means is that old drivers drive more slowly and have greater opportunities to avoid conditions that they find difficult, and take those opportunities, i.e., old drivers try to limit the amount

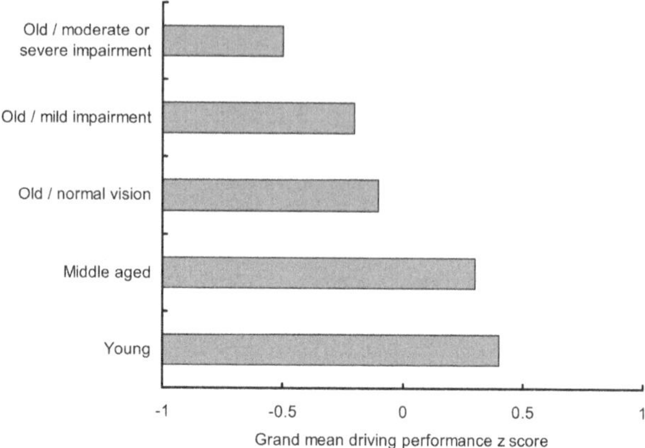

FIGURE 13.9 Grand mean composite driving performance z-scores for young, middle-aged, and old drivers, the old drivers being divided into those with normal vision, mild visual impairment, or moderate or severe visual impairment (after Wood, 2002).

of driving they do at night, in bad weather, in dense traffic, and on unfamiliar roads (Hakamies-Blomqvist, 1994). Such compensatory behaviour certainly has its advantages but it does have a negative side. Driving as a skill can be considered to have two components, a series of automatic routines that are used unconsciously and conscious control undertaken in unusual circumstances (Sivak et al., 1995). The automatic routines are the more resistant to the changes that occur with age. The automatic routines are developed and reinforced by continuous experience so if compensatory behaviour reduces experience of specific conditions, such as driving at night, when the driver has to drive at night, performance is likely to be degraded more than would have been the case – an example of use it or lose it in practice.

The consequences of low mileage are evident in Figure 13.10. This shows the number of crashes per million kilometres driven for five age groups, each of which has been divided into three annual mileage classes, based on data from The Netherlands (Langford et al., 2006). The interesting point is that for the intermediate and high annual mileage classes, there is no increase in the crash rate with increasing age. Rather, the reverse is the case with the crash rates for the high and intermediate annual mileages being highest for the youngest drivers. Indeed, the oldest age group has the lowest crash rate for both intermediate and high mileage groups. It is the lowest mileage class that shows the expected increase in crash rate for old drivers. A plausible reason for this is that it is drivers with the most degraded visual and cognitive systems or with other health problems that are most likely to limit their driving to the minimum. The problems caused by their physiological limitations are then likely to be compounded by the reduction in driving skill brought on by lack of practice. No matter whether this explanation is correct or not, there can be little doubt that age, per se, does not inevitably lead to higher crash rates (Hakamies-Blomqvist et al., 2002; Fontaine, 2003; Langford et al., 2006). Rather, within the population of old drivers, there are a group who are well down the steep decline of

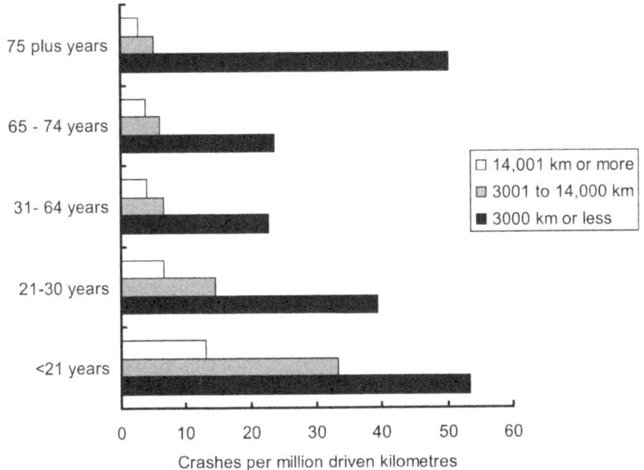

FIGURE 13.10   Crashes per million driven kilometres plotted against age group for three classes of different annual distances travelled (km) (after Langford et al., 2006).

visual, cognitive, and general health functions that degrade driving performance and eventually lead to cessation of driving (Hakamies-Blomqvist and Wahlstrom, 1998). The problem for society is to identify who these people are.

It is now necessary to consider what role lighting might have to play in reducing the crash rates for old drivers who drive low mileages. This needs to be considered for day and night. Old drivers with limited mileage are most likely to be on the road during the day. Given this situation it might be thought that lighting has no role to play. For road lighting, this is true but there are opportunities for vehicle lighting (Rumar, 1998). Crashes involving old drivers are dominated by a failure to yield the right of way at intersections or to see objects, people, and other vehicles (Sifrit et al. 2011). This suggests that the universal introduction of daytime running lights would be useful as they would increase the conspicuity of vehicles (see Section 7.14). The risk with this approach is that by increasing the conspicuity of vehicles, other road users without daytime lighting, such as cyclists and pedestrians, would decrease in conspicuity and so be more at risk. This risk could be reduced if all road users had some form of marking lighting attached or made more use of retro-reflective clothing elements.

Another form of lighting that could be improved is vehicle signal lighting, particularly brake and turn signals. As a general rule, old drivers need stronger signals to generate a response than young drivers. This suggests the use of higher luminous intensity signal lighting during the day, a proposal that implies two-level signal lighting if glare is to be avoided at night. This is technically feasible and allowed under the ECE recommendations but is not currently implemented. Yet another factor that might be used is the height of the marking and signal lights above road level. Old drivers have slower responses than young drivers so the longer notice they receive of a change ahead, the better. In traffic, signals about the intentions of vehicles some way ahead may be masked by intervening vehicles. By providing repeater signal lights at a high level, the likelihood of such obstruction occurring is reduced.

All these suggestions have the advantage of reinforcing the present system of vehicle marking and signal lights, a system that is familiar to most old drivers. This is an advantage because old people in general find it difficult to cope with very novel stimuli. There are many ideas for changing vehicle signal lighting so as to provide more information (see Section 7.15), but these should be carefully reviewed before being adopted because they may confuse old drivers rather than help them. This is also true for the idea of delivering more information to the driver by using sensors and displays. One such form of driver assistance that should be useful to older drivers is the automatic display of side views at intersections. There is definitely a place for such driver assistance systems but care needs to be taken that they make life easier for the old driver rather than more difficult. The same principle applies to signs and signals on roads. Indeed, it could be argued that one of the most valuable things that could be done to make driving easier for the old driver is to reduce the number and variety of road signs and markings. The introduction of bus lanes or cycle lanes that stop and start at short intervals, the presence of unexpected speed bumps and chicanes for traffic calming, the widespread use of mini-roundabouts, and the increasing variety of road markings with different meanings have all served to increase the complexity of the driver's task (Figure 13.11). The introduction of a set of road signs and markings designed with the objective of creating simple and consistent guidance to the old driver would be a marked improvement over the present trend of ever-increasing complexity designed to satisfy the demands of various special interest groups.

At night, the greatest improvement would be produced by a more widespread use of road lighting. Good road lighting enables the road ahead to be seen from a greater distance, diminishes glare from headlamps and provides visual guidance.

FIGURE 13.11 A complex road layout by day consisting of two mini-roundabouts in close proximity but with one offset relative to the other and a pedestrian crossing between.

Unfortunately, road lighting is expensive in both financial and environmental terms so its universal adoption is unlikely. What might be financially possible is to light all the locations where old drivers have problems. One such class of locations is rural intersections (see Section 11.6). In rural areas, the very fact that an intersection is lit while the nearby road network is not would draw attention to the decisions that will need to be made there and, provided the approaches to the intersection are also lit, make the decisions easier to reach. In suburban and urban areas, where all roads are lit, the lighting of the intersections could be different to the lighting of the adjacent roads, in either the amount or colour of light.

Another aspect of the road environment that could be improved for old drivers is the visibility of road signs from a distance (Sivak et al., 1981). This is basically a matter of sign size, luminance contrast, and luminance, all of which could be enhanced (Kline, 1995). At night, providing individual sign lighting rather than relying on the illumination from vehicle headlamps and retro-reflection would be a step forward. Other possibilities are the wider use of anti-glare screens where there is a median between the carriageways and the use of self-luminous road studs for visual guidance. For vehicle forward lighting, an adaptive forward lighting system should be of advantage as it changes the headlamp light distribution to match the environment automatically (see Section 6.7.3). If this is too exotic, Rumar (1998) suggests that old drivers would benefit from having a low beam headlamp luminous intensity distribution that provide more light to the sides of the road, more light on the road beyond 50 m and a softer cut-off. Other helpful changes would be more widespread use of automatic systems for changing between low- and high-beam headlamps, for reducing the reflectance of interior and side mirrors when a vehicle comes close behind, and for cleaning and levelling headlamps.

All these suggestions are technically feasible and would not have any deleterious effects on younger drivers but the question that needs to be asked is if the expenditure can be justified. From Figure 13.10 it appears that the old drivers who would benefit most are those who drive the fewest miles. Whether or not attempting to encourage drivers, whose visual and cognitive systems are in decline, onto the roads at night is a responsible use of resources is a matter of judgement. Even if the answer to this question were to be positive, it would still be necessary to carry out a number of small-scale demonstration projects to determine the effectiveness of the proposed suggestions. A much stronger case can be made for trying to make driving by day easier for old drivers. The proposed changes could be easily implemented and the expenditure limited. If a review of the current system of road signs and markings were to lead to a simplified system, there might even be some savings. The one certain thing about this discussion is that it will not go away. The demographics of many developed countries predict a greatly increased number of old people and hence a greatly increased number of old drivers. It would be a good idea to consider their needs now before fatalities and injuries start to rise.

13.5 RESTRICTIONS ON DRIVING

Some people are not allowed to drive, some have driven but are then forced to stop and some stop driving voluntarily. In modern society, not being allowed to drive imposes severe limitations on opportunities for employment and for socializing. Being forced to stop driving or stopping driving voluntarily is a blow to independence, not quite as extreme as going into a care home but a definite step in that direction. Different countries have different rules for determining who is allowed to drive and who should stop driving. These rules are based on age, level of training, behaviour on the road, and medical status. In this last category are visual capabilities.

13.5.1 STARTING DRIVING

In the UK, obtaining a licence to drive requires the applicant to demonstrate some basic visual capabilities, the requirements being more stringent for drivers of commercial vehicles than those of private vehicles. The requirements for people with normal vision involve visual acuity and visual field size. For private vehicle use, the applicant has to be able to read a UK vehicle number plate with letters and numbers 79 mm high and 50 mm wide (see Figure 3.18) at a distance of 20 m, in daylight, with spectacles or contact lenses if required. This ability is usually demonstrated in the field at the time of the practical driving test. This ability corresponds to a visual acuity of about 6/12 as measured using a conventional Snellen chart (Drasdo and Haggerty, 1984). For bus and lorry drivers, the applicant has to have a visual acuity of at least 6/7.5, with correction, in the better eye, and at least 6/60 in the poorer eye (see Section 3.4.1). For private vehicle use, the applicant's visual field has to be at least 120 degrees horizontally, extending at least 50 degrees on either side of the fixation point. Further, there should be no significant defect in the binocular visual field that encroaches within 20 degrees above and below the horizontal meridian. For bus and lorry drivers, the applicant's visual field has to be at least 160 degrees horizontally, extending at least 70 degrees on either side of the fixation point, and at least 30 degrees above and below the horizontal meridian. Further, there should be no significant defect in the binocular visual field within a circle subtending 30 degrees about the fixation point (Driver and Vehicle Licencing Agency, 2016).

There are two aspects of vision that might be expected to be required but are not. The complete loss of vision in one eye is acceptable for driving private vehicles provided the remaining eye can meet the acuity and visual field requirements given above but only after clinical assurance of successful adaptation to the condition is received. Monocular vision is not acceptable for driving buses or lorries. Similarly, normal colour vision is not required for driving private vehicles or for driving buses and lorries (Driver and Vehicle Licencing Agency, 2016). In addition to the requirements for visual acuity and visual field size, there are a number of visual problems for which a specialist's opinion is required with regard to the extent of the problem the condition causes for driving, whether the problem is permanent, temporary, or progressive, and, ultimately, whether the individual should be allowed to drive. Other countries have similar but not identical requirements (Casson and Racette, 2000).

Few would disagree that some level of visual capability is necessary for driving, but the basis of the current requirements is unclear (Gilkes, 1988). Attempts to relate specific visual functions to the frequency of crashes have met with mixed success (Burg, 1967; Burg, 1971; Johnson and Keltner, 1983; Ball et al., 1993; Higgins and

Bailey, 2000) probably because crashes occur infrequently and have many different causes, and drivers who are aware of their limitations modify when and where they drive.

An alternative approach to establishing how good vision has to be to permit an individual to drive has been to examine the effect of specific limitations of vision on driving performance. A study by Higgins et al. (1996) examined the effect of reduced visual acuity on driving performance. As might be expected, drivers with reduced visual acuity drove more slowly and had more difficulty in identifying traffic signs and hazards in the road simulating speed bumps.

Racette and Casson (2005) made an assessment of drivers who had slight to moderate visual field loss. The extent of their loss was measured prior to their driving over a variety of public roads accompanied by a driving instructor who made an assessment of their ability to drive safely based on their performance. As might be expected, there was a lot of individual variability but there was also a trend for drivers with greater visual field loss to be graded as unsafe. Bowers et al. (2005) have used a similar approach of driving on public roads to examine the ability of drivers with mild to moderate reduction in visual field to make different manoeuvres. They found that drivers with more restricted visual fields were worse at maintaining position in a lane while going round a curve and matching speeds when changing lanes, activities that rely on peripheral vision to some extent. Conversely, they found no relationship between visual field size and lane keeping on a straight road, maintaining distance from the vehicle ahead and crossing intersections, activities that primarily rely on foveal vision.

McKnight et al. (1991) examined the visual capabilities and driving performance of truck drivers with binocular and monocular vision. There were clear differences between the two groups in visual capabilities. Specifically, monocular drivers were inferior to the binocular drivers in contrast sensitivity, visual acuity at low light levels and in the presence of glare, and depth perception. However, these differences in capabilities made little difference to the measured driving performance. There were no differences between monocular and binocular drivers for lane keeping, clearance judgement, gap judgement, and hazard detection. The only aspect of driving for which a statistically significant difference was detected was the distance at which a sign requiring action could be read. For this, the binocular drivers read the sign about 12% further away. Both groups of drivers showed wide individual differences in driving performance, with much bigger differences occurring within each group than between the groups.

Defective colour vision in some form affects about 8% of males and 0.4% of females (Boyce, 2014). Cole and Brown (1966) examined the difference in reaction time to a red traffic signal for people with normal colour vision and for protanopes, a group that do not have a long wavelength-sensitive cone photoreceptor (see Section 3.2.5). They found that the protanopes had longer reaction times than colour normals to the same red traffic signal and were much more likely to miss the signal at lower signal luminances. Nathan et al. (1964) also found longer reaction times for people with colour-defective vision as well as more mistakes in identifying the signal colour. That such observations can have behavioural consequences is shown by the finding that people with defective colour vision had more crashes involving failure to halt at a red traffic signal than people with normal colour vision (Neubauer et al., 1978).

In a sense, such results are a statement of the obvious. Drivers who have limited visual capabilities tend to have problems when driving, the nature of the problem depending on the nature of the visual limitation. While this is interesting, it does not get us much further forward in answering the basic question: What limits should be set for a licence to drive? What can be said is that the current system of vision testing used in most countries is very crude. How crude it is can be seen from the results of Wood (2002). As part of this study of the effects of age and visual impairment on driving performance, measurements were made of the individual driver's static and dynamic visual acuity, useful field of view, static and kinetic fields of view, contrast sensitivity, sensitivity to glare, and motion sensitivity. A step-wise multiple linear regression analysis revealed that 50% of the variance in the composite z-scores (see Section 13.4.5) could be explained by a combination of motion sensitivity, useful field of view, contrast sensitivity, and dynamic visual acuity. Interestingly, the visual capability measure most widely used to determine suitability to drive, static visual acuity, did not increase the variance explained by a statistically significant amount. Such findings suggest that a more sophisticated visual screening system for testing applicants for driving licences could be developed (Clay et al., 2002; Owsley and McGwin, 2010). Alternatively, given the developments in driving simulators, it should not now be too difficult to produce a programme to test driving potential, a programme that examines both the visual and cognitive abilities required of the driver (Ball et al., 1993). As long as the current crude test remains, there is a risk that people who are capable of driving safely will be refused permission to drive and people who are a menace on the road will be given permission to drive.

The question that now needs to be addressed is what is the role of lighting in obtaining a licence to drive? The answer is very little. Tests of visual function are made during daytime or in conditions simulating daytime. Tests of driving performance are also made during daytime so driving at night is largely ignored. It is interesting to consider that anyone driving at night on an unlit road has a narrower horizontal field of view than that required for the issuing of a licence. This neglect of vision at low light levels means that the main role of lighting in determining whether or not someone gets a licence to drive comes through signs and signals. Signs are designed to have a high luminance contrast and are sized so as to be visible from the required distance (see Section 5.3) by a driver with the normal visual acuity. Larger signs would make driving easier for someone with poor visual acuity. Traffic signals are designed to have colours within restricted colour boundaries set out in the CIE 1931 chromaticity diagram (see Figure 2.3). These boundaries are intended to make signal colours easier to discriminate by people with the more common forms of colour-defective vision. It would be better still if colour was not the only difference between traffic signals with different meanings. With present traffic signal designs, colour-defective people can often use the relative positions of the red, yellow, and green signals to supplement the colour information they find hard to extract. It would be interesting to incorporate other means of conveying the necessary information, such as the shape of the signal (Whillans, 1983). Then, as long as the driver could tell which signal was lit, the meaning would be clear.

What should be clear from the above is that the current methods of determining whether or not people have adequate visual and cognitive capabilities to drive are crude and inadequate. That there have to be some limits is obvious but where those limits should be and whether or not they should be based on crash rates or on driving performance are open questions. The demographics of many countries are leading to more old drivers so there is a great deal of pressure to permit drivers with a wide range of visual disabilities to continue to drive. This has generated advice on driving with visual impairment (Peli and Peli, 2002) and consideration of the extent to which different forms of visual impairment might be accommodated on the road (Owsley and McGwin, 2010; Driver and Vehicle Licencing Agency, 2016; Wood, 2022). There is a definite need to give more thought to the visual and cognitive aspects of issuing a licence to drive and the conditions under which those visual aspects are tested. At the very least, some attention should be given to driving at night.

13.5.2 STOPPING DRIVING

In modern society, most people seek to gain a driving licence as soon as they are legally permitted and to retain it as long as possible. There are those who lose their licence through bad or foolish behaviour on the road, but for most the factor that determines when driving is stopped is a decline in health associated with increasing age. It is for this reason that many countries have introduced regular reviews of an individual's ability to drive when they reach a set age, typically 70 years. The rigour of this review varies from country to country. Some countries, such as Finland, make older drivers undergo a medical screening process, while others, such as Sweden and the UK, do not but rather rely on self-certification of fitness to drive. McGwin and Owsley (2022) examined the benefits of vision screening for reducing the frequency of crashes involving older drivers and concluded that even extended measures of visual function had little value. Indeed, they argued that, at a population level, the mental health problems caused by forcing people to stop driving would outweigh any traffic safety benefits.

The significance of general ill health for the decision to stop driving is evident from a study by Hakamies-Blomqvist and Wahlstrom (1998) who conducted a survey of all the drivers in Finland who had not renewed their licence when they reached 70 years of age and a sample of those who had. Questions about the reasons for not renewing their licence revealed that most were suffering from poor health. The most common form of ill health that differentiated between those who renewed their driving licences and those who did not was neurological conditions, depression, and glaucoma. Glaucoma is a form of partial sight that reduces the visual field and, if untreated, ultimately leads to blindness (Silverstone et al., 2000). Interestingly, only 7% of people who had stopped driving had been advised to do so by the physician treating their underlying health problem. Most had stopped voluntarily. Such voluntary cessation of driving may also explain the results of Hakamies-Blomqvist et al. (1996) who found that the trends in population-adjusted crash rates for old drivers were the same in Finland and Sweden, the former with medical screening of old drivers and the latter without.

The picture that is emerging is of old drivers gradually retreating from what they perceive to be stressful and dangerous situations until they eventually surrender another part of their independence. This picture is certainly evident in the results of Freeman et al. (2005) who examined the relationship between visual function and time of stopping driving for a large sample of older drivers in the United States. They found that drivers with poorer visual acuity, reduced contrast sensitivity, and smaller visual fields at the time of measurement were more likely to have stopped driving eight years later while those with better vision had not. Freeman et al. (2006) tried to determine if particular types of visual failings could be linked to particular changes in driving behaviour two years after measurement. What they found was that poorer visual acuity, degraded contrast sensitivity, and reduced visual field all led to a lower mileage being driven. Poorer contrast sensitivity and more limited visual fields were associated with a cessation of night driving, and poor visual acuity alone was linked to an unwillingness to drive in unfamiliar areas.

Given that deterioration of vision is linked to at first modifying and then stopping driving, it is appropriate to ask if better lighting might have a role to play in keeping older drivers driving safely for longer. The answer is the same as that given earlier for old drivers in general (see Section 13.4.5). The main point to take from this discussion of the reasons why people stop driving is that it is not age, per se, that is the cause of driving cessation but rather the consequences of ageing for visual and cognitive abilities. This implies that attempts to reduce crash rates are more likely to be successful if attention is focused on the health and not simply on the age of the driver. As part of that focus, it would be interesting to consider the use of a graduated licence to drive to which various restrictions could be added according to the individual's health. For example, conditions limiting driving to daytime only or within a set distance from home could be added. At the moment, withdrawing a licence to drive is often an all-or-nothing matter. A graduated licence would have the advantage of formalizing what is actually happening informally.

13.6 SUMMARY

Some people are better equipped to deal with the demands of driving than others. Among the important factors that determine a driver's ability are the amount of practice and any limitations in visual or cognitive systems. The influence of these factors can be seen in the number of fatal and non-fatal crashes, occurring by day and night for different age groups. The highest numbers of crashes occur for young and old drivers but fewer for female than male drivers. Another difference between young and old drivers lies in the contributory factors assigned to them when they crash. Both young and old drivers are identified as failing to look properly, failing to judge another person's path, and making a poor turn or manoeuvre, but young drivers are more likely to be identified as inexperienced, driving in a careless or reckless manner, travelling too fast for the conditions, and indulging in sudden braking or exceeding the speed limit. Conversely, older drivers tend to be involved in more crashes where illness or disability, mental or physical, is considered a contributory factor.

Young drivers are not likely to be involved in crashes because of visual or cognitive limitations. Young drivers who are in their late teens or twenties will usually have a visual system at the peak of its powers and a cognitive system capable of rapid responses to simple and complex stimuli, but they may be inexperienced. While a lack of experience is a factor in crashes for some young drivers, not all young drivers are inexperienced. Even when they are experienced, the nature of the types of crashes in which they are involved suggests that young drivers drive with reduced margins of error, relying on their fast reactions to survive. Further, young drivers are more likely to be out late at night and more likely to be suffering from the effects of alcohol or drugs. There is also a social aspect to the involvement of young drivers in fatal crashes. The presence of passengers, particularly male passengers, tends to increase the probability of a fatal crash. There is little that lighting can do to overcome such behavioural excesses and social pressures other than to provide more time for decision-making by lighting the road further ahead.

Old drivers show increased crash rates because ageing affects the visual, cognitive, and motor functions required for driving. As the visual system ages, a number of changes in its structure and capabilities occur. With increasing years, the ability to focus over a wide range of distances is diminished, the amount of light reaching the retina is reduced, more of the light reaching the retina is scattered, the spectrum of the light reaching the retina is changed, and more straylight is generated inside the eye. Such changes with age are the best that can be expected. With increasing age comes a greater likelihood of pathological changes in the eye leading to visual impairment in which the normal deterioration in visual capabilities with age is accelerated. These are all optical changes in the eye, but there are also neural changes. This means that the compensation for visual system ageing that can be provided by lighting is inevitably limited.

The consequences of such optical and neural changes with age are reduced visual field size, increased threshold luminance, reduced visual acuity, reduced contrast sensitivity, reduced colour discrimination, and greater sensitivity to glare. The consequence of these changes in visual capabilities is a deterioration in driving performance, but not necessarily an increase in the number of crashes. This is because old drivers tend to compensate for their reduced capabilities by avoiding situations that they find difficult and stressful, usually driving in bad weather, in dense traffic, at night, and on unfamiliar roads. Such compensatory behaviour certainly has its advantages but it does have a negative side. Driving as a skill can be considered to have two components, a series of automatic routines that are used unconsciously and conscious control undertaken in unusual circumstances. The automatic routines are the more resistant to the changes that occur with age. The automatic routines are developed and reinforced by continuous experience. This may be why it is the old drivers who cover the lowest mileage who are responsible for the increase in crash rate ascribed to old drivers as a whole.

Lighting could be used to help old drivers. By day, there is no role for road lighting but vehicle lighting could be modified to advantage. The universal introduction of daytime running lights would be helpful, as would the provision of higher luminous intensity vehicle signal lamps for use by day. By night, a wider use of road lighting would be beneficial. Unfortunately, road lighting is expensive so its universal adoption is unlikely. What might be financially possible is to light all the locations where old drivers have problems, such as road intersections, and to light all road signs. As for vehicle lighting, an adaptive forward lighting system would appear to be of advantage as it changes the headlamp luminous intensity distribution to match the environment automatically. All these suggestions are technically feasible and would not have any deleterious effects on younger drivers but the question that needs to be asked is if the expenditure can be justified. One thing is certain and that is that the problem of what to do about old drivers will not go away. The demographics of many developed countries predict a greatly increased number of old people and hence a greatly increased number of old drivers.

This raises the question as to how permission to drive is given and withdrawn. Different countries have different rules for determining who is allowed to drive and who should stop driving. These rules are based on age, level of training, behaviour on the road, and medical status. Getting a licence to drive always involves some test of visual capability, usually some simple measure of static visual acuity. Unfortunately, attempts to relate individual visual functions to crash rates have met with mixed success, probably because crashes occur infrequently and have many different causes, and drivers who are aware of their limitations modify when and where they drive. What is clear is that people with worse motion sensitivity, a smaller useful field of view, lower contrast sensitivity, and poor dynamic visual acuity show poorer driving performance.

Lighting has little involvement in obtaining a licence to drive. Tests of visual function are made during daytime or in conditions simulating daytime. Tests of driving performance are also made during daytime so driving at night is largely ignored. It is clear that the current methods of determining whether or not people have adequate visual capabilities to drive are crude and inadequate. That there have to be some limits is obvious but where those limits should be and whether or not they should be based on crash rates or on driving performance are open questions.

As for stopping driving, most people seek to retain a driving licence as long as possible. The picture that emerges from studies of the cessation of driving is of old drivers gradually retreating from what they perceive to be stressful and dangerous situations until they stop driving altogether and surrender another part of their independence. In many situations, the decision to stop driving is taken voluntarily. It is not age, per se, that causes people to stop driving but rather the consequences of ageing for visual and cognitive abilities. This implies that attempts to reduce crash rates for old drivers are more likely to be successful if attention is focused on the health and not simply on the age of the driver. As part of that focus, it would be interesting to consider the more widespread use of a graduated licence to drive to which various restrictions could be added according to the individual's health. At the moment, withdrawing a licence to drive is often an all-or-nothing matter. A graduated licence would have the advantage of formalizing what is actually happening informally.

14 Constraints

14.1 INTRODUCTION

All forms of lighting are subject to constraints and lighting for driving is no exception. These constraints range from the legal through the financial to the environmental, some of them being weak and some strong. Vehicle lighting meeting specific photometric conditions is a legal requirement in almost all countries. Road lighting is not a legal requirement but rather a matter of policy, the nature of which will vary from country to country and from time to time. Both vehicle lighting and road lighting imply costs. For vehicles, the first costs of lighting are included in the price of the vehicle. Costs-in-use are borne by the driver and depend on the timing and hours of use of the vehicle. For road lighting, both first costs and life cycle costs can be involved in the decision whether or not to install road lighting and certainly influence the amount and form of road lighting provided. Recently, environmental factors have also entered into consideration, particularly for road lighting. These environmental factors can take four forms: the amount of energy consumed, the carbon dioxide emissions produced when generating the electricity used, the disposal of the materials used at the end of product life, and the problem of light pollution. This chapter is devoted to a consideration of these constraints as they apply to road lighting. The legal requirements applicable to vehicle lighting have been discussed in Sections 6.3 and 7.3. The financial and environmental aspects of vehicle lighting are not considered separately from those of the vehicle itself. Even the light pollution produced by vehicle lighting is ignored, probably because it is considered to be transient rather than permanent.

14.2 ASSESSING THE NEED FOR ROAD LIGHTING

The lighting of roads costs money. When the roads are privately owned, whether or not to install lighting is a matter for the owner, but the vast majority of roads are publicly owned, and their lighting is a matter for local authorities. Local authorities have the power to install road lighting but not an obligation to do so. To guide decisions, local authorities have adopted different approaches for determining when expenditure on road lighting is warranted. In the USA, many states use the guidelines set out by the American Association of State Highways and Transportation Officials (AASHTO, 2018). These guidelines consist of a series of quantitative and qualitative criteria. For example, in North Dakota, the quantitative guidelines for determining if a linear section of freeway should be lit require that the annual average daily traffic flow should be more than 30,000 vehicles or the average distance between three or more adjacent intersections should be 1.5 miles (2.4 km) or less, or the distance between lit intersections is 1.5 miles (2.4 km) or less. The qualitative guidelines concern the extent, nature, and lighting of the surroundings through which the freeway passes. Meeting the quantitative and qualitative criteria is a necessary but not sufficient condition to ensure the installation of road lighting. Even if the criteria are met, most public authorities allow themselves some discretion by stating that they may need to conduct a full safety evaluation of the specific site to determine the night/day crash ratio as well as considering what funds are available and where they are best spent.

An alternative approach used in the EU is a calculation method based on the costs and benefits of accident prevention, expressed in monetary terms. This is not as simple as it may seem (European Commission, 2009; Hauer, 2011). The first costs of road lighting are easy to obtain, but assumptions are needed when life cycle costs are used. The assumptions required relate to hours of use, electricity costs, maintenance costs, and disposal costs. Hours of use can be identified fairly exactly, given that the rising and setting of the sun is quite predictable so the only unknown is the variation in switch on or switch off times caused by cloud cover. The same is not true for electricity costs, maintenance costs, and disposal costs. The economics of road lighting are usually assessed over a period of between 25 and 40 years, so different assumptions about electricity costs, monetary inflation, and interest rates can make a large difference to life cycle costs.

Even more contentious is the cost of benefits. This is because it is necessary to make assumptions about the reduction in crashes likely to occur if the road lighting is installed and the savings to be made by preventing crashes. Rea et al. (2009) carried out an extensive review of the literature on the safety benefits of road lighting in different locations and concluded that the reduction in nighttime crash risk associated with installing road lighting converged into the range 20% to 30%, the percentage increasing with greater complexity of the road geometry, higher population densities, and a higher level of pedestrian activity.

Having obtained the reduction in crash risk that will be produced by installing lighting, it is then necessary to estimate the value of preventing these crashes. This includes both measurable elements, such as the cost of any hospital treatment, police costs, and material damage, and imponderable elements such as the costs of lost productivity and the value of human pain and suffering. Once the costs of lighting and the financial benefits associated with a reduction in accidents are estimated, it is possible to calculate if an investment in road lighting gives a reasonable rate of return. Unfortunately, Wijnen et al. (2019) have revealed that estimates of the costs associated with a road fatality vary widely for 31 European countries, from €700,000 to €3 million euros, which means the associated cost/benefit values are subject to considerable uncertainty.

A hybrid approach has also been suggested (Deans et al., 2003; Lambert and Turley, 2003). The first step is a monetary balance between the costs and benefits of installing road lighting based on the reduction in crashes occurring during the hours of darkness, the costs of these crashes and the installation, maintenance, and energy costs of the proposed road lighting. The second step is a prescription method in which the specific site is assessed for crash risk. Among the factors considered are the traffic density, the traffic mix, the brightness and use of the surroundings, the geometry of the road layout, and the speed limits.

Each of the three methods summarized has different virtues. The prescriptive approach using the AASHTO guidelines has the virtue of simplicity although how the prescription is derived is not at all clear. The cost/benefit approach presents a high precision façade but is largely smoke and mirrors. By manipulating the assumptions about interest rates, the monetary value of fatalities and injuries, and the percentage crash reductions, policymakers can reach almost any decision they desire. The hybrid approach is the most attractive option, being a mixture of the general and the specific. The general monetary balance can be used to weed out the hopelessly uneconomic locations while the risk assessment of the specific site can be used to determine where an investment in lighting is likely to be most effective.

A benefit/cost approach has also been used to determine the relative effectiveness of a wide range of possible road safety measures, covering investments in infrastructure, legislation, and enforcement, post-crash treatment, and vehicle equipment (Daniels et al., 2019). The best estimate of the benefit/cost ratios for different actions is given in Table 14.1, benefit/cost ratio being used rather than cost/benefit ratio because the differences between the various actions are easier to appreciate. A benefit/cost ratio greater than 1 means the monetary value of the benefits in terms of road crashes and their consequences is greater than the costs of the action, while a value less than 1 indicates the reverse.

Study of the benefit/cost ratios in Table 14.1 reveals why many of the features of modern traffic safety have appeared. For example, the encouragement of seat belt use is particularly effective, as is the installation of crash barriers and speed bumps. But there is also evidence that monetary benefit cost is not always the deciding factor. For example, the introduction of alcohol interlocks would be effective in reducing drunk driving, but they have not been widely introduced, probably because of legal complications. Conversely, the installation of automatic barriers on railway crossings is not an effective use of monetary resources, but they have been widely introduced probably because of public pressure resulting from the spectacular nature of collisions between trains and vehicles. As for installing road lighting where there has been no lighting before, the benefit/cost ratio is 0.7 which means that while such an action may be effective in reducing crashes and injuries, it is not necessarily cost-effective, even when the costs are spread over 25 years. This calculation says more about the limitations of benefit/cost ratios and their inverse than anything else. When advantage is taken of the confidence intervals around both the benefit measure and the cost measure to predict the best possible benefit/cost estimate (upper confidence interval for the effect and lower confidence interval for the cost), the benefit/cost ratio for installing road lighting rises to 1.8. Conversely, the worst possible benefit/cost estimate (lower confidence interval for the benefit and upper confidence interval for the cost), the benefit/cost ratio falls to 0.3. What this implies is that benefit/cost analyses may be useful for identifying the most promising out of a wide range of possible actions for improving traffic safety but of limited value for determining if road lighting should be installed.

TABLE 14.1

Best Estimates of the Monetary Benefit/Cost Ratio for Different Road Safety Actions (from Daniels et al., 2019)

Action	Benefit/Cost Ratio
Identification and treatment of locations with elevated crash risk	16.3
Dynamic speed limits that change with traffic or weather conditions	1.1
Installation of speed bumps	18.2
Implementation of 30 km.h^{-1} (18 mph) zones	2.1
Installing road lighting on unlit road	0.7
Installing rumble strips on road centreline	0.1
Installing chevron markers on curves	2.7
Installing turn lanes at crossroads	0.4
Installing automatic barriers at railway crossing	0.05
Introducing traffic calming measures over an area	0.1
Installing crash barriers	19.5
Converting junctions to roundabouts	9.2
Installing traffic signals at junctions	1.1
Introducing mandatory visual acuity tests for drivers over 45 years	0.5
Increasing police checks on seat belt wearing	28.7
Introducing an autolock system for drivers with drink-driving records	10.9
Introducing red traffic signal cameras	3.7
Undertake random breath tests to detect drunk drivers	7.7
Introduce average speed cameras	19.5
Enhance police enforcement of speed limits	1.1
Train children in good pedestrian behaviour	2.6
Encourage the use of seat belts and stronger enforcement	69.8
Carry out an advertising campaign to reduce drink driving	2.1
Carry out a campaign to promote the use of booster seats for children	4.6
Provide ambulance helicopters in sparsely populated areas	9.9
Provide restraints for children in cars	3.4
Fit vehicles with stability control	4.4
Fit autonomous emergency braking systems to vehicles	4.8
Develop antilock braking systems for two-wheeled vehicles	9.0

14.3 COSTS OF ROAD LIGHTING

Once it has been decided that some form of road lighting is required, it is necessary to calculate the costs of the project. These can take various forms depending on what aspects of the installation are taken into account, but the most frequently used are the first cost per kilometre and the life cycle cost per kilometre. Both of these measures will differ depending on the nature of the installation. Where there was no previous road lighting on the site, there are additional first costs that are largely absent when the installation is a replacement of an existing luminaire on an already installed column.

For a new installation the first cost per kilometre is calculated as the sum of the cost of the light source, luminaire, column, and connection to the electricity supply, multiplied by the number of columns along a kilometre of road together with the cost of the work required to install the equipment. Depending on the proximity of the electricity supply this item can easily dominate the first cost per kilometre for a new installation. For a replacement installation, the first cost per kilometre is limited to the cost of light source and luminaire and the labour required to make the change as the column and electrical connection are already available. One aspect of replacement lighting that requires careful consideration is to ensure that the light distribution from the luminaire will provide the road surface luminance distribution specified in the relevant standard. This is important because the spacing of the columns along the road will have been designed for the light distribution of the luminaire being replaced.

Calculating life cycle costs per kilometre is rather more complicated (Onaygil et al., 2012). In its simplest form, to calculate the life cycle cost per kilometre of a road lighting installation, you need to know the first cost per kilometre, the number and power demand of the luminaires, the hours of use, the cost of electricity at the time the installation is operating, the proposed maintenance cycle, and the discount rate. The power demand will vary if dimming is used. The hours of use are typically taken to be 4,100 hours, but this will be reduced if the lighting is only used for part of the night. The cost of electricity may vary depending on how it is charged over time. The maintenance cycle will vary according to the light sources, luminaires, and columns used. Typically, road lighting columns are assumed to last for 40 years. Luminaires in which the light source can be easily replaced are assumed to last for 20 years. High-pressure sodium light sources are usually replaced after 4 years, and metal halide light sources after 3 years. For LED light sources, the whole luminaire has to be replaced but their longer life means that only needs to happen about every 8 years. The discount rate, which is the difference between the actual interest rate and the rate of inflation, will depend on the state of the economy at the time the life cycle cost is being calculated. Thus, the life cycle cost per kilometre represents the sum of the capital cost of 1 kilometre of the installation, the cost of replacing lamps and luminaires, and the cost of electrical energy over 40 years, all future costs being converted to their present values. As if this was not complicated enough, involving as it does assumptions about economic conditions many years ahead, there is now interest in moving from a linear model of production (take–make–dispose) to a circular model of the economy where the costs of raw materials, manufacturing, and distributing products and recycling their components at the end of life are also considered (Dzombak et al., 2020). What this means is that life cycle costs should always be viewed with suspicion until it is clear what assumptions have been made.

As a result of the calculations of the cost of road lighting and the enthusiastic promotion of the high luminous efficacy and long life of LEDs, the major trend in road lighting practice over the last decade has been the replacement of high-pressure sodium lighting with LED lighting. How far this has gone can be seen from a survey of the light sources used for road lighting in the UK in 2020, based on the listings sent to the Distribution Network Operators who determine how much the

TABLE 14.2

Number of Different Light Sources Used for Road Lighting in the UK in 2020 (Chartered Institution of Highways and Transportation, 2021)

Light Source	Number
Light emitting diodes (LED)	3,968,664
High pressure sodium (HPS)	1,274,497
Low pressure sodium (LPS)	548,087
Metal halide (MH)	512,091
Low pressure mercury (Fluorescent)	446,747
High intensity discharge (HPS or MH)	316,641
High pressure mercury (Mercury vapour)	119,700
Incandescent	3,739

local authorities are charged for the supply of electricity for road lighting (Chartered Institution of Highways and Transportation, 2021). Table 14.2 shows the make-up of light sources in the UK road lighting estate in 2020. In the UK, LEDs constitute almost 55% of road lighting light sources but there are still significant numbers of other light sources, including a few that have been banned or are no longer being manufactured. Other countries will show similar patterns, the exact numbers depending on the enthusiasm with which they have pursued the LED trail.

Another trend has occurred in controls, either via individual luminaires or through central management systems. Almost all new LED luminaires have a programmable dimming capability. Local authorities have taken advantage of these features and the approval given by standards to reduce the light levels on roads late at night when traffic densities and pedestrian activity are low, usually between midnight and 05:00 (BSI, 2020). The same survey shows that for all the light sources in the UK road lighting estate, only 42% operate at full light output from dawn to dusk, 22% are dimmed, and 5% are switched off late at night. These values may be underestimates as 29% of luminaires are controlled through central management systems but no details are available on how these are used. Fortunately, although dimming road lighting has been shown to increase driver's reaction times and shorten visibility distances (Wood et al., 2018), reducing lighting late at night has been shown to make very little difference to road casualties (Steinbach et al., 2015).

14.4 ENERGY CONSUMPTION

There are three reasons why the amount of energy consumed might become a constraint on the design of road lighting. The first reason is the cost. The higher the price of electricity, the more likely it is that energy costs will become a constraint on road lighting. The second is the need to reduce the maximum demand on the electricity network. This is very unlikely to be relevant except in countries in high latitudes, as the maximum demand usually occurs at times of maximum activity by the population, which is usually during the day, but road lighting operates at night. For countries in high latitudes, for some part of the year, darkness may coincide with high levels of activity in which case, cutting road lighting to reduce maximum demand might be considered. The third is the desire to minimize the use of fossil fuels. Given this desire, any use of electricity should be avoided unless it is generated from renewable resources such as nuclear, sun, wind, or waves.

14.5 CARBON DIOXIDE EMISSIONS

Carbon dioxide is one of several gases that contribute to global warming and hence climate change. It is not the most potent but it does have the biggest impact because it is produced in large amounts by the burning of fossil fuels for heat and for the generation of electricity. Given a desire to limit global warming by really reducing carbon dioxide emissions rather than just covering embarrassment with the fig leaf of carbon offset payments, road lighting is an obvious target. It consumes electricity, it is conspicuous, and it is sometimes in use at times when there are few people about. However, before cutting road lighting, it is important to remember its value for traffic safety and that the amount of carbon dioxide emitted will depend on how the electricity used to power the road lighting is generated. Different countries have very different fuel mixes when it comes to electricity generation. New Zealand generates approximately 82% of its electricity from renewables such as hydro, thermal, and wind, most of the rest coming from natural gas. Its neighbour, Australia, generates almost 75% of its electricity by burning coal. France generates 78% of its electricity through nuclear reactors. The USA fuel mix for electricity generation is 19.5% coal, 39.8% natural gas, 18.2% nuclear, and 21.5% renewables. The UK fuel mix for electricity generation is 3.4% coal, 39.3% natural gas, 13.9% nuclear, 40.8% renewables, and 2.6% other, which includes biomass. The carbon emissions of different fuels, expressed in kilograms of CO_2 per kilowatt-hour of electricity produced, range from zero for nuclear and renewables to 0.371 for natural gas and 0.945 for coal. If reducing carbon dioxide emissions is the aim, there is little point in limiting road lighting in countries where the fuel mix used to generate electricity is dominated by nuclear or renewables, but where gas and coal are extensively used for electricity generation, the need for road lighting should be carefully considered. Ideally, the answer to this problem lies in using photovoltaic panels linked to batteries to store electricity during the day and to use the stored electricity to power the road lighting at night. Such a system is technically possible and commercially available for parking lots, but its use is presently limited to locations where connecting to the electricity grid is prohibitively expensive.

14.6 WASTE DISPOSAL

Another factor that is beginning to be considered as a constraint on road lighting is the ease of disposal at the end of life. Questions about waste disposal can take two forms: the hazard to human health posed by the materials in the product and the ease of recycling. Some elements of a road lighting installation present little problem. For example, aluminium and steel are widely used for columns but present no hazard and can be recycled. Luminaires are usually constructed of some combination of aluminium, glass, and plastic, all of which present little hazard and can be recycled. Where there are problems is in the light source and control gear. There are still a number of different light sources in use for road lighting (see Table 14.2) some of which contain hazardous substances such as mercury. As for control gear, there are still capacitors containing PCBs present in the oldest part of the road lighting stock and while electromagnetic control gear is no longer widely used the electronic control gear and LED drivers require professional recycling. Without getting into the specifics of individual products, it is not possible to identify problems with waste disposal so the best that can be done here is to point out some general trends. For example, light sources that contain mercury will be subject to stricter regulation than those that contain sodium. This is evident through the existence of EU Directive 2002/95/EC, which prohibits the use of mercury, but not sodium, in electrical and electronic equipment to less than a low threshold level. Similarly, electronic control gear is likely to be more expensive to dispose of than electromagnetic control gear because it contains more expensive materials that are more difficult to isolate and recycle. The same applies to LED light sources. EU Directive 2002/96/EC, widely known as the Waste Electrical and Electronic Equipment (WEEE) Directive applies to LEDs and their associated drivers. This Directive has been implemented in many EU countries in an attempt to reduce the size of the electrical and electronic waste stream. What this Directive requires is that anyone who manufactures, brands or imports electrical and electronic equipment be responsible for the collection, treatment, recycling, and disposal of the product at the end of life. In the UK and other countries, trade groups have set up schemes to ensure the collection and appropriate treatment of both lamps and control gear. How effective these are remains to be seen, but there can be little doubt that questions about waste disposal are now on the agenda for the designers of road lighting.

14.7 LIGHT POLLUTION

Light is essential for life but it can also be considered a form of pollution when delivered at the wrong time. Complaints about light at night can be divided into three categories: light trespass, sky glow, and glare. Light trespass is local in that it is associated with complaints from individuals in a specific location. The classic case of light trespass is a complaint about light from a road lighting luminaire entering a bedroom window and keeping the occupant awake.

Sky glow is more remote than light trespass in that it can affect people over great distances. Complaints about sky glow originate from many people, ranging from those who have a professional interest in a dark sky, i.e., optical astronomers (McNally, 1994, Mizon, 2002), through those who are concerned with the effect of too much light at night on flora and fauna (Rich and Longcore, 2006) to members of the general public who simply like to be able to see the night sky. For night sky viewers, both professional and amateur, the problem sky glow causes is that it reduces the luminance contrasts of all the features of the night sky. A reduction in luminance contrast means that features that are naturally close to visual threshold will be taken below threshold by the addition of the sky glow. As a result, as sky glow increases the number of stars and other astronomical phenomena that can be seen is much reduced. As for the flora and fauna, the effects of sky glow are widespread, ranging from blooming at the wrong time through reduced nocturnal pollination and interference with bird migration to a change to the balance between predator and prey and misguided behaviour that leads to population decline.

Glare has been a feature of outdoor lighting for many years and is associated with the presence of a luminaire projecting a high luminous intensity towards an observer. If a luminaire produces a desire to shield the eyes, glare may be said to be occurring. As far as light pollution is concerned, glare is local in that it is caused by a particular luminaire seen from a particular direction by someone outside the lit area. Glare can be produced by nearby luminaires or those far from the observer and may or may not be linked to light trespass. Glare can disable vision by reducing the luminance contrasts in the retinal image of the scene or it may simply cause discomfort (see Sections 4.3 and 6.6).

14.7.1　Light Trespass

Light trespass is not well defined. For some people, light trespass occurs when light enters the rooms of their dwelling. For others, light trespass occurs when light extends over their property boundary into their garden. In principle, light trespass can be avoided by the careful positioning and aiming of luminaires with appropriate luminous intensity distributions. For road lighting, this means choosing luminaires that direct the vast majority of the light emitted onto the road surface and minimize the amount of light emitted beyond that surface, on both sides of the road. If complaints of light trespass still occur, then the classic response is to fit what is called a house-side shield to the luminaire. This is a baffle that prevents light travelling direct for the luminaire to windows immediately behind the luminaire. Figure 14.1 shows two forms of baffle. One is a simple, external house-side shield fixed to a luminaire to prevent light trespass into the bedroom of an adjacent house. The other is a cylindrical baffle fitted to a flashing beacon installed to mark a pedestrian crossing. The regular flashing of the beacon makes any light trespass particularly disturbing.

Fitting baffles adds to the cost of the lighting installation so it is as well to have some guidance as to whether or not a complaint of light trespass is justified. Advice in the form of maximum illuminances that should be allowed to fall on the windows of a property is available (CIE, 2017b, ILP, 2020b). This advice is presented for five different zones. Where there is road lighting, different zones are necessary because the ambient light level is very different in different locations. It makes no sense to specify one maximum illuminance on a window for both a city centre and a rural village. A maximum illuminance limit based on the city will be too high for a village while a criterion based on a rural village will be unachievable in a city. Table 14.3 identifies the five environmental zones suggested by the Commission Internationale de l'Eclairage (CIE). Note that it is recommended that the zone E0 Protected should always be surrounded by an E1 Natural zone. Table 14.4 sets out the maximum vertical illuminance that should be allowed to fall on a property in each of the four environmental zones. where road lighting is allowed, before and during a curfew (see Section 14.7.2).

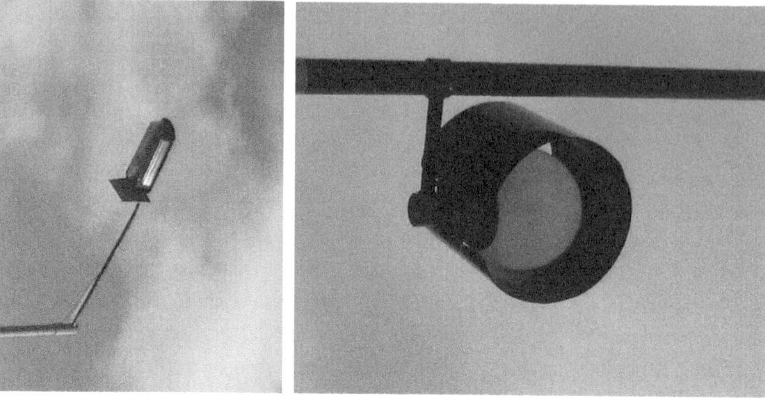

FIGURE 14.1   Two attempts to reduce light trespass, (a) an external house-side shield fitted to a deep bowl road lighting luminaire and (b) a cylindrical baffle fitted to a flashing beacon marking a pedestrian crossing.

TABLE 14.3
The Zoning System of the Commission Internationale de l'Eclairage (after CIE, 2017b)

Zone	Zone Description and Examples of Such Zones
E0 Protected	Areas with astronomically observable dark skies, UNESCO starlight reserves, IDA dark sky parks, and reserves (no road lighting)
E1 Natural	Areas with intrinsically dark landscapes: Relatively uninhabited rural areas, National Parks, areas of outstanding natural beauty (where roads are usually unlit)
E2 Rural	Areas of "low district brightness": Sparsely inhabited rural areas, outer suburban residential areas (where roads are lit to residential road standard)
E3 Suburban	Areas of "medium district brightness": Well-inhabited rural and urban settlements, small town centres (where roads are lit to traffic route standard)
E4 Urban	Areas of "high district brightness": Generally, urban areas having mixed recreational and commercial land use with high night-time activity

---

Specifying the maximum vertical illuminance on a window of a property is an appropriate criterion to determine if complaints of light trespass are justified. However, it is most likely to be used in a response to a complaint after the installation is complete, rather than at the design stage. An approach to light trespass based on property rights that can be used at the design stage has been developed (Brons et al., 2008). This approach, called the outdoor site-lighting performance

TABLE 14.4

Maximum Vertical Illuminance (lx) on a Window in a Property in Four of the Environmental Zones, before and during a Curfew. There Is no Recommendation for the E0-protected Zone because There Is Unlikely to Be any Road Lighting There (after ILP, 2020b)

Environmental Zone	Maximum Vertical Illuminance on a Property before Curfew (lx)	Maximum Vertical Illuminance on a Property during Curfew (lx)
E1 Natural	2	0.1
E2 Rural	5	1
E3 Suburban	10	2
E4 Urban	25	5

(OSP) method, uses a virtual, transparent "shoebox" surrounding the property to be lit. The virtual "shoebox" has vertical sides at the property boundary and a flat "ceiling" 10 m above the highest mounted luminaire in the installation or the highest point of the property illuminated. For road lighting, the "shoebox" is extended along a representative, straight section of road; the vertical planes being located along the edges of the right-of-way. Conventional lighting design software can be used to calculate the illuminances falling on these planes and a simple illuminance meter can be used to check the actual illuminances once the installation is complete. By identifying the location and magnitude of the maximum illuminance on a grid covering the vertical sides of the "shoebox", the potential for light trespass can be established. Based on 33 representative designs of road lighting conforming to either European or North American standards, a maximum illuminance of 200 lx is suggested as the limit to avoid light trespass (Brons et al., 2008). This will seem a very high value compared with the recommendations for the maximum illuminance falling on a window in an adjacent property (Table 14.4), but it should be noted that the 200 lx is at the boundary of the property containing the lighting and will decrease rapidly as distance from that boundary increases. Of course, this simply indicates where a light trespass problem is likely to occur, not what the solution might be. Possible solutions include choosing a luminaire with a more suitable luminous intensity distribution, moving the luminaire further into the site and away from the property boundary, planting screening vegetation, and fitting some form of baffle to the luminaire.

The OSP "shoebox" approach is consistent with the way people think about property rights. Owning property confers considerable freedom of action on the owner within the property boundaries, provided those actions do not impinge negatively on others nearby or on the public good. The OSP "shoebox" approach is both flexible and realistic. It is flexible in that different maximum illuminance limits can be set by different communities, using the CIE environmental zones, if desired. It is realistic in that it uses widely available software to make the necessary calculations;

decisions on actions necessary to avoid light trespass can be made at the design stage; it does not require detailed knowledge of what surrounds the site being lit, knowledge that is often not available to the designer; and it includes the contributions of both direct and reflected light.

14.7.2 SKY GLOW

Sky glow is evident over most cities and towns in the form of a glowing, flattened dome of light (Figure 14.2). Sky glow has two components, one natural and one due to human activity. Natural sky glow is light from the sun, moon, planets, and stars that is scattered by interplanetary dust, and by molecules and aerosols in the Earth's atmosphere, and light produced by a chemical reaction of the upper atmosphere with ultraviolet radiation from the sun. The luminance of the natural sky glow at zenith is of the order of 0.0002 cd.m^{-2}. The contribution of human activity is produced by light traversing the atmosphere and being scattered by the air molecules and aerosols therein. Aerosols are suspended water droplets and dust particles. Air molecules scatter light forward and back with a little to the side. This Rayleigh scattering is much greater for short visible wavelengths, which is why the sky appears blue. Aerosols scatter light predominantly forward. This Mie scattering is independent of wavelength in the visible region, which is why clouds appear white.

Where there are few aerosols and few air molecules, there is very little sky glow which is why major new optical telescopes are built in such areas as the Atacama Desert of the Chilean Andes, where the population is small, the air pollution is negligible, and the air is very thin and dry. Established optical telescopes at low altitude and near large cities are of diminishing value, despite rearguard actions fought to minimize sky glow.

FIGURE 14.2   Sky glow over the city of Cape Town, South Africa (iStock by Getty Images / janiebros).

The magnitude of the contribution of electric lighting from a city to sky glow at a remote location can be crudely estimated by Walker's Law (Walker, 1977). This can be stated as:

$$I = 0.01.P.d^{-2.5}$$

where I is the proportional increase in sky luminance relative to the natural sky luminance, for viewing 45 degrees above the horizon in the direction of the city, P is the population of the city, and d is the distance from the remote location to the city (km).

This empirical formula assumes a certain use of light per head of the population. Experience suggests that the predictions are reasonable for cities where the number of lumens per person is between 500 and 1,000 lumens. More sophisticated models based on the physics of light scatter have been used to generate light pollution maps, these models making allowances for the curvature of the earth and allowing predictions to be made for different altitudes and azimuths of viewing and different atmospheric conditions (Garstang, 1986; Baddiley and Webster, 2007; Kocifaj, 2007).

The problem in dealing with sky glow is not in measuring or predicting its effects on the visibility of the stars, but rather in agreeing what to do about it. The problem is caused by two facts. The first is that sky glow is caused by multiple light sources, some private, some public. The second is that what constitutes the astronomer's pollution is often the business owner's commercial necessity and sometimes the citizen's preference. Residents of cities like their streets to be lit at night for the feeling of safety the lighting provides. Similarly, many roads are lit at night to enhance the traffic safety. Businesses use light to identify themselves at night and to attract customers. Further, the floodlighting of buildings and the lighting of landscapes are methods used to create an attractive environment at night. The fundamental problem of sky glow is how to strike the right balance between these conflicting desires.

One of the earliest attempts to achieve a balance involved the use of low-pressure sodium light sources for road lighting in cities adjacent to observatories. This approach was effective because astronomers could easily filter out the very limited range of wavelengths produced by the low-pressure sodium light source (see Figure 2.5). Unfortunately, this approach has fallen by the wayside, for two reasons. The first is that the non-existent colour rendering properties of the low-pressure sodium light source make it an unattractive prospect for towns and cities. The second is that the growth in the use of light outdoors by commercial and residential property owners, using a wide range of light sources, has simultaneously increased the amount of light being emitted and undermined the effectiveness of the use of monochromatic light sources by local authorities. As a result, low-pressure sodium light sources are rapidly disappearing from use (see Table 14.2), a process that will accelerate as they are no longer being manufactured. However, a substitute using a monochromatic amber LED is sometimes considered.

Today, the most common advice given on how to reduce light pollution is to use what is called a full cut-off luminaire which has no light emitted above the horizontal (see Section 4.2). Following such advice is straightforward because luminaires of this type are widely available. But this advice is flawed, for two reasons: The first is because restricting the upward light output from luminaires only limits the proportion of light emitted directly upward but ignores the light scattered after reflection from the illuminated surfaces. Light scattered after reflection is a major contributor to sky glow when viewing the sky from a position close to the sources of light.

The second reason why simply using full cut-off luminaires may not be the best solution to the problem of light pollution is that the advice considers the luminaire in isolation and not as a part of a lighting system. Keith (2000) calculated the upward luminous flux produced by a roadway lighting installation, per unit area of road illuminated, including both light directly emitted upward and light reflected from the road and its surroundings. Figure 14.3 shows the upward luminous flux density for

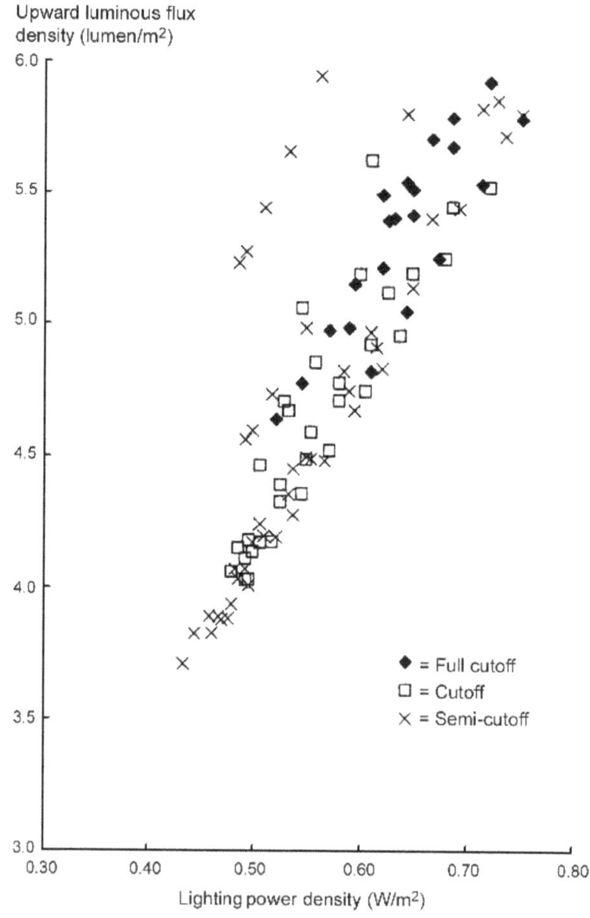

FIGURE 14.3   Upward luminous flux density (lm/m²) plotted against lighting power density (W/m²), for full cut-off, cut-off, and semi-cut-off luminaires. The values plotted are for a collector road of 10 m width, reflection class R3, lit by luminaires arranged in a staggered layout and using a 250 W high-pressure sodium light source (after Keith, 2000).

many different road lighting installations. The calculations were done for a collector road of 10 m width and of a specific reflectance, lit by lighting installations using a 250 W high-pressure sodium light source in full cut-off, cut-off, and semi-cut-off luminaires (see Section 4.2), arranged in a staggered pattern and spaced to meet average road surface luminance and luminance uniformity criteria.

What is interesting in Figure 14.3 is the fact that the upward luminous flux density increases as the lighting power density increases. What this implies is that if the use of full cut-off luminaires demands a closer spacing of the luminaires to meet the road lighting criteria, the lighting power density will be increased and, consequently, the upward luminous flux density will increase. Figure 14.3 also implies that a full cut-off luminaire is not necessarily the best option. In Figure 14.3, it appears that a semi-cutoff luminaire can give the lowest installed power density and the lowest upward luminous flux density, despite the fact that it may send a significant proportion of the lamp luminous flux directly up into the sky (Bullough, 2002a). This conclusion should not be too surprising given the small differences in the amount of uplight produced by different road lighting luminaires (see Table 4.1).

However, this conclusion assumes that all directions of light emitted above the horizontal produce the same level of scattered light. For a site remote from major sources of light, in clear sky conditions, which describes an observatory in operation, this assumption is not true (Baddiley and Webster, 2007). For these conditions, light emitted close to the horizontal causes more sky glow for most viewing directions than reflected light because reflected light is usually obstructed by the buildings around it and because light emitted close to the horizontal has a greater path length through an atmosphere that contains a greater density of molecules and aerosols. This is particularly significant for road lighting. Road lighting luminaires are designed to project light along the road at shallow angles. When the road surface is dry and diffusely reflecting, the reflected light will be evenly distributed over the upward hemisphere, but when the road is polished by traffic or wet from rain, strong specular reflections will occur at angles close to the horizontal. Thus, road lighting might make a greater contributor to sky glow than would be expected from the total light output, a contribution that is enhanced by the forward lighting of vehicles using the road. This is particularly important when a major road points directly towards a sensitive site such as an observatory (Waldram, 1972).

An alternative approach to limiting sky glow is to specify a maximum upward light percentage for a luminaire (ILP, 2020b). The upward light ratio of a luminaire is the percentage of the total luminous flux emitted by the luminaire that is emitted above the horizontal plane through the luminaire, i.e., that goes directly up into the sky. Table 14.5 shows the maximum upward light ratio recommended for a luminaire being used in the four CIE environmental zones where road lighting is permitted, expressed as a percentage of the total luminous flux emitted by the luminaire.

The maximum upward light percentage is an advance on the simple advice to use a full cut-off luminaire as a means of limiting sky glow in that it is quantitative. However, it is still inadequate in that it ignores the contribution of reflected light to sky glow. CIE (2017b) overcomes this limitation by considering the effect of the whole installation rather than just a luminaire. The metric is the upward flux ratio

TABLE 14.5

Maximum Upward Light Recommended for a Luminaire, Expressed as a Percentage of the Total Luminous Flux (lm) Emitted by the Luminaire, and the Maximum Upward Flux for an Installation, Expressed as a Percentage of the Total Luminous Flux Emitted by the Installation, in Different Environmental Zones. To Be Considered an Installation, There Must Be At Least Four Luminaires (after CIE, 2017b)

Environmental Zone	Maximum Permitted Upward Light Percentage for a Luminaire (%)	Maximum Permitted Upward Flux Percentage for a Lighting Installation (%)
E1 Natural	0	2.0
E2 Rural	2.5	5.0
E3 Suburban	5.0	8.0
E4 Urban	15.0	12.0

which is obtained by calculating the total upward luminous flux produced by the installation, which must have at least four luminaires, and expressing it as a percentage of the total luminous flux emitted by the installation. Table 14.5 also shows the maximum upward flux ratio for a road lighting installation expressed as a percentage for the four CIE environmental zones where road lighting is permitted.

Another approach that takes both direct and reflected light into account has been developed by the Association Francoise de l'Eclairage (AFE, 2006). This method calculates the maximum potentially lost luminous flux. This is given as the sum of three components: the luminous flux emitted above the horizontal plane through the luminaires, i.e., directly into the sky; the luminous flux reflected from the surface to be illuminated which for road lighting is the road surface; and the luminous flux reflected from the area surrounding the surface to be lit. The equation used for calculating the maximum potentially lost luminous flux for an installed luminaire is:

$$UPF = F_{la}\,(ULOR + p_1 u + p_2(DLOR - u))$$

where UPF is the maximum potentially lost luminous flux (lm), F_{la} is the luminous flux emitted by the light source in the luminaire (lm), $ULOR$ is the upward light output ratio, i.e., the proportion of luminous flux emitted by the light source that leaves the luminaire above the horizontal plane through the luminaire, p_1 is the reflectance of the surface to be illuminated, u is the utilization factor for the surface to be illuminated, i.e., the proportion of the luminous flux emitted by the light source that reaches the surface to be illuminated directly, p_2 is the reflectance of the surface surrounding the surface to be illuminated and $DLOR$ is the downward light output ratio i.e., the proportion of luminous flux emitted by the light source that leaves the luminaire below the horizontal plane through the luminaire.

The UPF is a maximum because it assumes that the area to be lit and its surroundings form a completely flat plane. It therefore ignores any reduction in luminous flux reaching the sky because light is absorbed or blocked by vegetation, buildings, or even the topography of the site and its surroundings. The *UPF* describes potentially lost flux because in urban situations, it may be that light reaching the surrounding area is itself useful in enhancing the safety of pedestrians and the quality of the visual environment.

The *UPF* is a minimum when the *ULOR* is zero and the utilization factor, u, is the same as the *DLOR*, i.e., all the luminous flux emitted by the lamp goes directly to the surface to be illuminated and nowhere else. The ratio of the actual *UPF* to the minimum *UPF* is called the upward flux ratio (*UFR*) and can be used as a measure of the efficiency of the installation.

Table 14.6 shows the results of calculations of *UPF* for a road lighting installation, using three different luminaire types representative of those used for road lighting in France. The road is 7 m wide, 1 kilometre long, and has a reflectance of 0.08. The surroundings have a reflectance of 0.10. The lighting is designed to produce an

---

TABLE 14.6

Installation Details, Achieved Photometric Conditions, Maximum Potentially Lost Luminous Flux (UPF) and Annual Electricity Consumption for Three Road Lighting Installations of the Same Kilometre of Road, Using Flat Lens, Shallow Bowl or Deep Bowl Luminaires (after Association Francoise de l'Eclairage, 2006)

Quantity	Flat Lens Luminaire	Shallow Bowl luminaire	Deep Bowl luminaire
Luminaire mounting height (m)	8	8	10
Luminaire spacing (m)	26	32	35
Tilt (degrees)	0	5	20
Overhang (m)	0	0	1
Utilization factor	0.38	0.43	0.30
ULOR	0	0.01	0.03
DLOR	0.79	0.81	0.79
Number of luminaires/km	38	31	29
Average initial illuminance (lx)	22	20	20
Overall luminance uniformity (minimum/average)	0.4	0.4	0.4
Longitudinal luminance uniformity (minimum/ maximum)	0.70	0.72	0.72
Threshold increment (%)	8	10	12
Maximum potentially lost luminous flux/km (lm/ km)	28,500	26,815	49,300
Direct upward luminous flux/km (lm/km)	0	3,255	14,355
Luminous flux reflected by road and surroundings/ km (lm/km)	28,500	23,560	34,945
Annual electricity consumption/km (kWh/km)	17,267	14,086	19,604

average initial illuminance of 20 lx, with an overall luminance uniformity ratio of 0.4, a longitudinal luminance uniformity ratio of 0.7, a maximum threshold increment of 15% and is to be used for 4,000 hours per year. Three different luminaire types are used, one with a flat lens, one with a shallow bowl, and one with a deep bowl, enclosures that approximate full cut-off, cut-off, and semi-cutoff luminaires, respectively. All three luminaires are used in a single-sided arrangement. The flat lens and shallow bowl luminaires are equipped with 100 W high-pressure sodium light sources, which have an initial luminous flux of 10,500 lm. The deep bowl luminaire is fitted with a 150 W high-pressure sodium light source, which has an initial luminous flux of 16,500 lm.

Close examination of Table 14.6 is worthwhile. Despite achieving very similar photometric conditions on the road, there are significant differences in maximum potentially lost luminous flux/km between the three luminaire types. Not surprisingly, the deep bowl luminaire with its 3% upward light output ratio (ULOR) produces the greatest maximum potentially lost upward luminous flux/km. What is interesting is that the shallow bowl produces the lowest maximum potentially lost upward luminous flux/km, lower than that of the flat lens luminaire. This is due to the fact that to achieve the same photometric conditions, a closer spacing is required with a flat lens luminaire than with a shallow bowl luminaire. As a result, the flat lens luminaire has more lamps per kilometre than the shallow bowl luminaire. This also explains why the annual electricity consumption/km is greater for the flat lens luminaire than for the shallow bowl luminaire. Also of interest are the relative amounts of upward luminous flux/km emitted directly and after reflection. The three luminaire types show different amounts of direct upward luminous flux, as would be expected from their different ULOR values, but, for all three, the amount of luminous flux reflected upward is greater than that emitted directly upward. The folly of ignoring reflected luminous flux when attempting to limit sky glow for observers close to the source of light should be apparent.

The OSP "shoebox" system described in the discussion of light trespass (see Section 14.7.1) can also be used to examine the amount of luminous flux lost from a property, directly and by reflection. For this purpose, the average illuminance is calculated on all five planes of the "shoebox". The average illuminance for each plane is multiplied by the area of the plane and the five resulting luminous flux values are summed. This sum gives an estimate of the total luminous flux leaving the property. This is a worst-case estimate of the contribution to sky glow because it is unlikely that all the luminous flux leaving the property will reach the sky because of obstruction and absorption by surrounding properties and landscape. Based on 33 representative designs of road lighting conforming to either European or North American standards, Brons et al. (2008) have shown that the total luminous flux leaving a roadway lighting installation (y) is linearly related to the average illuminance on the road (x) by the equation $y = 0.14x$. They suggest that this equation forms a sound basis for communities to decide on suitable limits to sky glow before any curfew. After curfew they suggest absolute limits to the amount of luminous flux leaving a property depending on the environmental zone. For zones E1 to E4, the limits are 1, 3, 10, and 30 *lx*, respectively.

All these metrics, upward flux ratio, maximum potentially lost luminous flux, and the total flux from the "shoebox", are attempts to quantify the amount of upward light generated by individual lighting installations. But if the number of exterior lighting installations continues to increase and the illuminances used in those installations continue to increase, then sky glow will inevitably increase. To effectively limit the contribution of human activity to sky glow, there are two complementary options.

The first option is to limit the amount of light used at night, accepting that if taken too far such limits will increase the risk to life and limb through decreases in visibility and consequent increases in vehicle crashes and/or crime. As would be expected, the CIE has produced a guide to limiting light pollution containing recommendations on technical matters such as limiting upward light ratio, minimizing upward flux ratio and setting maximum luminous intensities in specific directions (CIE, 2017b). The Illuminating Engineering Society of North America (IESNA) and the International Dark-Sky Association (IDA) have taken a rather different approach by producing a model lighting ordinance (IES, 2011). The difference lies not in the technical aspects but rather the assumed readership. The CIE document is aimed at the technically qualified while the IES document is aimed at local authorities who wish to reduce light pollution and who have the power to produce local regulations governing the use of light at night. An even simpler approach requiring little technical understanding of lighting would be to develop maximum lighting power densities for different activities in each zone. This is the approach adopted by the California Energy Commission (CEC, 2019). Maximum lighting power densities are specified for many outdoor lighting applications around buildings, such as parking lot lighting, retail display areas, façade and canopy lighting, and advertising signs. Unfortunately, the lighting of public roads is presently excluded from these limits, although why this should be so is not clear. Limiting power density will, unless there is a sudden major improvement in light source luminous efficacy, limit the installed luminous flux but it will not guarantee minimum upward light unless the design of the lighting is also controlled in some way. The Californian regulations attempt to do this in an ad hoc way by prescribing the equipment that should be used. A better approach, in that it would allow the lighting designer more freedom while still limiting sky glow, would be to use the OSP "shoebox" system to specify the maximum amount of luminous flux/unit lit area that should be allowed to leave a site, a different maximum being used for each of the different environmental zones.

The second option for reducing the sky glow is to pay careful attention to the timing of the use of light. Light pollution is unlike most other forms of pollution in that when the light source is extinguished, the pollution disappears very rapidly. This suggests that a curfew defining the times when lighting should not be used, or should only be used at a reduced level, could have a dramatic effect on sky glow. Curfews are commonly attached to the planning consents for sports lighting installations. The possibility of dimming road lighting in the middle of the night when traffic densities are low is a variation on this theme.

How successful these various approaches have been can be seen from the sky glow measured in recent years. The obvious place to start is the New World Atlas of Artificial Night Sky Brightness (Falchi et al., 2016). This claims to give a picture of the distribution of sky glow across the globe. Sadly, its title is a misnomer. Brightness is a perception of the human visual system. The data on which the atlas is constructed is taken largely from sensors on satellites. These sensors measure radiance rather than luminance so, unlike the human visual system, all wavelengths are given equal value. Further, the spectral sensitivity of the satellites' sensors cover the range 500–900 nm, which means they ignore the short wavelength end of the visible spectrum and extend into the near infrared. Given the increased use of LEDs for outdoor lighting, this is a serious deficiency for estimating sky glow as LEDs have a significant wavelength radiation below 500 nm (see Figure 2.5), wavelengths where Rayleigh scattering increases rapidly. Satellites are also insensitive to light emitted horizontally yet such light is more likely to interfere with the surrounding flora and fauna. Despite these limitations, the Atlas does show a broad picture of the relative distribution of one aspect of light pollution across the world. Countries with the most polluted skies based on the percentage of the population exposed include Singapore, South Korea, and Saudi Arabia. Those with the least polluted skies on the same basis include Chad, Somalia, and Madagascar. This division implies that light pollution is strongly influenced by a combination of population density and wealth per head of the population. Where both the population density and wealth are high, light pollution is high. Where both factors are low, light pollution is low. When the population density is low but the wealth is high, light pollution will be low because although the population can afford to use light at night, there are few to do so. Where the population density is high but the wealth is low, light pollution can again be low because the population cannot afford to use light at night and may not even have access to electricity. Of course, this is a very gross classification, different areas within the same country can have very different combinations of wealth and population density. For example, in Australia, the wealthy population is concentrated in the cities but the outback is largely unpopulated with the result that the skies in the outback are unpolluted. Similar variations in light pollution have been found in the USA and Europe (Falchi et al., 2019).

A very different approach to estimating sky glow has been used by Kyba et al. (2023). They used citizen scientists to estimate the visibility of the stars; 51,351 people received pictures of the clear night sky at different levels of light pollution and were asked to report which of the pictures best matched the visibility of the stars at their locality. From these data, it was concluded that in the decade since 2011 the night sky has been rapidly getting brighter at a rate of 7–10% per year.

Both sets of data suggest that the attempts to reduce sky glow worldwide have largely been unsuccessful although there are certainly local examples where dark skies have been achieved but these are usually in areas of low population density. It is now necessary to consider why this should be so. The obvious answer is that without any light, we humans are deprived of our principal sense, vision. With some light, we have some vision, although exactly what its capabilities are will depend on the amount, spectrum, and distribution of the light. Moonlight is sometimes enough to see the way ahead but rarely enough to identify colours. It is only when the sun rises that our vision achieves its full range of capabilities. Thus, the provision of light at night is designed to allow the visual system to function to some degree and this brings with it benefits (Boyce, 2019). Such benefits include greater safety for pedestrians and drivers, reduced fear of crime, more use of outdoor facilities after dark, enhanced economic growth, and the creation of built and natural environments that are a source of beauty and entertainment. This suggests that the use of light at night is linked to some very basic human motivations which in turn means that people value such benefits and will not willingly abandon them. This, in turn, suggests that the use of light at night will continue to grow as the world population continues to increase and as more people have access to lighting at a cost they can afford. This is probably why the Atlas (Falchi et al., 2016) shows major increases in light pollution in countries like India and China that are undergoing rapid economic growth. The best that is likely to be achieved is to limit the use of light at night rather than to abolish it. Fortunately, careful lighting design, soundly based outdoor lighting standards, the careful use of curfews and new lighting technology offer the possibility of providing the benefits of light at night while minimizing light pollution but it will require compromise and cooperation between people rather than conflict for this to be achieved. How successful such an approach can be, can be seen from a study of the contribution of road lighting to sky glow over the city of Tucson, Arizona (Kyba et al., 2021). In 2017, this city changed the light source used for road lighting from high-pressure sodium to LEDs. At the same time it reduced the total lumens emitted from 445 Mlm to 142 Mlm, installed a dimming capability and a central management system that allowed real-time power demand monitoring for each light source. A part-night lighting system is used, with the power demand being set to 90% of maximum for evening hours after dark but at midnight most of the road lighting is reduced to 60% of the power demand, except for intersections and pedestrian crossings. Over several successive nights in 2019, the city authorities agreed to dim the road lighting to 30% power after midnight, apart from intersections and pedestrian crossings, or to run it at 100% power throughout the night. With such controlled changes in light output of road lighting, it is possible to detect the contribution road lighting makes to the radiance detected by a satellite passing over the city. Specifically, with dimming to 30% power after midnight road lighting controlled by the City of Tucson contributed only 13% to the detected radiance. When such dimming was not applied, the contribution of road lighting to measured radiance was 18%. This is similar to that found in Flagstaff, also in Arizona (Luginbuhl et al., 2009), but lower than has been found in Reykjavik, Iceland (Hiscock and Gudmundsson, 2010), and two villages in Germany (Kyba et al., 2021). However, it does suggest that if the desire is to reduce light pollution, attention should be given to potential sources of light beyond road lighting, such as privately owned outdoor lighting and light escaping from buildings.

14.7.3 Glare

Both road lighting and vehicle lighting can cause disability and discomfort glare (see Sections 4.3 and 6.6). When glare is the cause of a complaint of light pollution, it is usually associated with road lighting rather than vehicle lighting and will be made by someone outside the lit area. Drivers, who are in the lit area, may complain about glare from the road lighting or from vehicles, but that glare is not light pollution.

TABLE 14.7

Maximum Luminous Intensity (cd) to Control Glare for the Four Environmental Zones, before and during a Curfew (after ILP, 2020b)

Environmental Zone	Maximum Luminous Intensity before Curfew (cd)	Maximum Luminous Intensity during Curfew (cd)
E1	2,500	0
E2	7,500	500
E3	10,000	1,000
E4	25,000	2,500

Bullough et al. (2008) have shown that discomfort glare from outdoor lighting is influenced by the illuminance received at the eye directly from the glare source, the ratio of this illuminance to the illuminance received at the eye from the immediate surround to the glare source, and the ambient illuminance received at the eye (see Section 4.3). The greater the illuminance received directly from the glare source, the higher the ratio of source/surround illuminances and the lower the ambient illuminance, the greater will be the discomfort. This understanding was incorporated into the OSP "shoebox", with a maximum de Boer discomfort glare rating of 4 being recommended (see Section 4.3). With this inclusion, the OSP "shoebox" method can deal with all the different aspects of light pollution; light trespass sky glow and glare (Brons et al., 2008).

The usual solution to a complaint of light pollution due to glare is to reduce the luminous intensity of the luminaire that is the subject of the complaint in the direction of the complainant. This can be done by changing the luminaire type, by re-aiming the luminaire or by fitting baffles to the luminaire to confine the luminous flux emitted by the luminaire more closely to the area that has to be lit. To indicate when such actions might be justified, maximum luminous intensities for each of the four environmental zones, before and during curfew, have been suggested by the Institution of Lighting Professionals (2020b). Table 14.7 shows these maximum luminous intensities. These maxima should be applied in whatever direction is relevant to the complaint, i.e., from the luminaire to the complainant.

14.8 SUMMARY

The constraints that limit lighting for driving range from the legal through the financial to the environmental. Vehicle lighting meeting specific photometric conditions is a legal requirement in almost all countries. Road lighting is not. Both vehicle lighting and road lighting imply costs. For vehicles, the first costs of lighting are included in the cost of the vehicle. For road lighting, both first costs and life cycle costs can be involved in the decision whether or not to install road lighting and certainly influence the amount and form of road lighting provided. As for environmental factors, these can take four forms: the amount of energy consumed, the carbon dioxide emissions produced when generating the electricity used, the disposal of equipment at the end of life, and light pollution. This chapter is devoted to a consideration of these constraints as they apply to road lighting.

The first decision that has to be made about road lighting is whether to install it or not. There are two approaches to making this decision. One is prescriptive and involves meeting a series of quantitative and qualitative criteria related to traffic flow, night/day crash ratio, road geometry, and ambient conditions. The other is a comparison of the costs and benefits of road lighting. The costs of road lighting come in two main forms, first cost and life cycle cost. For a new installation the first cost per kilometre is calculated as the sum of the cost of the light source, luminaire, column, and connection to the electricity supply, multiplied by the number of columns along a kilometre of road together with the cost of the work required to install the equipment. Depending on the proximity of the electricity supply this item can easily dominate the first cost per kilometre for a new installation. For a replacement installation, the first cost per kilometre is limited to the cost of light source and luminaire and the labour required to make the change as the column and electrical connection are already available. Calculating life cycle costs per kilometre is rather more complicated. In its simplest form, you need to know the first cost per kilometre, the number and power demand of the luminaires, the hours of use, the cost of electricity, the proposed maintenance cycle, and the discount rate. The life cycle of road lighting is usually assumed to lie between 25 and 40 years, so assumptions have to be made about the costs of electricity and interest rates many years ahead. As for the benefits, this is even more complicated because it involves the estimation of the reduction in number of crashes likely to occur if the road lighting is installed and the savings to be made by preventing cashes. These saving include both measurable elements, such as the cost of any hospital treatment, police costs, and material damage, and imponderable elements such as the costs of lost productivity and the value of human pain and suffering.

Despite the inevitable uncertainties, calculations of the costs and benefits of road lighting and the promotion of the high luminous efficacy and long life of LEDs have led to the widespread replacement of high-pressure sodium lighting with LED lighting on roads. Another trend has occurred in controls, either via individual luminaires or through central management systems. Almost all new LED road lighting luminaires have a programmable dimming capability. Local authorities have frequently used these capabilities to reduce light levels late at night by dimming when traffic densities and pedestrian activity are low, usually between midnight and 05:00.

Road lighting can also be constrained by four aspects of the environment: energy consumption, carbon dioxide emissions, waste disposal, and light pollution. There are three reasons why energy consumption might become a constraint on the design of road lighting. The first reason is the cost. The higher the price of electricity, the higher the cost of operating the installation. The second reason is the need to reduce the maximum demand on the electricity network. This is very unlikely to be relevant, except in countries at high latitudes, as the maximum demand usually occurs at times of maximum activity by the population, which is usually during the day, but road lighting operates at night. The third reason is the desire to minimize the use of fossil fuels. Given this desire, any use of electricity should be avoided unless it is generated from renewable resources such as nuclear, sun, wind, or waves.

A desire to reduce carbon dioxide emissions makes road lighting an obvious target for cuts. It consumes electricity, it is conspicuous, and it is often in use at times when there are few people about. However, before cutting road lighting, it is important to remember that the amount of carbon dioxide emitted will depend on how the electricity used to power the road lighting is generated. Different countries use very different fuel mixes for electricity generation. There is little point in limiting road lighting in countries where the fuel mix used to generate electricity produces little carbon dioxide, but where gas, oil, or coal are extensively used for electricity generation, then the need for road lighting should be carefully considered.

Another growing constraint on road lighting is the ease of disposal of equipment at the end of life. Questions about waste disposal can take two forms, the hazard to human health posed by the materials in the product and the ease of recycling. The elements of a road lighting installation that cause the most concern are the light source and the control gear. Light sources that contain mercury are likely to be subject to stricter regulation than those that contain sodium. Electronic control gear, including LED drivers, are likely to be more expensive to dispose of than electromagnetic control gear.

There is also growing concern about light pollution. Light pollution is evident as localized light trespass and glare and a more general increased sky luminance called sky glow. Light trespass and glare can be avoided by the careful positioning and aiming of luminaires with appropriate luminous intensity distributions. For road lighting, this means choosing luminaires that direct the vast majority of the light emitted onto the road surface and minimize the amount of light emitted beyond that surface, on both sides of the road. If complaints of light trespass still occur, the classic response is to fit what is called a house-side shield to the luminaire. If glare is still a problem, this can be dealt with by re-aiming the luminaire or fitting a baffle.

Sky glow is evident over mo t cities and towns in the form of a glowing, flattened dome of light. The effect of sky glow is to make many stars invisible because their luminance contrast is reduced to below threshold. There is a small natural contribution and a variable human contribution to sky glow. The magnitude of the human contribution is dependent on the density of air molecules and aerosols in the atmosphere and the amount of light traversing the atmosphere. The greater is the amount of scattering material and light in the atmosphere, the greater will be the sky glow.

There are a number of approaches being introduced to road lighting practice to minimize light pollution. The most common is the simplest, using only full cut-off luminaires that do not emit light above the horizontal. This approach is good for limiting glare but less useful for reducing sky glow because it ignores light reflected from the road surface. Another approach takes both direct and reflected light into account and measures the efficiency of a road lighting installation by calculating the proportion of light emitted by the luminaires that directly illuminates the road surface. By maximizing this proportion, the amount of wasted light that contributes to sky glow is minimized. Yet another approach is based on the idea of a virtual, inverted "shoebox" being placed around and over a property. For road lighting, the "shoebox" is extended along the axis of the road as far as desired. For roads, the idea is to specify the maximum luminous flux that should be allowed to cross the vertical planes parallel to the road and the "ceiling" plane over the road. This approach can be used to control sky glow, light trespass, and glare.

All three of these approaches are attempts to quantify the amount of upward light generated by individual lighting installations. But if the number of exterior lighting installations continues to increase and the illuminances used in those installations continue to increase, then sky glow will inevitably increase. Attempts have been made to limit the amount of light used outdoors through the planning system by classifying locations into different environmental zones and then setting either maximum luminous flux density limits or power density limits for each zone. An alternative approach that is sometimes used for sports facilities and could be used for road lighting is a curfew that restricts the use of the lighting to certain times. Long-term, limiting the amount of light that can be used, and the times when the lighting can be used are the only approaches to reducing sky glow that are likely to be successful.

All forms of lighting are subject to constraints. Road lighting is no exception. Cost constraints will always be present but the importance attached to the environmental constraints will vary over time as their political impact changes. Given the increasing evidence for the contribution of carbon emissions to climate change and the emotional appeal of such slogans as "Our children will not see the stars" it is reasonable to predict that, in the future, road lighting will be under greater pressure to justify its existence than in the past despite the evident benefits it bestows.

15 Envisioning the Future

15.1 INTRODUCTION

This chapter is different from all the others in that it is an attempt to envision the future, so it contains more opinions than facts. These opinions are based on a synthesis of the research and practice that has been the foundation of the previous chapters, but they are still opinions. The future can be envisioned at two levels: One attempts to predict what is most likely to happen; the other attempts to set out what should happen. The former is relatively straightforward because it is limited in time and is likely to be driven primarily by costs, benefits, and the opportunities presented by emerging technology. What these two factors suggest for lighting for driving is that over the next decade, vehicle forward lighting will become more sophisticated with elements of the adaptive forward lighting system most attractive to vehicle purchasers trickling down from up-market vehicles to more modestly priced vehicles. Vehicle signal lighting will continue to change as its design ceases to be just a simple means to convey information to other drivers and becomes an important part of the styling of the vehicle. Road lighting is likely to continue its transition to a more flexible service based on a combination of LED light sources and central management systems that enable dimming on demand. As for road markings and signs, these are likely to continue to grow in number and complexity as combinations that reinforce perceptual cues have been found to be effective in influencing drivers' behaviour (Charlton, 2007). In addition to these changes, it is likely that the mix of vehicles found on roads and sidewalks will change as mobility scooters and e-scooters become more widely used. These vehicles deliver great benefits in terms of mobility and convenience, but that can prove hazardous to their riders and pedestrians, and the laws concerning where, when, and how they are used vary widely. Further, their lighting is currently limited. It is to be hoped that public authorities will soon update lighting regulations to take such personal mobility vehicles into account.

Attempting to set out what should happen to lighting for driving is simultaneously more difficult and less likely to be successful. It is more difficult because it involves a longer time scale, so it is more open to unforeseen influences. It is less likely to be successful because what should happen is a fundamental review of current practice by vehicle manufacturers, road engineers and illuminating engineers working together, but these are groups that do not normally interact. Nonetheless, without ideas about how lighting for driving might be made more effective in the future, any action is unlikely. The approach taken here starts by identifying the fundamental problems facing the lighting used while driving and then seeks to identify solutions to those problems, solutions that are effective and without unintended consequences. Attention will be given first to vehicle lighting because it is used whenever the vehicle

DOI: 10.1201/9781003388906-15

is in motion after dark, and sometimes during the day, regardless of whether the road is lit or not. This will be followed by a discussion of the future of road lighting and possible developments in road markings, signs, and signals. Finally, the factors that may determine whether or not these opinions become facts are considered.

15.2 VEHICLE FORWARD LIGHTING

15.2.1 PROBLEMS

Vehicle forward lighting has changed dramatically over the last decade, but it still faces two fundamental problems. The first is the limited range of visibility available when using low-beam headlamps on unlit roads, which are the majority of roads. This is a problem because stopping distances at the speeds usually driven are much longer than the visibility distance (see Section 6.4). Whether or not stopping distances matter will depend on the opportunity for manoeuvre. For a pedestrian at the edge of a wide road, the driver may have the opportunity to swerve rather than stop, but on a narrow road or a road with heavy traffic, swerving may not be possible, and when encountering a large vehicle stationary in the lane of a busy road, stopping may be the only option. Given that an emergency stop is necessary but the stopping distance is longer than the visibility distance, a collision is inevitable although the consequences may be reduced from death to injury if the vehicle's speed can be lessened before impact.

It is also important to appreciate that visibility distances vary greatly with the size and luminance contrast of the target. Targets of small size and low luminance contrast have much shorter visibility distances than targets of large size and high contrast. This partly explains why the type of crashes which show the biggest effect of the absence of daylight are those involving small, low-contrast targets such as pedestrians wearing dark clothing and animals (see Section 1.5).

The number of miles driven on unlit roads where limited visibility distances occur is increased by the underuse of high-beam headlamps (see Section 6.4). This underuse arises from the second problem that vehicle forward lighting faces, glare between opposing vehicles. When facing an approaching vehicle drivers are taught to change from high-beam to low-beam headlamps in order to limit the extent to which the vision of the other driver is degraded. If such switching is required frequently, there is a tendency to keep the headlamps on low beam regardless of the resulting reduced visibility distances. Of course, the impact of glare from headlamps is not confined to the choice between low- or high-beam headlamps. The main effect of glare from oncoming headlamps is to reduce the luminance contrasts of all that lies ahead (see Section 6.6). This reduction is due to the veiling luminance created by light from the oncoming headlamps being scattered in the eye, at the windscreen and in the atmosphere. This is particularly a problem for old drivers who experience much more scatter in the eye than young drivers (see Section 13.4). Further, because the additional light from the oncoming headlamps changes the state of adaptation of the retina, immediately after the oncoming vehicle has passed, the driver is likely to be misadapted and it takes a finite period of time to recover the necessary sensitivity.

Recovery of sensitivity after exposure to glare is again a particular problem for old drivers who, because of increased absorption of light on passage through the eye, are effectively working at a lower light level than young drivers and who, hence, take longer to recover from exposure to glare. This conflict between visibility and glare has been a feature of the design of vehicle forward lighting since its inception.

15.2.2 SOLUTIONS

The first thing to say is that tinkering with existing headlamp luminous intensity distributions and spectra is unlikely to solve the problem of limited visibility distances caused by the use of low-beam headlamps when it is necessary to reduce glare to other drivers (Sivak, 2002). What is required are much more dramatic shifts in approach. There are four possible solutions to the problem of the short visibility distances available with low-beam headlamps. The first is to have road lighting on all roads. Good road lighting extends visibility distances, but this is a council of perfection. The cost of lighting every remote country lane would be colossal and the complaints of those concerned with light pollution would be deafening.

A more realistic option is to do nothing. This might be called the Darwinian solution because it effectively considers pedestrians foolish enough to walk on unlit roads at night in dark clothes as in the shallow end of the gene pool and as expendable. Such an approach is implicit in the system for deciding whether or not to invest in road lighting (see Section 14.2). The number of crashes involving death or injury occurring at night within a set time is a factor to be considered when deciding whether or not to install road lighting. In addition to the adverse consequences for pedestrians, doing nothing really does do nothing to reduce the risk of crashes involving other unexpected low-contrast obstacles such as animals, unmarked vehicles, potholes, or unusual, unlit road configurations.

Doing nothing is unlikely to be acceptable to nations devoted to reducing road deaths. Much more likely is a legislative solution. This could take two forms. One possibility would be to introduce lower speed limits on unlit roads after dark. Given a slow enough speed, the visibility distance of even low-contrast objects illuminated by low-beam headlamps can be made greater than the stopping distance. Even if the speed limits required to achieve this objective were to prove to be too low to be politically acceptable, any reduction in speed at night would reduce the damage to people and property should a collision occur. The disadvantage of setting lower speed limits on unlit roads after dark or in conditions of poor visibility would be the increase in journey times and the risk of increased traffic congestion although this might be offset for some authorities by the possibility of increased revenue derived from the speed cameras used to enforce the lower speed limits.

Another legislative possibility would be to demand that unlighted users of the roads, such as pedestrians and some cyclists, should wear clothing fitted with a minimum area of retro-reflective material at night. Luoma et al. (1995) measured the distances at which pedestrians in dark clothing with and without retro-reflective material could be recognized while walking beside or crossing the road. The observers were passengers in a car being driven with low-beam headlamps on unlit rural roads who had been instructed to respond only to pedestrians. Figure 15.1 shows the mean recognition distances for different forms of retro-reflective material for pedestrians walking towards the car at the edge of the road and crossing the road. Clearly, almost any form of retro-reflective material increases the recognition distances dramatically, but those mounted on parts of the body that are in frequent motion when walking are more effective than those that are not, especially for crossing the road. Such findings explain why those employed to work on and near the roads, by day or night, use high-visibility suits, jackets, or vests, clothing that contains fluorescent and retro-reflective materials. Such clothing has not yet been adopted by the general public for use at night, but given the example of the dramatically increased use of cycling helmets over recent years it is not beyond the realm of possibility that promotion of the idea would make legislation on this matter unnecessary.

The third solution would involve technology applied to the vehicle. Technology to automatically switch between high-beam and low-beam headlamps depending on the presence of approaching vehicles is now widespread. This technology eliminates the problem of underuse of high-beam headlamps, but does nothing for visibility distances when low-beam headlamps are in use because of the proximity of approaching vehicles. A technology solution that eliminates this problem is the adaptive headlamp (see Section 6.7.3). Using either an addressable LED array or an addressable micro-mirror illuminated by a small LED source to take out the part of the luminous intensity distribution that causes glare to an approaching driver enables the forward lighting to be maintained in high beam even as opposing vehicles pass (Bullough et al., 2016). But this is only one form of adaptive forward lighting. Using sensors and software to explore the road ahead, adaptive forward lighting can better match the luminous intensity distribution to the prevailing conditions, conditions

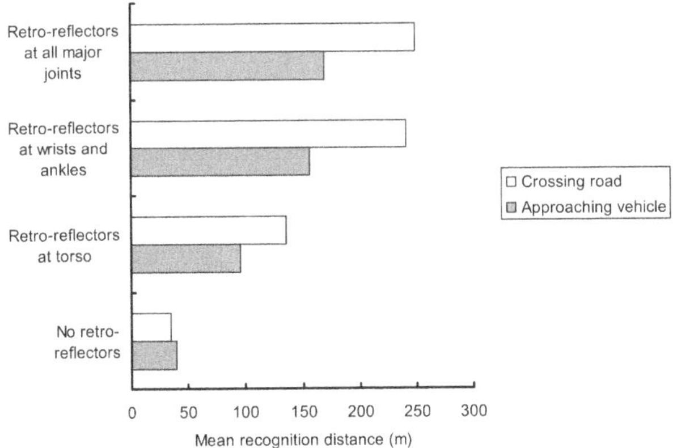

FIGURE 15.1 Mean recognition distances (m) for pedestrians approaching a vehicle while walking along the edge of the road or crossing the road. The recognition was made by passengers in a car being driven using low-beam headlamps, on an unlit rural road at night. The pedestrians wore dark clothing, some being fitted with retro-reflective material in various places (after Luoma et al., 1995b).

that can range from manoeuvring round curves, driving on motorways, in towns, on wet roads, or in fog. It is thought that a better match of the luminous intensity distribution of forward lighting to prevailing conditions will help reduce some types of crashes involving pedestrians (Sullivan and Flannagan, 2007).

Another possibility is to provide supplementary information to the driver, based on sensors. Both passive and active infrared systems can be used to generate an image of the road ahead and to display that image on a screen so that it is accessible to the human visual system. The passive infrared system uses the long-wavelength infrared radiation associated with the different temperatures of the various elements of the scene to create the image. The active infrared system emits short-wavelength infrared radiation and uses the differences in reflection from different elements of the scene to create the image. As the passive infrared system does not emit any additional electromagnetic radiation and the active infrared system only emits electromagnetic radiation outside the visible wavelength range, neither can cause glare to the human visual system, which means both can be kept perpetually in a state approximating high beam, even when other vehicles are approaching or close ahead.

How effective such supplementary information can be is shown by the work of Sullivan et al. (2004b). They had groups of young and old drivers drive a car equipped with a passive infrared system around a closed, unlit circuit using low-beam headlamps. The passive infrared system was mounted in either a heads-up position at the base of the windscreen or lower down and to the right, close to the central console. While driving round the circuit, there were targets of three types to be detected, pedestrians and warmed simulations of deer and small animals, all located at the edge of the road and all having a reflectance of 0.10. Table 15.1 shows the percentage of targets detected and the mean distance at which they were detected, using unassisted vision and vision assisted by the passive infrared system in the two positions.

Table 15.1 shows that the old drivers, whose mean age was 68 years, detected fewer of the targets than did the young drivers, whose mean age was 25 years, with and without the passive infrared system. When the targets were detected, the old

TABLE 15.1

Percentage Detection and Mean Detection Distance (m) of Pedestrians, Simulated Deer and Small Animals beside the Road by Young and Old Drivers Using Unassisted Vision and Using Assisted Vision, the Assistance Being Provided by a Passive Infrared System with a Head-Up Display (HUD) and a Display Mounted Down and to the Right (HDD) (from Sullivan et al., 2004b)

Driving Performance	Unassisted Vision		Assisted Vision with HUD Display		Assisted Vision with HDD Display	
	Young	Old	Young	Old	Young	Old
Per cent detection (%)	100	77	95	80	94	75
Mean detection distance (m)	57	32	85	51	115	48

drivers detected them at shorter distances than did the young drivers, with and without the passive infrared system. This is to be expected given the decline in visual and cognitive capabilities of older people. What is interesting is that, for both age groups, having the passive infrared system did not increase the percentage of detections but did increase the distance at which detection occurred, when it occurred. It is clear that providing supplementary information to the driver has the potential to alleviate the limitations of low-beam headlamps but not to overcome the effects of age. As for the form in which the head-up display delivers information, Table 15.1 shows little difference between the two locations of the display for the percentage detections for young or old drivers, but young drivers do seem to benefit from having the display placed down and by the central console rather than projected onto the windscreen for the mean detection distance. It should also be noted that such displays have to be easily read over a wide range of ambient conditions. Huang et al. (2023) found that black text on a white background was preferred during the daytime, but white text on a black background was preferred after dark.

A potential problem with delivering supplementary information to the driver is that of cognitive overload. Presenting information as a displayed image requires the driver to fixate that image during which time the driver will not be looking at the road. How long it takes to extract information from the display should depend on where it is positioned, how much information the display contains, and how much practice the driver has had with the system. However, Sullivan et al. (2004b) found no statistically significant differences in percentage detection nor mean detection distances for the two forms of passive infrared system that differed in where the display was located (see Table 15.1). Further, indirect measurements of workload, obtained from either the NASA Task Load Index questionnaire or the high-frequency component of steering, failed to show any difference in workload when the passive infrared imaging system was used to supplement the driver's vision, although there was an increase in workload when the driver was told to search for targets. This would suggest that, in this case, the workload was determined by the task defined rather than the means used to do it. Of course, these results were collected in undemanding conditions. There was no other traffic and the road layout was simple, both situations that would make performance insensitive to time spent looking at the display. Whether the position of the display becomes more important in more demanding conditions remains to be determined. As for the form of the display, the increasing use of satellite navigation displays has been beneficial as it has provided a lot of experience of the best ways to deliver information to drivers through displayed images. The images used for satellite navigation tend to be schematic rather than actual and the visual information is often augmented by auditory information. This suggests that there is a lot to be said for delivering the information about what lies ahead in a predigested form, in two sensory modes, rather than as a simple image that the driver has to analyse. Predigested information should be limited to what is essential even if it is restricted to a warning that there is something beyond the range of the headlamps that deserves attention (Tsimhoni et al., 2005).

Another development has been to link the use of passive infrared detection to an adaptive headlamp to deliver a narrow beam of light directly onto the object detected, typically a pedestrian or animal in or close to the road. This itself will draw the driver's attention to the pedestrian or animal, but pulsing the beam would provide an even stronger response because it causes a repeated change in the peripheral visual field, something that peripheral vision is designed to detect.

Another approach to solving the problems of vehicle forward lighting is to reduce the role of drivers and their visual systems in detection and decision-making. This is well on the way to being achieved in specific situations with the increased use of advanced driver-assistance systems such as autonomous emergency braking and autonomous cruise control (see Section 15.4), but the ultimate aim is a truly autonomous vehicle, one where there is no need for a driver at all. Such a vehicle would have to be able to find its way through traffic, anywhere, by day or night, using only its sensors, software, and actuators. Given the increase in the sophistication of vehicle systems that has occurred over the last 50 years, the driverless vehicle is not science fiction. Indeed, such vehicles are currently being field-tested as taxis in a number of cities in the USA.

15.3 VEHICLE SIGNAL LIGHTING

15.3.1 PROBLEMS

Vehicle signal lighting as currently practised has two problems. Vehicle signal lighting is designed to inform other drivers of the presence of the vehicle or changes in the movement of the vehicle. One problem is that of confusion. The implementation of LEDs as the light source of choice for vehicle signal lighting and the minimal requirements of the relevant standards have meant that some signals can be lost in a confusion of styling. This is not a problem for detecting the presence of a vehicle but rather of understanding the intention of the driver giving the signal. The other problem faced by vehicle signal lighting is adaptation to prevailing conditions. This is best exemplified by the use of rear fog lamps. Rear fog lamps have a much higher luminous intensity than rear-position lamps so as to ensure they are visible through dense fog (see Section 7.12). The problem occurs when drivers use their rear fog lamps in faint mist, the result being glare for drivers immediately behind.

15.3.2 SOLUTIONS

The problem of matching the signal lighting used to indicate the presence of the vehicle to prevailing conditions can be readily solved with existing technology. Many vehicles today have a photocell sensor that detects changes in ambient light levels so that forward lighting can be turned on or off as needed. It does not call for too great a leap of imagination to see the same system being used to provide a continuous adjustment of luminous intensity rather than to make a simple on/off switch. Such a system could be applied to front and rear-position lamps, side marker lamps, rear fog lamps, and daytime running lamps so as to ensure conspicuity without glare. Of course, a similar adjustment would need to be applied to intermittently lit signal lighting designed to convey an intended change in movement, such as turn lamps, hazard flashers, and stop lamps, to ensure they could still be easily discriminated from the continuously lit signal lighting.

One specific benefit of technology has been the ease of reversing. In their simplest form, reversing lights provide illumination of what lies behind the vehicle so the driver can use his or her visual abilities to avoid any obstacles. The first technology applied was reversing sensors that detected the proximity to any obstructions behind and warned the driver of the danger by emitting an intermittent audible tone, the frequency of the intermittency increasing as the distance to the obstruction shortened, ultimately becoming a continuous tone when very close. This is a help but still leaves the driver to use his or her visual abilities to detect the obstruction. A marked improvement on this has been the introduction of a rear-view camera which displays the view behind the vehicle on a screen. Even better is when other information collected from the vehicle controls is used to produce a framework indicating where the vehicle will be located as it reverses.

There are also a number of ways in which other forms of signal lighting might be used to enhance traffic safety. The easiest to implement would be lighting to ensure stable perception of movement in varying conditions. A vehicle moving on a road at night has multiple luminance contrasts. In fact, there are so many that it is very unlikely that the vehicle will ever become invisible. What is much more likely is that the pattern of luminance contrasts associated with the vehicle will change as it moves along the road, particularly if the road is lit and wet because then the background luminance against which the vehicle is seen will vary much more than when the road is unlit and dry. The opportunity is to use signal lighting to produce a stable pattern of luminance contrasts. There are two possible approaches. The first is to use contour lighting (Wordenweber et al., 2007). Contour lighting involves the outlining of the vehicle by self-luminous means, such as regularly spaced marker lamps or side-emitting fibre optics, or by retro-reflective material. Contour lighting is required on heavy trucks to make them easy to detect when pulling out onto another road in such a way that the truck is seen side on. Sadly, it is rarely used on smaller vehicles. The other approach is diffusing lighting placed under the vehicle. The idea behind this approach is to ensure that the luminance of the road against which the vehicle is seen is relatively stable so that the pattern of luminance contrasts associated with the vehicle is itself relatively stable. This is something that is already available in the after-market although it has acquired an unfortunate social connotation and has been banned in some jurisdictions. For under vehicle lighting to be effective, regulation would be required to fix the amount, distribution, colour, and stability of the lighting.

Both these approaches seek to provide a stable outline for the vehicle. The value of such a stimulus is that drivers can detect outlines faster through the magnocellular channel of the human visual system than information that passes through the parvocellular channel, such as colour (see Section 3.2). Further, an outline enables the type of vehicle to be identified and that, in turn, enables the experienced driver to have some idea of the type of movement of which the vehicle is capable. Of the two approaches, contour lighting is probably the more useful as it will be visible from many different directions over greater distances and, when provided by retro-reflective material, can signal the presence of the vehicle to other vehicles even when there is no power available for lighting. Such information may be useful in reducing the number of rear-end collisions at night, particularly collisions with vehicles broken down on the road (Morgan, 2001; Sullivan and Flannagan, 2004).

Another area where vehicle lighting could be used to provide more information important to other drivers is related to the perceptions of speed and distance. These perceptions require changes in the stimuli presented to the visual system relative to each other. By day, this is not a problem because the vehicle can be seen against a structured background. By night, on an unlit road, the perceptions of speed and distance are much more difficult, particularly when the vehicle is coming directly towards or away from the driver. In this situation, perceptions of speed and distance for vehicles approaching and vehicles ahead but moving in the same direction are governed by changes in the relative position of pairs of headlamps and rear-position lamps, respectively. The contour lighting discussed above would make speed and distance easier to judge as do well-separated pairs of headlamps and rear-position lamps. What is not helpful is having a single headlamp or rear-position lamp, as is effectively often the case for motorcycles, or a single rear fog lamp or a single rear reversing lamp as is allowed by some vehicle lighting regulations. Judging speed and distance from the expansion or contraction in the angular size of a single small lamp is much more difficult than using the change in separation between pairs of lamps. This suggests that single rear fog lamps and reversing lamps on cars and vans should be forbidden. For such vehicles, all signal lamps should be in pairs mounted on opposite sides of the vehicle.

Another aspect of signal lighting that deserves attention is the ability of drivers behind the immediately following vehicle to receive the signal. Vehicle signal lighting that is mounted close to the road may be masked from other drivers by the vehicle immediately following. One of the claimed advantages of the centre high-mounted stop lamp (CHMSL) is that it allows the operation of the brakes to be detected by drivers when there is another vehicle between because it can be seen through the windows of the immediately following vehicle. This is correct when the immediately following vehicle is a car but may not be true if it is a van or truck. This limitation could be alleviated somewhat if the conventional rear signal assemblies consisting of pairs of rear positions, stop, turn, fog, reversing, and hazard warning lamps were to be duplicated high up on the vehicle, at least on vehicles with a height of more than about two metres.

The main opportunity that is missed with current vehicle signal lighting is to provide other drivers with additional information. For example, the application of the brakes is shown by the lighting of the stop lamps, but that does not discriminate between a gentle touch and heavy braking. For that to be discriminated, the drivers of the vehicles behind have to perceive the magnitude of the deceleration of the vehicle ahead. It would certainly be possible to indicate the force with which the brakes had been applied by linking the area, luminous intensity, or flash rate of the stop lamps to that force. A similar but less common opportunity applies to reversing lamps. The intention to reverse is shown by the lighting of the reversing lamps. Modifying the CHMSL so that it would also operate as a high-level, white reversing lamp that would attract attention by flashing or pulsing would be advantageous. Pak et al. (2024) examined the effect of a turn signal guide lamp combined with a conventional turn signal on the time taken to recognize an intended change of direction. The turn signal guide lamp projects a luminous pattern on the road ahead and to the left or right of the vehicle. The results showed that when used with the conventional turn signal, the time taken to identify the intended change of course was reduced from what it was when the conventional turn signal alone was used without causing additional glare.

All these suggestions rely on lighting to convey the message, but automatic vehicle-to-vehicle communication is surely imminent. This will allow information of greater complexity to be conveyed by means other than looking through the windscreen. An example is a radio message broadcast over a limited range whenever a vehicle's hazard warning lamps are activated. Such a message could be integrated into a satellite navigation system to indicate to approaching drivers where the hazard is. Similar messages might be generated when a crash was detected either by those involved or by passing vehicles or through the emergency services. What this suggests is that the days of vehicle signal lighting as the only means to indicate the presence of a vehicle or the intentions of the driver are numbered. The fact is, as a means of conveying detailed information vehicle, signal lighting is rather primitive, closer to Morse code than the mobile phone. It cannot be long before additional sources of information operating through wireless networks are adopted. Such systems already exist for avoiding distant traffic congestion. What is required is for information to be exchanged between vehicles in close proximity about intentions such as braking or changing lanes. This is certain to occur as part of the development of autonomous vehicles but it should not be taken to mean that all signal lighting will disappear from vehicles. There are other people on and near the road, such as cyclists and pedestrians, who need to know about the presence of vehicles and the intentions of their drivers and who are unlikely to be equipped with the necessary technology. Rather, it means that vehicle-to-vehicle communication is most likely to supplement vehicle signal lighting than replace it entirely.

15.4 ADVANCED DRIVER-ASSISTANCE SYSTEMS

Driver-assistance systems such as automatic gear change and cruise control have been features of vehicles for many years. Their function has been to make the driver's task easier by eliminating some routine activities. To do this they use information from within the vehicle's machinery. Over the years, more sophisticated forms such as anti-lock brakes and electronic stability control have been introduced. Advanced driver-assistance systems are another step forward in that they use cameras, sensors, and software to examine what is happening ahead, behind, and at the sides of the vehicle. Advanced driver-assistance systems take two forms: those that deliver a warning to the driver but leave the driver to decide what to do and those that take over control from the driver. Among the former are blind spot warnings, forward collision warnings, and lane departure warnings. Among the latter are adaptive cruise control, automated emergency braking and intelligent speed assist. These systems are most effective when they interact. For example, when a forward collision warning is ignored by the driver, the automated emergency braking system takes over to avoid a collision or, at least, reduce the impact when the collision occurs. A subset of advanced driver-assistance systems concentrates on the state of the driver. The idea here is to identify drowsiness by monitoring physiological and behavioural signs and to warn the driver visually, audibly, or by seat vibration. If this fails and the driver falls asleep, an emergency assist programme may take over and stop and position the vehicle away from oncoming traffic and turn on the hazard flashing lights. The advanced driver-assistance systems fitted by different manufacturers to different vehicles in their range vary widely. Further, the extent to which they can be deactivated or engaged also varies from one vehicle to another. The result has been a call for standardization but this has yet to happen. What is clear is that technology has developed to such an extent that reliance on the performance of a human driver alone to ensure traffic safety is diminishing. This road leads ultimately to autonomous vehicles. But this is still some way away. Until that nirvana is reached, the interest in advanced driver-assistance systems is being driven by the belief that they offer the possibility of delivering a major improvement in traffic safety.

Wang et al. (2020) give estimates of the percentage reductions in crashes that advanced driver-assistance systems might produce, based on crash statistics. Table 15.2 gives the 95% confidence intervals of the mean percentage reductions attributable to different advanced driver-assistance systems. Such estimates are encouraging but fail to reflect the influence of the context in which the crash occurred. The context is important because different systems are likely to be more or less effective in different conditions. For example, automated emergency braking is likely to be more successful when the road is dry rather than when it is wet. Also, weather conditions, such as heavy rain or fog, and lighting conditions, such as daylight and darkness, may reduce the ability of the system to detect a hazard. Masello et al. (2022) have addressed this issue and provide estimates of the potential reductions in

---

TABLE 15.2

Ninety-five Per Cent Confidence Limits of the Percentage Reduction in Crashes When Different Types of Advanced Driver-Assistance Systems Are Available (from Wang et al., 2020)

Advanced Driver Assistance System	95% Confidence Interval for Percentage Reduction in Crashes (%)
Adaptive cruise control	5–14
Automated emergency braking	20–31
Blind spot warning	10–20
Forward collision warning	17–25
Intersection movement assist	40–57
Lane departure warning	21–30
Lane change warning	10–33
Electronic stability control	38–48

the number and severity of crashes due to some advanced driver-assistance systems across different road types, weather conditions, and lighting conditions. Specifically, the effectiveness of five advanced driver-assistance systems (adaptive cruise control; automated emergency braking; blind spot warning; forward collision warning; and lane departure warning) are categorized by driving context. The driving conditions are expressed in terms of road types (motorway, urban, and rural), weather conditions (clear, rain/fog, storm/snow) and lighting (daylight, darkness). Then the severity and frequency data of road casualties in these different driving contexts, in the UK, in 2019, are considered. The severity of a crash is quantified by the monetary loss associated with a fatality (£2,029,237), a serious injury (£228,029), and a slight injury (£17,579). Interestingly, out of the 18 combinations of driving context the five most severe all occur in darkness on motorways and on rural roads, often in adverse weather conditions (see Table 15.3). It should be noted that many kilometres of motorway and virtually all rural roads in the UK do not have road lighting.

The relationship between crash types and the advanced driver-assistance systems is then examined to identify which of the latter is most likely to reduce the former. For example, rear-end collisions are most likely to be avoided with adaptive cruise control, forward collision warning, or automated emergency braking, but blind spot warning will have little effect. Table 15.4 lists the crash types and the type of advanced driver-assistance system most likely to be effective for each type. Table 15.4 also shows the number of crashes of each type involving vehicles without the specified advanced driver-assistance system, in the UK, in 2019, and the mean predicted percentage reduction in the number of crashes when the specified advanced driver-assistance system is available. From Table 15.4 it is evident that advanced driver-assistance systems, particularly the automated emergency braking system, have the potential to make a significant improvement in traffic safety, both day and night. For rear-end collisions, it is worth noting that although automated emergency braking shows the biggest reduction, adaptive cruise control and forward collision warning systems also show benefits but to a lesser extent. Given the significant benefits to traffic safety evident from the availability of advanced driver-assistance

---

TABLE 15.3

Driving Contexts in which the Five Most Severe Crashes Happened in 2019 in the UK, Severity Being Based on the Mean Monetary Loss Associated with Such Crashes. The Driving Context Is Based on the Road Type, the Weather Conditions, and the Lighting Condition (from Masello et al., 2022)

Driving Context	Mean Monetary Loss (£)
Motorway–storm/snow–darkness	191,172
Rural–clear–darkness	175,921
Rural–storm/snow–darkness	174,829
Motorway–clear–darkness	151,521
Rural–rain/fog–darkness	136,974

TABLE 15.4

The Three Most Common Driving Contexts in which Five Different Types of Crash Occur, the Number of Crashes of the Type Occurring without the Most Effective Advanced Driver-Assistance System and the Predicted Percentage Reduction in Number of Crashes When the Advanced Driver-Assistance System Is Available (from Masello et al., 2022)

Type of Crash	Driving Contexts	Number of Crashes in Each Driving Context without an Advanced Driver-Assistance System	Advanced Driver-Assistance System	Predicted Percentage Reduction in Crash Count When Advanced Driver-Assistance System Available (%)
Collision with pedestrian	Urban, clear, daylight	4,536	Automated emergency braking	29
	Urban clear darkness	1,728		29
	Urban, rain/fog, darkness	763		27
Rear-end collision	Urban, clear daylight	8,694	Automated emergency braking	29
	Rural, clear, daylight	3,744		27
	Urban, clear darkness	2,016		29
Crashes at intersections	Urban, clear daylight	9,626	Automated emergency braking	29
	Urban, clear, darkness	3,709		29
	Rural, clear daylight	3,172		27
Leaving the road	Rural, clear, daylight	2,633	Lane departure warning	28
	Urban, clear,, daylight	1,558		26
	Rural, clear darkness	1,567		26
Lane change collision	Urban, clear, daylight	484	Blind side warning	18
	Rural, clear, daylight	322		14
	Motorway, clear daylight	230		18

systems, it is hardly surprising that regulations in Europe now require new vehicles to have a battery of such systems fitted.

15.5 ROAD LIGHTING

15.5.1 Problems

Road lighting, as currently practised, does not have a problem in the sense that it fails to do what it sets out to do. Rather, the problems it does have are whether or not what it sets out to do is what it should be doing and if what it sets out to do represents the best use of resources. The problems that surround road lighting can be considered under three headings: technical, financial, and environmental. The technical problems arise from the way road lighting practice has developed. The fact is that road lighting has evolved in isolation. This is most evident in the fact that road lighting does not consider the contribution of vehicle forward lighting to visibility (see Chapter 9). Where such metrics as resolving power and small target visibility (see Section 4.3) have been used as criteria for road lighting, the assumption is that road lighting alone should be sufficient to make small, low-reflectance targets visible. Such targets are rare in practice, apart from the occasional pothole and debris, and these represent a greater hazard to riders of two-wheeled than four-wheeled vehicles. Further, the distances at which these targets should be visible range from 80–100 m, only slightly more than should be possible by low-beam headlamps alone. This suggests that if vehicle forward lighting is taken into account and the role of road lighting is limited to ensuring the visibility of larger targets beyond the range of low-beam headlamps, standards for road lighting would change.

The other technical problem of road lighting arises from the growth in traffic densities. The studies that have examined the effectiveness of road lighting in making targets visible have mainly been concerned with an empty road. Given that traffic density is one of the factors considered when deciding whether or not to install road lighting, such a situation is unrealistic yet there has been little research concerning what road lighting may have to contribute to road safety at different levels of traffic density. It may be that in very dense traffic, road lighting has little contribution to make because the vehicle lighting dominates what needs to be seen, while in very light traffic, road lighting may only be significant beyond the range of low-beam headlamps so it is in medium-density traffic that road lighting is most important.

The financial problems associated with road lighting are concerned with all three costs relevant to lighting a road; first costs, operating costs, and life cycle costs. There is no doubt that road lighting makes life easier for the driver at night. Driving on a lit road enables the driver to see further ahead and hence to have more time to respond to what lies ahead. It also reduces glare from the headlamps of opposing drivers and provides some visual guidance. In urban and suburban areas, road lighting has ancillary benefits for pedestrians and householders in terms of crime reduction although this benefit might be more easily achieved through lighting specifically designed for crime reduction (Painter and Farringdon, 2001). Given these benefits, it might be thought that road lighting should be provided over the whole road network. Unfortunately, this would be prohibitively expensive with the result that there are established procedures for deciding where and when road lighting should be installed (see Section 14.2). These procedures are based on such variables as traffic flow, road geometry, crash history, and so on. Different countries strike different balances between the costs of road lighting and the benefits it provides and hence have different criteria for deciding where and when to install road lighting. If road lighting is installed without a central control system, it will remain unchanged over life apart from the inevitable decline in road surface illuminances and luminances brought about by the changes in luminous flux of the light source and dirt deposits on and in the luminaire. This means such road lighting is operating in the middle of the night when there are few drivers about to receive its benefits. Is this a good use of resources?

The environmental problems of road lighting are primarily those of light pollution (see Section 14.7). Light trespass has been an occasional problem with road lighting for many years but one that can be easily dealt with by shielding the luminaires. Sky glow is a more recent and more serious problem in that it is an inevitable consequence of road lighting.

15.5.2 Solutions

Advances in technology have the potential to reduce the financial and environmental problems of road lighting. One such technology is the photovoltaic generation of electricity during the day that is then stored in a battery and used to operate the road lighting at night. Such systems already exist but are largely confined to locations where connection to the electricity supply network is expensive. Further development of photovoltaic technology to increase its efficiency and reduce its cost will make such systems more attractive.

Another technology is remote monitoring and control of road lighting luminaires, using either mains signalling or wireless communication to connect a large number of luminaires to a local transmitter that in turn is linked to a central server through a mobile phone network or by landline. The central server provides a web portal through which authorized individuals gain access to monitor and control the luminaire network. Monitoring of the status of each luminaire allows light source failures to be identified quickly. Even when operating as expected, monitoring provides a record of the number of hours of operation of the light source so preventive maintenance can be planned (Valetti et al., 2023). In addition, monitoring supplies information on the amount of energy used and when it was used. Where real-time pricing of electricity is applied, knowing when the energy was used is as important as knowing how much has been used. As for control, the simplest control consists of adjusting the times when the luminaires are switched to different outputs depending on the expected traffic density (Ozadwicz and Grela, 2016).

LED road lighting luminaires offer another possibility because LEDs can be dimmed easily. This means there is the possibility of adjusting the amount of light used for road lighting according to the traffic flow and weather conditions. The longer the time spent operating in a dimmed state, the less will be the energy consumption. Using such remote-control systems has resulted in energy savings in the range 25–45% in the UK, China, and Finland (Guo et al., 2007; Walker, 2007). A similar approach for tunnel lighting where the luminances used are based on the luminance outside the tunnel and the traffic density and speed through the tunnel is claimed to have reduced energy consumption considerably (Xu et al., 2024).

An even more adventurous approach is to use the so-called smart road lighting luminaires. In addition to the usual equipment to generate and distribute light, these have sensors and software to detect the presence of pedestrians or vehicles so that the lighting can be operated at a low-light output when no pedestrian or vehicle is present and a higher-light output when they are, reverting to the low-light output when the pedestrian or vehicle has passed (Haans and de Kort, 2012; Mahoor et al., 2020). As long as the system adjusts several luminaires ahead, such a system would probably be acceptable to drivers safely cocooned in their vehicles, but pedestrians out in the open might find such a system less acceptable.

While switching, dimming, and adaptive control systems applied to road lighting are important for minimizing energy consumption and limiting operating costs, they do not address the major questions about road lighting: Where should it be provided and what form should it take? To answer these questions requires the identification of those areas where it will do the most good. The daylight saving time method can be used to identify the type of crashes most sensitive to a change in light level from day to night (see Section 1.5). These are crashes involving pedestrians. It would also be wise to include cyclists who are similarly vulnerable in collisions with vehicles (see Section 10.4). What would be appropriate would be to concentrate road lighting where pedestrians and cyclists are most common and to use it at its highest level when they are most active. Of course, this approach would leave rare pedestrians and cyclists on rural roads at night exposed to a higher level of risk, but the use of retro-reflective clothing and individual lighting would reduce that risk.

The first step towards such a rational application of road lighting would be to revise the road classification system giving greater emphasis to the level of pedestrian and cyclist activity. The existing road classification systems used in the UK and the USA do consider pedestrian and cyclist activity, but treat it as a subsidiary criterion after the road type has been identified according to the road layout and traffic density (see Section 4.4). The classification system should still be based on three types of road; residential roads, traffic routes, and conflict areas. In residential areas, where traffic is slow moving, on-street parking is common and pedestrians are primarily local residents, the lighting should be designed for the resident, although care should be taken to avoid glare to drivers. On traffic routes where there are very few pedestrians or cyclists, the lighting should be designed to serve the interests of the driver. In conflict areas, vehicular traffic, pedestrians, and cyclists all come together, hopefully not too intimately, the lighting should be designed to give good visibility for all three user groups. It is important to note that in this classification system, conflict areas extend beyond roundabouts and intersections to include all roads where there are significant numbers of cyclists and pedestrians, no matter whether the cyclists are in dedicated lanes or mixed with the traffic and the pedestrians are restricted or free range.

The basis for road lighting design would be different for these three road classes. For roads in residential areas, the road lighting is primarily there to meet the needs of pedestrians (BSI 2020). One of these needs is likely to be a feeling of safety at night. Early research in urban and suburban locations of the USA suggested that a median horizontal illuminance on the sidewalk of at least 10 lx is necessary to ensure a perception of safety at night by pedestrians (Boyce et al., 2000a). This illuminance is higher than the majority of the current recommendations for residential areas in the UK (see Table 4.9), but it does provide some flexibility to match the lighting to the ambient conditions. This matching is important because an illuminance that is much less than the ambient to which people are adapted will appear dark. Conversely, an illuminance that is much higher than the ambient to which people are adapted will appear bright. It may be that illuminances lower than 10 lx could provide a perception of safety provided the ambient lighting is lower than 10 lx and people have sufficient time to adapt. More recent research in the UK (Fotios et al. 2015) has shown that it is the minimum illuminance that is more closely related to pedestrian reassurance and recommends a minimum of 2 lx. How acceptable such a low illuminance will be is likely to be influenced by the illuminance uniformity (mean/minimum illuminance). Fortunately, the mean and minimum horizontal illuminances recommended for subsidiary roads in the UK imply an illuminance uniformity of 5.0 for five of the six P Classes (see Table 4.9), which means that a minimum illuminance of 2 lx corresponds to a mean illuminance of 10 lx. Despite this agreement, a minimum horizontal illuminance of 2 lx is only met or exceeded in two of the six P classes. Clearly, the lighting of subsidiary streets requires some revision (Fotios and Gibbons, 2018).

For traffic routes, mean road surface luminance is a reasonable criterion for road lighting design as the relevant directions of view are limited. The problem with lighting for traffic routes is the basis on which the average road surface luminance should be determined. At the moment, average road surface luminances are a matter of consensus rather than proof, based largely on experience rather than evidence, and are different in different countries (Fotios and Gibbons, 2018). There is no doubt that higher luminances would be better for vision but the judgement is that the average road surface luminances recommended are adequate. Most of the scientific evidence available is concentrated on the visibility of small, low-reflectance targets. The question that arises is whether or not this is the right sort of target on which to base road lighting recommendations for all traffic densities? At high traffic densities, it can be argued that detection of the relative movements of nearby vehicles matters much more than the ability to see detail. At the other extreme of traffic density, it may be more important to be able to estimate the speed and distance of what vehicles there are and to detect pedestrians, particularly pedestrians beyond the range of low-beam headlamps. The problem this approach poses is that the same traffic route may have a high traffic density at one time and a low traffic density at another. The remote monitoring and control systems discussed above could be used to adjust the road lighting to what is required for different traffic densities. The problem with this is that there is little evidence about what is required although there are a number of different methods that could be used to collect the relevant data (Fotios and Gibbons, 2018)

Finally, it is necessary to consider conflict areas. By definition, there are areas where many vehicles, cyclists, and pedestrians are trying to use the same piece of road at almost the same time, e.g., roundabouts, intersections, pedestrian crossings, bus, taxi, and bicycle lanes. The presence of drivers, pedestrians, and cyclists in different locations on and beside the road means there are multiple viewing directions and so average road surface illuminance is a reasonable metric for recommendations. Again, the problem is that there is little evidence as to what average illuminance is required to reduce crashes in conflict areas. The same approaches described above for traffic areas could be used but extending the performance measures to pedestrians and cyclists as well as drivers. However, given the diverse range of activities undertaken by drivers, pedestrians and cyclists, determining the illuminance required in conflict areas by this method is likely to take a long time and ultimately rely on some form of consensus over the weight to be attached to all the different measures. An alternative would be to assume that as conflict areas are where more people are taking more decisions more quickly than any other part of the road system, the lighting provided should be sufficient to allow the human visual system to operate without serious degradation. The question then changes to what background luminance is needed to avoid serious degradation of the human visual system. There are a number of ways of answering this question. One is to examine the vast literature on human vision so as to identify the changes that occur with decreasing background luminance (see Boff and Lincoln (1988) for a summary). The problem with this approach is that most visual functions decline gradually at first but at an increasing rate as luminance is reduced (see Section 3.4) which leaves anyone wishing to make a simple recommendation with the problem of deciding where to draw the line, a problem that also exists when visual performance is used as a basis for interior lighting recommendations (Boyce, 1996). An alternative would be to argue that there is an obvious decline in foveal visual functions where photopic vision changes to mesopic vision, which, by convention, is at an adaptation luminance of 3 cd m^{-2}. Another measure used to define this change is the illuminance provided at civil twilight, which is defined as that period from when the upper edge of the sun is at a tangent to the horizon to when the centre of the sum is six degrees below the horizon (Andre and Owens, 1999). The illuminance range for civil twilight in the centre of the USA is from approximately 320 to 3.2 lx, a range so large that it one is again left with the problem of deciding where to draw the line. Yet another approach would be to ask drivers their opinion of the adequacy of the amount of light. Measurements of observers' perceptions of brightness of a road lit by headlamps have been made in Germany (Wordenweber et al., 2007). These measurements led to the conclusion that a minimum average road surface luminance of 1 cd.m^{-2} is required for acceptable road illumination. Yet another approach is to look at the illuminances at which people consider they need additional light. Measurements of the illuminances on the road at which drivers in New York State turn on their headlamps have shown a wide distribution of values (Boyce and Fan, 1994). The illuminances at which 50% of drivers have their headlamps on ranged from 445 lx to 1119 lx, values that are much higher than the illuminances recommended for the lighting of intersections where pedestrians are present which range from 8 lx to 34 lx (see Table 4.5). This discrepancy may have occurred because the drivers are turning on their headlamps to be seen by rather than to see by, i.e., to increase their conspicuity. It would be interesting to determine if drivers of vehicles with daytime running lamps turn on their headlamps at the same ambient illuminance. Another possibility is to consider when existing road lighting is turned on. This is usually done by a photoelectric control unit that can be set to different illuminances, ranging from about 35 lx to 200 lx. The current most popular setting in the UK appears to be 70 lx. Experience of driving when road lighting is first lit suggests that such an illuminance is more than sufficient and using headlamps at this time makes little difference to visibility. By now it should be clear that in the absence of data on the impact of road lighting on crashes in conflict areas that there are many ways to identify a suitable amount of light, many of which lead to answers that differ significantly in their implications for the costs of road lighting. Again, there needs to be more research in this area if road lighting for conflict areas is to be put on a rational footing (Fotios and Gibbons, 2018).

15.6 MARKINGS, SIGNS, AND TRAFFIC SIGNALS

15.6.1 PROBLEMS

Road markings or road studs can be considered as the minimum required for visual guidance. Road signs can be considered as the simplest method to provide more detailed information to drivers. Road markings, road studs, and road signs that rely on retro-reflection of light from headlamps to be seen at night have a common problem. This is the decline in the luminances of the markings, studs, and signs in the presence of a layer of water or dirt. These reductions in luminance tend to reduce the luminance contrast of the marking or studs or the detail of the sign so that the distances from which the markings, studs, or signs are visible are reduced.

Road signs that are illuminated, either internally or externally, are much less sensitive to water or dirt than those that are not. The problem faced by some of these signs is how to match the luminances of the sign to the ambient conditions. The signs for which this problem occurs are those where light is used to convey the message, e.g., variable message signs using LEDs as pixels from which letters and numbers are formed. The problem is that much higher luminous intensities are needed for the message to be visible by day than at night. Unless, the luminous intensities of the sign are reduced at night, the sign itself may become a source of disability glare to approaching drivers. This is rarely a problem with variable message signs designed to provide information to drivers (He et al., 2021), but it certainly can be for video advertising displays placed near roads that are not reduced in luminance at night.

Disability glare is not a problem for road signs and bollards where the sign contains detail with inherent luminance contrast. This means that such signs need only be illuminated after dark and in adverse weather. Unfortunately for energy consumption, many such signs and bollards have the light source operating continuously because the cost of providing photoelectric control is more than the cost of the electricity consumed. Whether or not this wasteful practice will continue in the face of rapidly increasing electricity prices and demands to reduce carbon emissions remains to be seen.

A specific problem for externally illuminated road signs is the degree of light pollution produced by light that is reflected from the sign or that misses the sign altogether. As for traffic signals, these have no fundamental problems. They have been around in the same form for many years and are well understood by drivers. The biggest change to affect traffic signals in recent years has been the use of LEDs as the preferred light source. There is an opportunity to add shape to colour as a distinguishing feature between the different signals, thereby making life easier for drivers with colour-defective vision.

15.6.2 Solutions

There are three potential solutions to the problem of water layers on retro-reflective road markings. One is to increase the height of the retro-reflective elements so that the water layer has to be deeper to cover them. The negative side of this solution is that the wear is more rapid, resulting in a faster reduction in road marking luminance. Another is to use a porous asphalt pavement where rain is frequent. This material is constructed so as to produce faster draining of water from the road surface. Yet another is to change the materials used for the marking so that they are photoluminescent (Villa et al., 2023). Suitable materials can emit light for several hours after exposure to daylight. This means road markings using these materials can provide visual guidance beyond the range of headlamps for the first few hours after sunset, exactly how long depending on the duration of exposure to daylight, the ambient lighting at night, and the headlamps of passing traffic. This is an intriguing possibility that deserves further study.

As for road studs, one solution to the problems of rain and dirt would be to revert to the original design of the "cats-eye" in which a rubber housing deformed when a vehicle's wheel passed over it and thereby wiped the surface of the retro-reflector. Another would be to invest in active road studs that are self-powered. These produce guidance to the driver far beyond the range of headlamps. Their luminous intensity can be adjusted to match the environmental conditions (Villa et al., 2015).

One possible solution of the effect of rain or dirt on retro-reflective road signs would be to illuminate them so they no longer rely on retro-reflection alone to make them visible. Given the wide range of locations of such signs and the cost of bringing electricity to them, this course of action would be much more likely if sufficient power could be provided through a photovoltaic system. If the light pollution that can be caused by external illumination of road signs is a concern, then internally illuminated road signs are a solution. As for variable message signs, where it is necessary to adapt the luminous intensity of the elements of the sign to ambient conditions, there is advice available (Padmos et al., 1988; Soresen, 2011) and the technology required to implement that advice is relatively simple.

There is one other aspect of road markings, signs, and signals that can cause problems for drivers, but it has little to do with lighting. It is simply the ever-increasing number of markings, signs, and signals competing for a driver's attention. There is no doubt that duplicating signs or repeating signs increases the likelihood that they will be noticed, but ultimately, there is a risk of cognitive overload. All signs are not of equal importance so some could be omitted where cognitive overload is likely. What this means is that some discipline is required by road engineers to avoid excessive use of markings and signs.

15.7 WHY CHANGE?

This discussion of the future of lighting for driving has ignored one major factor, why should anyone bother to make any of the changes suggested? After all, most drivers get from A to B without a crash and, apart from the old or ill, think nothing of driving at night. The answer is that there are three major influences that are focussing attention on various aspects of lighting for driving. The first but not necessarily the most potent is the political will to reduce road deaths. Figure 15.2 shows that progress has been made on this objective in several countries since 2000, but there remain large differences in annual road deaths for different countries.

India is an exception to this downward trend, probably because of its rapid economic progress leading to many more vehicles appearing on its roads. Much of the downward trend has been due to improvements in the design of motor vehicles. Such features as side airbags, better crumple zones, child car seats, daytime running lights, as well as advanced driver-assistance systems like automated emergency braking have made the consequences of a collision less severe. Road design in the sense of separating vehicles from pedestrians and cyclists has also helped, as has more rigorous enforcement of driving laws. Moreover, recognition of the fundamental conflict between visibility and glare for headlamps is growing, as is the capability to do something about it. The result is a renewed interest in what contribution better vehicle lighting and advanced driver-assistance systems might make to reducing road deaths and injuries.

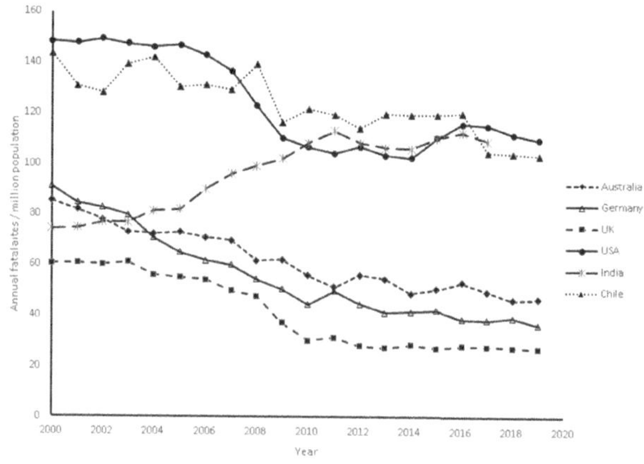

FIGURE 15.2 Trends in annual road fatalities per million population in Australia, Germany, the UK, the USA, India and Chile, from 2000 to 2019 (from OECD 2023).

Another factor influencing vehicle lighting is the desire of vehicle manufacturers to differentiate their product from others. The improvement in vehicle interior lighting over the last decade is mainly due to this. For many decades, the diffusing dome light was the only form of interior lighting, but once one manufacturer started to provide more sophisticated interior lighting, others soon followed. The same approach has now begun for vehicle forward lighting with the introduction of the adaptive forward lighting in luxury vehicles. No doubt some form of adaptive forward lighting system will soon begin to appear in less exotic vehicles alongside other forms of advanced driver-assistance systems.

Finally, road lighting is under two very different pressures. The first is that of costs. The increasing price of electricity in many countries is forcing the authorities responsible for road lighting to justify what they do. Typical of the questions being raised are why do they need to use so much light and why does it have to be operating at the same level throughout the night? Fortunately, the increasing use of LEDs as the primary light source for road lighting and the remote monitoring and control technology now available make it possible to adjust road lighting to match the traffic density. The second pressure is that of road lighting's impact on the environment. This is being discussed at two levels: national and local. Nationally, the concern about global warming is forcing people in many countries to look at how electricity consumption might be reduced. Road lighting is a very conspicuous example of electricity consumption and, as such, is certainly under scrutiny. Locally, there is an increasing awareness of light pollution and road lighting is seen as a major contributor to it. Taken together, these pressures are likely to influence where, when, and how much road lighting is used.

15.8 THE FUTURE

The one thing you can be certain about the future is that it will be different from the past. This is inevitably the case for lighting for driving. In a way, all the problems and solutions identified above can cynically be regarded as examples of "rearranging the deckchairs on the Titanic". To push this metaphor further, the iceberg here is the fully autonomous vehicle. If there is no human driver, what is the need for vehicle and road lighting? Actually, there is a driver in a fully autonomous vehicle but it is inanimate and consists of a series of sensors designed to identify what lies ahead, behind, and around the vehicle, coupled to decision-making software and actuators to control the movement of the vehicle. Given this system the need for road lighting and vehicle forward lighting will depend on the sensitivity and wavelength range of the sensors used. Road lighting and vehicle forward lighting will only be provided as long as they are necessary to ensure that the sensors and software can create an accurate and detailed image of the world around the vehicle. Road markings are another contributor to constructing this image. For road marking to be useful they will need to be much better maintained than they are at present.

Together, road lighting, vehicle forward lighting, and road markings still only cover part of what is needed for the safe operation of an autonomous vehicle. It should be remembered that any autonomous vehicle is likely to encounter humans, either on foot or on two wheels, who are using their visual systems to detect the presence of vehicles. This means that vehicle signal lighting will be required even on autonomous vehicles. This lighting should be designed to enable the autonomous vehicle to be easily detected and its intentions and movements recognized by the human visual system. A similar consideration also applies to road lighting in residential streets and urban areas. In these places, pedestrians are likely to be concerned with seeing more than vehicles. It may be that assessing the intentions of other pedestrians is their main concern. In this situation, the design of road lighting will have to be refocused on the needs of pedestrians rather than drivers. One final change that may be required is to improve the interior lighting of the vehicle. At present, interior lighting when the vehicle is in motion after dark is limited so that the driver can see out. In an autonomous vehicle, there is no need for the occupants to see out so the interior is a mobile room. This suggests that interior lighting generating the desired emotional response will have to be transferred from buildings to vehicles. Whether or not any of this comes about will depend on the rate at which fully autonomous vehicles are developed and accepted. It may be that more sophisticated driver-assistance systems will slow down growth in the adoption of fully autonomous vehicles or it may be that fully autonomous vehicles will be restricted to smart cities, where vehicle-to-vehicle and vehicle-to-infrastructure communication is available. One thing is certain, the role of lighting on roads and on vehicles is going to change in the future.

15.9 SUMMARY

This chapter seeks to envision what might be the future of lighting for driving, at two levels. One attempts to predict what is most likely to happen. The other attempts to set out what should happen. Over the next decade, what is most likely to happen to vehicle forward lighting is that it will become more sophisticated as elements of the adaptive forward lighting system trickle down from up-market vehicles to more modestly priced vehicles. Vehicle signal lighting will change little although the night-time appearance of vehicles, both inside and out, will continue to alter as LEDs deliver more opportunities for vehicle branding. Road lighting will see much greater use of LED light sources and much less use of discharge light sources. Road markings and signs are likely to continue to grow in number and complexity.

The approach used to determine what should happen with regard to lighting for driving is to identify fundamental problems and then to suggest solutions to those problems. Vehicle forward lighting faces two fundamental problems. The first is the limited range of visibility available when low beam headlamps are used on unlit roads. The second is the conflict between visibility and glare. There are several plausible solutions to these problems, ranging from the legal to the technical. The legal solution would be either to introduce slower speed limits on unlit roads after dark or to require pedestrians and cyclists on such roads to wear retro-reflective markers after dark. The technical solutions involve the delivery of supplementary information to the driver using infrared imaging systems to show what is beyond the range of the headlamps combined with advanced driver-assistance systems. Further into the future is the fully autonomous vehicle in which lighting to enable a human driver to see is no longer required, although some form of lighting will always be needed to enable other road users to detect the vehicle.

The main problem faced by vehicle signal lighting is adaptation to prevailing conditions, something that is well within the capabilities of modern technology to solve, but there are also missed opportunities. One type of missed opportunity relates to the use of lighting to ensure stable perception of movement in varying conditions. This could be done by using either contour lighting, as used on heavy trucks, to outline the vehicle or lighting under the vehicle to provide a constant pattern of luminance contrasts as the car moves along the road. Vehicle lighting could also be used to provide more information about the speed of and distance from other vehicles by insisting that all signal lamps on vehicles should be in pairs mounted on opposite sides of the vehicle. Another aspect of signal lighting that deserves attention is masking of the signal by an intervening vehicle. This limitation could be alleviated somewhat if the conventional rear signal assemblies were duplicated high up on the vehicle. However, the main opportunity that is missed with current vehicle signal lighting is to provide other drivers with additional information, such as the force with which the brakes have been applied. This could be done by linking the area, luminous intensity, or flash rate of the stop lamps to that force. In the future, this may be unnecessary as vehicle-to-vehicle communication becomes commonplace.

Road lighting, as currently practised, does not have a problem in the sense that it fails to do what it sets out to do. Rather, the problems it does have are whether or not what it sets out to do is what it should be doing and if what it sets out to do represents the best use of resources. The problems that surround road lighting can be considered under three headings: technical, financial, and environmental. Recent advances in remote monitoring and control technology have the potential to reduce the financial and environmental problems of road lighting by dimming when less light is required. Unfortunately, asking when less light is required itself raises questions about where road lighting should be provided and what form it should take. To answer the question about where road lighting should be provided means identifying those areas where it will do the most good. Examination of the crash data shows that the type of crashes most sensitive to the change of light level from day to night are those involving the most vulnerable, such as pedestrians and cyclists. This implies that a rational approach would be to concentrate road lighting where pedestrians and cyclists are most common and to use it at its highest level when pedestrians and cyclists are most active.

The first step towards implementing such an approach would be to revise the road classification system giving greater emphasis to the level of pedestrian and cyclist activity based on three types of road: residential roads, traffic routes, and conflict areas. In residential areas, the lighting should be designed primarily for pedestrians and cyclists. On traffic routes, the lighting should be designed to serve the interests of the driver. In conflict areas, vehicular traffic, pedestrians, and cyclists all come together so the lighting should be designed to give good visibility for all three user groups. The problem with lighting for all these three road classes is that there is little evidence about what is required for reducing crashes or improving driving performance. There needs to be more research in this area if road lighting is to be put on a rational footing.

Road markings, road studs, and road signs that rely on retro-reflection of light from headlamps to be seen at night have a common problem. This is the decline in the luminances of the sign in the presence of a layer of water or dirt. Self-luminous road markings and road studs and illuminated road signs do not have this problem but where light is used to convey the message, different luminous intensities are required by day and night. Technology to achieve this is available. Where the sign contains detail with inherent luminance contrast, the signs need only be illuminated after dark and in adverse weather but often they are illuminated continuously.

There are two potential solutions to the problem of water layers on retro-reflective road markings. One is to increase the height of the retro-reflective elements so that the water layer has to be deeper to cover them. The other is to use a porous asphalt road which allows faster draining of water from the road surface. Retro-reflective road studs could also be made more resistant to dirt by reverting to the original "cats-eyes" design where a flexible rubber housing wipes the retro-reflector whenever a vehicle's wheel passes over. As for retro-reflective road signs, one possible solution would be to illuminate them so they no longer rely on retro-reflection alone to be visible although there are also concerns that such illumination contributes to light pollution. This concern can be alleviated by using internally illuminated signs.

There is one other aspect of road markings, signs, and signals that can cause problems for drivers but it has little to do with lighting. It is simply the number of markings, signs, and signals. There is no doubt that duplicating signs increases the likelihood that they will be noticed, but there is a risk of cognitive overload. Some discipline is required by road engineers to avoid excessive use of markings, signs, and signals.

Given all these possibilities, it is necessary to ask why anyone should bother to make any of the changes suggested. The answer is that there are three major influences that are focussing attention on various aspects of lighting for driving. The first is the political will to reduce road deaths and injuries. This raises questions about the effectiveness of both vehicle and road lighting and encourages the adoption of advanced driver-assistance systems. The second is the desire of vehicle manufacturers to differentiate their product from others, the trend being towards more sophisticated vehicle lighting. The third is the financial and environmental pressures being applied to road lighting. The increasing price of electricity in many countries is forcing the authorities responsible for road lighting to justify what they do. In addition, concern about climate change is forcing people in many countries to look at how electricity is generated and how its consumption might be reduced. Road lighting is a very conspicuous example of electricity use and, as such, is under scrutiny. Taken together, these pressures are likely to influence where, when, and how much road lighting is used.

Despite these possibilities, the future of all facets of lighting for driving is uncertain because the end point of the advanced driver-assistance systems now being introduced into vehicles is the fully autonomous vehicle, where there will be no human driver. If there is no human driver, what is the need for vehicle and road lighting? In this situation, it can be argued that road lighting and vehicle forward lighting will only be provided if they are necessary to ensure that the sensors and software used to make the vehicle autonomous can create an accurate and detailed model of the world around the vehicle. However, it should be remembered that any autonomous vehicle is likely to encounter humans, either on foot or on two wheels, who are using their visual systems to detect the presence of vehicles. This means that vehicle signal lighting will be required even on autonomous vehicles. This lighting should be designed to enable the autonomous vehicle to be easily detected and its intentions and movements recognized by the human visual system. A similar consideration also applies to road lighting in residential streets and urban areas. In these places, pedestrians may be more concerned with seeing other pedestrians than vehicles. This means the design of road lighting will have to be refocused on the needs of pedestrians rather than drivers. One final change that may be required is to improve the interior lighting of the vehicle. At present, interior lighting when the vehicle is in motion after dark is limited so that the driver can see out. In an autonomous vehicle, there is no need for the occupants to see out so the interior is a mobile room. Whether or not any of this comes about will depend on the rate at which fully autonomous vehicles are developed and accepted but one thing is certain, the role of lighting on roads and on vehicles is going to change in the future.

List of references

Abboushi, B., Fotios, S. and Miller, N.J., (2024) Predictig discomfort from glare with pedestrian-scale lighting: A comparison of candidate models using four indepedent datasets, *Lighting Research and Technology*, 56, 225-246.

Abdel-Aty, M., Ekram, A-A., Huang, H. and Choi, K., (2011) A study on crashes related to visibility obstruction due to fog and smoke, *Accident Analysis and Prevention*, 43, 1730–1737.

Abdul Khalid, M.S., Khamis, N.K., Abu Mansor, M.R. and Hanzah, A., (2021) Motorcycle conspicuity issues and intervention: A systematic review, *Iranian Journal of Public Health*, 50, 24–34.

Adrian, W., (1976) Method of calculating the required luminances in tunnel entrances, *Lighting Research and Technology*, 8, 103–106.

Adrian, W., (1982) Investigations on the required luminance in tunnel entrances, *Lighting Research and Technology*, 14, 151–159.

Adrian, W., (1987a) Visibility levels under night-time driving conditions, *Journal of the Illuminating Engineering Society*, 16, 3–12.

Adrian, W., (1987b) Adaptation luminance when approaching a tunnel in daytime, *Lighting Research and Technology*, 19, 73–79.

Adrian, W., (1989) Visibility of targets: Model for calculation, *Lighting Research and Technology*, 21, 181–188.

Adrian, W., (1991) Comparison between the CBE and CIE glare mark formula and earlier discomfort glare descriptions, *First International Symposium on Glare*, New York: Lighting Research Institute.

Adrian, W., (1995) Change of visual acuity with age, In W Adrian (ed) *Lighting for Ageing Vision and Health*, New York: Lighting Research Institute.

Adrian, W. and Bhanji, A., (1991) Disability glare. *Proceedings of the CIE 22nd Session, Volume 2*, Melbourne, Australia, Vienna: CIE.

Adrian, W. and Schreuder, D.A., (1970) A simple method for the appraisal of glare in street lighting, *Lighting Research and Technology*, 2, 61–73.

Adrian, W.K., Gibbons, R.G. and Thomas, L., (1993) Amendments in calculating STV: Influence of light reflected from the road surface on the target luminance, *Proceedings of the 2nd International Symposium on Visibility and Luminance in Roadway Lighting*, New York: Lighting Research Institute.

Akashi, Y. and Rea, M.S., (2002) Peripheral detection while driving under a mesopic light level, *Journal of the Illuminating Engineering Society*, 31, 85–94.

Akashi, Y., Dee, P., Chen, J., Derlofske, J. and Bullough, J., (2003) Interaction between fixed roadway lighting and vehicle forward lighting. In H-J Schmidt-Clausen (ed) *Progress in Automobile Lighting*, Technical University of Darmstadt, Darmstadt, Germany.

Akashi, Y., Rea, M.S. and Bullough, J.D., (2007) Driver decision making in response to peripheral moving targets under mesopic light levels, *Lighting Research and Technology*, 39, 53–67.

Akashi, Y., van Derlofske, J., Raghavan, R. and Bullough, J.D., (2008) *National Highways Traffic Safety Administration Report DOT HS 810 947 Assessment of Headlamp Glare and Potential Countermeasures: The Effect of Headlamp Mounting Height*, Washington DC: NHTSA.

Akashi, Y., Ohashi, R. and Uchida, T., (2023) The effect of luminance reduction on foveal task performance while implementing mesopic photometry in street lighting standards, *Journal of Science and Technology in Lighting*, IEIJ 220000656

Akerboom, S.P., Kruysse, H.W. and La Heij, W., (1993) Rear light configurations: The removal of ambiguity by a third brake light, In E.G. Gale (ed) *Vision in Vehicles- IV*, Amsterdam, The Netherlands: North Holland.

Aldred, R., Johnson, R., Jackson, C. and Woodcock, J., (2021) How does mode of travel affect risks posed to other road users? An analysis of English road fatality data, incorporating gender and road type, *Injury Prevention*, 27, 71–76.

Alferdinck, J.W.A.M., (1996) Traffic safety aspects of high-intensity discharge headlamps; discomfort glare and direction indicator conspicuity, In A.G. Gale, I.D. Brown, C.M. Haslegrave, and S.P. Taylor (eds) *Vision in Vehicles-V*, Amsterdam: North-Holland.

Alferdinck, J.W.A.M. and Padmos, P., (1988) Car headlamps: Influence of dirt, age and poor aim on glare and illumination intensities, *Lighting Research and Technology*, 20, 195–198.

Alferdinck, J.W.A.M. and Varkevisser, J., (1991) *Discomfort Glare from D1 Headlamps of Different Size,* Report IZF 1991 C-21, Soesterberg, The Netherlands: TNO Institute for Perception.

Allen, K., (2009) *The Effectiveness of Amber Rear Turn Signals for Reducing Rear Impact*, Report DOT HS 811-115, Washington DC: National Highway Traffic Safety Administration.

Alm, H. and Nilsson, L., (2000) Incident warning systems and traffic safety: A comparison between the PORTICO and MELYSSA test site systems, *Transportation Human Factors*, 2, 77–93.

Alshdaifat, A., Moadab, N.H. and Fotios, S., (2024) The impact of road lighting on road user alertness in the evening, *Lighting Research and Technology*, 56, 207-218.

American Automobile Association (AAA), (2019) *Comparison of European and U.S. Specification Automotive Headlamp Performance*, Heathrow, FL: AAA.

American Association of State Highway Transportation Officials (AASHTO), (2011) *Roadside Design Guide*, Washington, DC: AASHTO.

American Association of State Highway Transportation Officials (AASHTO), (2018) *A Policy on Geometric Design of Highways and Streets*, Washington, DC: AASHTO.

American National Standards Institute (ANSI), (2017) *American National Standard for Electric Lamps: Specifications for the Chromaticity of Solid-State Lighting Products*, ANSI C78.377-2017, Rosslyn, VA: National Electric Manufacturers Association.

American National Standards Institute (ANSI) / Illuminating Engineering Society (IES), (2018) *IES Method for Evaluating Light Source Color Rendition* TM-30-18, New York City, NY: Illuminating Engineering Society of North America.

American Society for Testing and Materials (ASTM), (1996) *Standard Practice for Specifying Colors by the Munsell System*, D1535-1596, Philadelphia: ASTM.

American Traffic Safety Services Administration (ATSSA), (2013) *Nighttime Lighting Guidelines for Work Zones*, Fredericksburg, VA: ATSSA.

Andersson, K. and Nilsson, G., (1981) *The Effects on Accidents of Compulsory Use of Running Lights During Daylight in Sweden*, VTI –Report 208A, Linkoping, Sweden: National Road and Traffic Research Institute.

Andre, J.T. and Owens, D.A., (1999) The twilight envelope: An alternative approach toward defining roadway visibility at night, *Proceedings of the Fourth International Lighting Research Symposium: Vision at Low Light Levels*, Palo Alto, CA: Electric Power Research Institute

Arnold, E.D., (2004) *Development of Guidelines for In-Roadway Warning Lights*, Richmond, VA: Commonwealth of Virginia.

Aroke, O.M., Onuchukwu, I.S., Esmaeili, B. and Flintsch, A.M., (2022) Countermeasures to reduce truck-mounted attenuator (TMA) crashes: A state-of-the-art review, *Future Transportation*, 2, 425–452.

Arora, H., Collard, D., Robbins, G., Welbourne, E.R. and White, J.G., (1994) *Effectiveness of Daytime Running Lights in Canada*, Report TP 12298 (E), Ottawa: Transport Canada.

Ashley, W.S., Strader, S., Dziubla, D.C. and Haberlie, A., (2015) Driving blind: Weather related vision hazards and fatal motor vehicle crashes, *Bulletin of the American Meteorological Society*, 96, 755–778.

Association Francoise de l'Eclairage (AFE), (2006) *Les Nuisances Dues a la Lumiere*, Paris: AFE.

Assum, T., Bjornskau, T., Fosser, S. and Sagberg, F., (1999) Risk compensation - the case of road lighting, *Accident Analysis and Prevention*, 31, 545–553.

Attwood, D.A., (1979) The effects of headlight glare on vehicle detection at dusk and dawn, *Human Factors*, 21, 35–45.

Australian Institute of Health and Welfare (AIHW), (2019) *Mobility Scooter-related Injuries and Deaths*, Canberra, Australia: AIHW.

Ayama, M. and Ikeda, M., (1998) Brightness-to-luminance ratio of coloured light in the entire chromaticity diagram, *Color Research and Application*, 23, 274–287.

Bacelar, A. (2004) The contribution of vehicle lights in urban and peripheral urban environments, *Lighting Research and Technology*, 36, 69–78.

Bacelar, A., Cariou, J. and Hamard, M., (1999) Calculational visibility model for road lighting installations, *Lighting Research and Technology*, 31, 177–180.

Bacero, R., To, D., Aresta, J.P., Dela Cruz, M.K., Villaneva, J.P. and Uy, F.A., (2015) Evaluation of strontium aluminate in traffic paint pavement markings for rural and unilluminated roads. *Journal of the Eastern Asia Society for Transportation Studies*, 11, 1726–1744.

Baddiley, C.J. and Webster, T., (2007) *Towards Understanding Skyglow*, Rugby, UK: Institution of Lighting Engineers.

Bailey, I.L. and Bullimore, M.A., (1991) A new test for the evaluation of disability glare, *Optometry and Vision Science*, 68, 911–917.

Bajorski, P., Dhar, S. and Sandhu, D., (1996) Forward-lighting configurations for snowplows, *Transportation Research Record*, 1533, 59–66.

Balazsi, A.G., Rootman, J., Drance, S.M., Schulze, M. and Douglas, G.R., (1984) The effect of age on the nerve fiber population of the human optic nerve, *American Journal of Ophthalmology*, 97, 760–766.

Ball, K., Owsley, C., Sloane, M., Roeker, D. and Bruni, J., (1993) Visual attention problems as a predictor of vehicle crashes in older drivers, *Investigative Ophthalmology and Vision Science*, 34, 3110–3123.

Bargary, G., Furlam, M., Raynham, P.J., Barbur, J.L. and Smith A.T., (2015) Cortical hyperexcitability and sensitivity to discomfort glare, *Neuropsychologia*, 69, 194–200.

Becker, N, Rust, H.W. and Ulbrich U., (2022) Weather impacts on various types of road crashes: A quantitative analysis using generalized additive models, *European Transport Research Review*, 14, 37.

Bergkvist, P., (2001) Daytime running lamps (DRLs) – a North American success story, *Proceedings of the 17th International Technical Conference on the Enhanced Safety of Vehicles*, Washington, DC: Department of Transportation.

Berman, S.M., (1992) Energy efficiency consequences of scotopic sensitivity, *Journal of the Illuminating Engineering Society*, 21, 3–14.

Berman, S.M., Fein, G., Jewett, D.L., Saika, G. and Ashford, F., (1992) Spectral determinants of steady-state pupil size with a full field of view, *Journal of the Illuminating Engineering Society*, 21, 3–13.

Berson, D.M., (2007) Phototransduction in ganglion cell photoreceptors, *European Journal of Physiology*, 454, 849–855.

Berson, D.M., Dunn, F.A. and Takao, M., (2002) Phototransduction by retinal ganglion cells that set the circadian clock, *Science*, 295, 1070–1073.

Besenecker, U.C., Bullough, J.D. and Redetsky L.C., (2016) Spectral sensitivity and scene brightness at low to moderate photopic light levels, *Lighting Research and Technology*, 48, 676–688.

Bever, S., (2022) *An Updated Guide to Cycle Lighting Regulations*, Guildford, UK: CyclingUK.

Bhagavathula, R. and Gibbons, R.G., (2020) Light level for parking facilities based on empirical evaluation of visual performance and user perceptions, *Leukos*, 16, 115–136.

Bhagavathula, R. and Gibbons, R.G., (2022) *Initial Investigation of Intersection Lighting*, Blacksburg, VA: National Surface Transportation Safety Center for Excellence.

Bhagavathula, R. and Gibbons, R.G., (2023) Lighting strategies to increase nighttime pedestrian visibility at midblock crosswalks, *Sustainability*, 15, 1455.

Bhuvana, K.P., Bensingh, R.J., Kader, M.A. and Nayak, S.K., (2018) Polymer light emitting diodes: Materials, technology and device, *Polymer Plastics Technology and Engineering*, 57, 1784–1800.

Bil, M., Bilova, M. and Muller, J., (2010) Critical factors in fatal collisions of adult cyclists with automobiles, *Accident Analysis and Prevention*, 47, 1632–1636.

Billmeyer Jr., F.W., (1987) Survey of color order systems, *Color Research and Application*, 12, 173–186.

Blackwell, H.R., (1959) Development and use of a quantitative method for specification of interior illumination levels on the basis of performance data, *Illuminating Engineering*, 54, 317–353.

Blackwell, H.R. and Blackwell, O.M., (1971) Visual performance data for 156 normal observers of various ages, *Journal of the Illumination Engineering Society*, 1, 3–13.

Blackwell, H.R. and Blackwell, O.M., (1977) A basic task performance assessment of roadway luminous environments, In *Measures of Road Lighting Effectiveness*, Berlin: Lichttechnische Gesellschaft LiTG.

Blackwell, H.R., Schwab, R.N. and Pritchard, B.S., (1964) Visibility and illumination variables in roadway visual tasks, *Illuminating Engineering*, 59, 277–308.

Blankenbach, K., Hertlein, F. and Hoffman, S., (2020) Advances in automotive interior lighting concerning new LED approach and optical performance, *Journal of the Society of Information Display*, 28, 655–667.

Blazer, P. and Dudli, H., (1993) Tunnel lighting: Method of calculating luminance of access zone L_{20}, *Lighting Research and Technology*, 25, 25–30.

Bodmann, H.W. and Schmidt, H.J., (1989) Road surface reflection and road lighting: Field investigations, *Lighting Research and Technology*, 21, 159–170.

Boff, K.R. and Lincoln, J.E., (1988) *Engineering Data Compendium: Human Perception and Performance*, Wright-Patterson AFB, OH: Harry G. Armstrong Aerospace Medical Research Laboratory.

Bornstein, M.H., Kessen, W. and Weiskopf, S., (1976) The categories of hue in infancy, *Science*, 191, 201.

Bourdy, C., Chiron, A, Cottin, F. and Monot, A., (1988) Visibility at a tunnel entrance: Effect of temporal luminance variation, *Lighting Research and Technology*, 20, 199–200.

Bowers, A., Peli, E., Elgin, J., McGwin, G. Jr. and Owsley, C., (2005) On-road driving and moderate visual field loss, *Optometry and Vision Science*, 82, 657–667.

Bowman, K.J., Collins, M.J. and Henry, C.J., (1984) The effect of age on performance on the Panel D-15 and desaturated D-!5: A quantitative evaluation, In G. Verriest (ed), *Colour Vision Deficiencies VII*, The Hague, The Netherlands: W. Jumk Publishers.

Box, P.C., (1981) Parking lot accident characteristics, *ITE Journal*, 51, 12–15.

Boyce, P.R., (1996) Illuminance selection based on visual performance - and other fairy stories, *Journal of the Illuminating Engineering Society*, 25, 41–49.

Boyce, P.R., (2014) *Human Factors in Lighting*, Boca Raton, FL, CRC Press.

Boyce, P.R., (2019) The benefits of light at night, *Building and Environment*, 151, 356–367.

Boyce, P.R. and Fan, J., (1994) Drivers' headlight use: A lighting factoid, *Lighting Research and Technology*, 26, 23–27.

Boyce, P.R. and Stampfli, J.R., (2019) LRT Digest 3: New colour metrics and their uses, *Lighting Research and Technology*, 51, 657–681.

Boyce, P.R., Eklund, N.H., Hamilton, B.J. and Bruno, L.D., (2000a) Perceptions of safety at night in different lighting conditions, *Lighting Research and Technology*, 32, 79–91.

Boyce, P., Bierman, A., Carter, B., Hunter, C., Bullough, J., Figueiro, M. and Conway, K., (2000b) *The Color Identification of Traffic Signals*, Troy, NY: Lighting Research Center.

Boyce, P.R., Fotios, S. and Richards, M., (2009) Road lighting and energy saving, *Lighting Research and Technology*, 41, 245–260.

Boynton, R.M. and Clarke, F.J.J., (1964) Sources of entoptic scatter in the human eye. *Journal of the Optical Society of America*, 54, 110–119.

Boynton, R.M. and Gordon, J., (1965) Bezold-Brucke hue shift measured by a color naming technique, *Journal of the Optical Society of America*, 55, 78–86.

Boynton, R.M. and Purl, K.F., (1989) Categorical colour perception under low pressure sodium lighting with small amounts of added incandescent illumination, *Lighting Research and Technology*, 21, 23–27.

British Standards Institution (BSI), (1982) BS 6102-Part 2, *Cycles – Specification for Photometric and Physical Requirements of Reflective Devices*, London: BSI.

British Standards Institution (BSI), (1986) BS 6102-Part 3, *Cycles – Specification for Photometric and Physical Requirements of Lighting Equipment*, London: BSI.

British Standards Institution (BSI), (2013) BS 5489-1:2013, *Road Lighting – Part 1: Lighting of Roads and Public Amenity Areas*, London: BSI.

British Standards Institution (BSI) (2014) BS EN 12966: 2014 + A1:2018, *Road Vertical Signs- Variable Message Traffic Signs*, London: BSI.

British Standards Institution (BSI), (2015) BS EN 13201-2:2015, *Road Lighting – Part 2: Performance Requirements*, London: BSI.

British Standards Institution (BSI), (2019) BS EN 12767:2019, *Passive Safety of Support Structures for Road Equipment*, London: BSI.

British Standards Institution (BSI), (2020) BS 5489-1:2020, *Design of Road Lighting – Part 1: Lighting of Roads and Public Amenity Areas – Code of Practice*, London: BSI.

Brock, M.A., (1991) Chronobiology and aging, *Journal of the American Geriatrics Society*, 39, 74–91.

Brons, J.A., Bullough, J.D. and Rea, M.S., (2008) Outdoor site-lighting performance (OSP): A comprehensive and quantitative framework for assessing light pollution, *Lighting Research and Technology*, 40, 201–224.

Brooks, J.O., Tyrrell, R.A. and Frank, T.A., (2005) The effect of severe visual challenges on steering performance in visually healthy young drivers, *Optometry and Vision Science*, 82, 689–697.

Brown, T.M., Brainard, G.C., Cajochen, C., Czeisler, C.A., Hanifin, J.P., Lockley, S.W., et al., (2022) Recommendations for daytime, evening and nighttime indoor light exposure to best support physiology, sleep and wakefulness in healthy adults, *PLoS Biology*, 20(3), e3001571.

Brown, W.R.J., (1951) The influence of luminance level on visual sensitivity to color differences, *Journal of the Optical Society of America*, 41, 684–688.

Brumbelow, M.L., (2022) Light where it matters: IIHS headlight ratings are correlated with nighttime crash rate, *Journal of Safety Research*, 83, 379–387.

Bruneau, J-F. and Morin, D., (2005) Standard and non-standard roadway lighting compared with darkness at rural intersections, *Transportation Research Record*, 1918, 116–122.

Brusque, C., Paulmier, G. and Carta, V., (1999) Study of the influence of background complexity on the detection of pedestrians in urban sites, *Proceedings of the 24th Session of the CIE*, Warsaw, Poland, Vienna: CIE.

Buck, J.A., McGowan, T.K. and McNelis, J.F., (1975) Roadway visibility as a function of light source color, *Journal of the Illuminating Engineering Society*, 5, 20–25.

Bullough, J.D., (2002a) Interpreting outdoor luminaire cutoff classification, *Lighting Design and Application*, 52, 44–46.

Bullough, J.D., (2002b) Modeling peripheral visibility under headlamp illumination, *Proceedings of the TRB16th Biennial Symposium on Visibility and Simulation*, Lowa City, IA, Washington, DC: Transportation Research Board.

Bullough, J.D., (2022) *Understanding Glare in Exterior Lighting, Display and Related Applications*, Bellingham, Washington: SPIE Press.

Bullough, J. and Rea, M.S., (1997) A simple model of forward visibility for snow plow operators through snow and fog at night, *Transportation Research Record*, 1585, 19–24.

Bullough, J. and Rea, M.S., (2000) Simulated driving performance and peripheral detection at mesopic and low photopic light levels, *Lighting Research and Technology*, 32, 194–198.

Bullough, J.D. and Rea, M.S., (2001a) Forward vehicle lighting and inclement weather conditions, *Proceedings of the Symposium Progress in Automobile Lighting 2001*. Munich, Germany: Technical University of Darmstadt.

Bullough, J.D. and Rea, M.S., (2001b) Driving in snow: Effects of headlamp color at mesopic and photopic light levels, *Lighting Technology Developments for Automobiles*, Warrendale, PA: Society of Automotive Engineers.

Bullough, J.D. and Rea, M.S., (2010) *Visibility from Vehicle Headlamps and Roadway Lighting in Urban, Suburban and Rural Locations*, SAE Technical Paper 2010-01-0298, Warrendale, PA: SAE International.

Bullough, J.D. and Skinner N.P., (2009) *Pedestrian Safety Margins Under Different Types of Headlamp Illumination*. Troy, NY: Lighting Research Center.

Bullough, J.D. and Skinner, N.P., (2015) *Demonstrating Urban Outdoor Lighting for Pedestrian Safety and Security*, New York: University Transportation Research Center – Region 2.

Bullough, J.D. and Skinner, N.P., (2016) Dynamic signal lighting: Off-axis detection and directional signal identification, *Proceedings of the Vehicle and Infrastructure Safety Improvement in Adverse Conditions and Night Driving Congress*, Suresnes, France: Societe des Ingenieurs de l'Automobile.

Bullough, J.D., Boyce, P.R., Bierman, A., Conway, K.M., Huang, K, O'Rourke, C.P., Hunter, C.M. and Nakata, A., (2000) Response to simulated traffic signals using light emitting diode and incandescent sources, *Transportation Research Record*, 1724, 39–46.

Bullough, J.D., Boyce, P.R., Bierman, A., Hunter, C.M., Conway, K.M., Nakata, A. and Figueiro, M.G., (2001a) Traffic signal luminance and visual discomfort at night, *Transportation Research Record*, 1754, 42–47.

Bullough, J.D., Rea, M.S., Pysar, R.M., Nakhla, H.K. and Amsler, D.E., (2001b) Rear lighting configurations for winter maintenance vehicles, *Proceedings of the IESNA Annual Conference*, Ottawa, Canada, New York: IESNA.

Bullough, J.D., Zu, F. and Van Derlofske, J., (2002) *Discomfort and Disability Glare from Halogen and HID Headlamp Systems*, SAE Paper 2002-01-1-0010, Warrendale, PA: Society of Automotive Engineers.

Bullough, J.D., Van Derlofske, J., Fay, C.R. and Dee, P., (2003) *Discomfort Glare from Headlamps: Interactions among Spectrum, Control of Gaze and Background Light Level*, SAE Paper 2003-01-0296, Warrendale, PA: Society of Automotive Engineers.

Bullough, J.D., Yuan, Z. and Rea, M.S., (2007a) Perceived brightness of incandescent and LED aviation signal lights, *Aviation, Space and Environmental Medicine*, 78, 893–900.

Bullough, J.D., Van Derlofske, J. and Kleinkes, M., (2007b) *Rear Signal Lighting: From Research to Standards, Now and in the Future*, SAE Paper 2007-01-1229, Warrendale, PA: Society of Automotive Engineers.

Bullough, J.D., Brons, J.A., Qi, R. and Rea, M.S., (2008) Predicting discomfort glare from outdoor lighting installations, *Lighting Research and Technology*, 40, 225–242.

Bullough, J.D., Donnell, E.T. and Rea, M.S., (2013) To illuminate or not to illuminate: Roadway lighting as it affects traffic safety at intersections, *Accident Analysis and Prevention*, 53, 65–77.

Bullough, J.D., Skinner, N.P. and Plummer T.T., (2016) Assessment of an adaptive driving beam headlighting system, visibility and glare, *Transportation Research Record*, 2555, 81–85.

Bullough, J.D., Skinner, N.P. and Rea, M.S., (2019) *Impacts of Flashing Emergency Lights and Vehicle-Mounted Illumination on Driver Visibility and Glare*, SAE Technical Report 2019-01-0847. Warrendale, PA: Society of Automotive Engineers.

Burg, A., (1967) *The Relationship Between Vision Test Scores and Driving Records: General Findings*, Report 67-24, Los Angeles, CA: Department of Engineering, University of California.

Burg, A, (1971) Vision and driving: A report on research, *Human Factors*, 13, 79–87.

Burghout, F., (1979) On the relationship between reflection properties, composition and texture of road surfaces, *Proceedings of the CIE, 19th Session*, Kyoto, Japan, Vienna: CIE.

Caberletti, L., Elfmann, K., Kummel, M. and Schierz, C., (2010) Influence of ambient lighting in a vehicle interior on the driver's perceptions, *Lighting Research and Technology*, 42, 297–311.

Caird, J.K. and Hancock, P.A., (2002) Left turn and gap acceptance crashes, In R.E. Dewar and P.L. Olson (eds) *Human Factors in Traffic Safety*, Tucson, AZ: Lawyers and Judges Publishing Company.

Cairney, P. and Catchpole, J., (1996) Patterns of perceptual failures at intersections of arterial roads and local streets, In A.G. Gale, I.D. Brown, C.M. Haslegrave and S.P. Taylor (eds) *Vision in Vehicles-V*, Amsterdam, The Netherlands: North-Holland.

California Energy Commission (CEC), (2019) *Title 24, Part 6 of the California Code of Regulations*, Sacramento, CA: CEC.

Casson, E.J. and Racette, L., (2000) Vision standards for driving in Canada and the United States: A review for the Canadian Ophthalmological Society, *Canadian Journal of Ophthalmology*, 35, 192–203.

Castro, C., (2009) Visual demands and driving. In C. Castro (ed), *Human Factors of Visual and Cognitive Performance in Driving*, Boca Raton, FL, CRC Press.

Cavallo, V., Colomb, M. and Dure, J., (2001) Distance perception of vehicle rear lights in fog, *Human Factors*, 43, 442–451.

Cavallo, V. and Pinto, M., (2012) Are car daytime running lights detrimental to motorcycle conspicuity? *Accident Analysis and Prevention*, 49, 78–85.

Chandler, D., (1949) *The Rise of the Gas Industry in Britain*, London: Kelly and Kelly.

Chandra, D., Sivak, M., Flannagan, M.J., Sato, T. and Traube, E.C., (1992) *Reaction Times to Body-Colour Brake Lamps*, UMTRI-92-15, Ann Arbor, MI: University of Michigan Transportation Research Institute.

Chang, Y-M. and Wang, L.L., (2007) *The Visual Power of the Dashboard of a Passenger Car by Applying Eye-Tracking Theory*, SAE Paper 2007-01-0425, Warrendale, PA: Society of Automotive Engineers.

Chapparo, A., Wood, J.M. and Carberry, T., (2005) Effect of age and auditory and visual dual tasks on closed-road driving performance, *Optometry and Vision Science*, 82, 747–754.

Charlton, S.G., (2007) The role of attention in horizontal curves: A comparison of advance warnings, delineation and road marking treatments, *Accident Analysis and Prevention*, 39, 873–885.

Charman, W.N., (1997) Vision and driving – a literature review and commentary, *Ophthalmic and Physiological Optics*, 57, 371–391.

Chartered Institution of Highways and Transportation (CIHT), (2021) *State of the Nation: 2020 Street Lighting Survey*, London: CIHT.

Chatterjee, K. and Mcdonald, M., (2004) Effectiveness of using variable message signs to disseminate dynamic traffic information from field trials in European cities, *Transport Reviews,* 24, 559–585.

Clarke, D.D., Ward, P., Bartle, C. and Truman, W., (2006) Young driver accidents in the UK: The influence of age, experience and time of day, *Accident Analysis and Prevention*, 38, 871–878.

Clay, O.J., Wadley, V.G., Edwards, J.D., Roth, D.L, Roenker, D.L. and Ball, K.K., (2002) Cumulative meta-analysis of the relationship between useful field of view and driving performance in older adults: Current and future implications, *Optometry and Vision Science*, 82, 724–731.

Cobb, J., (1990) *Roadside Survey of Vehicle Lighting 1989,* Research Report 290, Crowthorne, UK: Transport and Road Research Laboratory.

Cobb, J., (1992) *Daytime Conspicuity Lights*, Report WP/RUB/14), Crowthorne, UK: Transport Research Laboratory.

Codling, P.J., (1971) *Thick Fog and its Effects on Traffic Flow and Accidents*, RRL Report LR 397, Crowthorne, UK: Road Research Laboratory.

Cole, B.L. and Brown, B., (1966) Optimum intensity of red road-traffic signal lights for normal and protanopic observers, *Journal of the Optical Society of America*, 56, 516–522.

Collins, J.J. and Hall, R.D., (1992) Legibility and readability of light reflecting matrix variable message road signs, *Lighting Research and Technology*, 24, 143–148.

Commission Internationale de l'Eclairage (CIE), (1976) *Calculation and Measurement of Luminance and Illuminance in Road Lighting*, CIE Publication 30, Vienna: CIE.

Commission Internationale de l'Eclairage (CIE), (1978) *Light as a True Visual Quantity*, CIE Publication 41, Vienna: CIE.

Commission Internationale de l'Eclairage (CIE), (1979) *Road Lighting for Wet Conditions*, CIE Publication 47, Vienna: CIE

Commission Internationale de l'Eclairage (CIE), (1983) *The Basis of Physical Photometry*, CIE Publication 18.2, Vienna: CIE.

Commission Internationale de l'Eclairage (CIE), (1984) *Road Surfaces and Lighting*, CIE Publication 66, Vienna: CIE.

Commission Internationale de l'Eclairage (CIE), (1986) *Colorimetry*, CIE Publication 15.2, Vienna: CIE.

Commission Internationale de l'Eclairage (CIE) (1987) *Guide to the Properties and Uses of Retro-reflectors at Night*, CIE Publication 72, Vienna: CIE.

Commission Internationale de l'Eclairage (CIE), (1989) *Mesopic Photometry: History, Special Problems and Practical Solutions*, CIE Publication 81, Vienna: CIE.

Commission Internationale de l'Eclairage (CIE), (1990) *CIE 1988 2° Spectral Luminous Efficiency Function for Photopic Vision*, CIE Publication No 86, Vienna: CIE.

Commission Internationale de l'Eclairage (CIE) (1992a) *Fundamentals of the Visual Task of Night Driving*, CIE Publication 100, Vienna: CIE.

Commission Internationale de l'Eclairage (CIE), (1992b) *Road Lighting as an Accident Countermeasure* CIE Publication 93, Vienna: CIE.

Commission Internationale de l' Eclairage (CIE), (1993) *Daytime Running Lights (DRL)* Report CIE 104, Vienna: CIE.

Commission Internationale de l'Eclairage (CIE), (1994) *Review of the Official Recommendations of the CIE for the Colors of Signal Lights*, CIE Technical Report 107, Vienna: CIE.

Commission Internationale de l'Eclairage (CIE), (1995) *Method of Measuring and Specifying Color Rendering Properties of Light Sources*, CIE Publication 13.3, Vienna: CIE.

Commission Internationale de l'Eclairage, (1998) *Road Traffic Lights – 200mm Round Signals Photometric Properties*, CIE Standard S006-1/E-1998, Vienna: CIE.

Commission Internationale de l'Eclairage (CIE), (1999) *Design Method for Lighting of Roads*, CIE Publication 132-1999, Vienna: CIE.

Commission Internationale de l'Eclairage (CIE), (2001) *Road Surface and Road Marking Reflection Characteristics*, CIE Publication 144-2001, Vienna: CIE.

Commission Internationale de l'Eclairage (CIE), (2002) *CIE Collection on Glare*, CIE Publication 146-2002, Vienna: CIE.

Commission Internationale de l'Eclairage (CIE), (2003) *The Maintenance of Outdoor Lighting*, CIE Publication 154:2003 Vienna: CIE.

Commission Internationale de l'Eclairage (CIE), (2004) *Guide for the Lighting of Road Tunnels and Underpasses*, CIE Publication 88:2004, Vienna: CIE.

Commission Internationale de l'Eclairage (CIE), (2010a) *Recommended System for Visual Performance Based Mesopic Photometry*, CIE Publication 191:2010, Vienna: CIE.

Commission Internationale de l'Eclairage (CIE), (2010b) *Lighting of Roads for Motor and Pedestrian Traffic*, CIE Publication 115:2010, Vienna: CIE.

Commission Internationale de l'Eclairage (CIE), (2010c) *Emergency Lighting in Road Tunnels*, CIE Publication 193:2010, Vienna: CIE.

Commission Internationale de l'Eclairage (CIE), (2017a) *Interim Recommendation for Practical Application of the CIE System of Mesopic Photometry in Outdoor Lighting*, CIE Technical Note TN007:2017, Vienna: CIE

Commission Internationale de l'Eclairage (CIE), (2017b) *Guide on the Limitation of the Effects of Obtrusive Light from Outdoor Lighting Installations*, CIE Publication 150:2017, Vienna: CIE.

Cooper, J., Stafford, K., Owlett, P. and Mitchell, J., (2008) *Review of the Lighting Requirement for Traffic Signs ad Bollards*, Crowthorne, Berkshire, UK: Transport Research Laboratory.

Copinschi, G. and van Cauter, E., (1995) Effects of aging on modulation of hormonal secretions by sleep and circadian rhythmicity, *Hormone Research*, 43, 20–24.

Crawford, B.H., (1949) The scotopic visibility function, *Proceedings of the Physical Society*, 62, 321.

Crawford, B.H., (1972) The Stiles-Crawford effects and their significance in vision, In D. Jameson and L.M. Hurvich (eds), *Handbook of Sensory Physiology, Vol. VII/4 Visual Psychophysics*, Berlin: Springer-Verlag.

Croft, T.A., (1971) Failure of visual estimation of motion under strobe, *Nature*, 231, 397.

Crosley, J. and Allen, M.J., (1966) Automobile brake light effectiveness: An evaluation of high placement and accelerator switching, *American Journal of Optometry and Archives of the American Academy of Optometry*, 43, 299–304.

Crundall, P. and Underwood, G., (1998) Effect of experience and processing demand on visual information acquisition in drivers, *Ergonomics*, 41, 448–458.

Czeisler, C.A., Rios, C.D., Sanchez, R., Brown, E.N., Richardson, G.S., Ronda, J.M. and Rogacz, S., (1988) Phase advance and reduction in amplitude of the endogenous circadian oscillator correspond with systematic changes in sleep/wake habits and daytime functioning in the elderly, *Sleep Research*, 15, 268.

Czeisler, C.A., Duffy, J.F., Shanahan, T.L., Brown, E.N., Mitchells, J.F., Rimmer, D.W. et al., (1999) Stability, precision and near 24 hour period of human circadian pacemaker *Science*, 284, 2177–2181.

Daniels, S., Martensen, H., Schoeters, A., van den Berghe, W., Papadimitiriou, E., Ziakopoulos, A., Kaiser, S., Aigner-Breuss, E., Soteropoulos, A, Wijnen, W., Weijermars, W, Carnis, L., Elvik, R. and Perez, O.M., (2019) A systematic cost-benefit analysis of 29 road safety measures, *Accident Analysis and Prevention*, 133, 105292.

Davidse, R., van Driel, C. and Goldenbeld, C., (2004) *The Effect of Altered Road Marking on Speed and Lateral Position: A Meta-Analysis*, Leidschendem, The Netherlands: SWOV.

Davoodi, S.R. and Hossayni, S.M., (2015) Role of motorcycle running lights in reducing motorcycle rashes during daytime: A review of the current literature, *Bulletin of Emergency and Trauma*, 3, 73–78.

De Boer, J.B., (1951) Fundamental experiments on visibility and admissible glare in road lighting, *Proceedings of the CIE, 12th Session*, Stockholm, Sweden, Paris: CIE.

De Boer, J.B., (1967) *Public Lighting*, Eindhoven, The Netherlands: Philips Technical Library.

De Boer, J.B., (1974) Modern light sources for highways, *Journal of the Illuminating Engineering Society*, 3, 142–152.

De Boer, J.B. and Westermann, H.O., (1964a) Characterisation and classification of road surfaces from the point of view of luminance in public lighting, *Lux*, 30, 385.

De Boer, J.B. and Westermann, H.O., (1964b) The discrimination of road surfaces depending on the reflection properties and its meaning for road lighting, *Lichttechnik*, 16, 487.

De Boer, J.B., Onate, V. and Oostrijk, A., (1952) Practical methods for measuring and calculating the luminance of road surfaces, *Philips Research Reports*, 7, 54–76.

De Boer, J.B., Burghout, F. and van Heemskerck Veekens, J.F.T., (1959) Appraisal of the quality of public lighting based on road surface luminance and glare, *Proceedings of the 14th Session of the CIE*, Brussels, Belgium, Vienna: CIE.

De Lange, H., (1958) Research into the dynamic nature of the human fovea cortex systems with intermittent and modulated light. 1. Attenuation characteristics with white and colored lights, *Journal of the Optical Society of America*, 48, 777–789.

DeLucia, P.R., (2013) Effect of size on collision perception and implications for perceptual theory and transportation safety, *Psychological Science*, 22, 199–204.

Deans, R.L., Miller A.R., Murrill, J.K., Sanders, J.R., Turley, T.C., Lambert, J.H., Bridewell, T.A. and Cottrell, B.H., (2003) Screening needs for roadway lighting by exposure assessment and site parameters, In M.H. Jones, B.E. Tawney and K. Preston White Jr (eds), *Proceedings of the 2003 IEEE Systems and Information Engineering Design Symposium*, Charlottesville, VA: Department of System and Information Engineering, University of Virginia.

Dee, P., (2003) *The Effect of Spectrum on Discomfort Glare*, MS Thesis, Troy, NY: Rensselaer Polytechnic Institute.

Department for Transport (DfT), (2005) *The Design of Pedestrian Crossings*, Local Transport Note 2/95, London: Department for Transport.

Department for Transport (DfT), (2007) *Local Transport Note 1/07 Traffic Calming*, London: DfT.

Department for Transport (DfT), (2013) *Safety at Street Works and Road Works: A Code of Practice*, London: DfT.

Department for Transport (DfT), (2015a) *Contributory Factors to Reported Road Accidents 2014*, London: DfT.

Department for Transport (DfT), (2015b) *Transport Advisory Leaflet 01/15 Variable Message Signs*, London: DfT.

Department for Transport (DfT), (2015c) *Facts on Pedestrian Casualties*, London: DfT.

Department for Transport (DfT), (2015d) *Facts on Young Car Drivers*, London: DfT.

Department for Transport (DfT), (2021a) *Reported Road Casualties Great Britain: Motorcycle Factsheet, 2020*, London: DfT.

Department for Transport (DfT), (2021b) *Reported Road Casualties Great Britain: Pedal Cycle Factsheet, 2020*, London: DfT.

Department for Transport (DfT), (2021c) *Reported Road Casualties Great Britain: E-scooter Factsheet 2020*, London: DfT.

Department for Transport (DfT), (2022a) *Traffic Sign Manual*, London: DfT.

Department for Transport DfT), (2022b) *Reported Road Casualties for Severity and Road User Type; Great Britain for Ten Years up to 2021*, London: DfT.

Department for Transport DfT), (2022c) *Reported Road Casualties Great Britain: Pedestrian Factsheet, 2021*, London: DfT.

Department for Transport DfT), (2022d) *Reported Road Casualties Great Britain: Older Driver Factsheet, 2021*, London: DfT.

Desimone, R., (1991) Face-selective cells in the temporal cortex of monkeys, *Journal of Cognitive Neuroscience*, 3, 1–8.

Devaney, K.O. and Johnson, H.A., (1980) Neuron loss in the ageing visual cortex of man, *Journal of Gerontology*, 35, 836–841.

Devonshire, J.M.,and Flannagan, M.J., (2007*) Effects of Automotive Interior Lighting on Driver Vision*, UMTRI -2007-1, Ann Arbor, MI: University of Michigan Transportation Institute.

DiLaura, D.L., (2006) *A History of Light and Lighting*, New York: The Illuminating Engineering Society of North America.

Distephano, N. and Leonardi, S., (2019) Evaluation of the benefits of traffic calming on vehicle speed reduction, *Civil Engineering and Architecture*, 7, 200–214.

Djuretic, A. and Kostic, M., (2018) Actual energy savings when replacing high pressure sodium with LED luminaires in street lighting, *Energy*, 157, 367–378.

Doyle, M., Edwards, A. and Avery, M., (2015) AEB real world validation using UK motor insurance claims data, *Proceedings of the 24th ESV Conference*, Gothenburg, Sweden.

Drasdo, N., (1977) The neural representation of visual space, *Nature*, 266, 554–556.

Drasdo, N. and Haggerty C.M., (1984) A comparison of the British number plate and Snellen vision test for car drivers, *Ophthalmic and Physiological Optics*, 1, 39–54.

Driving and Vehicle Licensing Agency (DVLA), (2016) *Visual Disorders: Assessing Fitness to Drive*, Swansea, UK: DVLA.

Duke-Elder, W.S., (1944) *Textbook of Ophthalmology, Vol. 1*, St Louis, MO: C.V.Mosby & Co.

Dumont, E. and Paumier, J-L., (2007) Are standard r-tables still representative of the properties of road surfaces in France? *Proceedings of the CIE, 26th Session*, Beijing China, Vienna: CIE.

Dunbar, C., (1938) Necessary values of brightness contrast in artificially lighted streets, *Transactions of the Illuminating Engineering Society (London)*, 3, 187–195.

Dzombak, R., Kasikaralar, E. and Dillon, H.E., (2020) Exploring cost and environmental implications of optimal technology management strategies in the street lighting industry, *Resources, Conservation and Recycling X*, 6, 100022.

Eastman, A.A. and McNelis, J.F., (1963) An evaluation of sodium, mercury and filament lighting for roadways, *Illuminating Engineering*, 58, 28–34.

Edewaard, D.E., Fekety, D.J.,Szubski, E.C., Tyrrell, R.A. and Rosopa, P.J., (2017) The conspicuity benefits of dynamic and static bicycle taillights at night, *Proceedings of the Human Factors and Ergonomics Society Annual Meeting*, 61, 1567–1568.

Edewaard, D.E., Szubski, E.C., Tyrell, R.A. and Duchowski, A.T., (2019) The conspicuity benefits of bicycle taillights in daylight, *Proceedings of the Driving Awareness Conference*, 24–27 June, Santa Fe, NM.

Edwards, C.S. and Gibbons, R.B., (2007) *The Relationship of Vertical Illuminance to Pedestrian Visibility in Crosswalks*, TRB Visibility Symposium, College Station, TX: Transportation Research Board.

Eklund, N.H., (1999) Exit sign recognition for color normal and color deficient observers, *Journal of the Illuminating Engineering Society*, 28, 71–81.

Eklund, N.H., Rea, M.S. and Bullough, J., (1997) Survey of snowplow operators about forward lighting and visibility during nighttime operations, *Transportation Research Record*, 1585, 25–29.

Ekrias, A., Eloholma, M. and Halonen, L., (2007) Analysis of road lighting quantity and quality in varying weather conditions, *Leukos*, 4, 89–98.

Elander, J., West, R. and French, D., (1993) Behavioral correlates of individual differences in road-traffic crash risk: An examination of methods and findings, *Psychological Bulletin*, 113, 279–294.

Electric Power Research Institute (EPRI) (2005) *Real World Background Luminance for Objects Viewed by Night Drivers*, Palo Alto, CA: EPRI.

Ellis, R.F., Amos, S. and Kumar, A., (2003) *Illumination Guidelines for Nighttime Highway Work*, NCHRP Report 498, Washington, DC: Transportation Research Board.

Eloholma, M. and Halonen, L., (2006) New model for mesopic photometry and its application to roadway lighting, *Leukos*, 2, 263–293.

Eloholma, M., Halonen, L. and Ketomaki, J., (1999) The effects of light spectrum on visual performance at mesopic light levels, *Proceedings of the CIE Symposium: 75 years of CIE Photometry*, Vienna: CIE.

Elvik, R., (1993) The effects on accidents of compulsory use of daytime running lights for cars in Norway, *Accident Analysis and Prevention,* 25, 383–398.

Elvik, R., (1995) Meta-analysis of evaluations of public lighting as accident countermeasure, *Transportation Research Record*, 1485, 112–123.

Elvik, R., (1996) A meta-analysis of studies concerning the safety effects of daytime running lights on cars, *Accident Analysis and Prevention,* 28, 685–694.

Eom, M. and Kim, R., (2020) The traffic signal control problem for intersections: A review, *European Transport Research Review*, 12, 50.

Erbay, A., (1974) A new method for the characterisation of the reflection properties of road surfaces, *Lichttechnik*, 26, 239.

Erkan, A., Hoffmann, D., Singer, T., Schikowski, J.M., Kunst, K., Peier, M.A. and Khan, T.Q., (2023) Influence of headlight level on object detection in urban traffic at night, *Applied Sciences*, 13, 2668.

European Commission, (2009) *SafetyNet: Cost Benefit Analysis*, Retrieved 7 August 2023. Brussels: European Commission, Directorate-General Transport and Energy.

European Commission, (2021) *Road Safety Thematic Report – Fatigue*, European Road Safety Observatory, Brussels: European Commission, Directorate General for Transport.

European Committee for Standardization (CEN), (2006) *European Standard: Traffic Control Equipment - Signal Heads*, EN 12368, Brussels: CEN.

European Road Safety Observatory (ERSO), (2007) *Traffic Safety Basic Facts, 2006*, Brussels: ERSO.

Ewing, R., (2019) *U.S. Traffic Calming*, London: Taylor and Francis.

Falchi, F., Cinzano, C.D., Duriscoe, D., Kyba, C.C.M., Elvidge, C.D., Baugh, K., Portnov, B.A., Rybnikova, N.A. and Eurgon, C., (2016) The new world atlas of artificial sky brightness, *Science Advances*, 2, e1600377.

Falchi, F., Furgoni, R., Gallaway, T.A., Rybnikova, N.A., Portnov, B.A., Baugh, K., Cinzano, P. and Elvidge, C.D., (2019) Light pollution in USA and Europe: The good, the bad and the ugly, *Journal of Environmental Management*, 258, 109227.

Falkmer, T. and Gregersen, N.P., (2005) A comparison of eye movement behaviour in experienced and inexperienced drivers in real traffic environment, *Optometry and Visual Science*, 82, 732–739.

Farmer, C.M. and Williams, A.F., (2002) Effects of daytime running lights on multiple vehicle daylight crashes in the United States, *Accident Analysis and Prevention*, 34, 197–203.

Farnsworth, D., (1947) *The Farnsworth Dichotomous Test for Color Blindness, Panel D-15 Manual*, New York: The Psychological Corporation.

Fayard, G.M., (2008) Work-related fatal injuries in parking lots, 1993–2002. *Journal of Safety Research*, 39, 9–18.

Federal Highways Administration (FHWA), (2003) *Manual on Uniform Traffic Control Devices*, Washington, DC: FHWA.

Federal Highways Administration (FHWA), (2022a) *Methods for Maintaining Pavement Marking Reflectivity*, FHWA-SA-14-017, Washington, DC: FHWA.

Federal Highways Administration (FHWA), (2022b) *FHWA Work Zone Facts and Statistics*, Washington, DC: FHWA.

Federal Highways Administration (FHWA), (2023) *Manual on Uniform Traffic Control Devices for Streets and Highways*, Washington, DC: FHWA.

Feeny-Burns, L., Burns, R.P. and Gao, C.L., (1990) Age-related macular changes in humans over 90 years old, *American Journal of Ophthalmology*, 109, 265–278.

Fekety, D.K., Edewaard, D.E., Stafford Sewall, A.A. and Tyrrell R.A., (2006) Electroluminescent materials can further enhance the nighttime conspicuity of pedestrians wearing retro-reflective materials, *Human Factors*, 58, 976–985.

Ferguson, B. and Blampied, N.M., (1991) Unenlightened: An unsuccessful attempt to promote the use of cycle lights at night, *Accident Analysis and Prevention*, 23, 561–571.

Ferguson S.A., Preusser, D.F., Lund, A.K., Zador, P.L. and Ulmer, R.G., (1995) Daylight saving time and motor vehicle crashes: The reduction in pedestrian and vehicle occupant fatalities, *American Journal of Public Health*, 85, 92–95.

Fildes, B.N., Corben, B., Kent, S., Oxley, J., Le, T.M. and Ryan, P., (1994) *Older Road User Crashes*, Clayton, Australia: Monash University Accident Research Centre.

Filipovic, F., Mladenovic, D., Lipovac, K., Das, D.K. and Todosijevic, B., (2022) Determining risk factors that influence cycling crash severity for the purpose of setting sustainable cycling mobility, *Sustainability*, 14, 13091.

Flannagan, M.J., (2001) *The Safety Potential of Current and Improved Front Fog Lamps*, Report UMTRI-2001-40, Ann Arbor, MI: University of Michigan Transportation Research Institute.

Flannagan, M.J., (2019) *A Market-weighted Description of Tungsten-halogen and LED Low-beam Headlighting Patterns in the US*, Report UMTRI-2019-5, Ann Arbor, MI: University of Michigan Transportation Research Institute.

Flannagan, M., Sivak, M., Ersing, M. and Simmon, C.J., (1989) *Effect of Wavelength on Discomfort Glare for Monochromatic Sources*, UMTRI-89-30, Ann Arbor, MI: University of Michigan Transportation Institute.

Folks, W.R. and Kreysar, D., (2000) Front fog lamp performance, *Human Factors in 2000, Driving, Lighting, Seating Comfort and Harmony in Vehicle Systems*, SP-1539, Warrendale PA: SAE.

Fontaine, H., (2003) Age des conducteurs de voiture et accidents de la route: Quel risqué pour les seniors, *Rescherche- Transports- Securite*, 79–80, 107–120.

Forbes, T.W., (1972) Visibility and Legibility of Highway Signs, In T.W. Forbes (ed) *Human Factors in Highway Safety Traffic Research*, New York: Wiley Interscience.

Foster R., (2022) *Life Time: The New Science of the Body Clock and How It Can Revolutionize Your Sleep and Health*, London: Penguin Random House.

Fotios, S. and Castleton, H.F., (2017) Lighting for cycling in the UK: A review, *Lighting Research and Technology*, 49, 381–395.

Fotios, S.A. and Cheal, C., (2007) Lighting for subsidiary streets: Investigations of lamps of different SPD, Part 2 – Brightness, *Lighting Research and Technology*, 39, 215–232.

Fotios, S. and Cheal, C., (2009) Obstacle detection: A pilot study investigating the effects of lamp type, illuminance and age, *Lighting Research and Technology*, 41, 321–342.

Fotios, S. and Gibbons, R., (2018) Road lighting research for drivers and pedestrians: The basis of luminance and illuminance recommendations, *Lighting Research and Technology*, 50, 154–186.

Fotios, S. and Yao, Q., (2019) The association between correlated colour temperature and scotopic / photopic ratio, *Lighting Research and Technology*, 51, 803–813 and Corrigendum, 814.

Fotios, S., Boyce, P. and Ellis, C., (2005a) *The Effect of Pavement Material on Road Lighting Performance*, London, UK Roads Liason Group, Department for Transport.

Fotios, S., Cheal, C. and Boyce, P.R., (2005b) Light source spectrum, brightness perception and visual performance in pedestrian environments: a review, *Lighting Research and Technology*, 37, 271–294.

Fotios, S. Unwin, J. and Farrall, P., (2015) Road lighting and pedestrian reassurance after dark: A review, *Lighting Research and Technology*, 47, 449–469.

Fotios, S., Cheal, C., Fox, S. and Uttley, J., (2018) The effect of fog on detection of driving hazards after dark, *Lighting Research and Technology*, 50, 1024–1044.

Fotios, S., Cheal, C., Fox, S. and Uttley, J., (2019) The transition between lit and unlit roads and detection of driving hazards after dark, *Lighting Research and Technology*, 51, 243–261.

Frederiksen, E. and Rotne, N., (1978) *Calculation of Visibility in Road Lighting*, Report 17, Lyngby, Denmark: The Danish Illuminating Engineering Laboratory.

Frederiksen, E. and Sorensen, K., (1976) Reflection classification of dry and wet road surfaces, *Lighting Research and Technology*, 8, 175–186.

Freedman, M., Janoff, M.S., Kuth, B.W. and McCunney, W., (1975) *Fixed Illumination for Pedestrian Protection*, Report FHWA-RD-76-8, Washington, DC: Federal Highways Administration.

Freedman, M, Zador, P. and Staplin, L., (1993) Effects of reduced transmittance film on automobile rear window visibility, *Human Factors*, 35, 535–550.

Freeman, M., Mitchel, J. and Coe, G.A., (2003) *Safety Performance of Traffic Management at Major Motorway Road Works*, TRL Report 595, Crowthorne, UK: Transport Research Laboratory.

Freeman, E., Munoz, B., Turano, K. and West, S.K., (2005) Vision loss and time to driving cessation in older adults, *Optometry and Vision Science*, 82, 765–775.

Freeman, E.E., Munoz, B., Turano, K.A. and West, S.K., (2006) Measures of visual function and their association with driving modification in older adults, *Investigative Ophthalmology and Visual Science*, 47, 514–520.

Freight Transport Association, (2013) *Conspicuity Marking Requirements on Goods Vehicles*, Tunbridge Wells, UK: Freight Transport Association.

Fry, G.A. and King, V.M., (1975) The pupillary response and discomfort glare, *Journal of the Illuminating Engineering Society*, 4, 307–324.

Gagliardi, G., Lupia, M., Cario, G., Tedesco, F., Cicchello, F., Lo Scudo, F. and Casavola, A., (2020) Advanced adaptive street lighting systems for smart cities, *Smart Cities*, 3, 1495–1512.

Gallagher, V.P. and Meguire, P.G., (1975) *Contrast Requirements of Urban Driving, Driver Visual Needs in Night Driving*, Transportation Research Board, Report 156: Washington, DC: TRB.

Gallagher, V.P., Janoff, M.S. and Farber, E., (1974) Interaction between fixed and vehicular illumination systems on city streets, *Journal of the Illuminating Engineering Society*, 4, 3–10.

Garland, J.A., (1971) Some fog droplet size distributions obtained by an impaction method, *Quarterly Journal of the Royal Meteorological Society*, 97, 483–494.

Garstang, R.H., (1986) Model for artificial night sky illumination, *Publications of the Astronomical Society of the Pacific*, 98, 364–375.

Gibbons, R. and Hankey, M., (2006) Influence of vertical illuminance on pedestrian visibility in crosswalks, *Transportation Research Record*, 1973, 105–112.

Gibbons, R., Guo, F., Medina, A., Terry, T., Du, J., Lutkevich, P. and Li, Q., (2014) *Design Criteria for Adaptive Roadway Lighting*, Report FHWA-HRT-14, 051, McLean, VA: Federal Highway Administration.

Gibbons, R.B., Bhagavathula, R., Warfield, B., Brainard, G.C. and Hanifin, J.P., (2022) Impact of solid state roadway lighting on melatonin in humans, *Clocks and Sleep*, 4, 633–657.

Gibson, J.J., (1950) *The Perception of the Visual World*, Boston MA: Houghton-Mifflin.

Gibson, K.S. and Tyndell, E.P.T., (1923) Visibility of radiant energy, *Bulletin of the Bureau of Standards*, 19, 131–191.

Gilkes, M.J., (1988) The basis of the medical recommendations for driver's visual standards in the United Kingdom, In A.G. Gale, M.H. Freeman, C.M. Haslegrave, P. Smith and S.P. Taylor (eds) *Vision in Vehicles-II*, Amsterdam: North Holland.

Girard, J., Villa, C. and Bremond, R., (2022) Discomfort glare from a cyclic source in outdoor lighting conditions, *Leukos*, 18, 459–474.

Girasole, T., Roze, C., Maheu, B., Grehan, G, and Menard, J., (1998) Visibility distances in a foggy atmosphere: Comparison between lighting installations by Monte Carlo simulation, *Lighting Research and Technology*, 30, 29–36.

Gomes-Franco, K., Rivera-Irquierdo, M., Martin-delosReyes, L.M., Jimenez-Mejias, E. and Martinez-Ruiz, V., (2020) Explaining the association between driver's age and the risk of causing a road crash through mediation analysis, *International Journal of Environmental Research and Public Health*, 17, 9041.

Goodman, T., Forbes, A., Walkey, H., Eloholma, M., Halonen, L., Alferdinck, J., Freiding, A., Bodrogi, P., Varady, G. and Szalmas, A., (2007) Mesopic visual efficiency IV: A model with relevance to nighttime driving and other applications, *Lighting Research and Technology*, 39, 365–392.

Gould, M, Poulter, D., Helman, S. and Wann, J.P., (2012) Judgments of approach speed for motorcycles across different lighting levels and the effect of an improved tri-headlamp configuration, *Accident Analysis and Prevention*, 48, 341–345.

Graham, C.H. and Kemp, E.H., (1938) Brightness discrimination as a function of the duration of the increment in intensity, *Journal of General Physiology*, 21, 635–650.

Green, M., (2000) "How long does it take to stop?" Methodological analysis of driver perception-brake times, *Transportation Human Factors*, 2, 195–216.

Green, J. and Hargroves, R.A., (1979) A mobile laboratory for dynamic road lighting measurement, *Lighting Research and Technology*, 11, 197–203.

Groupe de Travail Bruxelles (GTB), (1999) *Rationale of Harmonized Dipped (Low) Beam Pattern*, Report No. C.E. 3160, Geneva, Switzerland: GTB.

Guler, O. and Onaygil, S., (2003) The effect of luminance uniformity on visibility level in road lighting, *Lighting Research and Technology*, 35, 199–215.

Guo, L., Eloholma, M. and Halonen, L., (2007) Lighting control strategies for telemanagement road lighting control systems, *Leukos,* 4, 157–171.

Guo, F., Klauer, S.G., Fang, Y., Hankey, J.M., Antin, J.F., Perez M.A., Lee, S.E. and Dingus, T.A., (2016) The effects of age on crash risk associated with driver distraction, *International Journal of Epidemiology*, 46, 258–265.

Haans, A. and de Kort, Y.A.W., (2012) Light distribution in dynamic street lighting: Two experimental studies on perceived safety, prospect, concealment and escape, *Journal of Environmental Psychology*, 32, 342–352.

Habekost, M., (2013) Which color difference equation should be used? *International Circular of Graphic Education and Research*, 6, 20–33.

Haegestrom-Portnoy, G., Bryban, J.A., Schneck, M.E. and Jampolsky, A., (1997) The SKILL Card. An acuity test of reduced luminance and contrast. Smith-Kettlewell Institute Low Luminance, *Investigative Ophthalmology and Visual Science*, 38, 207–218.

Haegsetrom-Portnoy, G., Schneck, M.E. and Bryban, J.A., (1999) Seeing into old aqe: Vision function beyond acuity, *Optometry and Vision Science,* 76, 741–758.

Hagita, K. and Mori, K., (2014) The effect of sun glare on traffic accidents in Chiba prefecture. Japan, *Asian Transport Studies*, 3, 205–219.

Hakamies-Blumqvist, L., (1994) *Older Drivers in Finland: Traffic Safety and Behaviour*, Report 40/1994, Helsinki, Finland: Liikenneturva.

Hakamies-Blumqvist, L., Johansson, K. and Lundberg, C, (1996) Medical screening in a comparative Finnish–Swedish evaluation study, *Journal of the American Geriatrics Society*, 44, 650–653.

Hakamies-Blumqvist, L. and Wahlstrom, B., (1998) Why do older drivers give up driving? *Accident Analysis and Prevention*, 30, 305–312.

Hakamies-Blumqvist, L., Raitanen, T. and O'Neill, D., (2002) Driver ageing does not cause higher accident rates per km, *Transport Research*, Part F5, 271–274.

Hale, J., (1989) *Snowplow Lighting Study: Final Report*, Report No MN/RD-89/03, St. Paul, MN: Minnesota Department of Transportation.

Hall, R.R. and Fisher, A.J., (1977) Measures of visibility and visual performance in road lighting, In *Measures of Road Lighting Effectiveness*, Berlin: Lichttechnische Gesellschaft LiTG.

Hallet, P.E., (1963) Spatial summation, *Vision Research*, 3, 9–24.

Hamm, M., (2002) *Adaptive Lighting Functions History and Future – Performance Investigations and Field Test for User's Acceptance*, SAE Paper 2002-01-0526, Warrendale, PA: Society of Automotive Engineers.

Hamm, M., Keil, M. and Huhn, W., (2016) Analysis and overview on scientific research for adaptive driving beam and matrix beam systems, *Proceedings of the 4th International Forum on Automotive Lighting*, Shanghai, China.

Hankey, J.M., Kiefer, R.J. and Gibbons, R.B., (2005) *Quantifying the Pedestrian Detection Benefits of the General Motors Night Vision System*, SAE Technical Paper 2005-01-0443, Warrendale, PA: Society of Automotive Engineers.

Hansen, E.R. and Larsen, J.S., (1979) Reflection factors for pedestrian's clothing, *Lighting Research and Technology*, 11, 154–157.

Hare, C.T. and Hemion, R.H., (1968) *High Beam Usage on US Highways*, Report QR-666, Washington DC: Bureau of Roads.

Hargroves, R.A., (1981) Road lighting – as calculated and as in service, *Lighting Research and Technology*, 13, 130–136.

Hargroves, R.A., (2001) Lighting for pleasantness outdoors, *Proceedings of the CIBSE National Conference*, London: Chartered Institution of Building Services Engineers

Hargroves, R.A. and Scott, P.P., (1979) Measurements of road lighting and accidents - the results, *Public Lighting*, 44, 213–221.

Hartmann, E., Finsterwalder, J. and Muller, M., (1986) Kinetic luminance measurement and assessment of road tunnels, *Lighting Research and Technology*, 18, 28–36.

Hasson, P., Lutkevich, P., Ananthanarayanan, B., Watson, P. and Knoblauch, R., (2002) Field test for lighting to improve safety at pedestrian crosswalks, *Proceedings of the 16th Biennial Symposium on Visibility and Simulation*, Washington, DC: Transportation Research Board.

Hassan, M.M., Odeh, I. and El-Rayes, K., (2010) Glare and light characteristics of conventional and balloon lighting systems, *Transportation Research Board 89th Annual Meeting*, Washington, DC: TRB.

Hauer, E., (2011) Computing what the public wants: Some issues in road safety cost benefit analysis, *Accident Analysis and Prevention*, 43, 151–164.

Haun, A.M. and Peli, E., (2013) Perceived contrast in complex images, *Journal of Vision*, 13, e3817960.

Hawkins, R.K., (1988) Motorway traffic behaviour in reduced visibility conditions, In A.G. Gale, M.H. Freeman, C.M. Haslegrave, P. Smith, and S.P. Taylor (eds), *Vision in Vehicles-II*, Amsterdam: North Holland.

He, Y., Rea, M.S., Bierman, A. and Bullough, J., (1997) Evaluating light source efficacy under mesopic conditions using reaction times, *Journal of the Illuminating Engineering Society*, 26, 125–138.

He, Y., Li. and Zhang, X., (2021) Influence of text luminance, text colour and background luminance of variable message signs on legibility in urban areas at night, *Lighting Research and Technology*, 53, 263–279.

Hellier-Symons, R.D. and Irving, A., (1981) *Masking of Brake Lights by High-Intensity Rear Lights in Fog*, TRRL Report 998, Crowthorne, UK: Transportation and Road Research Laboratory.

Helman, S., Palmer, M, Haines, C. and Reeves, S.C., (2014) *The Effect of Two Novel Lighting Configurations on the Conspicuity of Motorcycles*, Crowthorne, UK: Transport Research Laboratory.

Helmers, G. and Rumar, K., (1975) High beam intensity and obstacle visibility, *Lighting Research and Technology*, 7, 38–42.

Hentschel, H.J., (1971) A physiological appraisal of the revealing power of street lighting installations for large composite targets, *Lighting Research and Technology*, 3, 268–273.

Higgins, K.E. and Bailey, I.L., (2000) Visual disorders and performance of specific tasks requiring vision, In B. Silverstone, M.A. Lang, B.P. Rosenthal and E.E. Faye (eds), *The Lighthouse Handbook on Vision Impairment and Vision Rehabilitation*, New York: Oxford University Press.

Higgins, K.E., Wood, J.M. and Tait, A., (1996) Closed road driving performance: The effect of degradation of visual acuity, *Vision Science and Its Applications*, Washington, DC: Optical Society of America.

Highway Loss Data Institute (HLDI), (2015) Volvo city safety loss experience – a long-term update, *HLDI Bulletin*, 32, 1–24.

Highways Agency, (2010) *The Use of New Materials to Reduce Traffic Sign Lighting*, Guildford, UK: Highways Agency.

Hills, B.L., (1975a) Visibility under night driving conditions, Part 1 Laboratory background and theoretical considerations, *Lighting Research and Technology*, 7, 179–184.

Hills, B.L., (1975b) Visibility under night driving conditions, Part 2 Field measurements using disc obstacles and a pedestrian dummy, *Lighting Research and Technology*, 7, 251–258.

Hills, B.L., (1976) Visibility under night driving conditions, Part 3 Derivation of (ΔL, A) characteristics and factors in their application, *Lighting Research and Technology*, 8, 11–26.

Hills, B.L., (1980) Vision, visibility and perception in driving, *Perception*, 9, 183–216.

Hilz, R. and Cavonius, C.R., (1974) Functional organization of the peripheral retina: Sensitivity to periodic stimuli, *Vision Research*, 14, 1333–1337.

Hiscocks, P.D. and Gudmundsson, S., (2010) The contribution of street lighting to light pollution, *Journal of the Royal Astronomical Society of Canada*, 104, 190.

Holladay, L.L., (1926) The fundamentals of glare and visibility, *Journal of the Optical Society of America*, 12, 271–319.

Holz, M. and Weidel, E., (1998) Night vision enhancement system using diode laser headlamps, *Electronics for Trucks and Buses*, Report SP-1401, Warrendale, PA: Society of Automotive Engineers.

Hopkinson, R.G., (1940) Discomfort glare in lighted streets, *Transactions of the. Illuminating Engineering Society (London)*, 5, 1–29.

Hoque, M.M., (1990) An analysis of fatal bicycle accidents in Victoria (Australia) with a special reference to nighttime accidents, *Accident Analysis and Prevention*, 22, 1–11.

Horne, J. and Reyner, L., (1999) Vehicle accidents related to sleep: A review, *Occupational and Environmental Medicine*, 56, 289–294.

Horowitz, A.D., (1994) Human factors issues in advanced rear signaling systems, *Proceedings of the Fourteenth International Technical Conference on Enhanced Safety of Vehicles*, Washington, DC: Department of Transportation.

Huang, H., Zegeer, C. and Nassi, R., (2000) *Innovative Treatments at Unsignalized Pedestrian Crossing Locations*, ITE Annual Meeting Compendium, Washington, DC:Institute of Transportation Engineers.

Huang, H.F. and Cynecki, M.J., (2001) *The Effects of Traffic Calming Measures on Pedestrian and Motorist Behavior*, Report FHWA-RD-00-104, Federal Highway Administration, McLean, VA.

Huang, H-P., Li, H-C., Wei, M. and Li, G-H., (2023) Investigation of text-background lightness combination on visual clarity using a head-up display under various surround conditions and different age groups, *Applied Sciences*, 13, 6037.

Hughes, P.K. and Cole, B.L., (1984) Search attention conspicuity of road traffic control devices, *Australian Road Research*, 14, 1–9.

Huhn, W., Ripperger, J. and Befelein, C., (1997) *Rear Light Redundancy and Optimized Hazard Warning Signal – New Safety Functions for Vehicles*, SAE Technical Paper 970656, Warrendale, PA: Society of Automotive Engineers.

Hunter, W.W., Stutts, J.C., Pein, W.E. and Cox, C.C., (1996) *Pedestrian and Bicycle Crash Types of the Early 1990s*, Report FHWA-RD-95-163, Washington DC: Federal Highways Administration.

Hutchinson, T.P. and Lindsay, V.L., (2009) *Pedestrian and Cyclist Crashes in the Adelaide Metropolitan Area*, Report CASR055, Adelaide, Australia: Centre for Automotive Safety Research

Hutt, D.L., Bissonnette, L.R., St Germain, D. and Oman, J., (1992) Extinction of visible and infrared beams by falling snow, *Applied Optics*, 31, 5121–5132.

Ikeda, M. and Shimozono, H., (1981) Mesopic luminous efficiency functions, *Journal of the Optical Society of America*, 71, 280–284.

Illuminating Engineering Society of North America (IESNA) (1980) Roadway Lighting Subcommittee, *Visual Comfort: The CBE Recommendation*, New York: IESNA.

Illuminating Engineering Society of North America (IESNA), (2000) *The IESNA Lighting Handbook, 9th Edition*, New York: IESNA.

Illuminating Engineering Society of North America (IESNA), (2001) Recommended Practice RP-19-01 *Roadway Sign Lighting*, New York: IESNA.

Illuminating Engineering Society of North America (IESNA), (2005) Recommended Practice RP-22-05 *Recommended Practice for Tunnel Lighting*, New York: IESNA.

Illuminating Engineering Society of North America (IESNA), (2006) Technical Memorandum TM-12-06 *Spectral Effects of Lighting on Visual Performance at Mesopic Light Levels*, New York: IESNA.

Illuminating Engineering Society of North America, (2007) *Technical Memorandum TM-15-07: Luminaire Classification System for Outdoor Luminaires*, New York: IESNA.

Illuminating Engineering Society of North America, (2011) *Model Lighting Ordinance with Users' Guide*, New York: IESNA.

Illuminating Engineering Society of North America (IESNA), (2020) *Technical Memorandum TM-15-20: Luminaire Classification System for Outdoor Luminaires*, New York: IESNA.

Illuminating Engineering Society of North America (IESNA), (2021a) *Recommended Practice: RP-8-21: Lighting Roadways and Parking Facilities, Part 1 Fundamentals*, New York: IESNA.

Illuminating Engineering Society of North America (IESNA), (2021b) *Recommended Practice: RP-8-21: Lighting Roadways and Parking Facilities, Part 2 Design*, New York: IESNA.

Institute of Transportation Engineers (ITE), (1985) *Vehicle Traffic Control Signal Heads, A Standard of the Institute of Transportation Engineers*, Washington, DC: ITE.

Institute of Transportation Engineers (ITE), (2001) *Equipment and Material Standards of the Institute of Transportation Engineers*, Washington, DC: ITE.

Institute of Transportation Engineers (ITE), (2005) *Specification for Traffic Control Signal Heads: LED Circular Signals*, Washington, DC: ITE.

Institute of Transportation Engineers (ITE), (2008) *Specification for Traffic Control Signal Heads: LED Arrow Signals*, Washington, DC: ITE.

Institute of Transportation Engineers (ITE), (2011) *Specification for Pedestrian Traffic Control Signal Indicators – Light Emitting Diodes*, Washington, DC: ITE.

Institution of Lighting Professionals (ILP), (2002) *Technical Report 25: Lighting for Traffic Calming Features*, Rugby, UK: ILP.

Institution of Lighting Professionals (ILP), (2007) *Technical Report 28: Measurement of Road Lighting Performance on Site*, Rugby, UK: ILP.

Institution of Lighting Professionals (ILP), (2012) *Lighting of Subsidiary Roads – Using White Light Serves to Balance Energy Efficiency and Visual Amenity*, Rugby, UK: ILP.

Institution of Lighting Professionals (ILP), (2020a) *Guidance Note 11: Determination of Maintenance Factors*, Rugby, UK: ILP.

Institution of Lighting Professionals (ILP), (2020b) *Guidance Note 01/20: Guidance Notes for the Reduction of Obtrusive Light*, Rugby, UK: ILE

Insurance Institute for Highway Safety (IIHS), (2018) *Headlight Test and Rating Protocol, Version 111*, Ruckerville, VA: IIHS.

International Dark-Sky Association (IDA), (2002) *Outdoor Lighting Code Handbook*, Version 1.14, http://www.darksky.org (accessed March 21st, 2007).

International Standards Organization (ISO), (1999) *ISO 16508:1999 Road Traffic Lights – Photometric Properties of 200 mm Roundel Signals*, Geneva, ISO

Isebrands, H., Hallmark, S., Hans, Z., McDonald, T., Preston, H. and Storm, R., (2004) *Safety Impacts of Street Lighting at Isolated Rural Intersections, Part II, Year 1 Report*, St Pauls, MN: Minnesota Department of Transportation.

Isebrands, H., Hallmark, S., Hans, Z., McDonald, T., Preston, H. and Storm, R., (2006) *Safety Impacts of Street Lighting at Isolated Rural Intersections, Part II*, St Pauls, MN: Minnesota Department of Transportation.

Jackett, M. and Frith, W., (2013) Quantifying the impact of road lighting on road safety – A New Zealand Study, *IATSS Research*, 36, 139–145.

Jagerbrand, A.K. and Sjobergh, J., (2016) Effects of weather conditions, light conditions and road lighting on vehicle speed, *SpringerPlus*, 5, 505.

Jalink, C.J., (2002) *Distributive Lighting Systems for Interior Applications*, SAE Paper 2002-01-0979, Warrendale, PA: Society of Automotive Engineers.

Janke, M.M., (1991) Accidents, mileage and the exaggeration of risk, *Accident Analysis and Prevention*, 23, 183–188.

Janoff, M.S., (1990) The effect of visibility on driver performance: A dynamic experiment, *Journal of the Illuminating Engineering Society*, 19, 57–63.

Janoff, M.S., (1992a) The relationship between visibility level and subjective ratings of visibility, *Journal of the Illuminating Engineering Society*, 21, 98–107.

Janoff, M.S., (1992b) Effect of headlights on small target visibility, *Journal of the Illuminating Engineering Society*, 21, 46–53.

Janoff, M.S. and Havard, J.A., (1997) The effect of lamp color on visibility of small targets, *Journal of the Illuminating Engineering Society*, 26, 173–181.

Janoff, M.S., Freedman, M. and Koth, B., (1977) Driver and pedestrian behavior - the effect of specialized crosswalk illumination, *Journal of the Illuminating Engineering Society*, 6, 202–208.

Janssen, W.H., Michon, J.A. and Lewis, O.H., (1976) The perception of lead vehicle movement in darkness, *Accident Analysis and Prevention*, 8, 151–166.

Johannson, G., (1973) Visual perception of biological motion and a model for its analysis, *Perception and Psychophysics*, 14, 201–211.

Johannson, O., Wanvik, P.O. and Elvik, R., (2009) A new method for assessing the risk of accident associated with darkness, *Accident Analysis and Prevention*, 41, 809–815.

Johnson, C. and Keltner, J., (1983) Incidence of field loss in 20,000 eyes and its relationship to driving performance, *Archives of Ophthalmology*, 101, 371–375.

Johnson, M., Charlton, J., Oxley, J. and Newstead, S., (2010) Naturalistic cycling study: Identifying risk factors for on-road commuter cyclists, *Annals of Advances in Automotive Medicine*, 54, 275–283.

Jones, H.V. and Heimstra, N.W., (1964) Ability of drivers to make critical passing judgments, *Journal of Engineering Psychology*, 3, 117–122.

Judd, D.B., (1951) Report of the US Secretariat Committee on Colorimetry and Artificial Daylight, *Proceeding of the CIE 12th Session*, Stockholm, Vienna: CIE.

Kahane, C.J., (2015) *Lives Saved by Vehicle Safety Technologies and Associated Federal Motor Vehicle Safety Standards: 1960–2012*, DOT HS 812-009, Washington, DC: National Highway Traffic Safety Administration.

Kahane, C.J. and Hertz, E., (1998) *The Long Term Effectiveness of Center High-Mounted Stop Lamps in Passenger Cars and Trucks*, DOT HS 808 696, Washington, DC: National Highway Traffic Safety Administration.

Kaiser, P.K. and Boynton, R.M., (1996) *Human Color Vision*, Washington, DC: Optical Society of America.

Keall, M.D., Fildes, B. and Newstead, S., (2017) Real-world evaluation of the effectiveness of reversing camera and parking sensor technologies in preventing backover pedestrian injuries, *Accident Analysis and Prevention*, 99, 39–43.

Keith, D.M., (2000) Roadway lighting design for optimization of UPD, STV and uplight, *Journal of the Illuminating Engineering Society*, 29, 15–23.

Kelly, D.H., (1959) Effects of sharp edges in a flickering field, *Journal of the Optical Society of America*, 49, 730–732.

Kelly D.H., (1961) Visual response to time-dependent stimuli, 1. Amplitude sensitivity measurements, *Journal of the Optical Society of America*, 51, 422–429.

Kelly, K.L. and Judd, D.B., (1965) *The ISCC-NBS Centroid Color Charts*, Washington, DC: National Bureau of Standards.

Kimlin, J.A., Black, A.A. and Wood, J.M., (2017) Nighttime driving in older adults: Effects of glare and association with mesopic visual function, *Investigative Ophthalmology and Visual Science*, 58, 2796–2803.

Kircher, K. and Niska, A., (2021) Testing of bicycle lighting: Method development and evaluation, *Transportation Research Interdisciplinary Perspectives*, 10, e100349.

Kirkpatrick, M., Baker, C.C. and Heasly, C.C., (1987) *A Study of Daytime Running Light Design Factors. Final Report*, Report DOT/HS 807-193, Washington DC: Department of Transportation

Kitsinelis, S., (2015) *Light Sources: Basics of Lighting Technologies and Applications*, 2nd Edition, Boca Raton, FL: CRC Press.

Kline, D., (1995) Visual requirements for the aging driver, In W. Adrian (ed) *Lighting for Aging Vision and Health*, New York: Lighting Research Institute.

Kocifaj, M., (2007) Light pollution model for cloudy and cloudless night skies with ground based light sources, *Applied Optics*, 46, 3013–3022.

Kocmond, W.C. and Perchonok, K., (1970) *Highway Fog*, National Cooperative Highway Research Program, Report 171, Washington, DC: The National Research Council.

Konyukhov, V.V., Koroleva, Y.E., Novakovsky, L.G. and Novikova, L.A., (2006) Motorcycle headlamps featuring a light beam stabilization dynamic adjuster during turning, *Light and Engineering*, 14, 50–61.

Koornstra, M.J., (1993) *Daytime Running Lights: Its Safety Revisited*, Report D-93-25, Leidschendam, The Netherlands: SWOV Institute for Road Safety Research.

Kosnik, W., Winslow, L., Kline, D., Rasinski, K. and Sekular, R., (1988) Visual changes throughout adulthood, *Journal of Gerontology: Psychological Sciences*, 43, 63–70.

Koth, B.W., McCunney, W.D., Duerk, C.P., Janoff, M.S. and Freedman, M., (1978) *Vehicle Fog Lighting: An Analytical Evaluation*, Report DOT HS-803-442, Washington, DC: National Highway Traffic Safety Administration.

Kraus, A., Benter, N. and Boerner, H., (2007) *OLED Technology and Its Possible use in Automotive Applications*, SAE Paper 2007-01-1230, Warrendale, PA: Society of Automotive Engineers.

Kwan, I. and Mapstone, D.E., (2006) Interventions for increasing pedestrian and cyclist visibility for the prevention of death and injuries, *Cochrane Database of Systematic Reviews*, 4, CD003436.

Kwong, R.C., Michaiski, L., Nugent, M., Rajan, K., Ngo, T., Brown, J.J., Lamansky, S., Djurovich, P., Murphy, D., Abdel-Razzaq, F., Brooks, J., Thompson, M.E., Adachi, C., Baldo, M. and Forrest, S.R., (2001) Recent advances in organic light emitting diodes, *Proceedings of the 9th International Symposium on the Science and Technology of Light Sources*, Ithica, NY: Cornell University Press.

Kyba, C.C.M., Ruby, A., Kuechly, H.U., Kinzey, B., Miller, N., Sanders, J., Barentine, J., Kleinodt, R. and Espey, B., (2021) Direct measurement of the contribution of street lighting to satellite observations of nighttime light emissions from urban areas, *Lighting Research and Technology*, 53, 189–211.

Kyba, C.C.M, Altintas, Y.O., Walker, C.E. and Newhouse, M., (2023) Citizen scientists report global rapid reduction in the visibility of stars from 2011 to 2022, *Science*, 379, 265–268.

Lahrmann, H., Madsen, T.K.O., Olesen, A.V., Masden, J.C.O. and Hels, T., (2018) The effect of a yellow bicycle jacket on cyclist accidents, *Safety Science*, 108, 209–217.

Lambert, J.H. and Turley, T.C., (2003) *Screening Methodology for Need of Roadway Lighting*, Report FHWA/ VTRC 03-CR14, Charlottesville, VA: University of Virginia.

Lamm, R., Kloeckner, J.H. and Choueiri, E.M., (1985) *Freeway Lighting and Traffic Safety: A Long Term Investigation*, TRB-1985-17, 64th TRB Annual Meeting, Washington, DC: Transportation Research Board.

Land, M.F. and Horwood, J., (1995) Which part of the road guides steering, *Nature (London)*, 377, 339–340.

Langford, J., Methorst, R. and Hakamies-Blumqvist, L., (2006) Older drivers do not have a high crash risk – a replication of low mileage bias, *Accident Analysis and Prevention*, 38, 574–578.

Lecocq, J., (1993) VL application for hemispherical multi-faceted targets, *Proceedings of the 2nd International Symposium on Visibility and Luminance in Roadway Lighting, Orlando, FL*, Cleveland OH: Lighting Research Office.

Lecocq, J., (1994) Visibility and lighting of wet road surfaces, *Lighting Research and Technology*, 26, 75–87.

Leibowitz, H.W. and Owens, D.A., (1975) Night myopia and the intermediate dark focus of accommodation, *Journal of the Optical Society of America*, 65, 1121–1128.

Leibowitz, H.W. and Owens, D.A., (1977) Nighttime driving accidents and selective visual degradation, *Science*, 197, 422–423.

Lestina, D.C., Miller, T.R., Knoblauch, R. and Nitzburg, M., (1999) Benefits and costs of ultraviolet fluorescent lighting, *Proceedings of the Association for the Advancement of Automotive Medicine*, Barcelona, Spain.

Lewis, A.L., (1999) Visual performance as a function of spectral power distribution of light sources at luminances used for general outdoor lighting, *Journal of the Illuminating Engineering Society*, 28, 37–42.

Lighting Energy Alliance, (2020) *Guide for Parking Lot Lighting*, Troy, NY: Rensselaer Polytechnic Institute.

Lighting Industry Association (LIA), (2013) Technical Statement TS 24 *SP Ratio and Mesopic Vision*, Telford, UK: LIA.

Lighting Research Center (LRC), (1996) *Evaluating Interior Lighting Schemes: General Motors Automobile Study*, Troy, NY: Lighting Research Center.

Lin, Y., Chen, W., Chen, D. and Shao, H., (2004) The effect of spectrum on visual field in road lighting, *Building and Environment*, 39, 433–439.

Lin, Y., Liu, Y., Sun, Y., Zhu, X., Lai, J. and Heynderickx, I., (2014) Model predicting discomfort glare caused by LED road lights, *Optics Express*, 22, 18056–18071.

Lingard, R. and Rea, M.S., (2002) Off-axis detection at mesopic light levels in a driving context, *Journal of the Illuminating Engineering Society*, 31, 33–39.

Liu, A., (1998) What the driver's eye tells the car's brain, In G. Underwood (ed), *Eye Guidance in Reading and Scene Perception,* Amsterdam: Elsevier.

Livingston, M.S. and Hubel, D.H., (1981) Effects of sleep and arousal on the processing of visual information in the cat, *Nature*, 291, 554–561.

Llewellyn, R., Cowie, J. and Maher, M., (2020) Active road studs as an alternative to lighting on rural roads: Driver safety perception, *Sustainability*, 12, e9648.

Lucas, R.J., Lall, G.S., Allen, A.E. and Brown, T.M., (2012) How rods, cones and melanopsin photoreceptors come together to enlighten the mammalian circadian clock. *Progress in Brain Research*, 199, 1–18.

Luginbuhl, C.B., Lockwood, G.W., Davis, D.R. Pick, K. and Selders, J., (2009) From the ground up 1: Light pollution sources in Flagstaff, Arizona, *Publications of the Astronomical Society of the Pacific*, 121, 185–203.

Luo, M.R., Cui, G. and Rigg, B., (2001) The development of the CIE2000 color difference formula, *Color Research and Application*, 26, 340–350.

Luoma, J., Schumann, J. and Traube, E.C., (1995) *Effect of Retroreflector Positioning on Nighttime Recognition of Pedestrians*, UMTRI-95-18, Ann Arbor, MI: University of Michigan Transportation Research Institute.

Luoma, J., Flannagan, M.J., Sivak, M., Aoki, M. and Traube, E.C., (1997) Effects of turn –signal color on reaction times to brake signals, *Ergonomics*, 40, 62–68.

Luoma, J., Sivak, M. and Flannagan, M.J., (2006) Effects of dedicated stop-lamps on nighttime rear-end collisions, *Leukos*, 3, 159–165.

Lynes, J.A., (1971) Lightness, colour and constancy in lighting design, *Lighting Research and Technology*, 3, 24–42.

Lynes, J.A., (1977) Discomfort glare and visual distraction, *Lighting Research and Technology*, 9, 51–52.

MacAdam, D.L., (1942) Visual sensitivity to color differences in daylight, *Journal of the Optical Society of America*, 32, 247–274.

Mace, D.J., Hosletter, R.S., Pollack, L.E. and Sweig, W.D., (1986) *Minimal Luminance Requirements for Official Highway Signs*, FHWA–RD-86-151, Washington, DC: Federal Highways Administration.

Mace, D., Garvey, P., Porter, R.J., Schwab, R. and Adrian, W., (2001) *Countermeasures for Reducing the Effects of Headlight Glare*, Washington, DC: AAA Foundation for Traffic Safety.

Maciej, J. and Vollrath, M., (2009) Comparison of manual and speech-based interaction with in-vehicle information systems, *Accident Analysis and Prevention*, 41, 924–930.

Madsen, J.C.O. Andersen, T. and Lahrmann, H.S., (2013) Safety effects of permanent running lights for bicycles: A controlled experiment, *Accident Analysis and Prevention*, 50, 820–829.

Mahoor, M., Hosseini, Z.S., Khpdaei, A., Paaso, A. and Kushner, D., (2020) State-of-the-art in smart streetlight systems: A review, *IET Smart Cities*, 2, 24–33.

Malin, F., Silla, A. and Mladenovic, M.N., (2020) Prevalence and factors associated with pedestrian fatalities and serious injuries: Case Finland, *European Transport Research Review*, 12, 29.

Mandlebaum, J. and Sloan, L.L., (1947) Peripheral visual acuity, *American Journal of Ophthalmology*, 30, 581–588.

Marshall, J., Grindle, J., Ansell, P.L. and Borwein, B., (1979) Convolution in human rods: An aging process, *British Journal of Ophthalmology*, 63, 181–187.

Martin, A., (2006) *Factors Influencing Pedestrian Safety: A Literature Review*, TRL Report PPR241, Crowthorne, UK: Transport Research Laboratory.

Masello, L., Castigani, G., Sheehan, B. and Murphy, F., (2022) On the road safety benefits of advanced driver assistance systems in different driving contexts, *Transportation Research Interdisciplinary Perspectives*, 15, 100670.

Massart, P., (1973) *Definition of synthetic surfaces in road lighting*, Thesis, Belgium: University of Liege.

Massie, D.L. and Campbell, K.L., (1993) *Analysis of Accident Rate by Age, Gender and Time of Day Based on the 1990 Nationwide Personal Transportation Survey*, UMTRI-93-7, Ann Arbor, MI: University of Michigan Transportation Research Institute.

Maycock, G., Lockwood, G.R. and Lester, J.F., (1991) *The Accident Liability of Drivers*, TRRL Report 315, Crowthorne, UK: Transport and Road Research Laboratory.

Mazzae, E.N., Scott Baldwin, G.H., Adrella, A. and Smith L.A., (2015) *Adaptive Driving Beam Headlighting System Glare Assessment*, NHTSA Technical Report DOT HS 812 174, Washington, DC: National Highways Traffic Safety Administration.

McColgan, M.W., Van Derlofske, J., Bullough, J.D. and Shakir, I., (2002) *Subjective Color Preferences of Common Road Sign Materials under Headlamp Bulb Illumination*, SAE Paper 2002-01-0261, Warrendale, PA: Society of Automotive Engineers.

McGuire, L. and Smith, N., (2000) Cycling safety: Injury prevention in Oxford cyclists, *Injury Prevention*, 6, 285–287.

McGwin Jr., G. and Owsley, C., (2022) Vision screening for motor vehicle collision involvement among older drivers, *Ophthalmology*, 129, 1022–1027.

McKnight, A.J., Shinar, D. and Hilburn, B., (1991) The visual and driving performance of monocular and binocular heavy duty truck drivers, *Accident Analysis and Prevention*, 23, 225–237.

McNally, D., (1994) *The Vanishing Universe*, Cambridge, UK: Cambridge University Press.

Middleton, W.E.K., (1952) *Vision through the Atmosphere*, Toronto: University Press.

Miller, J.W., (1958) Study of visual acuity during the ocular pursuit of moving test objects. II Effects of direction of movement, relative movement and illumination, *Journal of the Optical Society of America*, 48, 803–808.

Missikpode, C., Peek-Asa, C., Young, T. and Hamann, C., (2018) Does crash risk increase when emergency vehicles are driving with lights and sirens? *Accident Analysis and Prevention*, 113, 257–262.

Mitra, S., (2014) Sun glare and road safety: An empirical investigation of intersection crashes. *Safety Science*, 70, 246–254.

Mitran E., Codoe, J. and Edwards, E., (2020) *Impact of Crosswalk Lighting Improvements on Pedestrian Safety – A Literature Review*, Baton Rouge, LA: Louisiana Transportation Research Center.

Mizon, B., (2002) *Light Pollution, Responses and Remedies*, London: Springer.

Monahan, D.R., (1995) Safety considerations in parking facilities, *Proceedings of the International Parking Conference and Exposition*, Fredericksburg, VA: International Parking Institute.

Moon, P. and Hunt, R.M., (1938) Reflection characteristics of road surfaces, *Journal of the Franklin Institute*, 225, 1–21.

Moore, R.L., (1952) *Rear Lights of Motor Vehicles and Pedal Cycles*, Road Research Technical Paper 25, London: His Majesty's Stationery Office, London.

Moore, D.W. and Rumar, K., (1999) *Historical Development and Current Effectiveness of Rear Lighting Systems*, UMTRI-99-31, Ann Arbor, MI: University of Michigan Transportation Research Institute.

Morgan, C., (2001) *The Effectiveness of Retro-reflective Tape on Heavy Trailers*, NHTSA Technical Report DOT HS 809 222, Washington, DC: National Highway Traffic Safety Administration.

Mortimer, R.G., (1981) *Field Test Evaluation of Rear Lighting Deceleration Signals*, Safety Research Report 81-1, Champaign, IL: University of Illinois.

Mortimer, R.G. and Becker, J.M., (1973) *Development of a Computer Simulation to Predict the Visibility Distances Provided by Headlamp Beams*, Report UM-HSRI-IAF-73-15, Ann Arbor, MI: University of Michigan.

Mortimer, R.G. and Jorgeson, C.M., (1974) *Eye Fixations of Drivers in Night Driving with Three Headlight Beams*, Report UM-HSRI-74-17, Ann Arbor, MI: University of Michigan.

Murata, Y., (1987) Light absorption characteristics of the lens capsule, *Ophthalmic Research*, 19, 107–112.

Mure, L.S., (2021) Intrinsically photosensitive retinal ganglion cells of the human retina, *Frontiers in Neurology*, 12, 636330.

Murray, J., Feather, J. and Carden, D., (2002) Dynamic discomfort glare and driver fatigue, *Lighting Journal*, 67, 20–23.

Muzet, V., Greffier, F. and Verney, P., (2019) Optimization of road surface reflection properties and lighting: Learning of a three-year experiment, *Proceedings of the 29th Session of the CIE*, Washington DC, Vienna: CIE.

Muzet, V., Bernesconi, J., Lacomussi, M., Liandrat, S., Greffier, F., Blattner, P., Reber, J. and Lindgren, M., (2021) Review of road surface photometry and devices – Proposal for new measurement geometries, *Lighting Research and Technology*, 53, 213–229.

Nair, G. and Dhoble, S.J., (2020) *The Fundamentals and Applications of Light Emitting Diodes*, Sawston, Cambridge: Woodhead Publishing.

Nambisan, S.S., Pulugurtha, S.S., Vasudevan, V., Dangeti, M.R. and Virupuksha, V., (2009) Effectiveness of automatic pedestrian detection device and smart lighting for pedestrian safety, *Transportation Research Record*, 2140, 27–34.

Narendran, N., Vasconez, S., Boyce, P. and Eklund, N., (2000) Just-perceivable color difference between similar light sources in display lighting applications, *Journal of the Illuminating Engineering Society*, 29, 78–82.

Narendran, N., Bullough, J.D., Maliyagoda, N. and Bierman A., (2001) What is useful life for white light LEDs? *Journal of the Illuminating Engineering Society*, 30, 57–67.

Narendran, N., Freyssinier, J.P. and Zhu, Y., (2016) Energy and user acceptability benefits of improved illuminance uniformity in parking lot illumination, *Lighting Research and Technology*, 48, 789–809.

Narisada, K., (1995) Perception in complex fields under road lighting conditions, *Lighting Research and Technology*, 27, 123–131.

Narisada, K. and Karasawa, Y., (2001) Re-consideration of the revealing power on the basis of visibility level, *Proceedings of the International Lighting Congress, Istanbul*, 2, 473–480.

Narisada, K. and Yoshikawa, K., (1974) Tunnel entrance lighting - effect of fixation point and other factors on the determination of requirements, *Lighting Research and Technology*, 6, 9–18.

Narisada, K., Karasawa, Y. and Shirao, K., (2003) Design parameters of road lighting and revealing power, *Proceedings of the 23rd Session of the CIE*, San Diego, CA, Vienna: CIE.

Nathan, J., Henry, G. and Cole, B., (1964) Recognition of colored traffic light signals by normal and color vision defective observers, *Journal of the Optical Society of America.*, 54, 1041–1045.

National Academies of Sciences, Engineering and Medicine, (2020) *NCHRP 940 Solid State Roadway Lighting Design Guide, Vol 1: Guidance*, Washington, DC: National Cooperative Highway Research Program.

National Bureau of Standards, (1976) *Color: Universal Language and Dictionary of Names*, Special Publication 440, Washington, DC: National Bureau of Standards.

National Highway Traffic Safety Administration (NHTSA), (2006) *Traffic Safety Facts 2005*, Washington, DC: US Department of Transportation.

National Highway Traffic Safety Administration (NHTSA), (2021a) *Traffic Safety Facts 2019: A Compilation of Motor Vehicle Crash Data*, Washington, DC: US Department of Transportation.

National Highway Traffic Safety Administration (NHTSA), (2021b) *Traffic Safety Facts: Motorcycles*, Washington, DC: US Department of Transportation.

National Highway Traffic Safety Administration (NHTSA), (2021c) *Traffic Safety Facts: Pedestrians*, Washington, DC: US Department of Transportation.

National Highway Traffic Safety Administration (NHTSA), (2022a) *Fatality and Injury Reporting System*, Washington, DC: US Department of Transportation.

National Highway Traffic Safety Administration (NHTSA), (2022b) *Traffic Safety Facts: Bicyclists and Other Cyclists*, Washington, DC: US Department of Transportation.

National Safety Council, (2023) *Injury Facts- Emergency Vehicles*, Chicago: National Safety Council.

Neitz, J., Carroll, J. and Neitz, M., (2001) Color vision: Almost reason enough for having eyes, *Optics and Photonics News*, January, 26–33.

Neubauer, O., Harrer, S., Marre, M. and Verriest, G., (1978) Colour vision and traffic, In G. Verriest (ed), *Modern Problems in Ophthalmology*, Basel, Switzerland: Karger.

Nickerson, D., (1957) Horticultural color chart names with a Munsell key, *Journal of the Optical Society of America,* 47, 619–621.

Nilsson, L. and Alm, H., (1996) Effects of a vision enhancement system on drivers' ability to drive safely in fog, In A.G. Gale, I.D. Brown, C.M. Haslegrave and S.P. Taylor (eds), *Vision in Vehicles - V,* Amsterdam: North-Holland.

Novellas, F. and Perrier, J., (1985) New lighting method for road tunnels, *CIE Journal,* 4, 58–70.

Nygardhs, S. and Helmers, G., (2007) *VMS – Variable Message Signs – A Literature Review,* Linkuping, Sweden, VTI.

O'Boyle, C., (2010) *Are London's cyclists visible,* MSc Thesis, University College London, London.

Ohashi, R., Akashi, Y. and Uchida, T., (2020) Comparisons between off-axis detection and on-axis recognition to implement mesopic photometry in roadway lighting standards, *Lighting Research and Technology,* 52, 540–553.

Olson, P.L. and Sivak, M., (1983) Comparison of headlamp visibility distance and stopping distance, *Perceptual and Motor Skills,* 57, 1177–1178.

Olson, P.L., Cleveland, D.E., Fancher, P.S. and Schneider, L.W., (1984) *Parameters Affecting Stopping Sight Distances,* Report UMTRI-84-15, Ann Arbor, MI: University of Michigan Transportation Research Institute.

Olson, P.L., Battle, D.S. and Aoki, T., (1989a) *The Detection Distance of Highway Signs as a Function of Color and Photometric Properties,* Report UMTRI-89-36, Ann Arbor, MI: University of Michigan Transportation Research Institute.

Olson P.L., Halstead-Nussloch, R. and Sivak, M., (1989b) The effect of improvements in motorcycle / motorcyclist conspicuity on driver behavior, *Human Factors,* 23, 237–248.

Omran, Y.H., Sadechi-Bagamani, H. and Yarmohammadian, M.H, (2023) Driving hazard perceptions tests: A systematic review, *Bulletin of Emergency and Trauma,* 11, 51–68.

Onaygil, S., Guler, O. and Erkin, E., (2012) Cost analyses of LED luminaires in road lighting, *Light and Engineering,* 20, 39–45.

Organization for Economic Co-operation and Development (OECD), (1998) *Safety of Vulnerable Road Users,* Paris: OECD.

Organization for Economic Co-operation and Development (OECD), (2023) *Road Accidents,* Paris: OECD.

Ouimet, M.C., Pradhan, A.K., Brooks-Russell, A., Ehsani, J.P., Berbiche, D. and Simons-Morton, B.G., (2015) Young drivers and their passengers: A systematic review of epidemiological studies on crash risk, *Journal of Adolescent Health,* 57, S24–S35.

Owens, D.A., Atntonoff, R.J. and Francis, E.L., (1994) Biological motion and nighttime pedestrian conspicuity, *Human Factors,* 36, 718–732.

Owens, D.A. and Tyrrell, R.A., (1999) Effects of luminance, blur and age on nighttime visual guidance: A test of the selective degradation hypothesis, *Journal of Experimental Psychology: Applied,* 5, 1–14.

Owsley, C. and McGwin G., (2010) Vision and driving, *Vision Research,* 50, 2348–2361.

Owsley, C., Sekular, R. and Siemsen, D., (1983) Contrast sensitivity throughout adulthood, *Vision Research,* 23, 689–699.

Owusu, V., Taffour, Y.A., Obeng, D.A. and Salifu, M., (2018) Degradation of retroreflectivity of thermoplastic pavement marking: A review, *Open Journal of Civil Engineering,* 8, 301–311.

Oya, H. Ando, K. and Kanoshima, H., (2002) A research on interrelation between illuminance at intersections and reduction in traffic accidents, *Journal of Light and Visual Environment,* 26, 29–34.

Ozadowicz, A. and Grela, J., (2016) Energy saving in the street lighting control system – A new approach based on the EN-15252 standard, *Energy Efficiency,* 10, 563–576.

Padmos, P. and Alferdinck, J.W.A.M., (1983a) *Glare in Tunnel Entrances II: The Effect of Stray Light on the Windscreen*, IZF 1983 C-10. Soesterberg, The Netherlands: IZF-TNO.

Padmos, P. and Alferdinck, J.W.A.M., (1983b) *Glare in Tunnel Entrances III: The Effect of Atmospheric Stray Light*, IZF 1983 C-9. Soesterberg, The Netherlands: IZF-TNO.

Padmos, P., Van den Brink, T.D.J., Alferdinck, J.W.A.M. and Folles, E., (1988) Matrix signs for motorways: System design and optimum photometric features, *Lighting Research and Technology*, 20, 55–60.

Painter, K., (1999) The social history of street lighting (part 1), *Lighting Journal*, 64, 14–24.

Painter, K., (2000) The social history of street lighting (part 2), *Lighting Journal*, 65, 24–30.

Painter, K.A. and Farrington, D.P., (2001) The financial benefits of improved street lighting based on crime reduction, *Lighting Research and Technology*, 33, 3–12.

Pak, H., Kang, H., Song, D-W., Lee, J-H., Park, H. and Lee, C-S., (2024) Evaluation of direction judgments based on turn signal guide lamp and behavioural responses, *Lighting Research and Technology*, 56, 56–71.

Palmer, D.A., (1968) Standard observer for large-field photometry at any level, *Journal of the Optical Society of America*, 58, 1296–1299.

Pammer, V., Sabadas, S. and Lentern, S., (2018) Allocating attention to detect motorcycles: The role of inattentional blindness, *Human Factors*, 60, 5–19.

Peli, E., (1990) Contrast in complex images, *Journal of the Optical Society of America A*, 7, 2032–2040.

Peli, E. and Peli, D., (2002) *Driving with Confidence: A Practical Guide to Driving with Low Vision*, Hackensack, NJ: World Scientific Publishing.

Pelli, D.G., Robson, J.G. and Wilkins A.J., (1988) The design of a new letter chart for measuring contrast sensitivity, *Clinical Vision Sciences*, 2, 187–199.

Perel, M., Olson, P.L., Sivak, M. and Medlin Jr, J.W., (1983) *Motor Vehicle Forward Lighting*, SAE Technical Paper 830567, Warrendale, PA: Society of Automotive Engineers.

Plainis, S., Murray, I.J. and Charman, W.N., (2005) The role of retinal adaptation in night driving, *Optometry and Vision Science*, 82, 682–688.

Plainis, S., Murray I.J. and Pallikaris, I.G., (2006) Road traffic casualties: understanding the night-time death toll, *Injury Prevention*, 12, 125–128.

Polus, A. and Katz, A., (1978) An analysis of nighttime pedestrian accidents at specially illuminated crosswalks, *Accident Analysis and Prevention*, 10, 223–228.

Ponziani, R.L., (2006) *Electronic Intelligent Turn Signal System*, SAE Paper 2006-01-0714, Warrendale, PA: Society of Automotive Engineers.

Purves, D. and Beau Lotto, R., (2010) *Why We See What We Do Redux. A Wholly Empirical Theory of Vision*, Oxford, UK, Oxford University Press.

Racette, L. and Casson, E.J., (2005) The impact of visual field loss on driving performance: Evidence from on-road assessments, *Optometry and Vision Science*, 82, 668–674.

Range H.D., (1972) A simplified method for the characterisation of road surfaces for lighting, *Lichttechnik*, 24, 608.

Raynham, P., (2004) An examination of the fundamentals of roads lighting for pedestrians and drivers, *Lighting Research and Technology*, 36, 307–316.

Rea, M.S., (2018) The what and where of vision lighting research, *Lighting Research and Technology*, 50, 14–37.

Rea, M.S. and Bullough, J.D., (2007) Move to a unified system of photometry, *Lighting Research and Technology*, 39, 393–408.

Rea, M.S. and Bullough, J.D., (2016) Toward performance specifications for flashing warning beacons, *Transportation Research Part F: Traffic Psychology and Behavior*, 43, 36–47.

Rea, M.S. and Ouellette, M.J., (1991) Relative visual performance: A basis for application, *Lighting Research and Technology*, 23, 135–144.

Rea, M.S., Bierman, A., McGowan, T., Dickey, F. and Havard, J., (1997) A field study comparing the effectiveness of metal halide and high pressure sodium illuminants under mesopic conditions, *Proceedings of the CIE Symposium on Visual Scales: Photometric and Colourimetric Aspects, Teddington, UK*, Vienna: CIE.

Rea, M.S., Bullough, J.D., Freyssinier-Nova, J-P. and Bierman, A., (2004) A proposed unified system of photometry, *Lighting Research and Technology*, 36, 85–111.

Rea, M.S., Bullough, J.D., Foy, C.R., Brons, J.A., van Derlofske, J. and Donnell, E.T., (2009) *Effect of the Safety Benefit and Other Effects of Roadway Lighting*, Final Report for the National Cooperative Highway Research Program, Washington, DC: Transportation Research Board of the National Academies.

Rea, M.S., Bullough, J.D., Radetsky, L.C., Skinner, N.P. and Bierman, A., (2018) Toward the development of standards for yellow flashing lights used in work zones, *Lighting Research and Technology*, 50, 552–570.

Reagan, I.J. and Brumbelow, M., (2015) Perceived discomfort glare from adaptive driving beam headlight system compared with three low beam lighting configurations, *Procedia Manufacturing*, 3, 3214–3221.

Reagan, I.J., Brumbelow, M.L., Flannagan, M.J. and Sullivan, J.M., (2017) High beam headlamp use rate: Effects of rurality, proximity of other traffic and roadway curvature, *Traffic Injury and Prevention*, 18, 716–723.

Recarte, M.A. and Nunes L.M., (2009) Driver Distractions. In C. Castro (ed), *Human Factors of Visual and Cognitive Performance in Driving*, Boca Raton, FL: CRC Press.

Reed, M.P. and Flannagan, M.J., (2003) *Geometric Visibility of Mirror-Mounted Turn Signals*, UMTRI-2003-18, Ann Arbor, MI: University of Michigan Transportation Research Institute.

Regan, D.M. and Beverley, K.I., (1982) How do we avoid confounding the direction we are looking with the direction we are moving? *Science*, 213, 194–196.

Regev, S., Rolison, J.J. and Moutari, S., (2018) Crash risk by driver age, and time of day using a new exposure methodology, *Journal of Safety Research*, 66, 131–140.

Reynolds, C.C.O., Harris, M.A., Teschke, K.K., Cripton, P.A. and Winters, M., (2009) The impact of transportation infrastructure on bicycling injuries and crashes: A review of the literature, *Environmental Health*, 8, 47.

Rich, C. and Longcore, T., (2006) *Ecological Consequences of Artificial Night Lighting*, Washington, DC: Island Press.

Richter, M. and Witt, K., (1986) The story of the DIN color system, *Color Research and Application*, 11, 138–145.

Robbins, C.J., Allen, H.A. and Chapman, P., (2018) Comparing drivers' gap acceptance for cars and motorcycles at junctions using an adaptive staircase methodology, *Transportation Research Part F*, 58, 944–954.

Robertson, A.R., (1977) The CIE 1976 color difference formulae, *Color Research and Application*, 2, 7–11.

Robger, L. and Lennie, M.G., (2017) *Increasing Motorcycle Conspicuity; Design and Assessment of Interventions to Enhance Rider Safety*, Boca Raton: CRC Press.

Robger, L., Krzywinski, J., Muhlbauer, F. and Schlag, B., (2015) Design studies on improved frontal light configuration for powered two-wheelers and testing in laboratory experiments. In L. Robger and M.G. Lenne (eds), *Increasing Motorcycle Conspicuity*, London: CRC Press.

Roch, J. and Smiatek, G., (1972) The q_s-method for the characterisation of road surfaces for lighting, *Lichttechnik*, 24, 329.

Rockwell, T.H. and Safford, R.R., (1968) *An Evaluation of Automotive Rear Signal System Characteristics in Night Driving*, Report EES-272 B, Columbus, OH: The Ohio State University.

Rogers, J.G., (1972) Peripheral contrast thresholds for moving images, *Human Factors,* 14, 199–205.

Roper, V.J. and Howard, E.A., (1938) Seeing with motor car headlamps, *Transactions of the Illuminating Engineering Society (London)*, 33, 417-438.

Rumar, K., (1990) The basic driver error: Late detection, *Ergonomics*, 33, 1281–1290.

Rumar, K., (1998) *Vehicle Lighting and the Aging Population*, UMTRI-98-9, Ann Arbor, MI: University of Michigan Transportation Research Institute.

Rumar, K., (2000) *Relative merits of the US and ECE high beam maximum intensities and of two and four headlamp systems*, UMTRI 2000-41, Ann Arbor, MI: University of Michigan Transportation Institute.

Rumar, K., (2003) *Functional Requirements for Daytime Running Lights*, UMTRI-2003-11, Ann Arbor, MI: University of Michigan Transportation Research Institute.

Rumar, K. and Marsh II, D.K., (1998) *Lane Marking in Night Driving: A Review of Past Research and of the Present Situation*, Report UMTRI-98-50, Ann Arbor, MI: University of Michigan Transportation Research Institute.

Rutley, K.S. and Mace, D.G.W., (1969) *An Evaluation of a Brakelight Display Which Indicates the Severity of Braking*, Report LR-287, Crowthorne, UK: Transport and Road Research Laboratory.

Sabey, B.E. and Staughton, E.C., (1975) Interacting roles of road environment, vehicle and road user in accidents, *Proceedings of the 5th Conference of the International Association for Accident and Traffic Medicine*, London.

Sagawa, K. and Takahashi, Y., (2001) Spectral luminous efficiency as a function of age, *Journal of the Optical Society of America*, 18, 2659–2667.

Sagawa, K. and Takeichi, K., (1992) System of mesopic photometry for evaluating lights in terms of comparative brightness relationships, *Journal of the Optical Society of America, A,* 9, 1240–1246.

Sanders, A. and Scott, A., (2008) *A Review of Luminaire Maintenance Factors*, Southampton, UK, Mott McDonald.

Sayer, J.R., Mefford, M.L., Flannagan, M.J. and Sivak, M., (1996) *Reaction Time to Center High-Mounted Stop Lamps: Effects of Context, Aspect Ratio, Intensity and Ambient Illumination*, UMTRI-96-3, Ann Arbor, MI: University of Michigan Transportation Research Institute.

Sayer, J.R., Mefford, M.L. and Blower, D., (2001) *The Effects of Rear-Window Transmittance and Back-Up Lamp Intensity on Backing Behavior*, UMTRI-2001-6, Ann Arbor, MI: University of Michigan Transportation Research Institute.

Schieber, F., Schlorholtz, B. and McCall, R., (2009) Visual Requirements of Vehicular Guidance. In C Castro (ed), *Human Factors of Visual and Cognitive Performance in Driving*, Boca Raton, FL: CRC Press.

Schivelbusch, W., (1988) *Disenchanted Night: The Industrialization of Light in the Nineteenth Century*, Oxford, UK: Berg Publishing.

Schmidt, T.M., Chen, S.K. and Hattar, S., (2011) Intrinsically photosensitive retinal ganglion cells: many subtypes, diverse functions, *Trends in Neuroscience*, 34, 572–580.

Schmidt-Clausen, H.J., (1985) Optimum luminances and areas of rear-position lamps and stop lamps, *Proceedings of the 10th International Technical Conference on Experimental Safety Vehicles*, Washington, DC: National Highway Traffic Safety Administration.

Schmidt-Clausen, H.J., (2000) *Contour Marking of Vehicles, Final Report*, Darmstadt, Germany: Darmstadt University of Technology.

Schmidt-Clausen, H.J. and Bindels, J.H., (1974) Assessment of discomfort glare in motor vehicle lighting, *Lighting Research and Technology*, 6, 79–88.

Schmidt-Clausen, H.J. and Finsterer, H., (1989) Large scale experiment about improving the night-time conspicuity of trucks, *Proceedings of the 12th International Technical Conference on Experimental Safety Vehicles*, Washington, DC: National Highway Traffic and Safety Administration.

Schoettle, B., Sivak, M. and Flannagan, M.J., (2002) *High-Beam and Low-Beam Headlighting Patterns in the US and Europe at the Turn of the Millenium*, SAE paper 2002-01-0262, Warrendale, PA: Society of Automotive Engineers.

Schreuder, D.A., (1964) *The Lighting of Vehicular Traffic Tunnels*, Thesis, Centrex, The Netherlands: Techische Hochschule, Eindhoven.

Schreuder, D.A., (1976) *White or Yellow Light for Vehicle Head-Lamps*, Voorburg, The Netherlands: Institute for Road Safety Research.

Schreuder, D.A., (1993) Energy saving in tunnel entrance lighting, *Proceedings of Right Light, the Second European Conference on Energy-Efficient Lighting*, Arnhem, The Netherlands: Netherlands Institute of Illuminating Engineering.

Schreuder, D.A., (1998) *Road Lighting for Safety*, London: Thomas Telford.

Schumann, J., Sivak, M., Flannagan, M.J. and Schoettle, B, (2003) *Conspicuity of Mirror-Mounted Turn Signals*, UMTRI-2003-26, Ann Arbor, MI: University of Michigan Transportation Research Institute.

Schwab, R.N. and Mace, D.J., (1987) Luminance measurements for signs with complex backgrounds, *Proceedings of the CIE, 21st Session,* Venice, Vienna: CIE.

Scott, P.P., (1980) *The Relationship between Road Lighting Quality and Accident Frequency,* TRRL Laboratory Report 929, Crowthorne, UK: Transport and Road Research Laboratory.

Sekular, R. and Blake, R., (1994) *Perception*, New York: McGraw-Hill.

Setiawan, P., (2009) *The Use of Lights on the Bicycles: Cyclists' Perceptions of Safety*, PhD Thesis, Lund, Sweden: Lund University,.

Shahar, A., Bremond, R. and Villa, C., (2018) Can light emitting diode-based road studs improve vehicle control in curves at night? A driving simulator study, *Lighting Research and Technology*, 50, 266–281.

Shlaer, S., (1937) The relation between visual acuity and illumination, *Journal of General Physiology*, 21, 165–168.

Sifrit, K.J., Stutts, J. and Martell, C., (2011) *Intersection Crashes Among Drivers in their 60's. 70's ad 80's*, NHTSA Technical Report DOT HS 811 495, Washington, DC, National Highway Traffic Safety Administration.

Silverstone, B., Lang, M.A., Rosenthal, B.P. and Faye, E.E., (2000) *The Lighthouse Handbook on Vision Impairment and Vision Rehabilitation,* New York: Oxford University Press.

Simons, R.H. and Bean, A.R., (2000) *Lighting Engineering,* London: Butterworth-Heinemann.

Sivak, M., (2002) How common sense fails us on the road: contribution of bounded rationality to the annual worldwide toll of one million traffic fatalities, *Transportation Research, Part 5*, 259–269.

Sivak, M. and Flannagan, M.J., (1993) A fast rise brake light as a collision-prevention device, *Ergonomics*, 36, 391–395.

Sivak, M. and Olson, P.L., (1985) Optimal and minimal luminance characteristics for retro-reflective highway signs, *Transportation Research Record*, 1027, 53–57.

Sivak, M., Olson, P. and Pastalan, L., (1981) Effect of driver's age on nighttime legibility of highway signs, *Human Factors*, 23, 59–64.

Sivak, M., Simmons, C.J. and Flannagan, M., (1990) Effect of headlamp area on discomfort glare, *Lighting Research and Technology*, 22, 49-52.

Sivak, M., Flannagan, M.J., Ensing, M. and Simmons, C.J., (1991) Discomfort glare is task dependent, *International Journal of Vehicle Design*, 12, 152–159.

Sivak, M., Flannagan, M. and Gellatly, W., (1993) Influence of truck driver eye position on effectiveness of retro-reflective traffic signs, *Lighting Research and Technology*, 25, 31–36.

Sivak, M., Campbell, K., Scheider, L., Sprague, J., Streff, F. and Waller, P.F., (1995) The safety and mobility of older drivers: What we know and promising research issues, *UMTRI Research Reviews*, 26, 1–24.

Sivak, M, Flannagan, M.J., Traube, E.C., Hashimoto, H. and Kojima, S., (1996) *Fog Lamps: Frequency of Installation and Nature of Use*, Report UMTRI-96-31, Ann Arbor, MI: University of Michigan Transportation Research Institute.

Sivak, M., Flannagan, M.J., Traube, E.C. and Kojima, S., (1998) Automobile rear signal lamps: Effects of realistic levels of dirt on light output, *Lighting Research and Technology*, 30, 24–28.

Sivak, M., Flannagan, M.J. and Miyokawa, T., (2001) A first look at visually aimable and internationally harmonized low-beam headlamps, *Journal of the Illuminating Engineering Society*, 30, 26–33.

Sivak, M., Schoettle, B., Flannagan, M.J. and Minoda, T., (2005a) Optimal strategies for adaptive curve lighting, *Journal of Safety Research*, 36, 281–288.

Sivak, M., Schoettle, B., Minoda, T. and Flannagan, M.J., (2005b) Short wavelength content of LED headlamps and discomfort glare, *Leukos*, 2, 145–154.

Sivak, M., Schoettle, B. and Flannagan, M.J., (2006a) Mercury-free HID headlamps: Glare and colour rendering, *Lighting Research and Technology*, 38, 33–40.

Sivak, M., Schoettle, B. and Flannagan, M.J., (2006b) *Mirror-Mounted Turn Signals and Traffic Safety*, UMTRI-2006-23, Ann Arbor, MI: University of Michigan Transportation Research Institute.

Sivak, M., Schoettle, B., Flannagan, M.J. and Minoda, T., (2006c) Effectiveness of clear-lens turn signals in direct sunlight, *Leukos*, 2, 199–209.

Sliney, D., Fast, P. and Ricksand, A., (1995) Optical radiation hazards analysis of ultraviolet headlamps, *Applied Optics*, 34, 4912–4922.

Smith, F.C., (1938) Reflection factors and revealing power, *Transactions of the Illuminating Engineering Society (London)*, 3, 196–200.

Society of Automotive Engineers (SAE), (1990) *Lighting Committee Summary of DRL Tests*, Warrendale, PA: SAE.

Society of Automotive Engineers (SAE), (1995) *Harmonized Vehicle Headlamp Performance Requirements*, SAE J1735, Warrendale, PA: SAE.

Society of Automotive Engineers (SAE), (2014) *Directional Flashing Optical Warning Devices for Authorized Emergency, Maintenance and Service Vehicles*, SAE Standard J595, Warrendale, PA: SAE.

Society of Light and Lighting (SLL), (2006) *Factfile 2: Car Park Lighting – Dilemma Solved*, London: Chartered Institution of Building Services Engineers (CIBSE).

Sorensen, K., (1974) *Description and Classification of Light Reflection Properties of Road Surfaces*, Report 7, Lyngby, Denmark: The Danish Illuminating Engineering Laboratory.

Sorensen, K., (1975) *Road Surface Reflection Data*, Report 10, Lyngby, Denmark: The Danish Illuminating Engineering Laboratory.

Sorensen, K., (1977) *Road Lighting for Wet Conditions, Measures of Road Lighting Effectiveness*, Berlin: Lichttechnische Gesellschaft LiTG.

Sorensen, K., (2011) *Legibility of LED-based Variable Message Traffic Signs*. Lingby, Denmark: Trafitec.

Stahl, F., (2004) *Kongruenz des blickverlaufs, bei virtuellen und raelen autofahrten – Validierung eines nachtfahrsimulators*, Diplomarbeit, Ilmenau, Germany: Ilmenau Technische Universitat.

Starkey, N.J. and Charlton, S.G., (2020) Driver's use of in-vehicle information systems and perceptions of their effect on driving, *Frontiers in Sustainable Cities*, 2, 39.

STATS19 *Road Accident Injury Statistics*, London: Department for Transport.

Steinbach, R., Perkins, C., Tompson, L., Johnson, S., Armstrong, B., Green, J., Grundy, C., Wilkinson, P. and Edwards, P., (2015) The effect of reduced street lighting on road casualties and crime in England and Wales: Controlled interrupted time series analysis, *Journal of Epidemiology and Community Health*, 69, 1118–1124.

Stiles, W.S., (1930) The scattering theory of the effect of glare on the brightness difference threshold, *Proceedings of the Royal Society (London)*, 105B, 131–146.

Stiles, W.S. and Crawford, B.H., (1937) The effects of a glaring light source on extrafoveal vision, *Proceedings of the Royal Society (London)*, 122B, 255–280.

Styliois, K., Woxlin, A, Siljefalk, L., Heimersson, E. and Soderberg, R., (2020) Understanding light, A study on the perceived quality of car exterior lighting and interior illumination, *Procedia CIRP*, 93, 1340–1345.

Sullivan, J.M. and Flannagan, M.J., (2001) *Reaction Time to Clear-Lens Turn Signals Under Sun-Loaded Conditions*, UMTRI-2001-30, Ann Arbor, MI: University of Michigan Transportation Research Institute.

Sullivan, J.M. and Flannagan, M.J., (2002) The role of ambient light level in fatal crashes: Inferences from daylight saving time transitions, *Accident Analysis and Prevention*, 34, 487–498.

Sullivan, J.M. and Flannagan, M.J., (2004) *Visibility and Rear-End Collisions Involving Light Vehicles and Trucks*, UMTRI-2004-14, Ann Arbor, MI: University of Michigan Transportation Research Institute.

Sullivan, J.M. and Flannagan, M.J., (2007) Determining the potential safety benefit of improved lighting in three pedestrian crash scenarios, *Accident Analysis and Prevention*, 39, 638–647.

Sullivan, J.M., Adachi, G., Mefford, M.L. and Flannagan, M.J., (2004a) High-beam headlamp usage on unlighted rural roadways, *Lighting Research and Technology*, 36, 59–67.

Sullivan, J.M., Bargman, J., Adachi, G. and Schoettle, B., (2004b) *Driver Performance and Workload Using a Night Vision System*, UMTRI 2004-8, Ann Arbor, MI: University of Michigan Transportation Institute.

Sumner, R., Baguley, C. and Burton, J., (1977) *Driving in Fog on the M4*. Supplementary Report 281, Crowthorne, UK: Transport Research Laboratory.

Tan, H., Zhao, F., Han, H., Liu, Z., Amer, A.A. and Babiker, H., (2020) Automatic emergency braking (AEB) system impact on fatality and injury reduction in China, *International Journal of Environmental Research and Public Health*, 17, 917.

Tanner, J.C. and Harris, A.J., (1956) Comparison of accidents in daylight and in darkness, *International Road Safety and Traffic Review*, 4, 11–14, 39.

Tefft, B.C., (2013) Impact speed and a pedestrian's risk of severe injury or death, *Accident Analysis and Prevention*, 50, 871–878.

Tefft, B.C., (2017) *Rates of Motor Vehicle Crashes, Injuries and Deaths in relation to Driver Age, United States, 2014–2015*, Washington, DC, AAA Foundation for Traffic Safety.

Teichner, W. and Krebs, M., (1972) The laws of simple reaction time, *Psychological Review*, 79, 344–358.

Tenkink, E, (1988) Lane keeping and speed choice with restricted sight, In T. Rothengatter and R. de Bruin (eds) *Road User Behaviour: Theory and Research*, Assen / Maastricht, The Netherlands: Van Gorcum.

Theeuwes, J. and Alferdinck, J.W.A.M., (1996) *The Relation Between Discomfort Glare and Driver Behavior*, Report DOT HS 808 452, Washington, DC: US Department of Transportation.

Theeuwes, J. and Riemersma, J., (1995) Daytime running lights as a vehicle collision countermeasure: The Swedish evidence reconsidered, *Accident Analysis and Prevention*, 27, 633–642.

Thompson, P.A., (2003) *Daytime Running Lamps (DRLs) for Pedestrian Protection*, SAE Technical Paper 2003-01-2072, Warrendale, PA: Society of Automotive Engineers.

Tomczuk, T., Wytrykowska, A. and Chizanonicz, M., (2023) Analysis of luminance contrast values on illuminated pedestrian crossings in urban conditions, *Energies*, 16, 8031.

Trezona, P.W., (1991) A system of mesopic photometry, *Color Research and Application*, 16, 202–216.

Trick, L.M. and Enns, J.T., (2009) A Two-Dimensional Framework for Understanding the Role of Attentional Selection in Driving. In C. Castro (ed), *Human Factors of Visual and Cognitive Performance in Driving*, Boca Raton, FL: CRC Press.

Troutbeck, R. and Wood, J.M., (1994) Effect of restriction of view on driving performance, *Journal of Transportation Engineering*, 120, 737–752.

Tsimhoni, O., Flannagan, M.J. and Minoda, T., (2005) *Pedestrian Detection with Night Vision Systems Enhanced by Automatic Warnings*, UMTRI-2005-23, Ann Arbor, MI: University of Michigan Transportation Research Institute.

Tuaycharoen, N. and Tregenza, P.R., (2005) Discomfort glare from interesting images, *Lighting Research and Technology*, 37, 329–341.

Turner, D., Nitzburg, M. and Knoblauch, R., (1998) Ultraviolet headlamp technology for nighttime enhancement of roadway markings, *Transportation Research Record*, Paper 98-1187, Washington, DC: Transportation Research Board.

Turner, P.L., van Someren, E.Z.J.W. and Mainster, M.A., (2010) Effects of ocular aging and cataract surgery, *Sleep Medicine Reviews*, 14, 269–280.

TUV Rheinland Group, (2004) *Conspicuity of Heavy Goods Vehicles: Final Report*, Brussels: European Commission.

Ueki, K., Ohkubo, N. and Fujimura, H., (1992) Motorway accidents in tunnels in relation to lighting, *Road Lighting as an Accident Counter-Measure*, CIE Publication No. 93, Vienna: CIE.

Uttley, J. and Fotios, S., (2017) The effect of ambient light conditions on road traffic collisions involving pedestrians on pedestrian crossings, *Accident Analysis and Prevention*, 108, 189–200.

Uttley, J., Fotios, S. and Lovelace, R., (2020) Road lighting density and brightness linked with increased cycling rates after-dark, *PLoS ONE*, 15, e0233105.

Uttley, J., Fotios, S., Lovelace, R. and Vidal-Tortosa, E., (2023) Cyclist fatalities increase on unlit roads, *Proceedings of the 30th Session of the CIE Part 1*, Ljubljana, Slovenia, Vienna: CIE.

Valetti, L., Piccablotto, G., Taraglio, R. and Pellegrino, A., (2023) Long-term monitoring campaign of LED street lighting systems: Focus on photometric performances, maintenance and energy savings, *Sustainability*, 15, 16910.

Van den Berg, T.J.T.P., (1993) Quantal and visual efficiency of fluorescence in the lens of the human eye. *Investigative Ophthalmology and Vision Science*, 34, 3566–3573.

Van den Berg, T.J.T.P., IJspeert, J.K. and de Waard, P.W.T., (1991) Dependence of intraocular straylight on pigmentation and transmission through the ocular wall, *Vision Research*, 31, 1361–1367.

Van Bommel, J.M. and Tekelenburg, J. (1986) Visibility research for road lighting based on a dynamic situation, *Lighting Research and Technology*, 18, 37–39.

Van Bommel, W., (1976) Optimization of the quality of roadway lighting installations, especially under adverse weather conditions, *Journal of the Illuminating Engineering Society*, 55, 95–104.

Van Bommel, W., (2015) *Road Lighting: Fundamentals, Technology and Application*, Dordrecht: Springer.

Van Derlofske, J. and Bullough, J.B., (2003) *Spectral effects of high-intensity discharge automotive forward lighting on visual performance*, SAE Technical Paper Series 2003-01-0559, Warrendale, PA: Society of Automotive Engineers.

Van Derlofske, J. and Bullough, J.B., (2006) *Spectral effects of LED forward lighting: Visibility and glare*, SAE Technical Paper Series 2006-01-0102, Warrendale, PA: Society of Automotive Engineers.

Van Derlofske, J., Bullough, J.B. and Hunter C.M., (2001) *Evaluation of high-intensity discharge automotive forward lighting*, SAE Technical Paper Series 2001-01-0298, Warrensburg, PA: Society of Automotive Engineers.

Van Derlofske, J., Boyce, P.R. and Hunter C.M., (2003) Evaluation of in-pavement warning lights on pedestrian crosswalk safety, *IMSA Journal*, 20 and 54.

Van Derlofske, J., Bullough, J.B., Dee, P., Chen, J. and Akashi, A., (2004) *Headlamp parameters and glare*, SAE Technical Paper Series 2004-01-1280, Warrendale, PA: Society of Automotive Engineers.

Van Derlofske, J., Bullough, J.B. and Gribbin, C., (2007) Comfort and visibility characteristics of spectrally-tuned high intensity discharge forward lighting systems, *European Journal of Scientific Research*, 17, 73–84.

Van Derlofske, J., Chen, J., Bullough, J.D. and Akashi, Y., (2005) *Headlight glare exposure and recovery*, SAE Paper 05B-269, Warrendale, PA: Society of Automotive Engineers.

Van Houten, R., Healy, K., Malenfant, J.E.L and Retting, R., (1998) *The Use of Signs and Signals to Increase the Efficacy of Pedestrian-Activated Flashing Beacons at Crosswalks*, Proceedings of the 77th Transportation Research Board, Washington, DC: Transportation Research Board.

Van Nes, F.L. and Bouman, M.A., (1967) Spatial modulation transfer in the human eye, *Journal of the Optical Society of America*, 47, 401–406.

Varady, G., Freiding, A., Eloholma, M., Halonen, L., Walkey, H., Goodman, T. and Alferdinck, J., (2007) Mesopic visual efficiency III: Discrimination threshold measurements, *Lighting Research and Technology*, 39, 355–364.

Viikari, M., Eloholma, M. and Halonen, L., (2005) 80 years of V (λ) use: A review, *Light and Engineering*, 13, 24–36.

Villa, C., Bremond, R., Eymond, F. and Saint-Jaques, E., (2015) Visibility and discomfort glare of road studs, *Lighting Research and Technology*, 47, 945–963.

Villa, C., Bremond, R., Eymond, F. and Saint-Jaques, E., (2023) Characterization of luminescent road markings, *Lighting Research and Technology*, 55, 459–473.

Voevodsky, J., (1974) Evaluation of a deceleration warning light in reducing rear end collisions, *Journal of Applied Psychology*, 59, 270–273.

Vogel, S., Fiedelak, S., Niedling, M. and Volker, S., (2024) Influence of automotive headlamp systems on the visibility of targets under different road lighting conditions, *Lighting Research and Technology*, 56, 37–55.

Vos, J.J., (1984) Disability glare - a state of the art report, *CIE Journal*, 3, 39–53.

Vos, J.J., (1999) Glare today in historical perspective: Towards a new CIE Glare observer and a new glare nomenclature, *Proceedings of the CIE 24th Session*, Warsaw, Vienna: CIE.

Vos, J.J., (2003) Reflections on glare, *Lighting Research and Technology*, 35, 163–175.

Vos, J.J. and Boogaard, J., (1963) Contribution of the cornea to entoptic scatter, *Journal of the Optical. Society of America*, 53, 869–873.

Vos, J.J. and Padmos, P., (1983) Straylight, contrast sensitivity and the critical object in relation to tunnel lighting, *Proceedings of the CIE 20th Session*, Amsterdam: Vienna: CIE.

Wada, T., Kurokawa, K., Itozawa, K. and Saito, H., (2007) *Development of a New Instrument Cluster with Electrochromic Device (ECD)*, SAE Paper 2006-01-0945,, Warrendale, PA: Society of Automotive Engineers.

Wald, G., (1945) Human vision and the spectrum, *Science*, 101, 653–658.

Waldram, J.M., (1938) The revealing power of street lighting installations, *Transactions of the Illuminating Engineering Society (London)*, 3, 173–186.

Waldram, J.M., (1972) The calculation of sky haze luminance from street lighting, *Lighting Research and Technology*, 4, 21–26.

Walker, M.F., (1977) The effects of urban lighting on the brightness of the night sky, *Publications of the Astronomical. Society of the Pacific*, 89, 405–409.

Walker, T., (2007) Remote monitoring systems assessed, *Lighting Journal*, 72, 49–53.

Walraven, P.L., (1974) A closer look at the tritanopic convergence point, *Vision Research*, 14, 1339–1343.

Wang, J-S., (2008) *The Effectiveness of Daytime Running Lights for Passenger Vehicles*, Report DOT-HS-811-029, Washington, DC: National Highway Traffic Safety Administration.

Wang, J-S. and Knipling, R.R., (1994) *Lane Change / Merge Crashes: Problem Size Assessment and Statistical Description*, Report DOT-HS-808075, Washington, DC, National Highway Traffic Safety Administration.

Wang, Y., Yang, X., Liang, M. and Liu, Y., (2018) A review of the self-adaptive traffic signal control system based on future traffic environment, *Journal of Advanced Transportation*, 1096123.

Wang, L., Zhong, H., Ma, W., Abdel-Aty, M. and Park, J., (2020) How many crashes can connected vehicle and automated vehicle technologies prevent: A meta-analysis, *Accident Analysis and Prevention*, 136, 105299.

Wanvik, P.O., (2009) *Road Lighting and Traffic Safety*, PhD Thesis, Trondheim, Norway: Norwegian University of Science and Technology.

Watanabe, B.L., Patterson, G.S., Kempema, J.M., Magallanes, O. and Brown, L.H., (2019) Is use of warning lights and sirens associated with increased risk of ambulance crashes? A contemporary analysis using National EMS Information System (NEMSIS) data, *Annals of Emergency Medicine*, 74, 101–109.

Watkinson, J.C., (2005) *Additivity of Discomfort Glare*, MS Thesis, Troy, NY: Rensselaer Polytechnic Institute.

Weale, R.A., (1992) *The Senescence of Human Vision*, Oxford, UK: Oxford University Press.

Weirich, C., Lin, Y. and Khanh, T.Q., (2022) Evidence for human-centric in-vehicle lighting: Part 1, *Applied Sciences* 12, 552.

Wells, S., Mullin, B., Norton, R., Langley, J., Connor, J., Lay-Yee, R. and Jackson, R., (2004) Motorcycle rider conspicuity and crash related injury: case-control study, *British Medical Journal*, 328, 857–862.

Werner, J.S., Peterzell, D.H. and Scheetz, A.J., (1990) Light, vision and aging. *Optometry and Visual Science*, 67, 214–229.

Wertheim, A.H., (1981) On the relativity of perceived motion, *Acta Psychologica*, 48, 97–110.

Whillans, M.G., (1983) Colour-blind driver's perception of traffic signals, *Canadian Medical Association Journal*, 128, 1187–1189.

White, M.E. and Jeffrey, D.J., (1980) *Some Aspects of Motorway Traffic Behavior in Fog*, Report LR 958, Crowthorne, UK: Transport Research Laboratory.

Whitlock and Weinberger Transportation, (1998) *An Evaluation of a Crosswalk Warning System Utilizing In-Pavement Flashing Lights*, Santa Rosa, CA: Whitlock and Weinberger Transportation.

Whittaker, J., (1996) An investigation in to the effects of British summer time on road traffic accident casualties in Cheshire, *Journal of Accident and Emergency Medicine*, 13, 189–192.

Wierda, M., (1996) Beyond the eye: Cognitive factors in drivers' visual perception, In A.G. Gale, I.D. Brown, C.M. Haslegrave and S.P. Taylor (eds) *Vision in Vehicles-V*, Amsterdam: North-Holland.

Wijnen, W., Weijermars, W., Schoeters, A., van den Berghe, W., Bauer, R., Carnis, L., Elvik, R. and Martensen, H., (2019) An analysis of official road crash cost estimates in European countries, *Safety Science*, 113, 318–327.

Williams, G.L., Kennedy, J.V. and Beesley, R., (2008) *The Use of Passively Safe Signposts and Lighting Columns*, Crowthorne, UK: Transport Research Laboratory.

Willis, M.L., Palermo, R. and Burke, D., (2011) Judging approachability on the face of it: The influence of face and body appearance on the perception of approachability, *Emotion*, 11, 514–523.

Wolf, E. and Gardiner, J.S., (1965) Studies on the scatter of light in the dioptric media of the eye as a basis of visual glare, *Archives of Ophthalmology*, 74, 338–345.

Wolfe, J.M., Kluender, K.R., Levi, D.M., Bartoshuk, L.M., Herz, R.S., Klatzky, R.L. and Lederman, S.J., (2006) *Sensation and Perception*, Sunderland, MA: Sinauer Associates Inc.

Wood, J.M., (2002) Age and visual impairment decrease driving performance as measured on a closed-road circuit, *Human Factors*, 44, 482–494.

Wood, J.M., (2022) Vision impairment and on-road driving, *Annual Review of Vision Science*, 8, 195–216.

Wood, J.M., Tyrrell, R.A. and Carberry, T.P., (2005) Limitations in drivers' ability to recognize pedestrians at night. *Human Factors*, 47, 644–653.

Wood, J.M., Tyrrell, R.A., Marszalek, R, Lacherez, P, Carberry, T., Chu, B.S. and King, M.J., (2010) Cyclist visibility at night: Perceptions of visibility do not necessarily match reality. *Journal of the Australian College of Road Safety*, 21, 56–60.

Wood, J.M., Tyrrell, R.A., Marszalek, R, Lacherez, P, Carberry, T. and Chu, B.S., (2012) Using reflective clothing to enhance the conspicuity of bicyclists at night, *Accident Analysis and Prevention*, 45, 726–730.

Wood, J., Isoardi, G., Black, A. and Cowling, I., (2018) Night-time driving visibility associated with LED streetlight dimming, *Accident Analysis and Prevention*, 121, 295–300.

Wood, J.M., Black, A.A. and Tyrell, R.A., (2022) Increasing the conspicuity of cyclists at night by using bicycle lights and clothing to highlight their biological motion to oncoming drivers, *Transportation Research Part F: Traffic Psychology and Behavior*, 90, 326–332.

Wordenweber, B., Wallaschek, J., Boyce, P. and Hoffman, D.D., (2007) *Automotive Lighting and Human Vision*, Berlin: Springer.

Wu, H., Zhang, X.M. and Ge, P., (2016) A freeform reflector for a light emitting diode bicycle head lamp, *Lighting Research and Technology*, 48, 642–652.

Wulf, G., Hancock, P.A. and Rahirni, M., (1989) Motorcycle conspicuity: An evaluation and synthesis of influential factors, *Journal of Safety Research*, 26, 153–176.

Wyszecki, G., (1981) Uniform color spaces, *Golden Jubilee of Colour in the CIE*, Bradford, UK: The Society of Dyers and Colourists.

Xu, R., Ye, H., Hu, B., Jin, S., Lu, M., Huang, Z. and Chen, L., (2024) Intelligent dimming control and energy consumption monitoring system of tunnel lighting, *Lighting Research and Technology*, 56, 72–86.

Yannis, G., Papadimitrou, E., Lejeune, P., Treny, V., Hemdorff, S., Bergel, R., Haddak, M., Hollo, P., Cardoso, J., Bijleveld, F., Houwing, S. and Bjornskau, T., (2005) *State of the Art Report on Risk and Exposure Data*, Deliverable 2.1 Project 506723 Building the European Road Safety Observatory, Brussels: European Union Directorate-General Transport and Energy.

Yerrell, J.S., (1971) *Headlamp Intensities in Europe and Britain, LR 383*, Crowthorne, UK: Road Research Laboratory.

Yerrell, J.S., (1976) Vehicle headlamps, *Lighting Research and Technology*, 8, 69–79.

Ying S-P., Chen B-M., Fu, H-K. and Yeh, C-Y., (2021) Single headlamp with low and high beam light, *Photonics*, 8, 32.

Zein, S.R., Geddes, E.A., Hemsing, S. and Johnson, M., (1997) Safety benefits of traffic calming, *Transportation Research Record*, 1578, 3–10.

Ziakopoulos, A., Theofilatos, A., Papadimitriou, E. and Yannis, G., (2019) A meta-analysis of the impact of operating in-vehicle information systems on road safety, *AATSS Research*, 43, 185–196.

Zulkefli, M.T., Shafiai, N.I.A. and Azizan, H.A., (2021) Automotive interior: A study on the dashboard touch screen panel and its impact to the driver, *Design Decoded 2021, Proceedings of the 2nd International Conference on Design Industries and Creative Culture*, Kedan, Malaysia.

Translation

George Bernard Shaw described America and England as two nations divided by a common language. This is certainly the case for the words used to describe roads and vehicles. In this book I have used terms from both countries according to my everyday speech. To avoid confusion, the list below offers a translation.

English	American
Amber traffic signal	Yellow traffic signal
Autumn	Fall
Bonnet	Hood
Boot	Trunk
Breakdown Recovery	Tow truck
Car park	Parking lot
Dustcart	Garbage truck
Hire car	Rental car
Multi-story car park	Parking garage
Crossroads	Intersection
Dipped beam	Low beam
Junction	Intersection
Level crossing	Grade crossing
Lorry	Truck
Main beam	High beam
Motorway	Freeway
MPV	Minivan
Pavement	Sidewalk
Pedestrian crossing	Crosswalk
Railway	Railroad
Silencer	Muffler
Road surface	Pavement
Road works	Work zone
Roundabout	Rotary
Waistcoat	Vest
Windscreen	Windshield
4 X 4	SUV

As well as words, there are also differences in the measures used for different quantities in different countries. For example, both the USA and the UK use miles per hour for speed, but most of the rest of the world uses kilometres per hour. Conversion factors for some commonly used measures are given below.

1 mile per hour (mph) = 1.6093 kilometres per hour (km.h^{-1})
1 kilometre per hour (km.h^{-1}) = 0.6214 miles per hour (mph)
1 footcandle (Fc) = 10.76 lumens.metre^{-2} (lx)
1 lumen.metre^{-2} (lx) = 0.0929 footcandles (Fc)
1 mile = 1.6093 kilometres
1 kilometre = 0.6214 miles
1 metre (m) = 3.278 feet (ft)
1 foot (ft) = 0.305 metres (m)
1 degree = 60 minutes of arc
1 minute of arc = 0.0166 degrees

Index